Proceedings

Ein stetig steigender Fundus an Informationen ist heute notwendig, um die immer komplexer werdende Technik heutiger Kraftfahrzeuge zu verstehen. Funktionen, Arbeitsweise, Komponenten und Systeme entwickeln sich rasant. In immer schnelleren Zyklen verbreitet sich aktuelles Wissen gerade aus Konferenzen, Tagungen und Symposien in die Fachwelt. Den raschen Zugriff auf diese Informationen bietet diese Reihe Proceedings, die sich zur Aufgabe gestellt hat, das zum Verständnis topaktueller Technik rund um das Automobil erforderliche spezielle Wissen in der Systematik aus Konferenzen und Tagungen zusammen zu stellen und als Buch in Springer.com wie auch elektronisch in Springer Link und Springer Professional bereit zu stellen. Die Reihe wendet sich an Fahrzeug- und Motoreningenieure sowie Studierende, die aktuelles Fachwissen im Zusammenhang mit Fragestellungen ihres Arbeitsfeldes suchen. Professoren und Dozenten an Universitäten und Hochschulen mit Schwerpunkt Kraftfahrzeug- und Motorentechnik finden hier die Zusammenstellung von Veranstaltungen, die sie selber nicht besuchen konnten. Gutachtern, Forschern und Entwicklungsingenieuren in der Automobil- und Zulieferindustrie sowie Dienstleistern können die Proceedings wertvolle Antworten auf topaktuelle Fragen geben.

Today, a steadily growing store of information is called for in order to understand the increasingly complex technologies used in modern automobiles. Functions, modes of operation, components and systems are rapidly evolving, while at the same time the latest expertise is disseminated directly from conferences, congresses and symposia to the professional world in ever-faster cycles. This series of proceedings offers rapid access to this information, gathering the specific knowledge needed to keep up with cutting-edge advances in automotive technologies, employing the same systematic approach used at conferences and congresses and presenting it in print (available at Springer.com) and electronic (at Springer Link and Springer Professional) formats. The series addresses the needs of automotive engineers, motor design engineers and students looking for the latest expertise in connection with key questions in their field, while professors and instructors working in the areas of automotive and motor design engineering will also find summaries of industry events they weren't able to attend. The proceedings also offer valuable answers to the topical questions that concern assessors, researchers and developmental engineers in the automotive and supplier industry, as well as service providers.

More information about this series at http://www.springer.com/series/13360

Euroforum Deutschland GmbH
Editor

CTI SYMPOSIUM 2019

18th International Congress and Expo 9–12 December 2019, Berlin, Germany

Springer Vieweg

Editor
Euroforum Deutschland GmbH
Düsseldorf, Germany

ISSN 2198-7432 ISSN 2198-7440 (electronic)
Proceedings
ISBN 978-3-662-61514-0 ISBN 978-3-662-61515-7 (eBook)
https://doi.org/10.1007/978-3-662-61515-7

Responsible Editor: Markus Braun
This Springer Vieweg imprint is published by the registered company Springer-Verlag GmbH, DE part of Springer Nature.
The registered company address is: Heidelberger Platz 3, 14197 Berlin, Germany

Preface

Dear reader,

The transformation to the age of electromobility and automation demands enormous investments. The goal is clear, but the timeline is still uncertain – and above all, region-specific. While China sets higher quotas each year for electric vehicles, in the US electromobility still takes a back seat. In Europe, stricter CO_2 limits – 81 g/km from 2025 and 59 g/km from 2030 – will further boost electrified drives and accelerate the transition to sustainable mobility. As a result, in a few years' time nobody will be making drives without an e-motor – starting with the companies producing premium vehicles.

Commercial vehicle manufacturers are also required to electrify their drives, not just for the "last mile". In this sector, recurring urban usage scenarios offer totally new opportunities for electromobility – for example by adapting batteries and charging infrastructures to range requirements. For heavy commercial vehicles and buses, entirely different concepts, such as overhead power lines, are also being discussed and tested.

So drive and transmission developers must prepare for multiple scenarios, with factors such as automation and vehicle connectivity creating high levels of diversity and complexity. Currently, variants range from conventional drives through 12 + 12 V or 12 + 48 V mild hybridization up to high-voltage 800 V electric drives. With electromobility and automation transforming the markets, no manufacturer wants to be left behind. In a race that may not be wholly justified at a time when consumers are hesitant about EVs, we are seeing rivalry among manufacturers in electric drive concepts.

The primary goal for automotive suppliers and manufacturers is to safeguard their future by offering more attractive products.

For this, they need to understand customer expectations and their impact on new drive concepts. Many companies, and particularly small and medium enterprises, could apply their skills here to add value for their customers – either by extending the functionality of existing products, or by using existing facilities to build completely new ones. Conversely, automotive manufacturers need to hear what existing or potential new opportunities the products suppliers can offer them. Never before has communication at all levels been so important between automotive companies and suppliers as well as sub-suppliers.

This year's symposium addressed these issues in numerous contributions and statements:

- Development of markets and regulations and their influence on drives.
- New hybrid and electric drives for passenger cars, commercial vehicles and buses, networked and automated.
- 48 V mild hybridization for various drives and markets.
- New AT, DCT, CVT and MT concepts with enhanced comfort and efficiency, plus their modular hybridization.
- Diverse DHTs for new hybrid drives.
- Single and multi-speed transmissions for EV drives, high-rpm concepts.
- Conventional and electrified four-wheel drives, torque vectoring.
- Semiconductors, electric motors and batteries as key technologies in electrified drives.
- New generation of oils and lubricants for electrified drivetrains.
- Virtual, customer-focused development using AI to determine efficiency, driveability and durability.
- Electro-mechanical actuators, sensors
- Compact, efficient launch and shift elements, jaw clutches, parking locks
- Production: quality assurance, lightweight battery housings, fuel cell stacks, ring gear manufacturing, laser welding in EM production, surface treatment of friction linings.

To address the broad range of topics outlined above, the extensive programme included around **100 specialist lectures** in plenary and in **16 parallel sessions**. In the panel discussion, the question **"Do we have the technology to meet customer expectations while meeting legal requirements?"** was discussed by vehicle, mobility and drive experts. All this was accompanied by **CTI SYMPOSIUM EXPO**, our "hands-on technology market for innovative products" with **120 exhibitors**.

The 18th CTI Symposium in Berlin was an excellent forum for international drive and transmission specialists to exchange opinions and experiences, get valuable updates on the industry status quo, and gain a clearer view of the challenges and solutions that lie ahead.

Thank you for your participation. I hope you had plenty of useful dialogue and inspiration.

Yours sincerely,
Prof. Dr. Ferit Küçükay
Chairman of the CTI Symposium

Contents

48 V Hybrid Manual Transmission

Konstantin Baron[1](✉), Matthias Werra[2], Arno Ringleb[2], and F. Küçükay[2]

[1] Opel Automobile GmbH, Bahnhofsplatz, 65423 Ruesselsheim, Germany
konstantin.baron@opel-vauxhall.com
[2] Institute of Automotive Engineering Braunschweig, Hans-Sommer-Str. 4, 38106 Brunswick, Germany
m.werra@tu-braunschweig.de

Abstract. The automotive industry faces significant changes. Main driver for those changes are legal requirements, new technology trends, new customer preferences and their mobility behaviour. Propulsion electrification could be part of solution to solve new challenges. While combustion engine technology will be part of this future propulsion architecture e.g. in hybrids, manual transmissions are frequently claimed to become obsolete.

But, Opel as well as other car manufacturers still have a high penetration of manual transmissions in their current portfolio for good reasons like high efficiency, low cost, low weight, its proven technology and lower complexity compared to automatic transmissions. Combination of those attributes with a propulsion electrification technology leads to a hybrid version of a manual transmission that could be the successor of the conventional version.

Keywords: Hybrid · Manual transmission · CO_2

1 Introduction

Opel Automobile GmbH and Automotive Engineering of the Technical University Braunschweig (TUB) have performed an investigation to understand the hybridization impact on manual transmissions more in detail. Opel's long experience and deep knowledge in the transmission sector combined with scientific method of the university facilitated in-depth analysis for an appropriate setup, expected fuel economy improvement and new requirements for the transmission. Results and conclusions of those activities are provided in this publication.

2 Reason for Hybrid Manual Transmission

Worldwide, the manual transmission share is dropping while total transmission volume is increasing in same time [1]. But even if it is true, the manual transmissions volume is still be significant in the overall production volume and in South Asia it is even increasing [1] (Fig. 1).

Due to this fact, developing a hybrid version of a manual transmission is a good way to reduce the CO_2 emissions of passenger cars [2] worldwide. This paper presents

Euroforum Deutschland GmbH (Ed.): *CTI SYMPOSIUM 2019*, pp. 1–16, 2021.
https://doi.org/10.1007/978-3-662-61515-7_1

the outcome of analysis done for a new type of transmission – it is a combination of a conventional manual transmission with a 48 V electrical motor and an electric actuated clutch. Other components like DC/DC converter and 48 V battery are part of the system but not in scope of the analysis described here.

Similar to applications available with hybrid automated transmissions (AT, DCT, CVT), also the hybrid manual transmissions can improve powertrain efficiency. But compared to automatics, the electrified manual transmission takes advantage of lower cost, higher efficiency and lower complexity. In addition, this hybrid concept makes this manual transmission more attractive to customers, since it supports new comfort features that have not been feasible before with the conventional version.

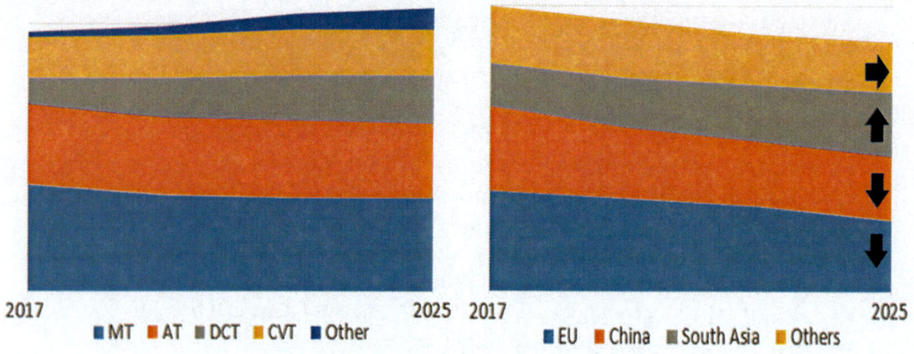

Fig. 1. Overall transmission share worldwide

Major attributes:

- Electric driving - Launch and drive by electric motor, change driving direction by electrical motor for autonomous parking
- Efficiency improvement - Recuperation during vehicle deceleration maneuver, load point shifting for combustion optimization, support combustion engine with electrical boost
- Comfort improvement - No clutch pedal (optional), torque fill during shifts (depends on topology)

Without going into details in this paper, there is an additional aspect. The introduction of 48 V power grid enables the usage of new electrical components with higher efficiency like e-turbocharger, e-AC compressor or e-brake booster.

3 Transmission Setup Definition

Many parameters have to be investigated for concept selection of a hybrid transmission. The major focus in this analysis is on fuel economy improvement. But also the customer satisfaction needs to be considered in the system setup.

The most dominant parameter is the position of the electrical motor (EM) in the power-flow. A hybrid solution for a transmission requires the EM position to be at the input-shaft or output-shaft. Therefore, the options can be limited to P2 or P3 topology.

Further factors have to be carefully analyzed and the optimum solution has to be a trade-off between different factors:

- Fuel economy (on-cycle, RDE, customer related)
- Drive quality (transition between different modes)
- Package (fit to target vehicles)
- Dimension (torque capacity, power)
- Design complexity (controls, reliability and durability)
- Cost (development, manufacturing)
- Feature support (el. driving, autonomous parking, torque fill etc.) (Fig. 2)

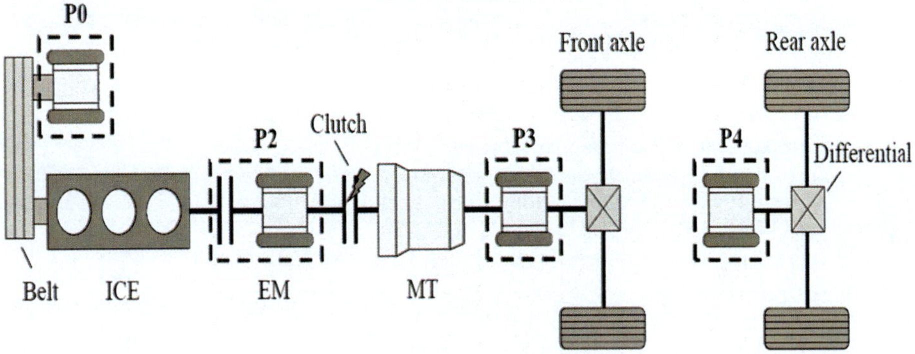

Fig. 2. Hybrid topology [2]

The A, B- and C-segments of passenger cars are the major domain of manual transmissions. In those applications the space in the engine compartment is often limited, vehicles are front wheel driven with transversely installed transmissions. For this analysis a C-segment vehicle was selected for reference.

Therefore, the additional distance required between engine and transmission becomes an important factor for the concept selection. In this aspect, the P2 topology is critical, especially with serial connection of the EM (coaxial to transmission input shaft). Due to package limitation, a parallel connection of the EM to the transmission output shaft was selected for this analysis. In this P3 configuration, the EM is very close to the wheels resulting in low torque losses in electrical drive mode and recuperation.

To keep the complexity of the gearbox on a minimum level, only one gear step (fixed ratio) for the EM connection was considered for the analysis. A fixed ratio between wheels and EM means that the EM speed is proportional to vehicle speed hence the use of the EM in less efficient operating areas cannot be avoided with a P3 architecture. This leads to a negative impact on recuperation and e-drive performance at certain vehicle speed operation areas. In this aspect, a P2 takes advantage of various gear steps of the transmission, which enables an operation of the EM in its highest efficient areas [2].

On the other hand, specific for manual transmissions with a P2 layout, the EM cannot remain connected to the driveline during gearshifts. The EM has to be disconnected in order to prevent an increase of inertia and associated synchronization effort. As a result, there is no recuperation and electrical boost possible during gearshifts. The overall system performance is linked to driver's shifting profile - it develops into a function of the number of gearshifts to be shifted and the gearshift duration itself.

Additional opportunity of the EM connected to the output shaft (P3) is the possibility to transfer torque to the wheels independently from gear actuation. The negative effect of propulsion torque loss during gearshifts can be reduced by EM torque applied to the wheels, resulting in higher driving comfort.

Not only the position of the EM in the power flow is an important factor but also the performance data of the electrical machine. Saving potential depends on EM recuperation performance, the higher the electrical power of the machine, the higher the potential for energy recuperation and thus for fuel saving. Figure 3 below shows that the relation between saving potential and electrical power of EM is not linear, the sweet point for a P3 architecture is at 20 kW peak (kWp). An additional increase of electrical power does not provide an appreciable saving improvement [2]. For the analysis described in this paper, an EM with a power of ~25 kWp (in generator mode) was selected in order to cover the upper end of the reasonable range.

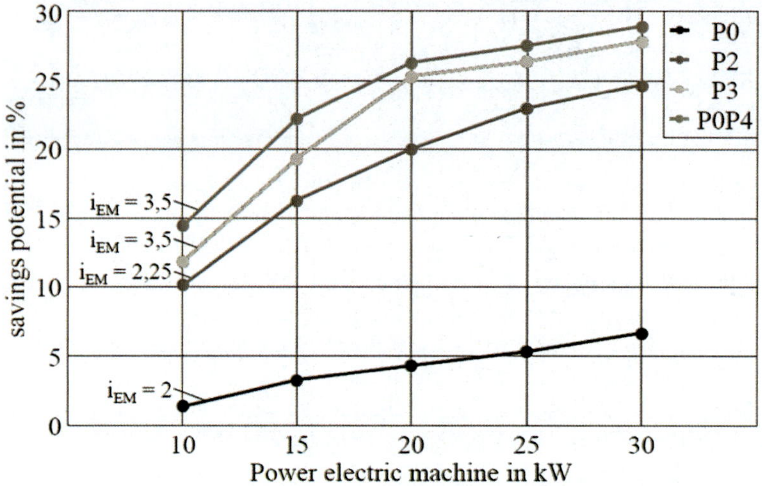

Fig. 3. Impact of different EM power on saving potential in WLTC [2]

There are more components in a hybrid vehicle which affect the entire system performance but are not further described in detail here. The 48 V battery is a big contributor to the overall performance. It needs to be carefully selected to make full use of EM performance in terms of power and energy. Within this project a range between 12 Ah and 20 Ah has been applied to identify impact on fuel consumption on- and off-cycle.

An electrically actuated clutch (e-Clutch) is part of the setup and used to disconnect the combustion engine according to the operation strategy if required.

4 Simulation Method

For the simulation, a 3D-model has been development to simulate the fuel consumption and the loads in the drivetrain as realistic as possible. The 3D-model is divided into three parts: the driven vehicle with all its drivetrain components, the driver and the driving environment. The drivetrain model contains all drivetrain components of the vehicle between the tank + combustion engine up to the wheels and all necessary vehicle parameters. In this model, the electric machine with the power electronics and the 48 V traction battery have been implemented in a way to model the hMT powertrain. At this point, the thermal behaviour of these electric components have been implemented based on real measured data. By doing so, the impact of hybridization is modelled closely to reality. As part of the model, the Institute of Automotive Engineering of the Technical University of Braunschweig has executed a driving study with 27 probands in different road profiles using a measurement car. This car has been equipped with a position and force sensor for the clutch pedal and the manual transmission gearshift lever. The clutch actuations as well as gearshift operations have been recorded, then analyzed and finally classified. To be able to model the gear shifting behaviour of the e-clutch in the model correctly, launch and gear shifting operations were implemented into the simulation based on the power requirement. Launch and gear shifting operations depending on the drivers driving behaviour regarding the driving style have been taken into account. A unique part of the modelling of the hybrid MT is the gear shifting operation. While clutch actuations are controlled and are optimized by the e-clutch system (no clutch pedal) the gear shifting is actuated by the driver.

The driver and the driven environment is based on the 3D method of the Institute of Engineering [4], which contains of 3 variations of each:

- Vehicle Load: low, medium and high
- Driving style: mild, average and sporty
- Road type: urban, rural and motorway

Every 3D-combination represents one driver. Each driver uses various representative statistics. For example, for gas and brake pedal end position and gradients as well as gear shifting points and shifting engine speed. By doing so, this represents the different driving behaviors. In sum, every combination of these variations adds up to 27 various customer related fuel consumption values [4].

Finally, a parameter study was executed with the aim of evaluating the influence on fuel consumption by varying for example EM power, gear ratio, ambient temperature or 12 V grid energy consumption. In particular, the influence of the gear ratio was of interest for the transmission design and layout. This ratio was increased to allow the electric machine to be connected up to 100 kph and then reduced for a spread up to 180 kph.

The advantage in regards to CO_2 of the hMT over the conventional MT drivetrain was investigated in WLTC (Worldwide harmonized Light Duty Test Cycle) for a C-segment vehicle. In addition, three RDE (Real Driving Emissions) cycles and customer related cycles have been analyzed. In general, RDE cycles can differ in terms of parameters and dynamics of the cycle. This is due to tolerances of RDE trips. This results in that a RDE cycle can be more or less demanding which ultimately leads to different effects on the emissions [3]. In this paper, three various valid RDE cycles are taken into account for the simulations. The three RDE cycles are shown in Fig. 4:

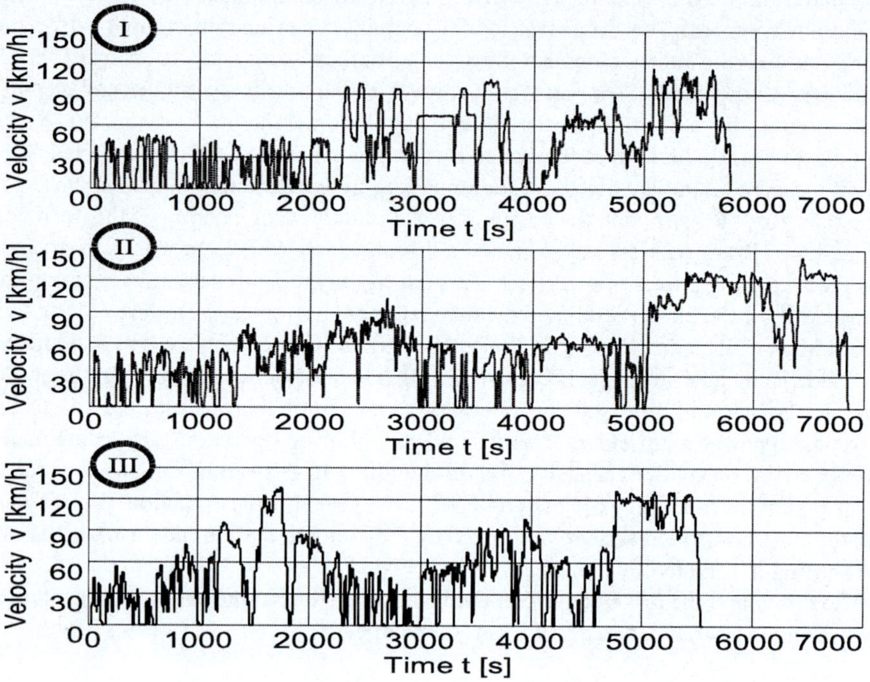

Fig. 4. Valid RDE cycle variant for the simulation [3]

The top profile (I) represents the RDE cycle with mild vehicle speed times acceleration. In the middle (II), the RDE is representative for an average case. The lowest course (III) represents for a more demanding RDE cycle.

5 Operation Strategy

The operation strategy in a hybrid vehicle is a major control parameter (beside battery size and EM performance) in order to achieve best fuel economy on the one side, but also to guarantee for a "charge sustaining system" on the other side. Moreover, it is also a goal to make use of the electric energy for comfort and customer facing features (like electric parking etc.). The operation strategy defines the coordination of torque between the combustion engine and the electric machine and the direction of electric (power) flow in any driving condition.

The available torque capacity of the components used in this project and the corresponding operating modes are shown in Fig. 5. As already described in "Concept Selection" chapter the gear ratio of the electric drive dictates the region were EM can be operated efficiently, in this analysis the limit is at 90 kph. Above this operation speed only the combustion engine will be used for propulsion - ICE solo mode. For lower torque demands and below 90 kph only EM can be used to propel the vehicle - EM solo mode. The maximum usable EM capacity depends on different factors like temperature (inverter and machine) and battery state of charge (SoC). It is obvious that there are conditions where EM solo mode is not available at all.

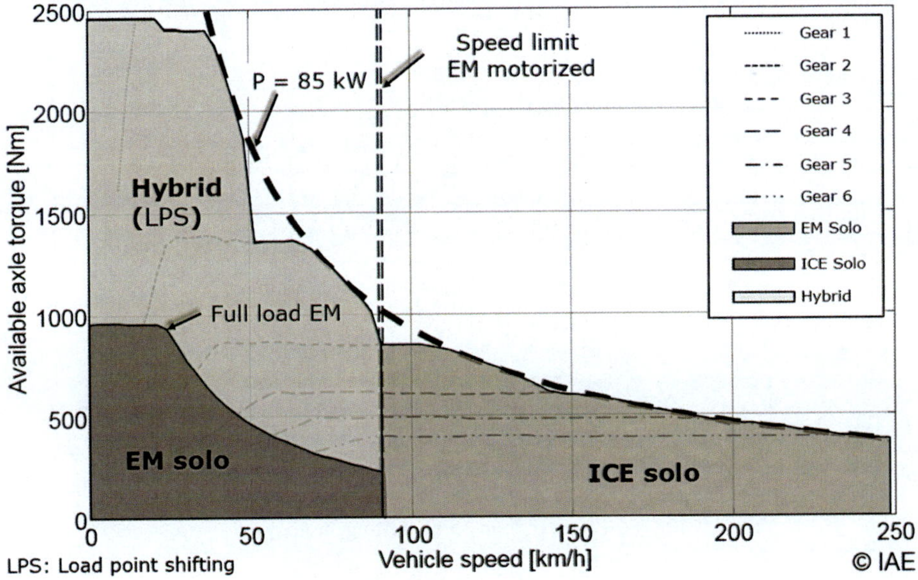

Fig. 5. Modes of operation strategy

Since the concept is based on manual transmission, the driver needs to shift gears even in EM solo mode. Therefore, a strategy needs to be developed to guide the driver (especially when the ICE is switched off), to ensure the transmission input shaft speed never underruns nor exceeds an appropriate value. This is crucial because at any time it might be required to re-couple the ICE to the drivetrain when changing from e-drive-mode to Hybrid- or ICE mode.

The area with higher torque requests is used as a Hybrid mode, that means the combustion engine and the electric machine are working together in order to achieve the best fuel economy. This operating mode is also called Load Point Shifting (LPS).

When the required propulsion torque is low and the ICE would be operating in a low efficient operating point, the electric machine can be used as a generator to charge the battery. This additional torque moves the operating point of the ICE to a fuel efficient point with higher torque. On the other way around, if high propulsion power is required, the EM can work as a motor and add power. With this, the ICE torque and thus CO_2 emissions can be reduced. For every ICE engine speed there exists an optimum torque regarding the efficiency. Based on the SoC, the requested torque by the driver as well as the torque capacity, the operating strategy algorithm calculates a factor called kopt. This factor determines the range the ultimate torque is shifted to between the ICE's requested torque and the optimum torque regarding the efficiency. The factor kopt is shown in Fig. 6.

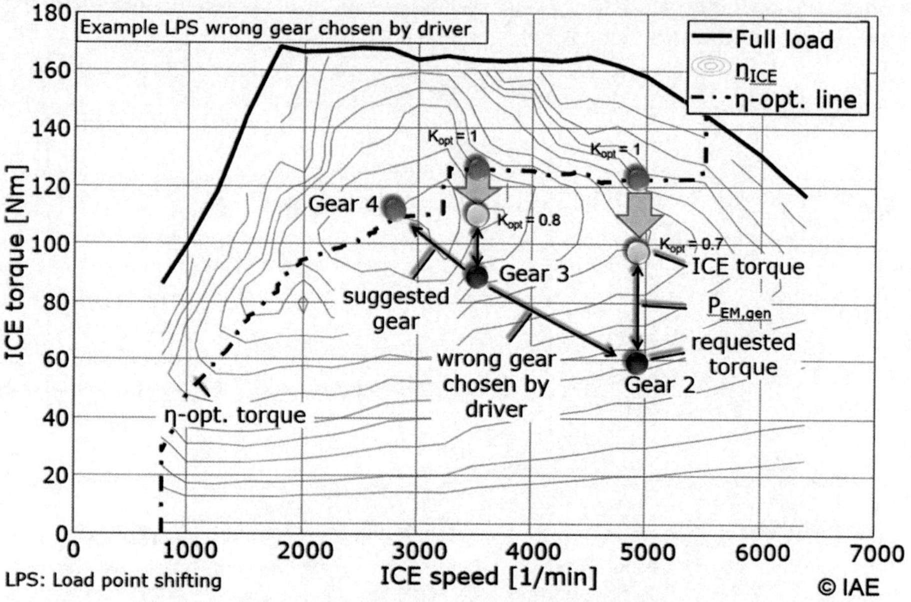

Fig. 6. Effect of load point shifting strategy

The transition process between all three modes – EM solo, Hybrid and ICE solo – is realized in combination with hysteresis in order to prevent high frequency shifting between the modes.

While combined with an automatic transmission, the operating point of the ICE can be optimized not only by shifting the load but also with changing the gear, in the manual transmission application the driver might not follow the shift indication command and skip the shifting operation or even shift into a wrong gear (e.g. to high engine speed). Thus the LPS function gains more importance for fuel consumption reduction in driver's hand.

Figure 6 shows an example where a driver selects 3rd gear. Shifting into 4th gear would move the ICE to the optimum load point and is indicated as recommended gear. Instead the 2nd gear is selected wrongly by the driver. With the Load Point Shifting functionality the engagement of the wrong gear is at least partially compensated.

5.1 Brake Blending

One major goal of a hybrid propulsion system is to make a maximum use of kinetic energy that would be wasted in a conventional powertrain during braking. Therefore, the aim is to apply generator torque at any time the driver uses the brake pedal to slow down the car.

The strategy for recuperation during breaking with an MT-hybrid basically is comparable to an AT application. Within this project, a simplified approach was used. The brake-system was calibrated in the simulation model in a way that there is a free travel in the first 15% brake pedal travel before the mechanical brakes are activated. Within this free travel the EM generating torque is applied. From driving quality perspective, this methodology is not a preferred one, but it ensures that in any maneuver the friction brakes are only active when the EM braking capacity is not sufficient to obtain the desired deceleration and thus a maximum energy recuperation is achieved (Fig. 7).

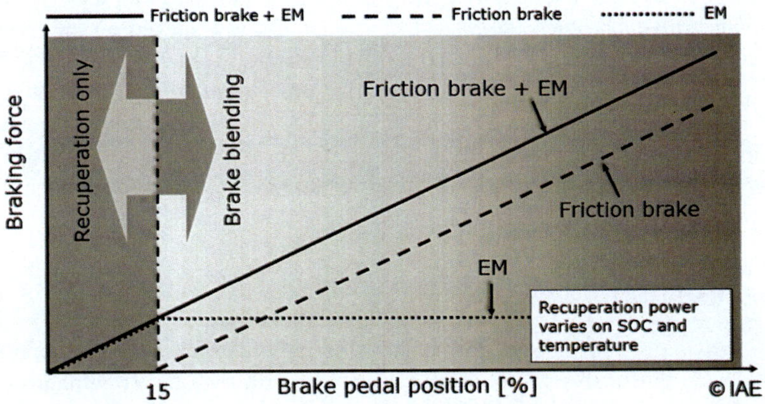

Fig. 7. Brake blending strategy

6 Fuel Consumption Results

6.1 Fuel Consumption Results

This section deals with the simulation results regarding the fuel consumption of the hMT compared to the conventional MT. The analysis shows the results in the WLTC, the RDE cycles, the customer based cycles and the variation of key parameters of the model listed before (Vehicle Load, driving style, road type). Figure 8 summarizes the results for the first three mentioned cases.

6.2 WLTC

The fuel consumption in the WLTC is shown by the solid horizontal line for the conventional MT and by the dashed horizontal line for the hMT. It can be seen that the hybridization reduces the fuel consumption by 21.1%. The conventional MT the fuel consumption is 5.61 L/100 km or 133.5 g CO_2 per km, respectively. This simulated value is very close to the real reference base car. Thus, the simulation quality is regarded as very high. The hybrid vehicle hMT achieves a value of 4.43 L/100 km or 105.4 g CO_2 per km. This relative improvement of the fuel consumption indicates a very promising capability of the 48 V hybrid system.

6.3 Customer Based Cycle

The results of the 3D customer based cycles are shown in Fig. 8, see three sections on the left side of the picture. Comparing the driving style with each other, the lowest fuel savings potential can be achieved by a driver with a mild driving style. This is because of the relatively low driving performances in terms of low drive and brake pedal actions. The highest savings potential is realised in the sporty driver. This results due to high driving requirements of the sporty driving style. The electric machine with its high power can meet the power requirements and eventually reduce the fuel consumption by the described hybrid functionalities in this paper. Comparing the different driving environments with each other, the lowest fuel reduction is achieved on the motorway. This is because of combination of relatively high vehicle speed and the characteristics of the electric machine. In this operating condition, the electric machine cannot operate as efficient and as often as it does at lower speeds. The highest savings potential is realised in urban areas. The reason for this are the brake and launch event in urban areas. Recuperation while braking and launch events in driving mode reduce the fuel consumption effectively.

6.4 RDE (Real Driving Emissions)

The RDE test is defined in order to measure the pollutants over on-road conditions, such as NOx. In this analysis three RDE cycles with different power demands applied not for pollutant emissions but as valid indicator for fuel consumption in customer use. Comparing these three cycles with each other, the cycle with higher loads is positively affected by the hybrid system and shows significantly higher reduction capability for CO_2 emission. In general, the absolute fuel consumption values are at a level between the 3D customer based cycles with a mild driving style and the WLTC.

6.5 Variation of Vehicle Key Parameters

In order to evaluate the impact of the various vehicle key parameters, the according variation is analyzed in the following. The results are shown in the following Fig. 9. The selected key parameters are:

- Gear ratio of the Torque Transmission Unit
- Number of gears
- Limit of recuperation only and brake blending
- Ambient temperature
- Power of auxiliaries
- Maximum power of the electric machine

The selected gear ratio of the Torque Transmission Unit indicates a very promising choice. By changing the gear ratio – longer and shorter – the relative fuel savings potential is higher compared to the simulated reference hMT vehicle described in this paper. A change in gear ratio leads to a change in speed for the electric machine and thus different

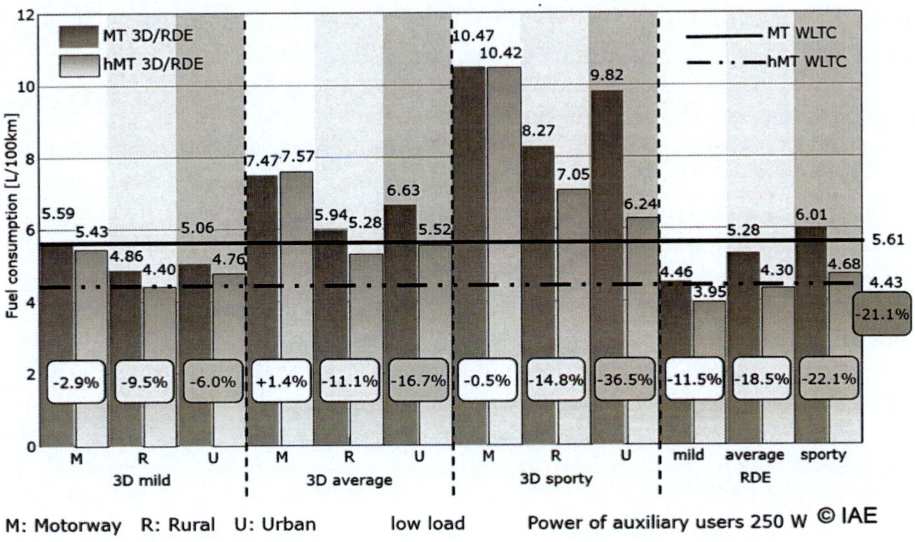

Fig. 8. Simulation result - Fuel Consumption WLTC/RDE/3D model

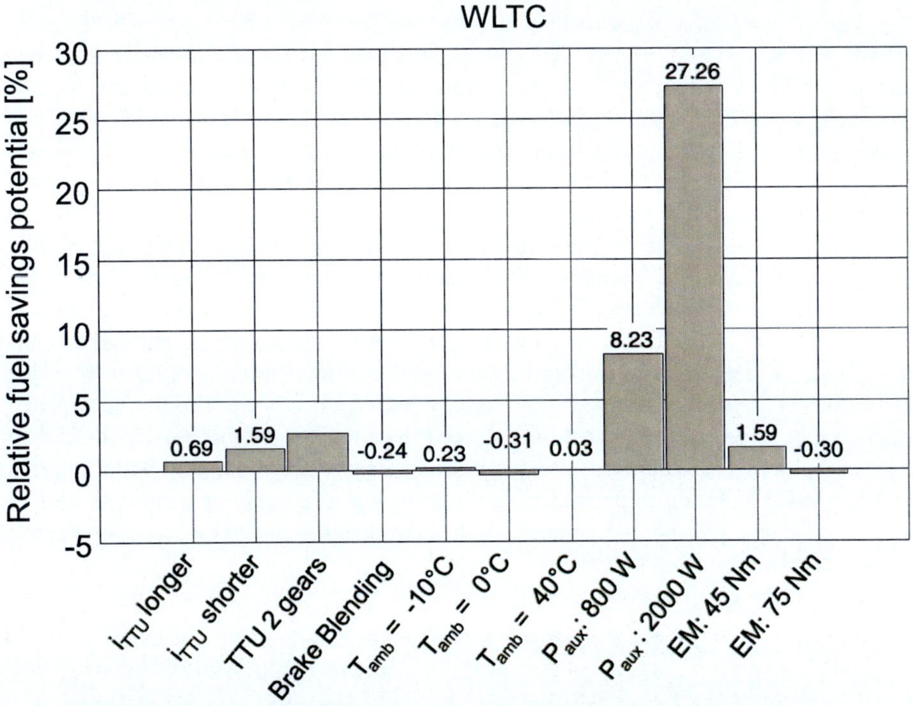

Fig. 9. Variation of selected key parameters – Relative fuel consumption change

operating points. All in all, these operating points are less efficient in the variation of the gear ratio.

One of the main points in vehicle development regarding electrified vehicles is the difference of one and two gears for the electric machine. At this point, a 2-gear-variant has been taken into account for the variation. The result shows a negative impact on the fuels savings potential. This is because of selected ratios of both gears and resulted speed profile for the electrical motor vs vehicle speed. Further analysis in terms of varying the gear ratios of the two gears can results in an improvement of the system efficiency.

The third analysis varies the limit for recuperation energy as function of driver brake request. This value has been increased in order to evaluate its impact on the system efficiency. The results in Fig. 9 show that the fuel saving can be improved by 0.24%. Higher recuperation limit does not necessarily provide a significant efficiency improvement in driving cycles used for this analysis. The already calibrated limit in the simulated reference hMT vehicle is capable of recuperating almost all of the potential recuperation energy.

The ambient temperature has an impact on the 48 V battery due to the battery's characteristics that depend on the temperature. The reference hMT vehicle is simulated with ambient conditions meeting the WLTC regulations that is at 23 °C. As expected, the battery performance decreases at colder environment and at −10 °C the fuel consumption gets worse by 0.23%.

A higher power of auxiliaries increases the fuel consumption significantly. To characterize this negative impact of fuel consumption, the power of auxiliaries has been enhanced to 800 W and 2000 W in the simulations. For the first enhancement, the relative fuels savings potential is increased by 8.23%. For the second enhancement, the increase is 27.26%. This result shows that the fuel saving potential does not increase linearly with the electrical load and the electrical consumption is the biggest contributor to fuel saving potential in this setup.

The last selected key parameter is the maximal power of the electric machine. Reduction of the maximal power down to 45 Nm has a negative impact on fuel savings of 1.59%. An enhancement of the electric machine's power up to 75 Nm can achieve an addition saving of 0.30%. The different machine power affects its efficiency in particular operation points as well as its capability in motor and generator mode. A higher maximal electric power supports the operating strategy with higher power requirement hence broader range were electrical machine can contribute to overall efficiency. With 75 Nm maximum power, the vehicle is operated more in electric driving mode that ultimately leads to a lower fuel consumption. But considering the small improvement with 75 Nm version on the one hand but higher costs on the other hand, the power increase of electric motor is considered as not beneficial

7 Mechanical Load on HMT

The mechanical stress for components on the conventional vs. hybrid manual transmission is expected to be different. The effort for the new design depends on scope of modifications due to new requirements coming from the operations strategy, torque vs. speed profile, new features.

In order to be able to compare the transmission loads of the hMT and MT with each other and to identify the most critical driver for the gearbox, the load spectra are taken into account. Thus, the Wöhler line plays an important role when looking at the damage resulted due to loads. In order to cover a wide range of failure mechanisms in the transmission two different Wöhler lines have been used, one with low exponential and another one with high exponential. In a double logarithmic diagram – overrolling cycles on the x-axis and torque on the y-axis – a low exponential results in a steep Wöhler line whereas high exponentials lead to a flat line. Thus, flatter course will focus on the higher torque parts and lower impact of low torque areas while steeper courses of the Wöhler line take the lower part torque into account as well.

This sensitivity study shows, that the range the Wöhler line varies in, does not influence the statements on the most critical driver. The following Fig. 10 shows the results for the damaging of the gear input shaft.

Generally, Fig. 10 shows that the minimum and maximum damaging differ in a wide percentage range for the gear input shaft. Both Wöhler lines are created in a way that the most critical driver causes a damaging of 100%. By doing so, it is possible to directly compare the 27 different driver behaviour investigated within the 3D method with each other as well as to compare the damaging of the drivers with the two Wöhler lines. Figure 10 shows that for both Wöhler lines the most critical driver does not change. In particular, it is the sporty driver on motorways with a vehicle with low load. Regarding the vehicle load, higher vehicle load results in lower damaging. This is due to the 3D statistics. For example, the gear shifting speed is different for a vehicle with low and high load.

The following Fig. 11 represents the damaging of the hMT compared to the MT with two different Wöhler lines for the gear output shaft. The most critical driver for the gear output shaft is the sporty driver in urban areas with a low load vehicle. One of the main aims of the hMT study was to identify whether the gearbox of the conventional vehicle can be used for the hMT without modifications for durability. Thus no boosting functionality was applied in the operating strategy. With the results shown in Fig. 11 the gear output shaft load behaviour is the basically the same for the hMT as it is for the MT.

8 Conclusion

There is a significant potential for efficiency improvement that can be reached by P3 architecture of a hybrid manual transmission. In combination with a 25 kWp electrical motor and the applied operation strategy, the fuel consumption improvement opportunity is more than 20% in the WLTC compared to a similar conventional C-segment vehicle. The forecasted improvement in customer hand is even higher, based on statistical data used for this analysis.

Use cases come along with hybridization do not lead to higher absolute loads. In certain operation conditions, like in generator mode, the load is different compared to a conventional manual transmission and needs to be assessed more in detail for the affected components.

Further advantage of this setup is the ability to drive electrically in both directions without the need to shift gears. Features like autonomous park assist can be supported

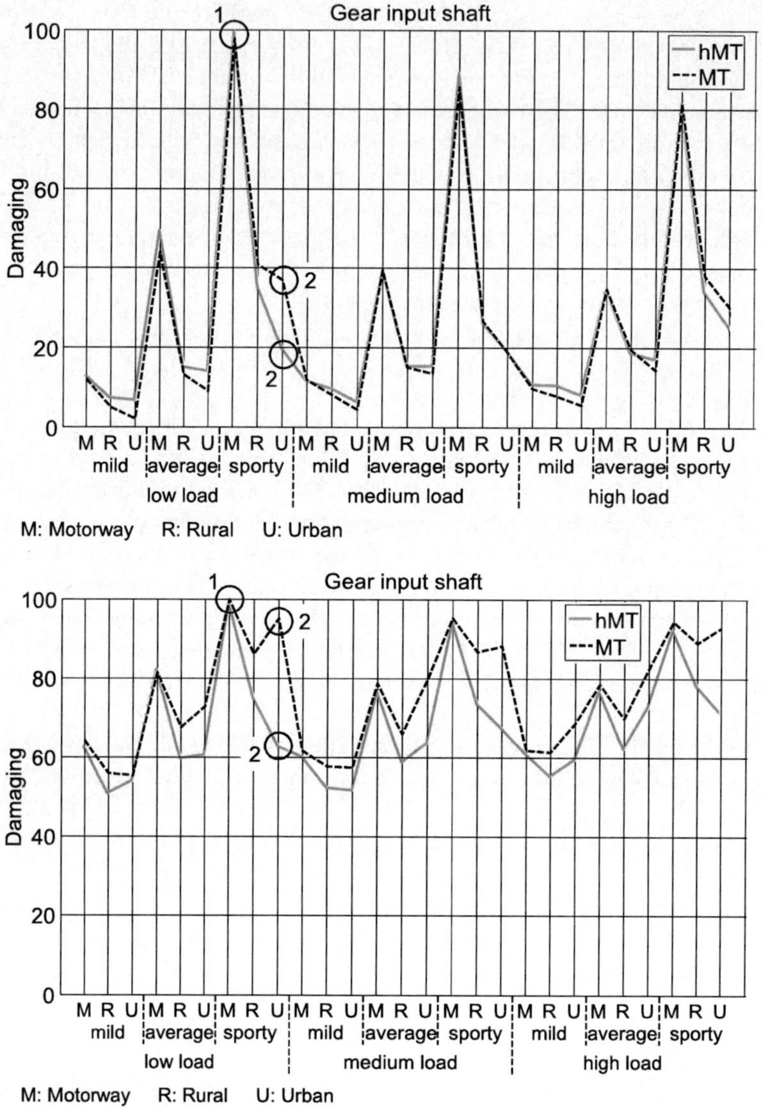

Fig. 10. Damage based on "Wöhler" curve with high exponential (left) vs low exponential (right)

even with a manual transmission, which makes this technology more attractive to customers. In addition, propulsion torque is possible during gear shifts, providing the opportunity for driving comfort improvement during gear shifts on manual and automated manual transmission (AMT) systems.

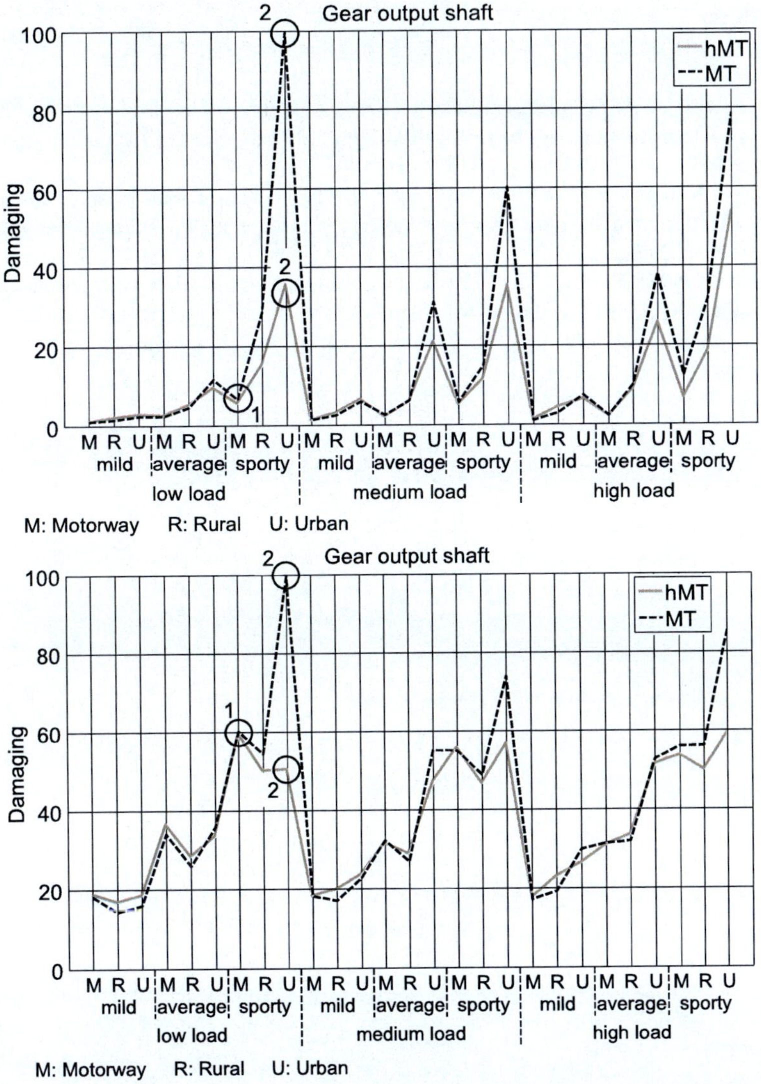

Fig. 11. Damage based on "Wöhler" curve with high exponential (left) vs low exponential (right)

References

1. IHS Markit statistic
2. Werra M, Ringleb A, Müller J, Küçükay F (2018) Topology comparison of 48 V drivetrains with manual transmission. In: 17th CTI symposium, TU Braunschweig, Institute of Automotive Engineering
3. Ringleb A, Küçükay F (2017) Representative driving cycles for real driving emissions. In: 14th symposium on hybrid and electric vehicles, Braunschweig. TU Braunschweig, Institute of Automotive Engineering
4. Küçükay F (1995) Repräsentative Erprobungsmethoden bei der Pkw-Getriebeentwicklung; VDI Berichte Nr. 1175

Torque Sensors for High Volume Production Applications

Julius Beck(✉)

Methode Electronics International GmbH, Rheinstr. 40, 55435 Gau-Algesheim, Germany
Julius.beck@methodegermany.com

Abstract. Magnetoelastic sensors are the first solution that allows measuring torque and other forces economically so they can be integrated into high volume applications. The current status quo for force measurements - strain gauge type sensors - require extensive manual labor during application, fault susceptible telemetries and recalibration due to aging effects. This makes them suitable for use in test benches and prototype vehicles, rather than high volume production applications. Magnetoelastic sensors have many advantages that make them ideal for high volume production applications. The sensors have small space requirements, the sensing object (e.g. a driveshaft) does not need to be modified and there is no telemetry, making the technology truly non-contact. Magnetoelastic sensors have excellent performance characteristics, exceptional long-term stability and have been proven in many high volume applications for more than 10 years, including automotive (e.g. electronic power steering, anti-roll stabilization), consumer products (e.g. eBikes), and agricultural equipment (e.g. power take-off shaft).

The magnetoelastic effect states that a ferromagnetic material will change its magnetic properties when subjected to mechanical stress (e.g. due to an applied torque). A standard magnetoelastic sensor consists of two circumferential magnetization bands that are encoded into the shaft during production. Sensing coils placed around the shaft pick up changes in the magnetic field when torque is applied to the shaft, which is the sensor electronics converts into the output signal.

An almost completely automated production process and multiple large-scale manufacturing facilities around the world allow Methode Electronics to produce over three million sensor units per year, satisfying even the highest volume orders.

Keywords: Torque sensor · Magnetoelastic · Drivetrain optimization

1 Overview

1.1 Introduction

Over the last decades, several sensor technologies to measure torque on load bearing components, such as a driveshaft, have been developed. These include Surface Acoustic Wave technology and torsion bar setups, where the torsion angle is measured either through magnetic or optical system [1]. The most used and well-known technology, however, is the strain gauge type sensor.

Euroforum Deutschland GmbH (Ed.): *CTI SYMPOSIUM 2019*, pp. 17–24, 2021.
https://doi.org/10.1007/978-3-662-61515-7_2

The common misconception still held by most people in the automotive industry, that measuring torque accurately and reliably is too expensive for high volume production applications, originates largely from the most commonly used technology in the automotive industry to measure torque - strain gauges. Strain gauge type sensors require extensive manual labour during application, fault susceptible telemetries and recalibration due to aging effects, making them expensive and difficult to integrate. Consequently, strain gauge sensors are more suitable for use in test benches and prototype vehicles, rather than high volume production applications. In many applications, direct torque measurements essential for closed loop control systems (e.g. transmissions) have been replaced by complex and often inaccurate mathematical models, which only estimate the true torque value, due to cost considerations.

1.2 Technology Description

Magnetoelastic torque sensors make use of a physical phenomenon called the Villari effect, discovered in 1865 by Italian physicist Emilio Villari (1836–1904), which states that a ferromagnetic material will change its magnetic properties when subjected to a mechanical stress (e.g. due to an applied torque), creating a magnetic anisotropy [2]. Ways to exploit this effect to measure torque have been researched for more than 30 years [3]. One way is to magnetise a region of a shaft with two circumferential bands as shown in Fig. 1.

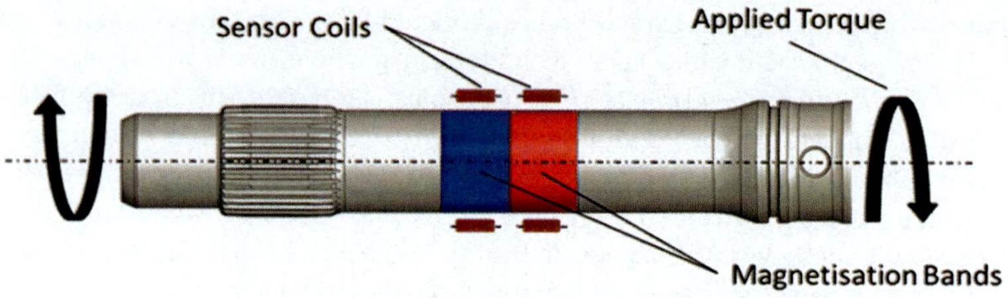

Fig. 1. Magnetization with two circumferential bands

This magnetic encoding takes place only once during production and remains inside the shaft for the lifetime of the sensor. In the stationary secondary part of the sensor, highly sensitive sensor coils are placed in close proximity of the magnetic bands around the shaft. These sensor coils pick up changes in the magnetic field emitted by the encoded shaft caused by an applied torque, creating the sensor signal. The relationship between applied torque and signal received by the sensor coils is linear, making any signal linearization unnecessary.

Having a sensor that consist only of a magnetised shaft and a stationary housing containing sensor coils and an electronics PCB creates a truly contactless torque sensor with unique benefits compared to other competing technologies:

- Non-contact: There is nothing attached to the shaft and no other modifications are necessary. In contrast to strain gauges, magnetoelastic sensors do not require any fault susceptible telemetry or adhesives that cause long-term instabilities and have a limited lifetime.

- Durability: The sensor can be completely encapsulated and is suitable for harsh operating environments. It can be submerged in caustic liquids, exposed to high temperatures (210°) and withstand continuous and strong vibrations. It is insensitive to dirt.
- Performance: Accuracy, repeatability and linearity are comparable to strain gauge sensors. The magnetoelastic sensors have an outstanding dynamic response in comparison to any other known torque or force sensing technology. This extremely high signal bandwidth enables detection of engine cylinder misfiring and accurately resolves any other high frequency signal.
- Minimal Packaging Requirements: Small packaging and a flexible design that can be optimised for axial or radial space constraints makes it easy to integrate the sensor.
- Certification: ISO 26262 ASIL levels are achievable and already in production for several years.
- No Cross-coupling: In magnetoelastic sensors the torque/force measurements are completely decoupled, so that, for example, a torque sensor is not influenced by shear forces on the shaft.
- Long-term Stability: The magnetisation is permanent, and the calibration will remain constant over the lifetime of the vehicle. Using a gearbox mounted magnetoelastic torque sensor, a long-term stability of 480.000 miles was demonstrated. The sensor ran without fault or change in performance characteristics. The test was only stopped due to the wearing out of gear teeth in the gearbox and torque could no longer be transferred.

2 Production and Performance

2.1 Industrial Scale Production

The key element that has allowed Methode to make magnetoelastic sensors a serious competitor for strain gauge sensors is the industrialisation of the technology. The industrialisation of a product that used to be assembled in largely manual fashion has allowed Methode to produce in high quantities and at very competitive prices. The technology allows for a level of automation not possible with strange gauge sensors. Current production facilities are located in Malta, Belgium, China, USA, Canada and Mexico (see Fig. 2), having a combined capacity of over three million sensors per year, equating to more than 10000 sensors per working day.

2.2 Cost and Performance Comparison

The industrial scale and automated production of Methode's magnetoelastic torque sensing technology have a beneficial effect when comparing it to other competing sensing technologies. Figure 3 shows that Methode's magnetoelastic sensing technology has the highest sensor performance at the lowest production cost. In the future, this lead is likely to increase as R&D work is continually further improving the sensor performance while slightly decreasing the production costs.

Fig. 2. Methode's worldwide manufacturing locations

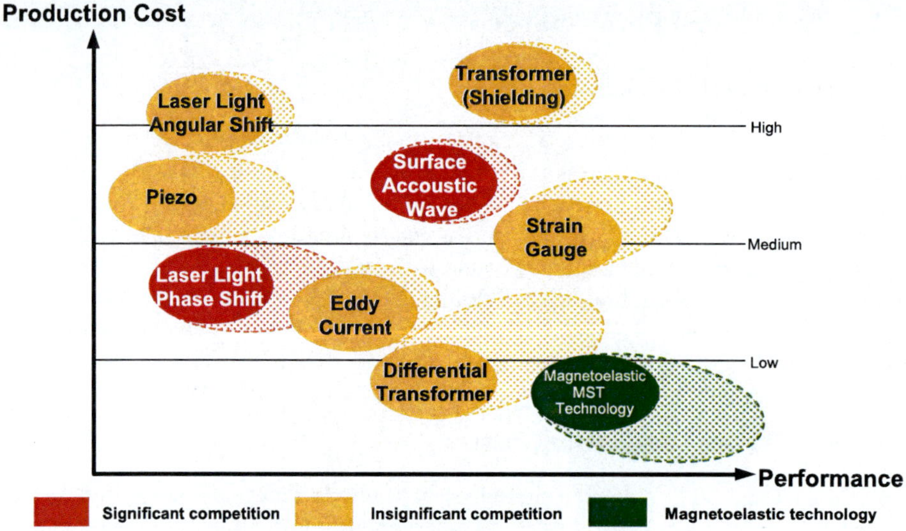

Fig. 3. Cost/Performance comparison chart

3 Examples of Torque Sensor Applications

3.1 Electronic Power Steering (EPS)

The magnetoelastic torque sensor technology is used in Electronic Power Steering (EPS) products with a total of several million units produced since 2009. Methode's proprietary Dual-Dual-Band technology provides the necessary redundancy in this safety critical application. Additionally, high overload requirements make this a very difficult application for any sensing solution and magnetoelastic is the only one known to provide a solution for a non-compliant sensor setup removing mechanical complexity and thereby providing a much more direct steering performance and improved driving experience.

3.2 Active Anti-roll Stabilizer

Methode has also worked together with a major German supplier for the automotive and mechanical engineering industry supporting the design of an electromechanical anti-roll stability control system. This system is designed to reduce the roll angle of large passenger vehicles at high cornering speeds, improving driving comfort whilst simultaneously increasing vehicle dynamics and safety. The system is mounted between the left and right half of the stabilising bars and can either stiffen or decouple the system depending on the driving conditions. The magnetoelastic torque sensor is the underlying sensor technology that updates the system's control unit with real-time torque values within milliseconds. The system is primarily used in larger vehicles such as SUVs and has been in production since 2015 with a maximum production capacity of 140 k units per year.

3.3 eBikes and Pedelecs

Another safety critical application where the magnetoelastic sensor technology has proven itself is eBikes and pedelecs. The direct and absolute measurement of the torque applied to the bottom-bracket by the rider, results in optimized motor control and smoothest riding experience. Methode is exclusive supplier of torque sensor for the current market leader as well as a large number of other eBike motor and system suppliers.

3.4 Engine Control

Methode's magnetoelastic sensor technology can be applied to components other than a shaft in order to measure torque. For example, through a patented process it is possible to magnetically encode a region of the Flexplate; there are no magnets or other sensing targets fitted to the Flexplate. A secondary sensor unit placed behind the Flexplate features a series of fluxgates sensors. Each time a cylinder fires it creates a unique torque signal that is propagated through the Flexplate. This torque applied to the Flexplate creates a specific change to the magnetic coding which in turn is detected by the fluxgates. The resultant measurement is conditioned and transferred to the ECU as a linear signal representative of the applied torque.

However, this is only half the story since the Flexplate mounted Methode torque sensor is also capable of measuring the torque exerted on a transmission (Fig. 4). Currently, electronically controlled transmissions rely on empirical based look-up tables to estimate engine torque based on throttle position and RPM. The transmission control unit (TCU) uses this information to help select the gear, shift points, and shift speed. However, it is often difficult to accommodate all the potential factors that can affect the efficient operation of the powertrain. New engine technologies, such as, cylinder deactivation, variable valve timing, start stop sequencing, and various environmental treatments all contribute to this complexity with service wear also affecting powertrain characteristics over time. Complex control algorithms, test evaluations, and qualification time can often be significantly reduced with a real-time torque sensor measurement.

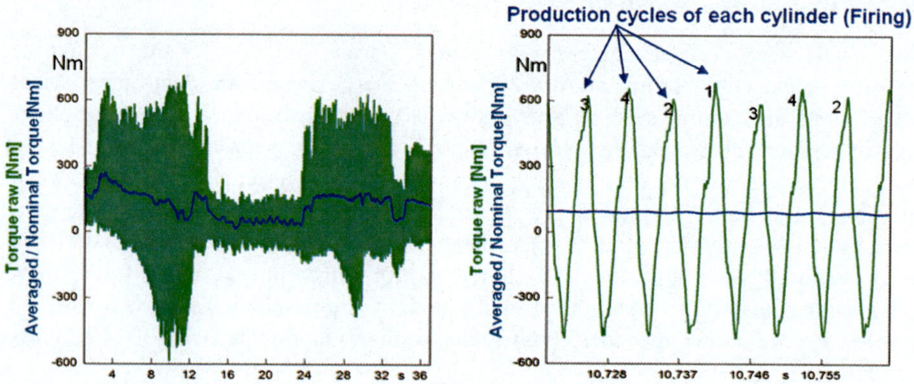

Fig. 4. Engine control vs. transmission control (Note: unique and repeatable peak torque)

3.5 Cylinder Misfire Detection

Figure 5 shows the effect of misfiring in the measurement signal. The signal resolution of the sensor means misfiring can be detected immediately after the first incidence. As long as cylinder 1 is working properly there is a clear peak visible. After shutting down this cylinder, the corresponding peak disappears. Current detection time required of the ECU is between 1 and 2 s. The real-time torque sensor is a good indicator of engine abnormalities.

3.6 Transmission in and Output Shaft

Figure 6 shows a transmission sensor based on the same dynamic torque sensing technology that has surpassed 480,000 miles without failure or change to its performance. The magnetisation is permanent, and the calibration will remain constant over the lifetime of the vehicle. Shown in Fig. 7 are test results regarding the deviation in sensitivity (±0.5%), offset (±0.05%) and hysteresis (±0.1%). The benefit of Methode's torque sensor arrangement is that it requires no alteration to the existing engine, has been proven in

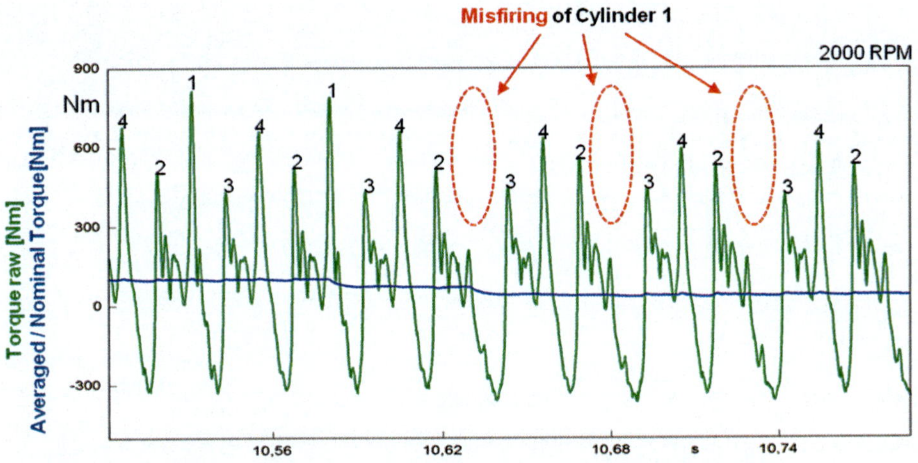

Fig. 5. Misfire signal by switching cylinder 1 ignition off

key life testing, requires no specific processes or after treatment of the powertrain components, and can withstand temperatures of up to 210 °C making it extremely suitable for harsh environments.

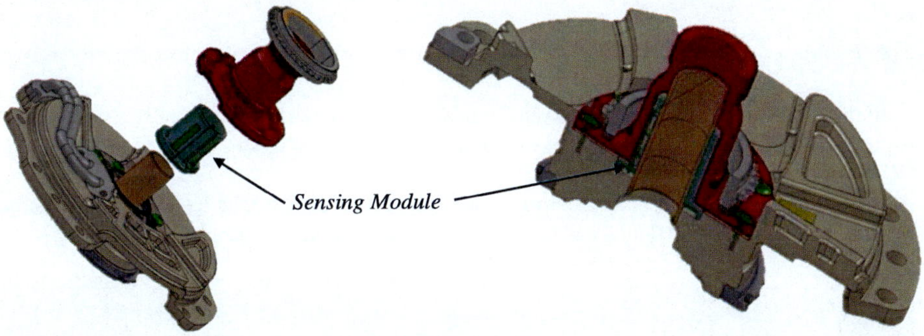

Fig. 6. Transmission torque sensor

4 Conclusion

There have been significant improvements to the torque sensing system since sensing torque in a shaft was best described as being a laboratory operation. Among various torque sensing methods, a magnetoelastic torque sensing technology appears the most promising for high volume automotive production applications, due to their superior performance and competitive cost. Summarising the magnetoelastic sensor technology:

- High Performance: Meets or exceeds strain gauge performance characteristics

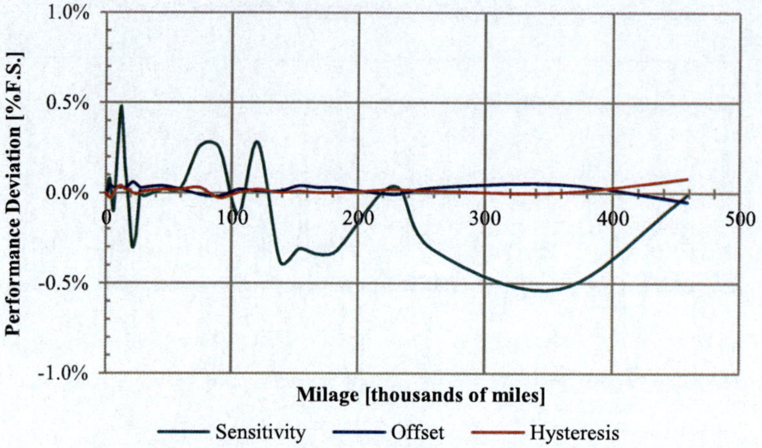

Fig. 7. Sensor performance deviation over 480.000 miles

- Low Cost: Highly automated manufacturing brings cost down
- Proven: multiple high volume automotive applications in the field today
- Easy Integration: Small packaging requirements & no telemetry for easy integration

References

1. Kalinin V (2011) IEEE international ultrasonics symposium proceedings, pp 212–221, IEEE
2. Boll R, Overshott KJ (2008) Sensors, a comprehensive survey, Magnetic sensors, vol 5. Willey-VCH
3. Fleming WJ (1990) SAE technical paper series: magnetostrictive torque sensors – comparison of branch, cross, and solenoidal designs. SAE, Warrendale, pp 51–78

Quick Start with AI for Diagnostics and Calibration

Ulrich Bodenhausen[1,2](✉)

[1] Vector Consulting Services GmbH, Ingersheimer Str. 24, 70499 Stuttgart, Germany
ulrich.bodenhausen@vector.com, info@grow-with-ai.com
[2] Ulrich Bodenhausen AI Coaching, Stuttgart, Germany

Abstract. AI offers great new opportunities. To become a successful player in AI driven business, companies need to find the right balance between investments for AI and benefits in their product strategy. The paper gives a short introduction to AI, especially Deep Learning Neural Networks (DLNN) and provides long-term experiences on the successful application to real-world applications. A scenario for the application of DLNN to drivetrain calibration and diagnostics as a very promising field of application with a beneficial balance of investment and benefits is described.

Who will be the right companies to drive AI in drivetrain calibration and diagnostics? These applications are "too special purpose" for AI giants like Google, Apple or Amazon and will therefore very likely not be a suitable field of application for them. This is a great opportunity for the automotive industry: With the application of AI, drivetrain calibration and diagnostics can provide very valuable benefits for the automotive industry by reducing the growing effort for calibration on the one hand, and on the other hand enhancing availability and reducing quality cost by predictive maintenance.

The paper concludes with an extract of experiences from 25 years of successful AI commercialization like speech recognition that will support a quick start with AI for automotive applications besides autonomous driving.

Keywords: Artificial Intelligence · Diagnostics · Calibration

1 AI is Important for Automotive Industry

Artificial Intelligence (AI) offers great potential not only for autonomous vehicles, but also for many other automotive applications, where an AI-based algorithm can be used to augment a car functionality by usage of available or new sensor data. Voice assistants are one example of an already common application of AI in cars. According to a recent report about the market in the U.S. [1], voice assistants are already used at least daily by approx. 27 million users. Concerning the usage of devices, consumers say that they are equally likely to have used voice through the voice assistant native to the car and a smartphone connected to the car via Bluetooth.

Euroforum Deutschland GmbH (Ed.): *CTI SYMPOSIUM 2019*, pp. 25–32, 2021.
https://doi.org/10.1007/978-3-662-61515-7_3

How can AI be used to augment automotive functionality? Existing ECUs can be extended by additional sensors combined with AI to enhance the functionality (see Fig. 1) or diagnostic functions can be extended by AI to improve diagnosis (see Fig. 2).

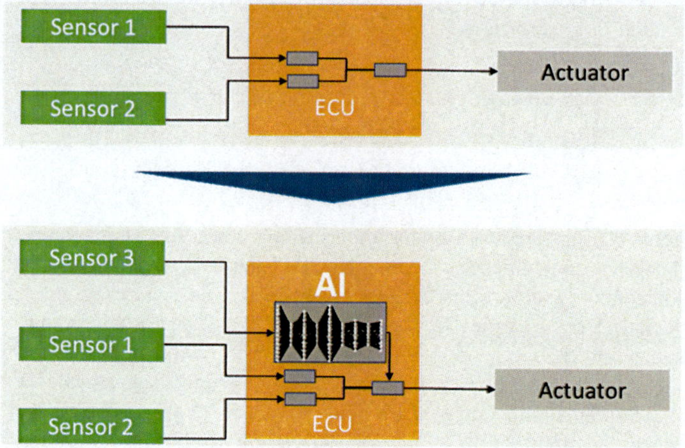

Fig. 1. AI offers great potential to enhance automotive functionality. Existing ECUs can be extended by additional sensors combined with AI to enhance the functionality.

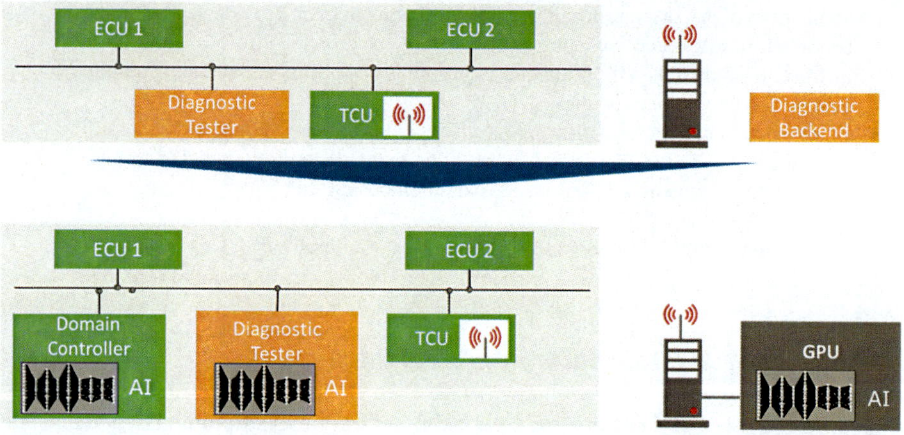

Fig. 2. AI offers great potential to enhance automotive functionality. Diagnostic functions can be extended by AI, both in the diagnostic tester as well over the air.

In practice of real-world AI applications, the optimization of the performance is very important to achieve a functionality that meets the expectations of the users. Example: In case of a voice assistant, the number of languages that are accepted, the size of the vocabulary, the recognition accuracy in various noisy environments are important criteria for user acceptance. To meet the expectations of the users, functional improvements of the AI module by optimization of internal architectural parameters (such as number of hidden layers and detailed connectivity of Deep Learning Neural Networks) and big data for the training are key success factors for excellent performance. This results in a big increase in the required resources [2].

In summary, AI provides great potential, but it will require specific AI-know how, intensive data collection and functional improvement as well as expertise and resources for safety assurance. It is likely that a collaboration of several partners is needed to bring together the required expertise and resources.

2 Machine Leaning, Neural Networks and Deep Learning

For the context of this paper, Machine Learning is defined as an approach using inductive learning techniques for system design, where the run-time system uses the results of a learning process to perform algorithmic operations (e.g. running a Deep Learning Neural Network having precomputed weights).

Machine Learning algorithms have been developed and applied for several decades. In case of application to real world data, the following design goals have been pursued:

- The run-time system performs with low deviation from desired system functionality in the defined application context
- The functionality is robust against variances (i.e. rotation, size, shape)
- The performance is uniform across classes of inputs (i.e. critical classes do not yield unacceptable poor performance)
- The system is robust against differences between its training and testing data sets

One class of Machine Learning approaches is Neural Networks (NN). A NN consists of simple processing units ("nodes") which are connected via weights. In the generic approach each node of one layer of nodes is connected to each node of the next layer. A NN is a Deep Learning NN, if there are several hidden layers between input and output for internal processing (see Fig. 3).

In practice over several decades of enhancement of Machine Learning, the approach of Deep Learning Neural Network NN and Deep Learning Convolutional NN have proven to be very powerful approaches to Machine Learning in real-world applications.

The steps for the successful development of AI-driven applications and the implications on resources and time [2] are shown in Fig. 4.

3 Deep Learning, Calibration and Diagnostics

The principle of application of an AI algorithm for diagnostics in the automotive context is to analyze data from the car to predict an unwanted event in future, see Fig. 5. The challenge is to detect the unwanted event early enough with a time interval Δ t to be big enough to be able to react with measure (e.g. service). In order to use Machine Learning for system design it is necessary to collect and label meaningful data, which means data of events leading to unwanted events.

In case of Deep Learning NN, the approach could be implemented with a moving window over data in time (realized by "time delays") and a specific Neural Network

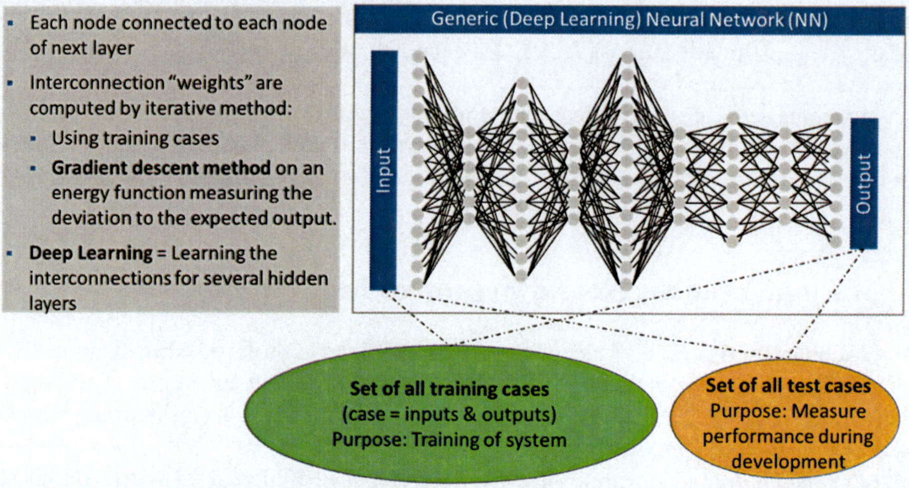

Fig. 3. Generic (Deep Learning) Neural Network. Each node is connected to each node of next layer. Deep Learning Neural Network refers to several internal (hidden) layers between input and output for internal processing.

Fig. 4. Seven steps for your success with development of AI applications. The diagram shows the resources over time. There are two phases: In the first phase, the team has to be effective to be fast. In the second phase before market entry, there is a high demand for resources.

architecture that is tailored to the task and the available amount of data. In practice, Big Data and computing power allows training of large Neural Networks today. However, the data has to be collected and prepared/labeled before it can be used for the training. As shown in Fig. 4, it is important to invest resources to enhance an AI application such that it is beneficial and attractive to the users.

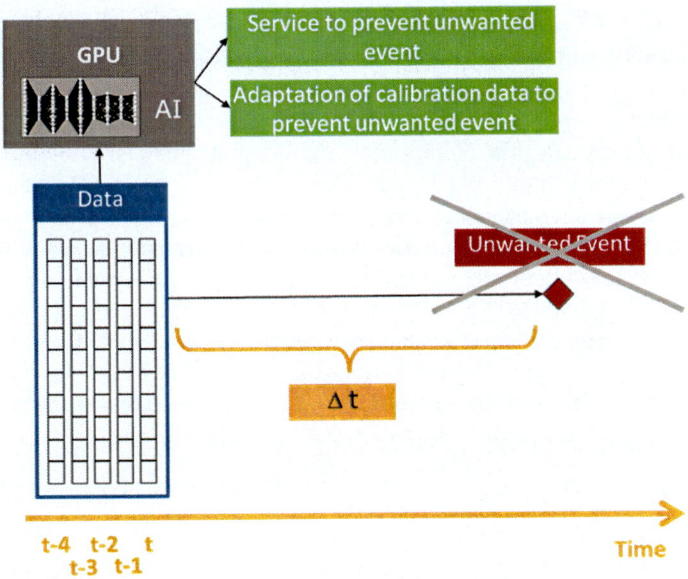

Fig. 5. The principle of application of an AI algorithm for diagnostics in the automotive context is to analyze data from the car to predict an unwanted event in future.

4 AI Landscape and Strategy

In the last chapters we have discussed the potential of AI in automotive, clarified what a Deep Learning NN is and how it could be applied to Calibration and Diagnostics. Now there are two questions:

1) Who will be the right companies to drive AI in drivetrain calibration and diagnostics?
2) Who will be interested in applying AI to calibration and diagnostics?

Without doubt there are a couple of large, major AI companies who have invested into AI for decades, have collected a huge experience and reputation with it and are making money with business models they have developed around it. These companies continue to further improve general purpose application like Alexa, Siri, Cortana, meaning that these applications are driven towards even more languages or bigger vocabulary. These applications are general purpose in a sense that there is a potentially very large group of users. English alone has 340 million native speakers. Google and to some extent Apple also invest in another general purpose application which needs to cover many different types of conditions and country specific adaptations: Autonomous cars.

What about Diagnostics and Calibration? These applications seem to be "too special purpose" for AI giants like Google, Apple or Amazon and will therefore very likely not be a suitable field of application for them.

This is a great opportunity for the automotive industry: With the application of AI, drivetrain calibration and diagnostics can provide very valuable benefits for the automotive industry by reducing the growing effort for calibration on the one hand,

and on the other hand enhancing availability and reducing quality cost by predictive maintenance. This could become a very attractive special purpose AI application with strong advantages for those companies who have detailed know about the diagnosed and calibrated products.

In addition, there is a third opportunity of AI application in automotive: Clever application of general purpose AI products like speech recognition for diagnostics and calibration to provide up to date user interfaces. Example: Instead of connecting a diagnostic tester to the car, the diagnostic car data could be potentially accessed via special spoken keywords. In summary there are three types of companies applying AI:

- Major AI companies who develop and drive general purpose applications
- Clever users of general purpose applications like speech recognition for their product
- Companies developing special purpose AI applications like car Diagnostics or Calibration to provide a great additional value to their customers

See Fig. 6 for the AI landscape.

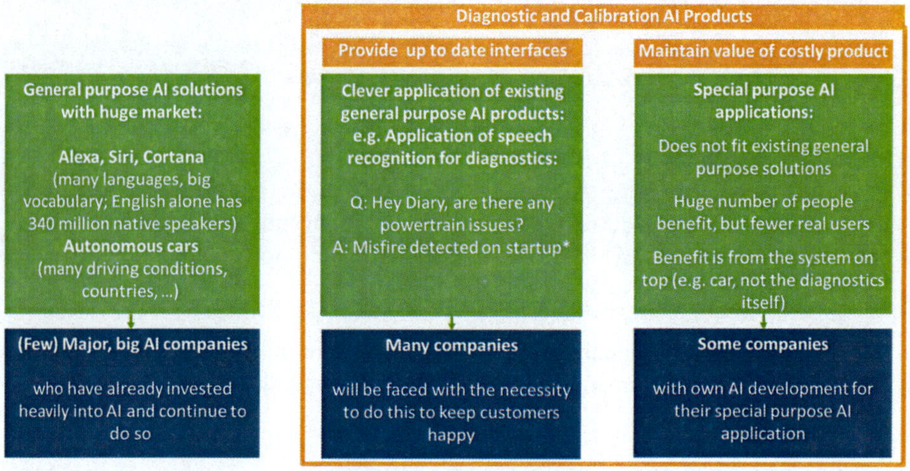

Fig. 6. AI landscape. Besides general purpose AI solutions like Alexa, Siri, Cortana or autonomous cars, which have a he market and are dominated by the major, big AI companies who have invested for a long time into AI, there are attractive applications in automotive, especially for diagnostics and calibration.

5 Case Studies for AI Development

As mentioned in the introduction, speech recognition is currently one of the most successful commercial applications of AI [1]. It has been developed for more than 25 years and there are many practical best practices available from that time. For this paper, two best practices have been selected which may be of importance to the development of automotive calibration and diagnostics development:

5.1 Practice 1: Development with High Pace

In a small speech recognition and translation team developing a SW that would run on a mobile device there was the challenge to integrate new team members with a high pace, in the need to improve many things in parallel and integrate many improvements to one solution.

The following development approach was implemented:

- Highly parallel work on advancements: Deep Learning, speech modeling, natural language understanding, data collection, system integration, verification
- Adding and integrating new team members at a high pace
- Frequent integration of solutions: Daily cycles of creation of improvement ideas, implementation, integration, verification.

Some elements of this approach would be called "Agile Development with a sprint duration of one day" today.

5.2 Practice 2: Market Entry to Get Data

A small Start Up company, which has grown to a company of several thousand employees be now, was developing speech recognition for clinical documentation. The challenge was to deliver the very high quality standards for clinical documentation. This could have been developed for decades before market entry, with a severe problem to find the funding for this. The approach was to enhance the system in several evolution steps over many years:

- Evolution step 1: Market entry with a combination of automatic speech recognition combined with manual correction by transcribers listening to complete recording Learning: You don't need to be perfect to enter the market!
- Evolution step 2: Manual correction by transcribers only listening to suspect parts.
- Evolution step 3: In-workflow feedback to the clinician, combining speech recognition and natural language understanding.

Further best practices are summarized in [2].

6 Summary

AI offers great new opportunities. To become a successful player in AI driven business, companies need to find the right balance between investments for AI and benefits in their product strategy. The paper gives a short intro-duction to AI, especially Deep Learning Neural Networks and provides long-term experiences on the successful application to real-world applications. A scenario for the application of DLNN to drivetrain calibration and diagnostics as a very promising field of application with a beneficial balance of investment and benefits is described.

Who will be the right companies to drive AI in drivetrain calibration and diagnostics? These applications are "too special purpose" for AI giants like Google, Apple or Amazon and will therefore very likely not be a suitable field of application for them. This is a great opportunity for the automotive industry: With the application of AI, drivetrain calibration and diagnostics can provide very valuable benefits for the automotive industry by reducing the growing effort for calibration on the one hand, and on the other hand enhancing availability and reducing quality cost by predictive maintenance.

The paper concludes with an extract of experiences from 25 years of successful AI commercialization like speech recognition that will support a quick start with AI for automotive applications besides autonomous driving.

References

1. Kinsella B, Mutchler A (2019) In-car voice assistant consumer adoption report, January 2019. voicebot.ai
2. Bodenhausen U (2018) Quick start with AI for businesses, ML conference, June 2018. www.grow-with-ai.com

Thermal Management System for High Performance Battery Based on an Innovative Dielectric Fluid

Nicolas Champagne(✉)

Total Marketing & Services, Solaize R&D Center, Solaize, France
nicolas.champagne@total.com

Abstract. Fast and ultra-fast charging can create tremendous thermal stresses on a battery pack and may cause battery failure or a reduction in its service life. To tackle this issue a new and efficient battery thermal system is proposed. The originality of this system lies in the use of a highly advanced fluid that allows excellent thermal performance by direct contact with the electrochemical cells. A bench test was designed to measure the cooling properties of this system at the cell level and then demonstrate its efficiency. Numerical show an excellent agreement with the experimental data. These results allow the simulation at the battery pack level to be scaled up. An ultra-fast charging scenario (350 kW) was simulated for a 60 kWh battery pack: temperature inhomogeneity or maximal temperature appears to be under control. Lastly, to assess the performance of this battery thermal management system a holistic approach is made by considering parameters for the pumps required for this system (weight and consumption), but also parameters concerning the types and sizes of heat exchangers. Ultimately this enables a global view of the system (performance, weight, power consumption) in order to compare it easily with current solutions.

Keywords: Battery · Battery pack · Thermal management · Fast-charging · Thermal runaway · Battery electric vehicles

1 Introduction

In recent years, transportation electrification has emerged as an important trend in support of reducing global CO_2 emissions [1]. Indeed, to achieve the foreseen CO_2 fleet targets for passenger cars and light duty vehicles as well as heavy-duty trucks and buses, high growth rates in electrification are needed, and vehicle electrification can already be observed to varying degrees across all markets. This electrification is only possible with high performance batteries. The key implications for acceptable battery performance are: longer service life to lower the Total Cost of Ownership (TOC), fast-charging capability and good operation over the extremes of external temperature. The battery thermal management system has therefore become an important aspect of the development of electric vehicles [2, 3].

Euroforum Deutschland GmbH (Ed.): *CTI SYMPOSIUM 2019*, pp. 33–42, 2021.
https://doi.org/10.1007/978-3-662-61515-7_4

Battery service life is known to be directly linked to its thermal history [4, 5]. Even a short spell of time at elevated temperature may have an impact, so maintaining the battery pack at constant temperature is crucial for its health. Moreover, any temperature inhomogeneity within a cell may damage it. The heat in a battery pack can come from outside but above all can be generated during use, notably fast-charging. Indeed, being an electrical component a battery has an internal electrical resistance, and the Joule Effect is then responsible for heat generation. When considering fast-charging or sport use, large currents up to hundreds of amperes may flow, leading to extreme heat and temperature inhomogeneities within the battery pack if the battery thermal management system is not properly designed. This problem is not just theoretical, and several issues linked to battery thermal management systems have been reported over the last few years. For instance, a "RapidGate" spread in social media because of the impossibility of fast-charging a car after several uses of this option on account of the rapid ageing of the battery.

Hence battery thermal management is a key issue for the acceptance of any Battery Electric Vehicle. Even if conventional systems are satisfactory for some uses, innovative systems are required for the next generation of BEVs. The next section will present current solutions, and then our innovative system. The aim of this paper is to give the reader enough information to compare our innovative solution with current ones.

2 Overview of Current Battery Thermal Management Solutions

Before getting into details of current battery thermal management system, it is worth considering how heat is generated during fast-charging. A 50 kWh battery pack, for example, contains 240 cells having a capacity of 70 A.h (3.6 V), a surface area of 0.120 m^2 and an internal resistance of 1 mΩ. This battery pack is intended to be charged with a 150 kW charger. This corresponds to a severe 3C charging scenario, which is not currently implemented.

For this scenario, the heat generated by a cell can be estimated as the product of the cell internal resistance with the square of the current in the cell during fast charging. This current is equal to the cell capacity divided by the time to be charged, hence

$$W_{cell} \sim r \times \left(\frac{Capacity}{Chargingtime} \right)^2$$

This leads to the generation of 44 W heat per cell (0.037 W/cm^2), leading to a total of 10 kW of heat generated during the fast charge. That represents a huge amount of heat to be transferred from the cell to the exterior. This figure is in line with those published by different component suppliers, and demonstrates the real challenge that thermal engineers have to deal with.

As stated in the introduction, the thermal management system should be capable not only of extracting this large quantity of heat but must also avoid any non-uniformities in temperature within the battery. The current existing solutions are described in Table 1. Basically, they all work in the same manner [6]. A dedicated fluid with specific properties will transport the heat from the cell to the exterior. Depending on the fluid properties, the flow will be managed to ensure temperature homogeneity in the battery cells. The fluid properties will also drive the design of the other components in the thermal loop (radiators, chillers, pump, etc.) so they then appear central to the whole battery thermal management system.

Table 1 Overview of current battery thermal management system.

Solution	Passive	Fan	Refrigerant	Water
Description				
Thermal fluid	Air	Air	R1234yf, R134a	Water/Glycol mixture
Comment	• Costless • Low efficency	• Low cost • Requires high power	• High cost • High power consumption	• Medium cost • Indirect cooling

In the first two systems described (passive and fan), air is the medium that extracts heat. The main disadvantage of this fluid is its poor density leading to a low heat transfer coefficient. In the first system, which is obviously the simplest, no external action is necessary. It leads to high thermal stress on the battery due only to the external climatic conditions. It may be suppressed in most cars for the future. In the second case, a fan allows the air to extract heat from the battery pack to the exterior. Good thermal management implies that it is necessary for the air to flow at high speed, leading to a large power consumption for the fan. For extreme heat (~fast-charge) or in a hot climate, this solution does not meet the requirements.

The third solution uses a refrigerant fluid. This fluid evaporates in the battery pack and extracts heat due to its latent heat of vaporisation. This leads to high thermal transfer. It needs to be conducted away in a specific channel to avoid any leakage, which in turn leads to a certain weight and cost penalty. Moreover, the fluid needs to be compressed, which can lead to high power consumption.

Lastly, the fourth thermal management system uses a water/glycol mixture as the coolant fluid [7]. A water/glycol mixture is perfectly suited to thermal management and has excellent thermal properties. In particular, it is used for extracting large quantities of heat such as those generated in internal combustion engines. Fluid channels are required to avoid any short circuit but it can be simpler than those for refrigerant, e.g. using cold plates. However, this configuration leads to large temperature inhomogeneities, causing some OEMS to add fins between cells, which are costly and heavy. Battery packs are also becoming ever larger and these cold plates can become costly to produce. Moreover, one or several thermal pads are necessary to ensure good thermal contact between the plate and the cells, which could be costly and heavy.

All these thermal management systems have their advantages and drawbacks, but a better option needs to be developed to allow fast charging while still ensuring battery longevity. An innovative thermal management system is proposed, based on the use of a dedicated fluid with enhanced dielectric and thermal properties, as described in the next sections.

3 An Innovative Battery Thermal Management System

3.1 Description

As explained previously, at the heart of this new thermal management system is the fluid. This innovative fluid has a core property: insulation. This insulation lasts longer than the battery service life, and confers a major advantage by allowing direct contact between the electrochemical cells and the fluid. This large contact area will lead to a high heat transfer coefficient, as will be shown in the next sections.

The fluid has been developed specially for this application. An in-depth analysis of the global approach to battery thermal management systems leads to improvements in several properties of this fluid. 5 main parameters were studied and improved:

1. Electrical Properties: the insulation properties of the fluid are essential and no compromise can be made here. Particular attention is given to the stability of these properties over time, whatever the climate.
2. Thermal properties: the main purpose of the fluid is to extract heat so its thermal conductivity, heat capacity and density have been optimized to this end.
3. Suitability for the whole system: The fluid properties may be adapted to the whole system such as, for instance, low viscosity for pumping or properties for heat exchangers.
4. Durability: the chemical structure of the fluid is designed to exceed the life of the battery to avoid any drain. Its compatibility with a large range of materials has also been assessed.
5. Safety: the fluid should create no accident in the event of thermal runaway of the battery cells, and may also be helpful in such a dramatic case. This feature will not be detailed here.

The aim of this paper is not to present all the properties of the fluid and the way in which to optimize them but to present the overall performance in an appropriate thermal management system. To this end, a bench test has been developed, simulations have been carried out and information concerning the whole system has been gathered.

3.2 Bench Testing and Simulation

The aim of this paper is not to present all the properties of the fluid and the way in which to optimize them but to present the overall performance in an appropriate thermal management system. To this end, a bench test has been developed, simulations have been carried out and information concerning the whole system has been gathered.

The heat inside the box is generated using 4 electrical resistors, type RH50 (Vishay). Each is positioned at the centre of a face and is encased by a thermal gel (Raytech) in order to homogenize the temperature within the cell. Using such "cells" also presents the major advantage that they can mimic easily and safely ultra-fast charging scenarios. The test bench is formed by the assembly of two parts, namely, the "box" and the "electrical cell", between which the fluid is circulated by a pump at a desired flow rate varying from 0.2 l/min to 2 l/min. The height of the electrical cell is 115 mm and its section outer dimensions are 55.5 mm by 64 mm. The gap between the electrical cell and the box is 4 mm, leading to a fluid velocity in the vicinity of the electrical cell of roughly 3 mm/s to 30 mm/s. When exiting the cell, fluid flows back to a reservoir of approximately 5 l, the temperature of which is regulated around 20 °C. The power dissipation can be controlled to generate heat fluxes of between 0.02 W/cm^2 and 0.1 W/cm^2. Remarkably, these values are higher than those found in vehicle batteries but the purpose of the experiment is to easily discriminate our fluids. Temperatures in the test bench are measured using Pt100 overmoulded temperature probes (Correge) located on the surface and inside the cell.

Figure 2a shows the effect of the fluid flow rate on the cooling performance when the heat flux is 0.23 W/cm^2 (larger than current heat flux generated in a vehicle battery pack). First, this test demonstrates that passive cooling (blue line) is not an option, the cell reaching temperatures above 40 °C in less than 10 min. By contrast, even at a low flow rate (0.33 L/min), the temperature increase within the cell is maintained below 25 °C (accuracy is around 1 °C). Figures 2b represent the cooling performance of our fluid in different heat flux scenarios at a given flow rate (0.33 L/min). Even at 0.4 W/cm^2 the temperature is still well controlled, with an increase of less than 10 °C.

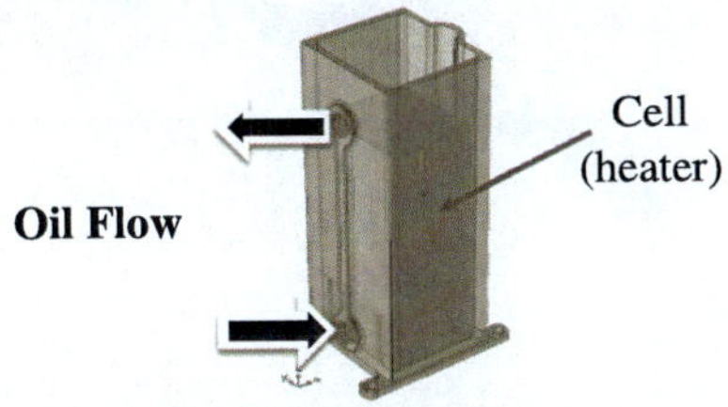

Fig. 1. Scheme of the Battery Thermal Bench Test.

These results demonstrate the performance of this solution at the cell level. However, to fully understand the benefit of this solution, it is necessary to scale this bench test up to pack level. One solution may require a larger bench test to be built but the preferred solution is to try to simulate the bench to extend the simulation at the battery pack level.

The numerical model has been implemented in the commercial software package COMSOL Multiphysics® and aims to reproduce the thermal behaviour of the test bench under different operating conditions (dissipated power and flow rates). The details of the simulation have been previously reported and are described elsewhere [8]. The comparison between simulation and experiment is depicted in Fig. 3.

Excellent agreement between simulation and experiment can be seen. The important point is that no external parameters have been added to the fitting parameters: the behaviour of the bench has been properly understood. In particular, this excellent agreement is due to the fact that liquid direct cooling is well captured by simulation. Flow is laminar (in contrast with thermal management based on air) and the contact surface is well defined (in contrast with indirect contact).

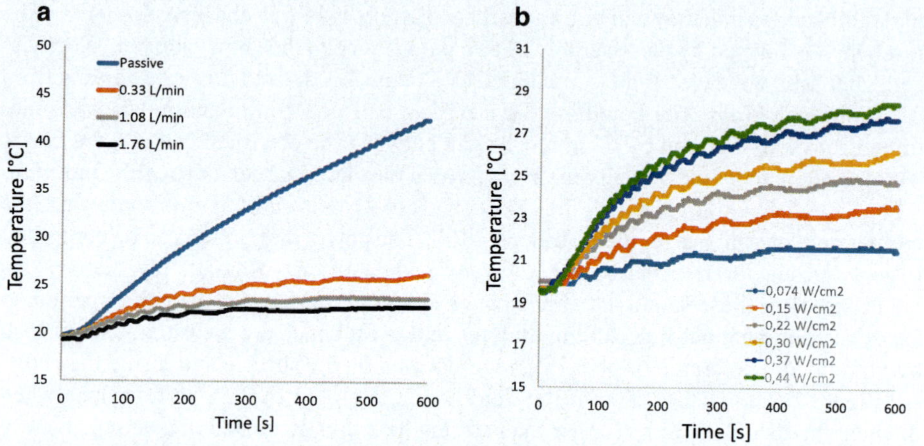

Fig. 2. Battery Thermal Bench test results. a) H = 0.23 W/cm^2. Cell surface temperature across time for different fluid flow rate. b) Q = 0.33 L/min. Cell surface temperature versus time for different heat fluxes.

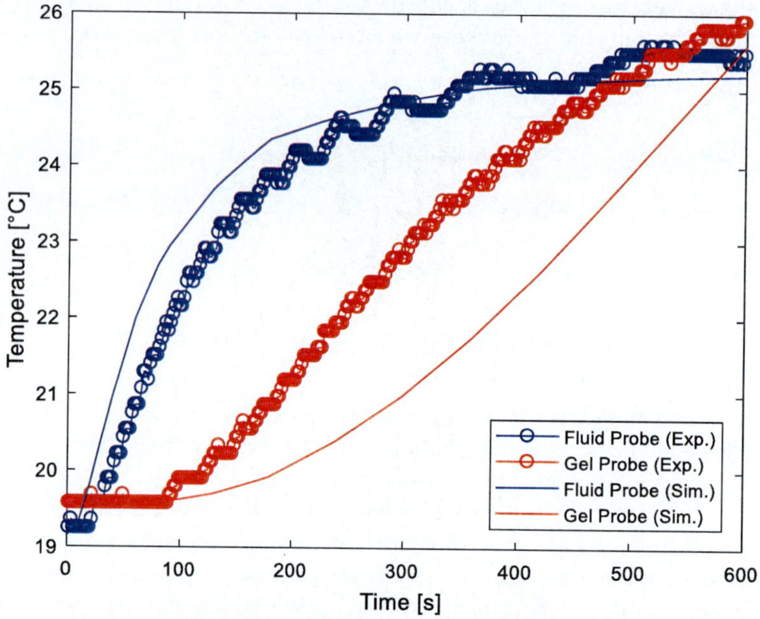

Fig. 3. Battery Thermal Bench test temperature versus time. Dots represent the experimental results and the lines the numerical simulation. Blue corresponds to the probe inside the cell and red the temperature on the surface of the cell.

This agreement allows us to scale up the simulation at the battery pack level. More precisely, the battery pack under consideration is composed of 2 blocks of 96 cells connected together as shown in Fig. 4.

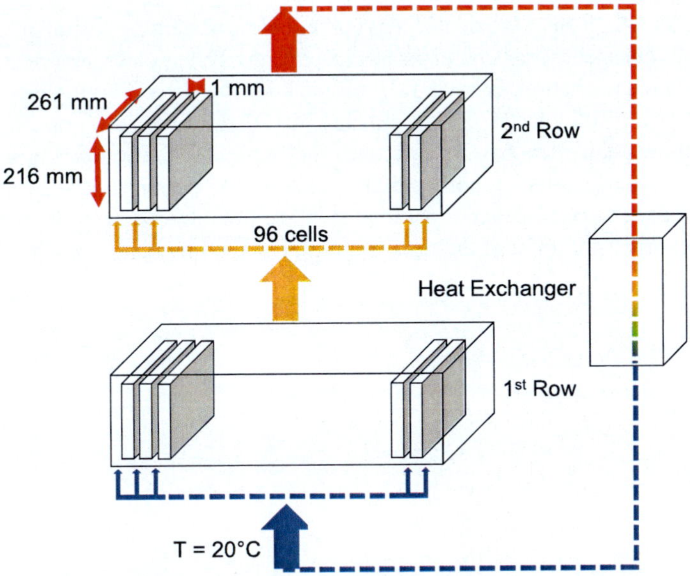

Fig. 4. Battery pack simulated. Flow rate, Gap between cells or area covered by the fluid are parameters of the simulations.

Each cell in the pack has dimensions of 261 mm * 216 mm * 7.91 mm for a capacity of 57 A.h, at a nominal voltage of 3.65 V. The ohmic resistance of the cells has been fixed at a typical value of 0.05 Ω.Ah, giving 880 μΩ per cell. The numerical model aims to characterize the overall performance of direct oil cooling of the cells under different fast charging scenarios. It therefore comprises a single parallelepiped domain in which our fluid flows in direct contact with the entire surface of the battery cell described above. The heat flux is in a plane at the centre of each cell. The simulation is versatile enough to generate a parametric study where the effect of each fluid characteristic can be evaluated for the different configurations as represented in Table 2. Concerning the system, fluid flow rate and the gap between cells (the fluid channels) can be modified. Moreover, the cell surfaces can be fully immersed or the fluid can be in contact only with a certain portion of the surface. The flow is located at the centre (not the bottom or the top) of the cell.

Table 2 Numerical parametric study. First row are system parameters and second corresponds to the fluid characteristics.

System	Fluid
Flow rate	Viscosity
Gap between cells	Density
% cells in contact with the fluid	Thermal conductivity
	Heat capacity

The results are given at the end of a hypothetical fast-charging scenario in which the battery pack is charged from 0% to 100% at full charge. This is clearly more severe than a real case in which the battery is never fully empty and where the charging rate diminishes when approaching full charge. The results are given with two main criteria: the first is the temperature peak seen by the cells and the second is the temperature difference within the cell (temperature inhomogeneity). These two parameters will influence the health of the battery cells. System parameters were first simulated in order to design a high-performance solution and the results are represented by the graphs in Fig. 5.

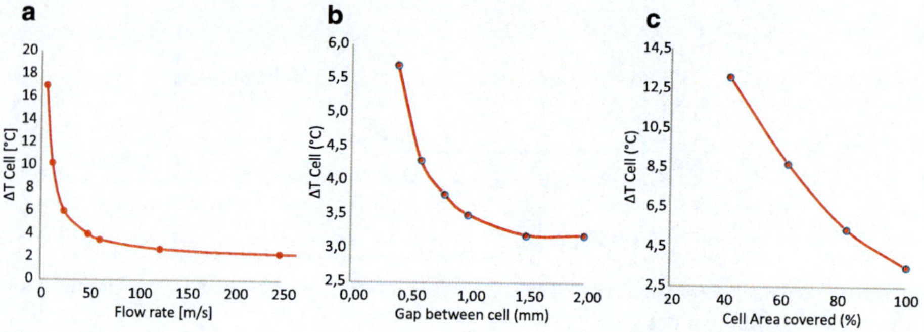

Fig. 5. Numerical parametric study: Results. a) Temperature inhomogeneity versus flow rate. b) Temperature inhomogeneity versus gap between cells. c) Temperature inhomogeneity versus cell area covered.

The first graph (Fig. 5a) demonstrates that the temperature inhomogeneity within a cell is dependent on the flow rate. This flow rate should be great enough to avoid inhomogeneity but small enough that the power required for the pump remains reasonable. The same analysis can be done when choosing the gap between cells (Fig. 5b). Too large a gap will increase the total fluid mass without any real benefit and the temperature will become too great if the gap is too small. The last point concerns the percentage of surface covered by the fluid. It can be seen that it is not essential to fully immerse the battery pack. The performance will decrease with the surface area in contact with the fluid but it could be an interesting compromise for certain configurations and is an efficient tool for reducing the weight of such a system. All these data demonstrate the importance of fully understanding the system when designing the fluid. Indeed, the same parametric study has been carried out for different fluids. The optimal fluid will depend on the configuration chosen.

The next section will describe a specific case, and a discussion concerning the other component of the battery thermal management system will follow.

3.3 Holistic Approach for an Optimal Solution

The simulations presented above enable the fluid and the system configuration to be optimized. The optimization process will not be detailed but the overall performance of a specific configuration is given. Table 3 presents this specific performance of the fluid in a configuration where the fluid is fully covering the cell and the gap between cells is 0.4 mm. A 3.75 C charging rate is considered, which is far above what is currently in the market; this leads to a severe scenario for thermal consideration.

Table 3 Performance indicator for the specific configuration chosen with our thermal fluid. The cells are 100% immersed and the gap between cells is 0.4 mm.

Performance indicator	Value	Unit
Total fluid mass	4, 5	Kg
Fluid flow rate	20	1/min
Total pressure loss	<300	mbar
ΔT cell (max)	8	°C

The temperature inhomogeneity in such a configuration remains below 10 °C, demonstrating the excellent performance in extracting heat from the cell to the fluid, even for this extreme scenario. Such an efficiency has not been clearly demonstrated for other current thermal management solutions. However, this performance at battery pack level is questionable at the whole system level. Indeed a high-performance system is not necessarily efficient: the weight of the system and power requirements are key parameters to be assessed. The next sections will attempt to provide information about the fluid in order to help answer this question.

The first question is the compatibility of the innovative fluid with the materials (polymer, seals, tubes) used in a battery pack. Indeed, materials considered for battery pack use could be different from those used in current lubrication systems. This point has been taken into account and materials compatibility tests have been developed. Even though care must be exercised, its use will not necessarily prevent such a technology from being employed.

The second question to be answered is the power required to circulate such a fluid, given that it is more viscous than a water/glycol mixture. The fluid has been optimized for that purpose. As described, the flow rate required is 20 L/min and the pressure loss induced by the flow within the battery pack is less than 300 mbar. Commercially available pumps can be found having a power consumption of 50 W and a mass less than 0.5 kg. Such a component appears to be relevant in a battery thermal management system.

The third point deals with the efficiency with which heat is transferred from fluid to the exterior. Indeed, the simulations assumed a perfect heat exchange with the exterior. Thermal loops usually use two types of component: radiators and chillers. These both have an exchange surface and work better if the flow is turbulent. Knowing that the fluid thermal conductivity is lower than for water/glycol, the surface area has to be increased. Moreover, in order to make the flow turbulent the channel may have to be redesigned to allow turbulence to occur. These two criteria (thermal conductivity and Reynolds Number) are well known and may lead to a controlled increase in terms of size, weight and price for these components. This disadvantage has to be balanced against the simplicity of a solution that does not require large cold plates or expensive thermal pads.

The last question concerns system weight. First, note that the quantity of fluid in the battery pack remains reasonable (less than 5 kg). The thermal loop will probably double this quantity leading to roughly 8 kg for the whole BTMS. This quantity can still be optimized, given that the scenario considered is extremely severe. This reasonable quantity is seen as an advantage when comparing current thermal loops using water/glycol.

3.4 Conclusion and Future Prospects

This article investigated the use of an innovative dielectric fluid as a major component for an efficient battery thermal management system. A dedicated bench test at cell level has been designed, demonstrating the great ability of this solution to extract heat from the cells. At the battery pack level, a simulation of a case study of a 40 kWh battery pack has been carried out. The fluid used in this configuration makes a 4C charging scenario possible with a temperature difference within a cell less than 8 °C, confirming the excellent performance measured at the cell level. Lastly, an analysis of the whole thermal management system (weight, power required, pumping system, heat exchangers) demonstrates the advantages of using an appropriate fluid with low viscosity and good thermal properties. This solution is clearly a viable option when compared with current solutions such as indirect cooling with water/glycol.

It has been demonstrated that the growing market for electric vehicles will require innovative solutions to be developed, notably in tackling fast charging issues. Due to its acceptable cost and excellent thermal performance, we believe that the combination of our innovative fluid and direct battery cooling represents one of these solutions.

References

1. Global EV Outlook, OECD, IEA 2019
2. Rao Z, Wang S (2011) A review of power battery thermal energy management. Renew Sustain Energy Rev 15(9):4554–4571
3. Pesaran AA (2001) Battery thermal management in EV and HEVs: issues and solutions. Battery Man 43(5):34–49
4. Leng F, Tan CM, Pecht M, Ramadass P, et al. (2002) „Capacity fade of Sony 18650 cells cycled at elevated temperatures: Part I. Cycling performance." J Power Sources 112(2): 606–613
5. Ramadass P et al (2002) Capacity fade of Sony 18650 cells cycled at elevated temperatures: part I. Cycling performance. J Power Sources 112(2):606–61
6. Chowdhury S et al (2018) Total thermal management of battery electric vehicles (BEVs). vol 1. No. DOE-MAHLE-06840-1. MAHLE, Lockport, NY
7. Ye B, Rubel RH, Li H (2019) Design and optimization of cooling plate for battery module of an electric vehicle. Appl Sci 9(4):754
8. Champagne R (2019) Innovative fluid allowing a new and efficient battery thermal management, to be published in SAE Journal

PUNCH Powerglide's Dedicated Hybrid Transmission (DHT)

Philippe Ramet and Wolfgang de Loth(✉)

PUNCH Powerglide Strasbourg, 81 rue de la Rochelle, 67100 Strasbourg, France
sales@punchpowerglide.com

Abstract. The Automotive industry is currently looking for efficient hybrid transmission concepts for use with a single electric machine. The aim is to find a cost effective alternative to current parallel and power-split hybrid concepts.

To meet this challenge, PUNCH Powerglide is introducing a novel hybrid concept that delivers electric and mechanical states using less number of gears and dog clutches to maximize efficiency.

The proposed DHT concept contains a unique patented gears and shafts arrangement, in a very compact environment to achieve very stringent package space of new OEM platform. This DHT offers a full-integrated high-voltage solution, with its own electric machine and electronic power.

The transmission operation requires a very precise and reliable operation of the dog clutches chosen to support the highest efficiency of the overall powerflow.

The synchronization of the dog clutch during the gear change is managed with the introduction of the EVT mode, which makes this DHT concept a powershift transmission.

The base variant has been developed to propose four mechanical speeds, which allow a good compromise between functionality and system cost. By applying slight changes within the modular design, the proposed concept can also evolve to a 2-gear transmission with EVT mode or even a single gear transmission to support lowest possible transmission cost while still being able to manufacture all the variants on the same manufacturing line.

Keywords: Innovation · Optimization · Flexibility

1 DHT Concept as Cost Reduction Opportunity

Car manufacturers around the globe have to hybridize their vehicle range to meet increasingly challenging CO_2 and emission standards. We all acknowledge that CO_2, emissions and energy transition will be prevalent for the years to come. This is one reasons why a holistic approach is required. Unavoidably, this challenge results is additional development effort and costs for all organizations involved in powertrain development.

Hybrid P0 and P1 versions are attractive solutions from a cost perspective. However, they provide limited potential to achieving future emission standards. Other approaches such as P2, P3 and P4 hybrid solutions have shown that significant vehicle alterations

Euroforum Deutschland GmbH (Ed.): *CTI SYMPOSIUM 2019*, pp. 43–50, 2021.
https://doi.org/10.1007/978-3-662-61515-7_5

to integrate an electric motor and power electronics onto the front or rear drivetrain is required.

The main drivers for DHT and DHE (Dedicated Hybrid Engines) drivetrains is to facilitate cost reductions for specific sub-systems. The benefits afforded by the electric motor to reduced emissions while maintaining the same vehicle performance, can also support a de-contenting of the combustion engine and transmission. This is one logical step to compensate for the on-cost of these electrified drivetrain solutions. Cost reduction will be instrumental in facilitating DHT's market penetration.

DHT is a radical transformation of the conventional transmission, bringing new philosophies that needs permanent electric power management. DHTs driven by an internal combustion engine and an electric machine, offer multiple running modes where the e-machine is used either as a generator or as a motor.

Given the high costs of batteries, amongst other items, further improvements in efficiency (reduction of losses) of the DHT is one of the key success factors.

1.1 Ambitious Goal for OEMs Needs

Having identified the technical issues, PUNCH Powerglide has developed the following requirements in order to identify a well suitable dedicated hybrid transmission concept.

A key objective is to identify technical solutions to optimize gearbox efficiency, while assuring a similar level of comfort and performance as today's solutions. A weakness of current power-shift transmissions on the market is the use of multiple friction plates for the shifting system. It is generally recognized that dog clutches represents the most efficient solution for this type of connection. However, in today's transmission market, only few have implemented dog clutches and expertise is still limited (ZF 9HP [1] and Renault E-Tech [2]).

In order to limit vehicle electrification cost, a de-contenting approach has been considered for the DHT definition. A solution with four mechanical gears states (ICE) and one electric gear state is sufficient to meet the emission targets. Apart from the significant reduction of components, a considerable benefit is realized in the reduction in installation space.

The project target is to use one single high-speed IM (induction machine) where torque, speed and packaging is adapted to suit various hybrid vehicles. The number of gears and ratios are to be flexible to cater for various vehicle requirements such as gradeability and top speed.

Even if the concept must utilize dog clutch, the DHT has to be a power-shift transmission and has to fulfill standard hybrid functions such as regeneration/recuperation, boost mode, and serial mode (Figs. 1).

1.2 PUNCH Powerglide's Concept: A DHT with Dog Clutches

PUNCH Powerglide's research outcome has resulted in a DHT with modular powerflow.

The core element of the powerflow is a single planetary gear set where the carrier is directly connected to the output shaft, the ring gear to the electrical machine, and the sun to the combustion engine through a C0 connection. This is the key element to enable an EVT mode in which the relative speed of the connection elements is controlled [3].

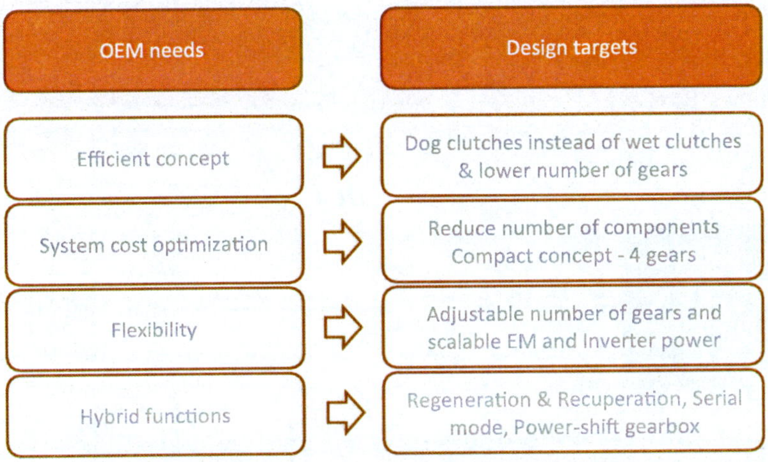

Fig. 1. PUNCH powerglide's integrated hybrid design targets

This powerflow provides 8 different modes comprising: 1 electric gear, 4 mechanical gears (ICE), 1 EVT mode, and 2 serial modes. Only 5 connection elements are needed to provide these states. In hybrid mode, the torque at the wheel is a combination of the combustion engine and electrical machine torques. In this mode, the electric machine is used either as a generator or as a motor, allowing for boost, regeneration (also at standstill) and recuperation functions.

The primary and intermediate shafts contains all the connection elements and gears. The advantage of this configuration is to limit the number of transmission components, i.e. main shafts, bearings and casings. This arrangement is significant in minimizing the transmission length and helps facilitating the transmission integration into what is typically a limited workspace in hybrid variants. Naturally, lowering the component count also results in cost reduction Fig. 2.

Activation of various connection elements enable the various mode and gear states. These are described in the table depicted in Fig. 3. The ratios shown in this table are only one example and are modifiable through different gear designs, i.e. number of teeth.

In hybrid mode, upshifts and downshifts are performed with the release of one connection element (dog clutch) and the application of another one. However, during this state change, C0 remains engaged putting the transmission into EVT mode. This mode is used to control the speed of the connection elements for shifting into the various states.

1.3 EVT Mode for Dog Clutch Engagement

The advantage of dog clutches is the negligible losses when closed, but especially when open. However, their use brings new constraints on transmission concepts. Engagement and disengagement of a dog clutch is possible only if the relative speed of elements as well as the torque are close to zero. The challenge is even more complex considering that synchronization of dog clutches should take place without wheel torque disruption, to maintain an acceptable shift comfort [4–6].

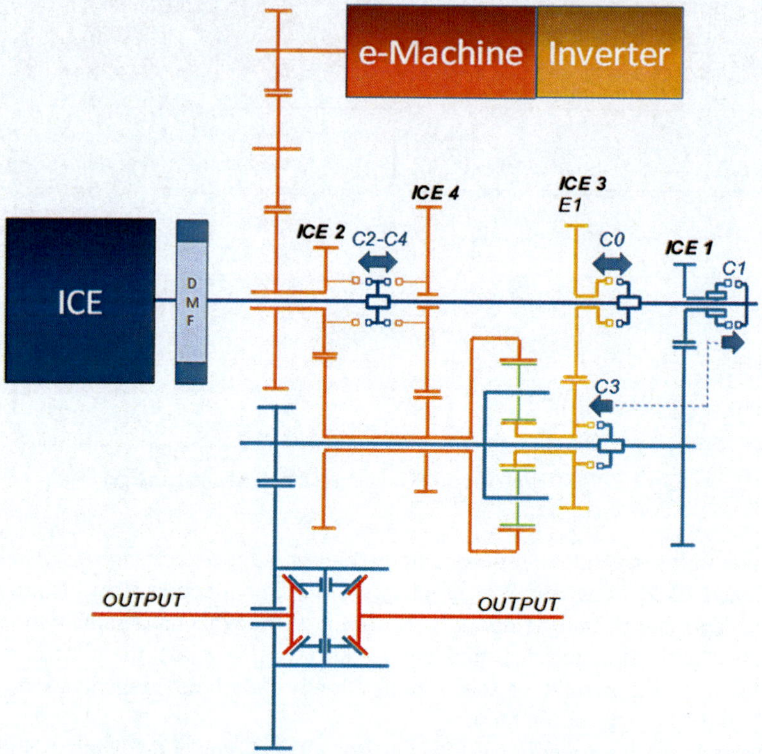

Fig. 2. PUNCH powerglide's DHT powerflow

1.3.1 Strategy for Dog Clutch Disengagement in Hybrid Mode

In hybrid mode, the reduction of torque required for tooth disengagement of the dog clutch is managed by adjusting the torque distribution between the combustion engine and electrical machine. The kinematics of this transmission has engaged dog clutches physically linked to the combustion engine and electrical machine. Therefore, there is always a torque state where the dog clutch transmitted torque becomes null, allowing the disengagement of the element.

1.3.2 Strategy for Dog Clutch Synchronization Before Engagement

The synchronization engagement speed of the dog clutch is achieved again via the EVT mode. In this mode, (C0 remains applied), the carrier is connected to the wheel while the electrical machine is connected to the ring gear, and the combustion engine to the sun gear. For a simple planetary gearset arrangement, there are infinite possibilities of motor speeds for a given output speed. Based on this relationship the combustion engine and the electrical speed are controlled until dog clutch tooth speeds reach the design threshold for engagement. The benefit of the EVT mode is that constant speed to the wheels is maintained while synchronizing the connection elements.

Mode	ICE ratio	EM ratio	C0	C1	C2	C3	C4
Hybrid drive 1	8.72	5.61	X	X			
Hybrid drive 2	6.07	8.55	X		X		
Hybrid drive 3	4.16	10.66	X			X	
Hybrid drive 4	2.90	12.06	X				X
E1	/	10.66				X	
EVT	13.79 ⇔ 15.26		X				
Serial mode 2	0.71				X		
Serial mode 4	0.24						X

Fig. 3. DHT actuation table – 8 modes

1.3.3 Results & Program Risks:

The introduction of an EVT mode support perfectly the dog clutch implementation. The result is a powershifting transmission (refer to Fig. 4 for gear 1 to gear 2 change). However, this specific mode requires precise control of speed and torque of both motors to prevent mis-engagement and to limit output torque disturbance affecting shift quality. There are tooth design and the actuation selections that improve this concept's implementation success.

1.4 EVT Mode for Vehicle Launch

Another benefit of the EVT mode is during vehicle launch. It consumes electric energy from the High Voltage battery but in such condition, the electric machine is acting as a reaction element, and so, as a generator.

The selection between electrical mode and EVT mode during vehicle launch depends on the battery SOC (State Of Charge). When the battery SOC is insufficient, EVT is selected; otherwise, pure electric driving is preferable.

As the torque ratio in EVT mode is higher than first gear, this mode can act as an additional gear to the transmission. Although this DHT concept has four mechanical gear, it can behave as a five-gear transmission if desired.

1.5 Combustion Engine Restart

During transitions from electric to hybrid mode, the combustion engine needs to be restarted. Transmissions using only dog clutches can only start the ICE via the electrical

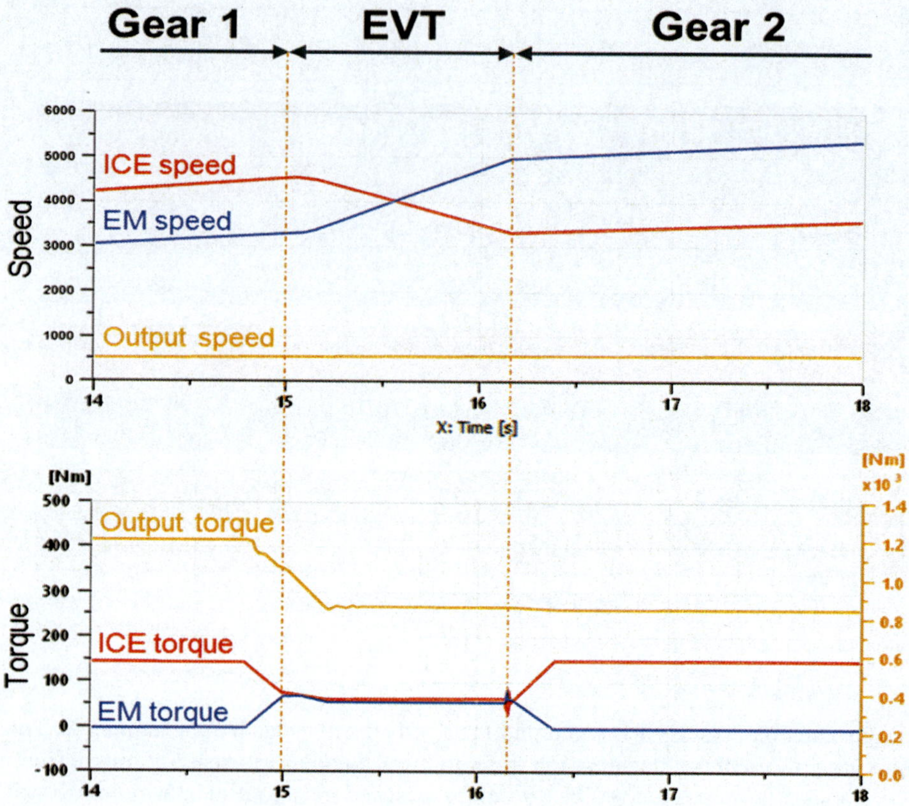

Fig. 4. Hybrid mode - gear 1 to gear 2 change

machine when the vehicle is stopped. While driving, this causes loss of vehicle traction and is thus not desirable. Several solutions are adaptable to the concept. These being: add a starter motor, a launch clutch, or a second electric machine (as the Renault Locobox). Replacement of a dog clutch by a friction clutch is also an option but is probably least desirable.

2 Performance and Flexibility

Figure 5 shows the traction force in each modes of PUNCH's DHT. Traction torque depends on ICE (Internal Combustion Engine) and EM (Electric Machine) characteristics, and values shown are only one application example. Here, the maximum available traction force is similar for EVT mode, 1st gear with EM boost, or second gear with EM boost.

The base kinematic variant has been developed to propose four mechanical speeds, which allow a good compromise between functionality and system cost. By applying slight changes within the modular design, the proposed concept can be adapted to a 2-gear transmission with EVT mode or even a single gear transmission. These may be

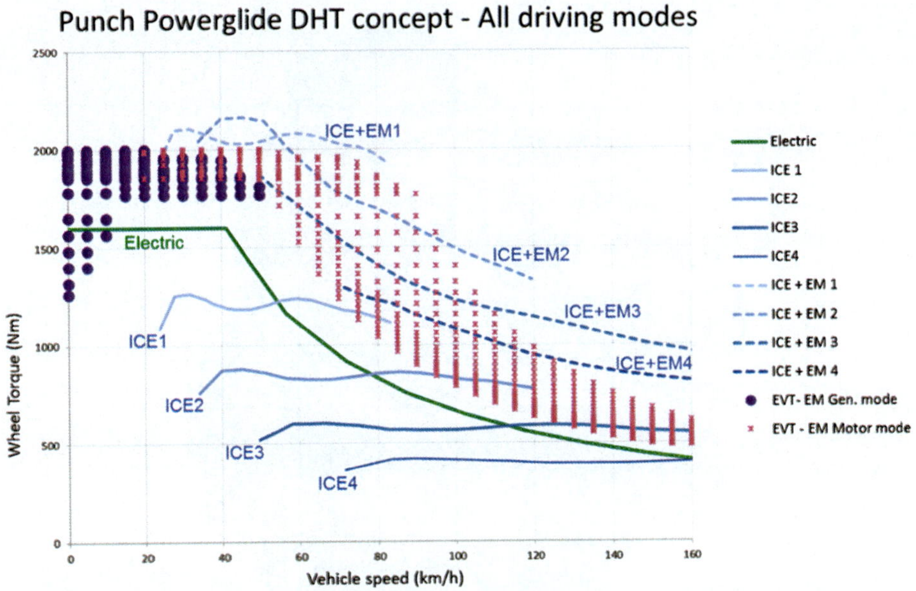

Fig. 5. Performance and traction diagram

advantageous in supporting a lowest transmission family cost due to common parts on the same manufacturing line.

For example, C1 and C4 (refer table in Fig. 3) could be optional connecting elements which are removed for cost optimization without compromising DHT functionalities.

2.1 Detailed Design

A 3D model (Fig. 6) has been developed for a powertrain configuration of 50 kW [150 Nm peak] for the electric motor and 70 kW/150 Nm for the ICE.

The transmission is designed for a transversal installation using four axes. The ICE is on the primary shaft while the planetary gear set on the intermediate shaft, the differential on the output shaft and the EM on an offset axis. For integration simplicity, reliability improvement and cost reasons, a common housing is used for the gearbox, the electrical machine and the inverter. As a result, the total length of the transmission does not exceed 350 mm.

3 Outlook

In the past years, OEMs have introduced HEV/PHEV vehicles using existing transmissions and adding electrification concepts such as P1, P2 or P3. The market trend is clearly to develop electric drivetrain solution with increased electrification content and power.

Consequently, more affordable solutions with combustion engines are still needed. DHTs that are cleverly de-contented are attractive concepts in delivering efficient compact solutions at reduced cost. The increasing market share of hybrids will provide the

Fig. 6. DHT 3D model with a 50 kW EM

needed volumes for such dedicated drivetrains and thereby help to reduce the overall fleet consumption. It is the logical next step.

References

1. Description of 9HP product using dog clutches available on ZF website
2. Description of E-TECH product using dog clutches available on Renault group website
3. Cheng Y, Cheng M (2014) EVT and E-CVT for Full Hybrid Electric Vehicles. En-cyclopedia of Automotive Engineering
4. Bataus M-V, Vasiliu N (2016) Modeling of a dual clutch transmission for real-time simulation. UPB Scientific Bulletin, Series D: Mechanical Engineering
5. Ambalavanan S (2018) Dynamic modelling of a Power Transfer Unit of All-Wheel Drive vehicle in a 1-D simulation environment: Master of Science Thesis: TRITA-ITM-EX 2018:367
6. Piracha MZ (2017) Improving gear shift quality in a PHEV DCT with integrated PMSM. 16th International CTI Symposium Automotive Transmissions, HEV and EV Drives, Berlin

Development and Prototyping of Lithium-Ion Cells for Demonstrator Drivetrains

Jan Diekmann(✉) and Sebastian Kraas

Custom Cells Itzehoe GmbH, Fraunhofer Str. 1b, 25524 Itzehoe, Germany
jan.diekmann@customcells.de

Abstract. The progressive drive train electrification of vehicles also leads to a technological and technical development of battery cells and systems.

Developments on the cell, module and system level have various objectives. The most common are: cost reduction, increase of energy density, high power charging (HPC).

The corresponding adaptations at cell level are naturally subject to physical limitations, but can achieve the desired cell performance by optimizing all cell components from the electrode active material combination to the tab design.

The development process starts with a basic cell design and selection of materials. During the subsequent development process, the electrochemically active components in particular are optimized. Recurring prototyping enables the verification on pack level and even in the actual application, which in turn leads to an efficient overall product development. Even radical system designs become testable in this way.

Keywords: Lithium-ion cell development · Cell prototypes · Fast charging

1 Introduction

The technological and technical developments in the field of lithium-ion batteries (LIB) in recent years have led and continue to lead to an increasing number of applications for which the use of these energy storage devices is of interest.

Herein, the automotive sector plays an important role. Various applications, from hybrid trucks to motor racing, result in a great variety of demands on modern battery systems. These requirements cannot always be met at the module or system level alone, so the cells must also be adapted in terms of performance, contacting or size and shape. This in turn means that these cells do not yet exist in the correct configurations during the earlier development phases of battery systems.

The necessary cell development always takes place in several stages. The performance of the target cell should always be checked at module or system level with the corresponding performance characteristics.

The development goals depend on the actual application, but can be summarized in the following main parameters in the area of automotive cells.

Euroforum Deutschland GmbH (Ed.): *CTI SYMPOSIUM 2019*, pp. 51–55, 2021.
https://doi.org/10.1007/978-3-662-61515-7_6

1.1 Development Aims

Of course, the development goals are always accompanied by a **cost reduction**. This cost reduction has already been observed in the past and is partly due to an **increase in energy density** (compare Fig. 1). The main drivers are new active materials, improved production processes and the further development of cell formats.

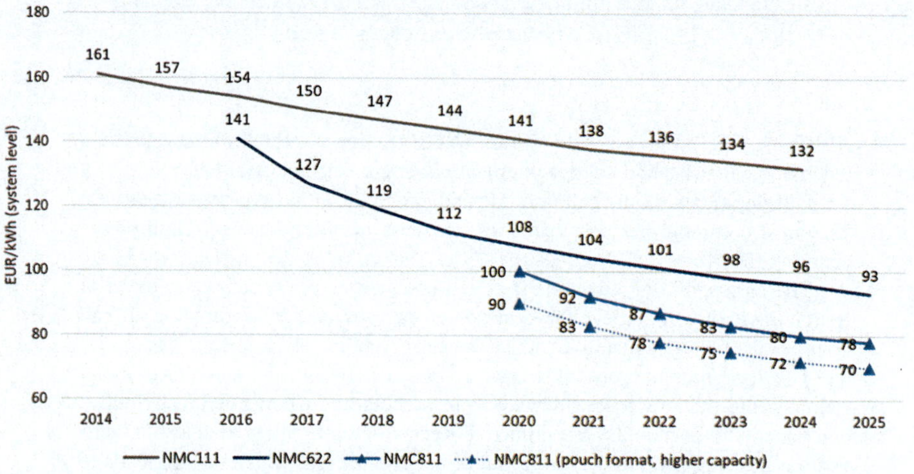

Fig. 1. Cost in EUR per kWh on battery system level for different active materials (NMC111, NMC622, NMC811) in one pouch cell type. Assumption: long-range battery electric vehicle (BEV) with a 90 kWh battery system, automotive system cost structure: ≈80% cells, ≈20% components. Data obtained from [1]

Another development goal that has gained in importance is **high power charging** and in special cases high C-rate stability.

With every new or further development, however, at least the existing **safety** standard must be maintained or a further safety improvement must be achieved.

In addition to high safety, cycle stability and thus service life are decisive factors.

2 Customizing Lithium-Ion Cells for Automotive Applications

Physical limitations ensure that many parameters cannot be fulfilled at the same time, e.g. very high energy densities at the same time as very high usable currents are excluded.

Thus, no lithium-ion cell can meet all requirements of all applications and the cells must be adapted to the required target performance. Certain key factors are particularly important.

2.1 Key Parameters Which Can Be Adapted

2.1.1 Active Materials.

As already mentioned, the choice of active material has a major influence on cell parameters such as energy density, lifetime or cost. Energy is stored in lithium-ion cells by reversible storage of lithium in host materials such as graphite or silicon on the anode and lithium nickel manganese oxide (LNMO), lithium nickel cobalt aluminium oxide (NCA) or lithium nickel manganese cobalt oxide (NMC) on the cathode. All cathode materials contain lithium ions in their structures, which are transferred into the anode material during charging and transferred back into the cathode material during discharging. The energy in a battery cell is thus generated by forcing lithium ions (by applying a current during the charging process) into a material that naturally contains no or lesser lithium ions. Therefore a potential difference between the anode and cathode which correlates with the amount of lithium ions in both materials is present. A large number of lithium ions per mass of anode material results in a high potential (state of charge is high) and a high lithium ion content in the cathode correlates with a low state of charge.

Typically, the average potential of a LIB containing a graphite or silicon anode and an NMC cathode is 3.7 V. The energy of such a cell is represented by the capacity, e.g. a 30 Ah cell would have 111 Wh. Important for many applications is not only the energy content of a cell, but the gravimetric energy density (Wh/Kg) or volumetric energy density (Wh/L). Here the different materials come into play, NMC is available in different compositions like NMC111, NMC532 or NMC622. All these materials result in the same average voltage, but have different specific capacities (mAh/g). Two 3.7 V and 30 Ah cells from the same materials, except for the cathodes, would weigh 10% less if switched from NMC111 to NMC622. Other materials such as NMC811 or LNMO also alter the average potential and the specific capacity, so that even lower energy densities are possible.

On the anode side, the use of silicon helps to increase the specific energy density (not the average voltage) of the electrode. However, implementing silicon in the electrode has some disadvantages, such as higher resistivity (silicon is not a conductor) or volume changes during charging and discharging.

Active materials are therefore chosen for the highest possible voltage with the highest possible specific energy density.

2.1.2 Cell Size and Inactive Materials

The influence of cell size is an often underestimated parameter. If the design goal is to achieve a high energy density, the weight of the inactive material in the battery cell must be reduced as much as possible. Changing from a 30 Ah cell to a 60 Ah cell can increase the energy density by about 10% to 15% by doubling the cell length, as the ratio of active materials to inactive materials is better.

Substrates also have a direct influence on the weight of the battery cell. For cathodes, the substrate foil of choice is aluminum and for anodes copper is used. One design goal is to reduce as much weight as possible in the cells, and therefore the weight of the substrate must be reduced. The only way to reduce the weight of the substrate is to make it thinner. Typically, the thickness of the substrate is between 6 μm and 15 μm.

2.2 High Power Charging

Charging a battery in a short time is a challenge at cell level for several reasons, such as overvoltage at higher currents. Furthermore increased resistance leads to more heat generation in LIB cells. One possible way to improve the high performance charging capabilities is to reduce the internal resistance of the electrodes by modifying the electrode recipes, such as adding more carbon black or other highly conductive additives, and simultaneously changing the morphology of the active material particles. The layer thickness and thus the mass per unit area load also has a direct influence on the fast charging capability. Electrodes with high loading generally have a much higher internal resistance and thus a lower rapid charge capability.

2.3 Safety

Cell safety is a prerequisite for every cell development and the corresponding applications. This, in turn, means that the influence on cell safety must be checked when the components and processes are adapted.

Cell defects can lead to a thermal runaway, self-reinforcing exothermic reactions. These result in the worst cases in a cell fire. The sources of error can be reduced to overcharging, deep discharge, high ambient temperatures or an external or internal short circuit. While the chance of all other failures can be minimized by an adequate battery management system and proper cell installation, a spontaneous internal short circuit can only be avoided by a safe cell design and high quality cell production. For example, lithium plating can be avoided by the appropriate electrode configurations during cell design and a sufficient placement accuracy of anode and cathode during cell production. The consideration of the future application in order to avoid bending effects or cell respiration damage also plays an important role.

It is often assumed that pouch cells in particular are less safe due to the thinner cell housing and the lack of passive safety devices. However, the Fraunhofer Battery Alliance has summarized in an evaluation that despite these features, pouch cells have a higher safety level than 18650, 21700 and PHEV2 cells (compare Fig. 2). This evaluation is based on a battery system of equivalent capacity.

In particular, the better monitoring by means of the battery management system due to the lower number of cells and the safe installation and contacting in battery modules and system are responsible. Accordingly, it is possible to achieve lower costs per kWh (see Fig. 1) while maintaining or improving the safety level.

3 Conclusion

This summary shows which goals are pursued in the development of cells for automotive applications and which parameters lead to a cost reduction, increase in energy density or the enabling of high power charging. However, cell safety always remains a basic requirement.

The development and production of cell prototypes according to application-specific requirements enables the verification of concepts at system level and even in actual application. Accordingly, the feedback of such tests in the early stages of the development of the cell may enable an efficient product development and highly optimized application at a later date.

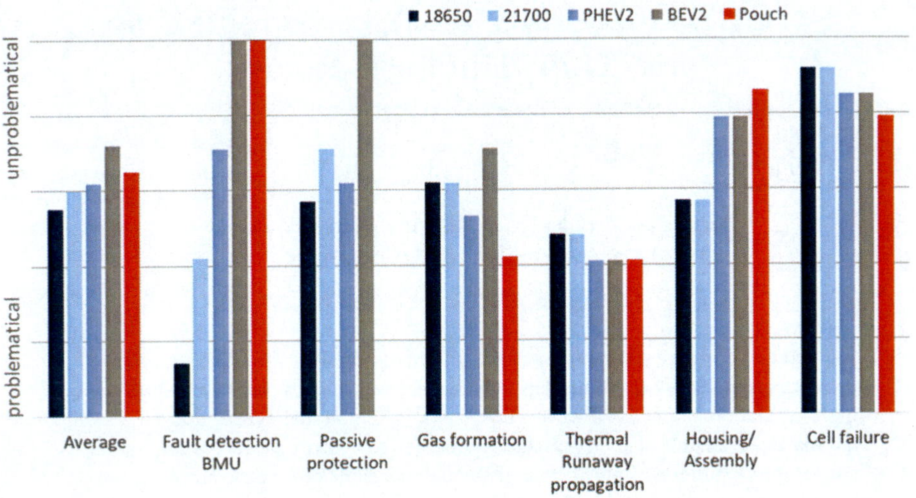

Fig. 2. Evaluation of safety relevant parameters of different cell types. Data obtained from [2].

References

1. Markus Woland: Differenzierungsmöglichkeiten und Kostenpotentiale für lokale Batterie- & Zellfertigungen, VDMA Jahrestagung 2018
2. Fraunhofer-Allianz Batterien; Entwicklungsperspektiven für Zellformate von Lithium-Ionenbatterien in der Elektromobilität (2017)

Spring Loaded Rotor Shafts as New Flexible Shaft Hub Joint for E-Rotors

Benjamin Dönges(✉), Maximilian Rolfes, and Stefan Buchkremer

Muhr und Bender KG, 57439 Attendorn, Germany
benjamin.doenges@mubea.com

Abstract. To improve the performance and range of electrified cars, the power density of the electric machines has to be increased, which is possible by increasing the rotational speed. However, the maximum rotational speed is limited by the radial expansion of the lamination stack due to centrifugal forces and radial temperature gradients. Improper centrifugal expansions may cause a loss of contact between shaft and lamination stack as well as a loss of radial centering of the lamination stack on the rotor shaft.

Hence, Mubea developed and presents a revolutionary spring loaded tubular rotor shaft, whose innovation is a defined radial spring stiffness. This radial spring-like behavior results in a significantly reduced sensitivity of the transmitted torque with regard to the rotational speed, since the spring loaded shaft compensates the centrifugal deformations of the stack. This ability is new in the state of the art including interlocking connections. In addition, the spring behavior results in lower sensitivities regarding tolerances, the needlessness of fine machining the press fit surfaces and the avoidance of thermal joining. Also, the mechanical load on the stack is on average reduced by approximately 50%, which results in better electromagnetic properties. Tests and simulations show the high robustness with respect to transmitted torque and radial centering of the stack, also beyond 25.000 rotations per minute. Inner cooling of the shaft is easily applicable due to a relatively large contact surface.

Keywords: Electric engine · E-Rotor · Shaft hub joint

1 Introduction

Due to air pollution and increasing scarcity of resources, the demand for electric driven cars increases. In order to secure suitability for daily use and economic feasibility of automobiles in the future, electric machines with high power densities are necessary. The power density of electric machines may be maximized by increasing the rotational speed. However, this may bring established technical solutions to the limit of their applicability – new solutions need to be developed.

One example for such a development is the shaft hub joint between rotor shaft and lamination stack. About 500 single sheets with a thickness of around 0.3 mm are joined to a rotor shaft. The required torque has to be transmitted without slipping and change of position – even under high temperature changes, frequently changing operating

Euroforum Deutschland GmbH (Ed.): *CTI SYMPOSIUM 2019*, pp. 56–61, 2021.
https://doi.org/10.1007/978-3-662-61515-7_7

conditions (driving, recuperating), high engine torque and rotational speeds [1]. Pure form fit connections are not suitable due to missing radial centering properties, which enables relative motion and thereby worsening of initial balancing quality [2].

Requirements regarding shaft hub joints of rotors for electric machines are shown in torque versus rotational speed diagrams for a press fit in Fig. 1a and for a form fit in Fig. 1b. The red line shows the required torque according to the characteristics map of an electric machine, which has to be transmitted between lamination stack and rotor shaft. Operating point 1 represents the transition between the basic speed range and the field weakening range at maximum torque, whereas operating point 2 shows the maximum operating speed $n_{\max}$. No plastic deformation is permitted up to the overspeed operating point 3 (typically $1.2 \times n_{\max}$) and no separation of parts is allowed up to the maximum operation speed, shown by operating point 4 (typically $1.4 \times n_{\max}$). Peak values of torque – occurring in extreme operation modes such as an ABS breaking situation – need to be considered. The green line shows the torque corresponding to the required axial fixing force between the lamination stack and the rotor shaft. Both values are proportional to each other. Moreover, the blue line illustrates the transferable torque by means of a form fit and the yellow line shows the transferable torque by means of a press fit. The latter decreases with increasing rotational speed due to centrifugal forces. The maximum temperature gradient causing an additional expansion of the lamination stack needs to be considered.

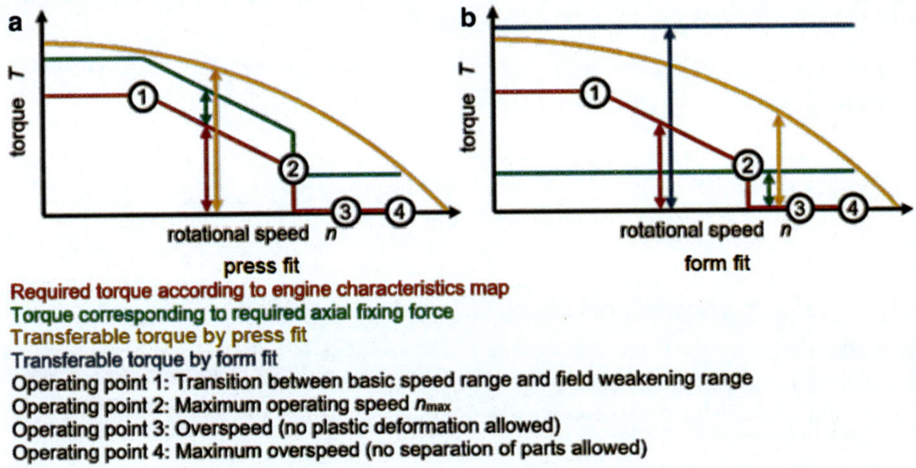

Fig. 1. Shaft hub joint with a) press fit and b) form fit.

In case of a press fit between lamination stack and rotor shaft (Fig. 1a), the vector sum of the required torque according to the engine characteristics map (red line) and the torque corresponding to the required axial fixing force (green line) has to be lower than the transferable torque by the press fit (yellow line). In case of a form fit between lamination stack and rotor shaft (Fig. 1b), the required torque according to the engine characteristics map (red line) has to be lower than the transferable torque by the form fit (blue line). Furthermore, the torque, which is corresponding to the required axial fixing force (green line), has to be lower than the transferable torque by the press fit (yellow line), which corresponds to the transferable axial fixing force. In order to serve for radial centering of the lamination stack on the rotor shaft, the oversize causing the press fit has to be ensured at least up to overspeed operating point 3.

2 New Spring Loaded Shafts for High-Speed Rotors of Electric Traction Machines

2.1 Rigid Press Fit

In order to fulfill the aforementioned requirements regarding the shaft hub joint of rotors for electric machines, press fit connections are often chosen. In comparison to solid shafts, tubular shafts serve for lightweight and enable inner cooling of the rotor. However, high assembly forces – which correlate with high transferable torques at idleness and may lead to a damaging of rotor parts – and moderate maximum rotational speeds limit the applicability (Fig. 2, dashed line). If applicable, complex and expensive assembly processes such as fine machining of the press fit surfaces, heating of the lamination stack and/or cooling of the rotor shaft are needed to achieve the necessary oversize between lamination stack and rotor shaft. Further disadvantages are relatively high mechanical stresses in the lamination stack (Fig. 3a) due to the rigid press fit, which leads to a degradation of magnetic flux and thereby to a lower efficiency of the electric machine and a lower range of the electrified automobile.

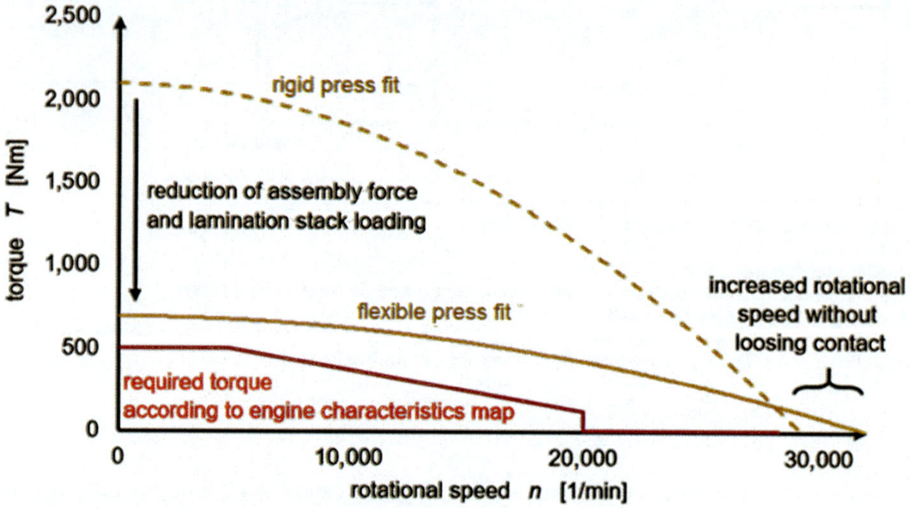

Fig. 2. Transferable torque between lamination stack and rotor shaft of a rigid press fit (dashed) and a flexible press fit (solid).

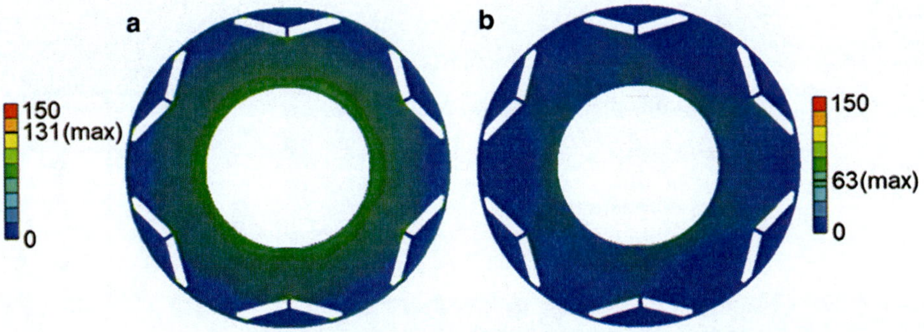

Fig. 3. Von Mises stress in lamination stack due to a) rigid press fit and b) flexible press fit.

2.2 Flexible Press Fit

A tubular rotor shaft was developed, which is characterized by a defined radial flexibility. This enables the transmission of the required torque, the axial fixation between lamination stack and rotor shaft and the radial centering of the lamination stack on the rotor shaft at rotational speeds of more than 25.000 rotations per minute (Fig. 2, solid line). Expensive assembly processes such as fine machining of the press fit surface of the lamination stack as well as heating of the lamination stack and/or cooling of the rotor shaft are not necessary due to a low sensibility concerning tolerances of shaft and lamination stack. The axial assembly force to join the rotor shaft to the lamination stack is considerably reduced in comparison to conventional rigid press fits. Inner cooling of the rotor is easily applicable due to a relatively large contact surface and a good heat transfer between lamination stack and tubular rotor shaft. Furthermore, the mechanical stress in the lamination stack due to the flexible press fit is significantly reduced (Fig. 3b) in comparison to a rigid press fit (Fig. 3a). This causes a high magnetic flux in the lamination stack and thereby a high efficiency of the electric machine and a higher range of the electrified automobile.

2.3 Comparison of Shaft Hub Joints

A comparison between shaft hub joints is shown in Table 1. Press fit connections are distinguished between rigid and flexible ones, whereas form fit connections are differentiated between gear teeth and groove-tongue connections. The assessment of the fulfillment of functional and customer requirements shows distinct technical and economical advantages of the developed spring loaded tubular shaft in comparison to the established shaft hub joint solutions.

3 Experimental Verification

In order to experimentally confirm the simulation based dimensioning of the developed spring loaded tubular shaft, static torque tests were executed (Fig. 4). The determined transmittable torque between lamination stack and rotor shaft exceeded the simulated torque by around 10%, which represents a very good accordance between the dimensioning method and the effectively transmittable torque.

Table. 1. Technological comparison between shaft hub joints.

● requirements **fulfilled**

◒ requirements fulfilled **under circumstances**

○ requirements **not fulfilled**

functional requirements	customer requirements	press fit: flexible	press fit: rigid	form fit: gear teeth	form fit: groove-tongue
transmission of torque	low assembly forces and no thermal treatment	●	○	●	●
	applicability for high speed drives	●	○	●	●
	no fine machining of shaft and lamination stack	●	○	●	●
radial centering of lamination stack	low assembly forces and no thermal treatment	●	○	○	○
	applicability for high speed drives	●	○	○	○
	no fine machining of shaft and lamination stack	●	◒	◒	◒
axial fixation of lamination stack	low assembly forces and no thermal treatment	●	○	○	○
	applicability for high speed drives without axial form fit (retaining ring)	●	○	○	○
	no fine machining of shaft and lamination stack	●	◒	◒	◒
cooling of lamination stack	large contact surface between shaft and lamination stack	◒	●	◒	●
	direct cooling of lamination stack possible	●	○	◒	○

Furthermore, spinning tests were successively conducted at 150 °C to confirm the centering function of the lamination stack up to 25.000 rotations per minute. No rigid body motion, worsening of initial balancing or improper plastic deformation were determined.

4 Conclusions

Powertrains of automobiles are more and more electrified by means of electric machines in order to reduce fuel consumption and emissions. Potential for reduction of weight, designed space and usage of expensive and environmentally critical materials lies in an enhancement of rotational speed of the electric machines. However, this may bring established technical solutions to the limit of their applicability. Shaft hub joints of rotors for electric machines are examples for such an issue. New technical solutions have to be developed in order to secure centering of the lamination stack on the rotor shaft

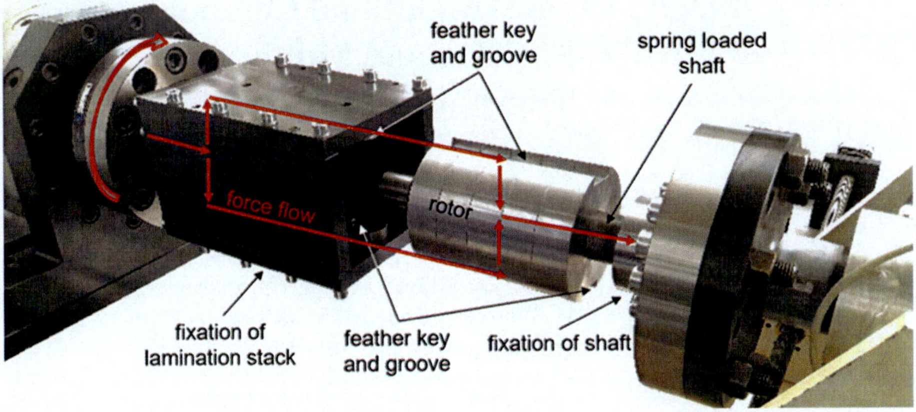

Fig. 4. Static torque test of a rotor with a spring loaded tubular shaft.

and to enable the required torque and axial fixation even under high expansions of the lamination stack due to high rotational speeds and associated centrifugal forces.

Hence, a tubular rotor shaft with a defined radial flexibility was developed, which is characterized by:

- high tolerance regarding expansions of the lamination stack due to high rotational speeds and radial temperature gradients,
- a secure centering function of the lamination stack on the rotor shaft,
- secure transmission of the required torque and axial forces even under high expansions of the lamination stack due to radial temperature gradients as well as high rotational speeds of 25.000 rotations per minute and more and associated centrifugal forces,
- significantly lower assembly forces to join the rotor shaft to the lamination stack at room temperature,
- needlessness of fine machining of the press fit surface of the lamination stack as well as heating of the lamination stack and/or cooling of the rotor shaft,
- suitability for inner cooling of a rotor due to relatively large contact areas and a good heat transfer between rotor shaft and lamination stack.

References

1. Braun A (2018) Neue Herausforderungen für Welle-Nabe-Verbindungen. In: Leidich E (ed) Welle-Nabe-Verbindungen 2018 – Dimensionierung - Fertigung - Anwendungen, VDI-Berichte, vol 2337. VDI-Verlag, Düsseldorf, pp 3–6
2. Bader M (2018) Einfluss alternativer Automotive-Antriebskonzepte auf Welle-Nabe-Verbindungen. In: Leidich E (ed) Welle-Nabe-Verbindungen 2018 – Dimensionierung - Fertigung - Anwendungen, VDI-Berichte, vol 2337. VDI-Verlag, Düsseldorf, pp 109–121

BEV Range Increase by Optimal Combination of 800 V E-Machine and Multispeed Transmission

Boris Dotz[1](✉), Camelia Jivan[2], Sebastian Waider[1], and Norberto Termenon[2]

[1] Valeo Siemens eAutomotive Germany GmbH, 97616 Bad Neustadt, Germany
{boris.dotz.jv,sebastian.waider.jv}@valeo-siemens.com
[2] Valeo Transmissions, 14 Avenue des Beguines, 95800 Cergy, France
{camelia.jivan,norberto.termenon}@valeo.com

Abstract. To cope with increasing customer demands on vehicle range and efficiency, Valeo and Valeo Siemens eAutomotive Germany develop electrical drives from the beginning as a fully integrated system. This work presents the newest seamless multispeed and 800 V eAxle designed for SUV premium class vehicles. The paper depicts the architecture and technical advantages of the 250 kW eAxle system combining a multispeed transmission with an 800 V electrical permanent magnet machine and a SiC Inverter. Advantages in efficiency and resulting range increase on the WLTC cycle are highlighted. Throughout the work, a comparison with a single speed transmission and a 400 V round wire machine illustrates decisive factors and key enablers of the newly developed eAxle.

Keywords: eAxle · Multispeed transmission · Hairpin permanent magnet machine · Range increase

1 Introduction

1.1 State of the Art for Battery Electric Vehicles

Today, battery electric vehicles (BEV) are mainly compounded by a high voltage electric machine, an inverter, a single speed transmission and a differential (Fig. 1). Traditional electric drives equipped with a single speed transmission are generally a compromise between high wheel torque and high maximum vehicle speed, which especially for Sport Utility Vehicles (SUVs) targeting the premium segment denotes a strong disadvantage. At higher speed, the electric machine as well as the transmission operate less efficiently thereby limiting the vehicle range considerably.

As future premium vehicles require a significantly higher vehicle range, next generation high efficient powertrains are developed by Valeo and Valeo Siemens eAutomotive Germany.

This section highlights the state of the art of electrical machines and transmissions, while Sect. 2 details the targeted system requirements. The following Sects. 3 and 4 illustrate the newly developed electrical machine and multispeed transmission. Section 4 shows the performance advantage of the optimized fully integrated eAxle, while Sect. 6 concludes the paper.

Euroforum Deutschland GmbH (Ed.): *CTI SYMPOSIUM 2019*, pp. 62–79, 2021.
https://doi.org/10.1007/978-3-662-61515-7_8

1.2 Electrical Machine – Limits and Future Needs

With the advantage in efficiency and size, the permanent magnet machine with interior magnets is now state of the art in nearly all drive systems across the BEV and HEV (Hybrid Electric Vehicle) market. Choosing the main machine parameters, designers have to choose an appropriate pole pair number and winding type. With the number of poles denoting a trade-off of machine size and electrical frequency, eight-pole machines are now seen as the classical electrical machine for BEVs [1–3]. This number of poles defines the optimum solution for round wire wounded stators, the slots of which carry a distributed round wire winding. Round wire windings are technically feasible with slot fill factors up 0.6. The winding is a single- or double layer winding with 48 slots and a number of slot per pole and phase equal to 2. To reach required vehicle speed, operation speed is around 16000 rpm, leading to an electrical frequency of approximate 1 kHz. Latter value denotes a reasonable limit for classical inverters with a switching frequency around 10 kHz.

Considering these parameters, the machine performance shows a clear trade-off between efficiency and size, i.e. as machine speed and pole pair number are fixed, smaller machines are less efficient than larger machines. Additionally, continuous torque and power performance is mainly limited by the air gap flux density of the rotor permanent magnets and the electric loading of the stator, which in turn is limited by the cooling system of the electric machine.

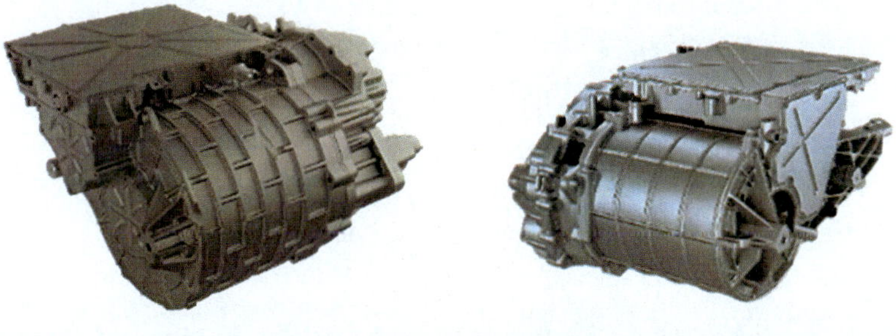

Fig. 1. Units of electric drives

One way to increase the efficiency without increasing the size of the machine is to aim at a high copper fill factor to reduce Joule losses. To overcome today's limitation of round wire windings, profile copper windings can be used. Profile copper, or hairpin windings can be manufactured with slot fill factors of up to 0.8, thereby reducing the DC joule losses. However, the main drawback of the profile copper windings are increased AC joule losses, especially at high operation speed. From the e-machine prospective, it is therefore advantageous to aim at a high slot fill factor while reducing the operation speed of the electrical machine. Thereby efficiency increases, while size and power requirements are kept or even exceeded. This is the main track, which is discussed in Sect. 3, where the machine optimization is described. To allow for a reduction of operation speed the full eAxle is optimized as a system aiming at the best powertrain efficiency. The optimization of the multispeed transmission is described in Sect. 4.

1.3 Transmission – Limits and Future Needs

Traditional single speed transmissions come with advantages regarding weight, packaging, number of components and cost. The disadvantages of single speed transmissions are coming with speed increase. In order to ensure a high vehicle speed single speed should have at least 16000 rpm as maximal functional speed. In this case, speed dependent losses (e.g. bearings, joints, churning) are increased exponentially and becoming dominant in the transmission efficiency. In order to minimize speed dependent losses and to increase powertrain efficiency, lowering input operation speed is targeted.

2 Targeted Application Description

2.1 Premium SUV Class

Based on benchmarks and different customer requirements, Valeo and Valeo Siemens eAutomotive Germany created an ideal vehicle target for the market segment premium SUV. The premium SUV should meet the requirements detailed in Table 1.

Table. 1. Golden car requirements for premium SUV class

	Requirements
Traction type	2WD or 4WD
Wheel torque	4500…6000 Nm/axle
Acceleration 0…100 km/h	<6 s
Maximum vehicle speed	>200 km/h
Powertrain peak power	>250 kW
Battery energy	>75 kWh

The needs for the future electric premium SUV are the following

- Increased powertrain efficiency
- Increased vehicle range
- Additional integrated functions
- Seamless multispeed

In order to increase the powertrain efficiency on driving cycles (WLTC, highway and urban), system optimization is focused on a torque of <200 Nm for the electrical machine while further components are optimized for maximum efficiency in part load condition.

2.2 Optimal System Overview

To assess new performances and to develop an optimal integrated eAxle, different set-ups regarding transmission and electrical machine are designed and analyzed. Table 2 illustrates different possibilities of speed and torque combinations. All systems considered, meet the golden car requirements discussed in previous section. The comparison on the overall eAxle efficiency for WLTC and highway operation points shows a combination of a rather low speed electrical machine with a maximum operation speed of 12000 rpm and a medium machine torque of 440 Nm to be the optimum choice. To simplify the inverter set-up as well as to decrease inverter related losses the voltage level is additionally increased to 800 V. Table 3 compares basic properties of the original baseline and the new multispeed system. The following Sects. 3 and 4 describe the individual components and their optimization in detail, while Sect. 5 gives the overall reduction of losses and related BEV range increase for the WLTC.

Table. 2. Different configurations comparison

	System 1: single speed		System 2: multispeed	
	High speed 14…16 krpm	High torque >500 Nm	Low speed 10…12 krpm	Medium torque 400…450 Nm
Packaging	Reference		+	
eMotor efficiency			++	
Transmission efficiency			+	
System efficiency			++	

Table. 3. Comparison of baseline and optimal system for 250 kW

	Baseline system	New optimal system
Targeted cycle	WLTC and additional highway driving	
Vehicle	SUV premium class	
Max. wheel torque in Nm	>4500	
Voltage in V	400	800
eMotor peak power in kW	250	250
eMotor winding type	Round wire	Hairpin
eMotor max. operation speed	16000 rpm	12000 rpm
Transmission	Single speed	Multispeed 12.3/6.8

Table 2 depicts two main configurations which are targeted. The main conclusion is that from efficiency point of view, it is preferable to target a low speed electrical machine and transmission configuration. Using the suitable ratios, both wheel torque and vehicle speed requirements are met.

3 Design and Optimization of the Electrical Machine

3.1 Decisive Factors for an Optimal E-Machine Design

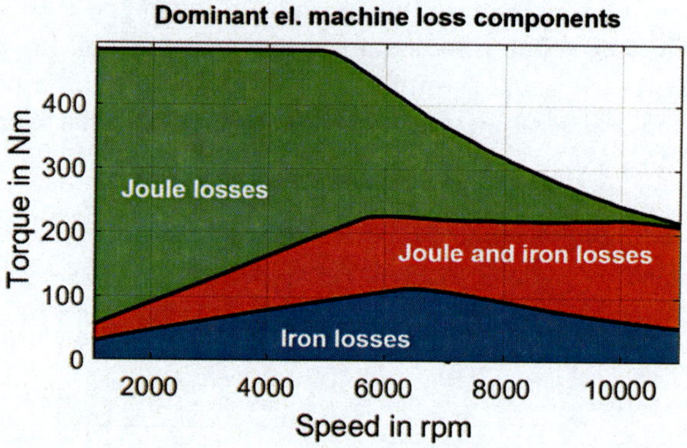

Fig. 2. Dominant loss components over speed and torque of electrical machines, depicted exemplarily for the machine under consideration.

Figure 2 shows an exemplary speed-torque map of a permanent magnet machine including the dominant loss components. As can be seen, in low load condition (blue area), mainly iron losses are dominant, while at high torque Joule losses become more important (green area). In mid load operation (red area) Joule and iron losses contribute equal parts to total machine losses. There are several decisive design factors for electrical machines, which can be used to minimize loss components. A reduction of iron losses is achieved by reducing the electrical frequency of the machine, which in turn is possible by reducing the operation speed as well as the pole pair number. A reduction of the electrical frequency is furthermore important to limit the AC Joule losses in the stator winding, which in turn facilitates the hairpin technology. With hairpin windings, the slot size can be reduced facilitating a lower flux density in the stator iron and therefore reducing iron losses further. Additionally, with a high slot fill factor of hairpin windings, DC joule losses are minimized. Figure 3 shows the interrelations of basic design measures.

As depicted in Fig. 3, a lower operation speed of the electrical machine comes with the advantage of lower iron and AC losses. The lower speed is made possible by the redesign of the transmission and the application of a two speed solution (refer to Sect. 4). To reduce the AC losses further, the number of poles is chosen to 6, reducing the electrical frequency by 25% compared to classical 8 pole designs. Using a hairpin winding further allows to reduce DC losses and enables a compact design of the electrical machine.

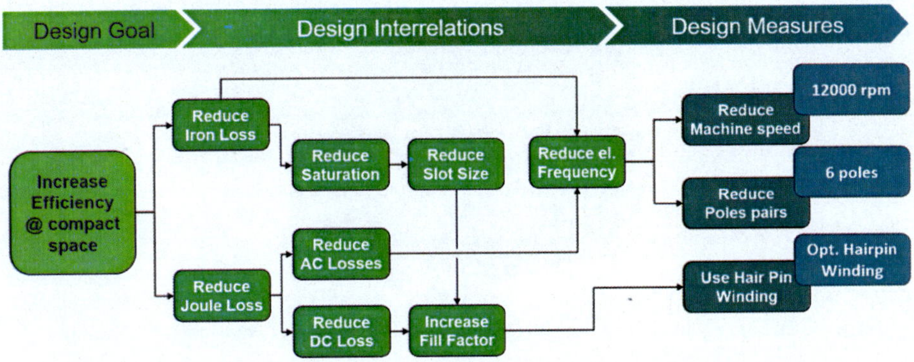

Fig. 3. Design measures and design interrelations to achieve an increased efficiency

3.2 Comparison of Round Wire and Hairpin Winding

Table 4 shows a comparison of basic properties of round wire and hairpin windings. An exemplary hairpin stator is shown in Fig. 4. Due to higher slot fill factor, machines with hairpin windings have significantly lower DC joule losses. For the same slot area, considering an increase of 20% of the fill factor, DC joule losses are reduced by 28%. As the slot area is classically rectangular, the hairpin technology is also used to decrease the cross-section of the slots, thereby limiting flux density in the iron core. Again, the goal is to balance out copper and iron losses as discussed in previous section.

Table. 4. General comparison of winding technologies

	Round wire winding	Profile copper winding
Slot fill factor	o	+
DC losses	–	+
AC losses	+	–
Phase connection	o	+
Thermal performance	o	+
Machine size	o	+

The biggest drawback of hairpin windings are high AC joule losses, which decrease the efficiency at medium to high speed considerably. Through an optimization of the multispeed transmission, the maximum operation speed is reduced by 25%, while the additional 6-pole design reduces the electrical frequency once again by 25%. In total, the resulting maximum electrical frequency of the new machine is only 55% of the base line machine, allowing the hairpin winding to be highly efficient up to maximum speed 12000 rpm. Figure 5 illustrates a comparison of the newly designed 6-pole hairpin

Fig. 4. Exemplary hairpin stator developed by Valeo Siemens eAutomotive Germany

machine compared to a classical 8-pole round wire machine. It is clear that the optimization lead to an improved efficiency especially for desired part load operation determining vehicle range.

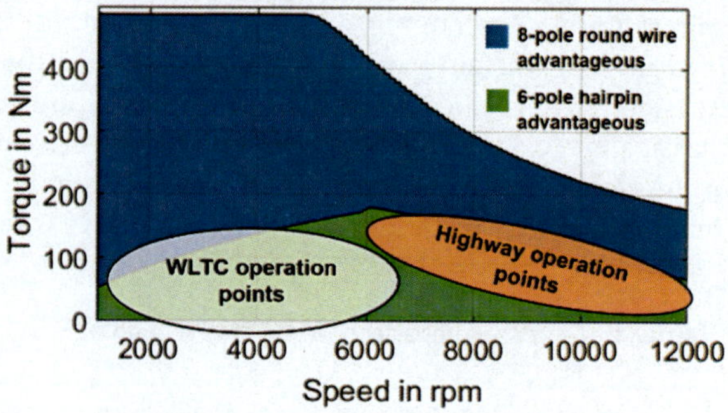

Fig. 5. Comparison of round wire and hairpin machine optimized for WLTC and highway operation. Green area indicates load conditions, where the new hairpin machine has higher efficiency compared to the baseline round wire machine.

3.3 Optimization of the Electrical Machine

The main challenge is to balance out iron and Joule losses of the electrical machine, assuring high torque capability. The fully optimized machine developed by Valeo Siemens eAutomotive Germany considers several load and driving conditions, achieving an optimum balance of iron and Joule losses. Figure 6 shows one exemplary load point, which is analyzed for more than 500 designs. It can be seen that several designs are suboptimal, while the Pareto front indicates optimal designs for considered operation point showing the trade-off between torque and efficiency. The second step is to provide a detailed weighting and analysis of all load conditions during a driving cycle, taking additional customer requirements into account. Figure 7 shows a comparison of exemplary 56 designs considering several driving conditions. The optimum design can now be identified easily.

From previous discussion, it can be seen that general design measures allow to find the optimum target design. As for hairpin stators, not only the slot geometry is of interest. Additionally, the hairpin winding itself must be carefully designed and optimized. Valeo Siemens eAutomotive Germany is advanced in hairpin windings with several slot and layer combinations. The number of layers can be exemplary four to eight conductors, while the number of slots ranges between 48 and 72 slots for classic designs. The final choice depends on customer requirements as well as on the design voltage, machine size, operation speed and torque requirements.

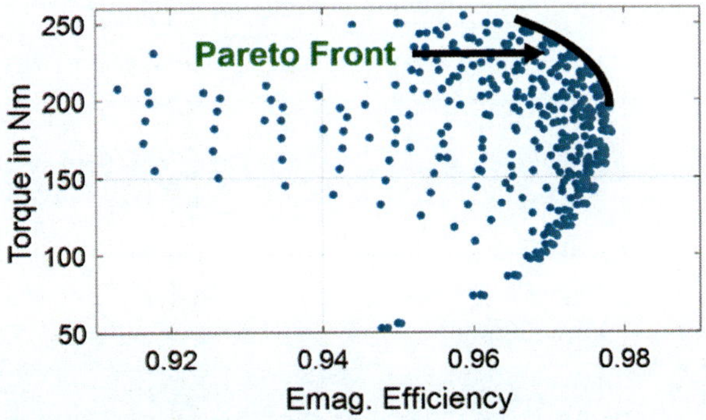

Fig. 6. Exemplary operation point simulated for more than 500 designs showing torque over electromagnetic efficiency. The black line indicates the Pareto front.

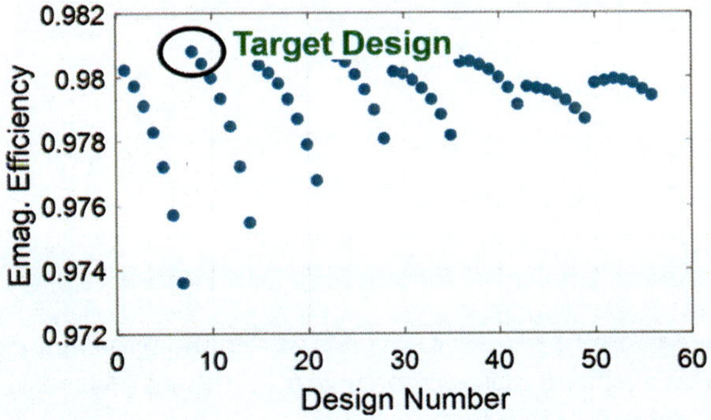

Fig. 7. Exemplary results of 56 designs evaluated over driving cycle requirements showing resulting electromagnetic efficiency.

Basic requirements for hairpin windings can be found in [4]. For the hairpin winding optimization Valeo Siemens eAutomotive Germany always considers the following goals

1. High slot fill factor
2. Low loss increase due to AC losses
3. Full winding symmetry
4. Simplified end winding and phase connection

Goals 1. and 2. as well as 3. and 4. are opposing each other at first. However through new winding topologies developed by Valeo Siemens eAutomotive Germany highly efficient hairpin machines are possible, without compromising one or the other. Through optimized insulation systems, increased fill factors are feasible. The phase connection is optimized to a high degree, resulting in a simple connection unit, which can be attached to the hairpin basket fully automatically. The key data for the electrical machine is summarized in Table 5.

Table. 5. Key machine data for baseline and new developed eAxle

	Baseline single speed eAxle	New Valeo TwinSpeed eAxle
DC voltage	400 V	800 V
Operation speed	16000 rpm	12000 rpm
Torque	530 Nm	440 Nm
Power	200 kW	250 kW
Max phase current	1200 A	600 A
Number of poles	8	6
Max. el. frequency	1100 Hz	600 Hz
Winding type	Round wire	Hairpin
Winding connection	Partly manual	Fully automated
WLTC Efficiency	+	++
Machine Size	Ref. size	+

4 Multispeed Transmission

4.1 Best Transmission Choice – Methodology Description

The transmission is chosen by using a specific methodology, which is detailed in this section. Based on targeted criteria (refer to Table 6), different transmission architectures are evaluated and compared (Fig. 8).

Three main configurations and transmission architectures are deeply evaluated in order to converge to the best solution for this application. The three architectures are

- DCT 2 (2 ratios transmission, with dual-clutch)
- DCT 3 (3 ratios transmission, with dual-clutch)
- Epicycles classical transmissions

Table 7 depicts the comparison as well as the overall results. For targeted application DCT2 is identified as the best solution.

Table. 6. Considered criteria

	Criteria
Customer requirements	Packaging
	Vehicle performances
	Parklock function
	System efficiency
	Powershift
Transmission evaluation	Control challenges
	NVH
	Cost

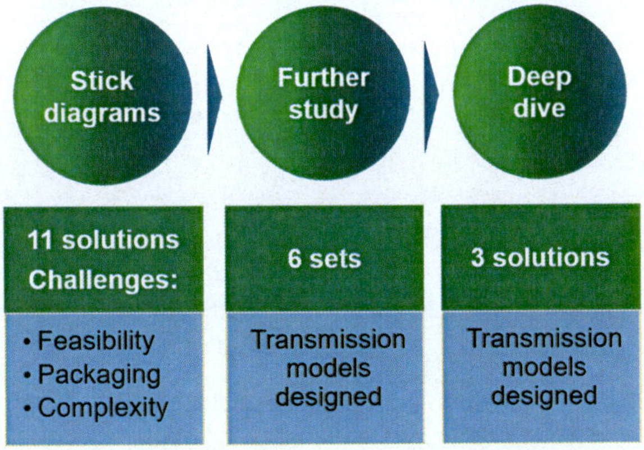

Fig. 8. Transmission choice methodology

4.2 DCT 2 Description

As a result of the analysis described in previous section a DCT2 transmission is chosen. The DTC2 topology denotes the best compromise between system efficiency, packaging and complexity 1 layshaft. The stick diagram is described in Fig. 9. The principle is shown in Table 8.

As the dual wet clutch is driven directly by the electrical machine, it is designed in order to meet the requirements of this application. The disconnected synchronizer has two main functions: to protect the dual wet clutch spinning at a very high speed when 2nd gear is engaged and also to reduce clutch drag torque on cycle. Figure 10 illustrates a DCT2 TwinSpeed. Figures 11 and 12 detail the transmission architecture and the torque path.

In first gear, the torque flow passes through the closed first clutch, the first gear stage, the closed synchronizer, the final gears stage and the differential. In second gear, the torque flow passes through the closed second clutch, the second gear stage, the final gears stage and the differential.

Table. 7. Comparison of transmission configurations for multispeed eAxles

	Criteria	DCT2	DCT3	Epicycles
Customer requirements	Packaging	++	+	+++
	Vehicle performances	+++	+++	+++
	Parklock function	+++	+++	+
	System efficiency	+++	++	+
	Powershift	+++	+++	++
Transmission evaluation	Control challenges	++	++	++
	NVH	++	++	+
	Cost	+++	++	++
Total		**21**	**18**	**15**

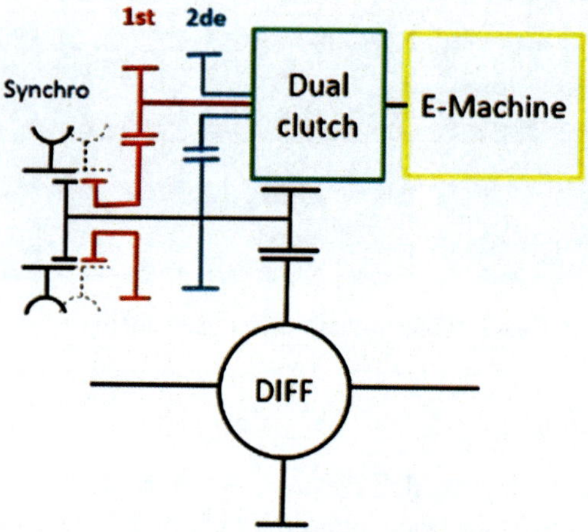

Fig. 9. TwinSpeed stick diagram

The system is designed in order to perform full powershift in all conditions, drive and regen the way it is shown in Fig. 13. Also, for reverse drive, it is possible to use both gears (1st and 2nd), based on most challenging technical requirements in terms of wheel torque.

4.3 Functions Integration Description

The TwinSpeed transmission concept proposes different functions such as

Table. 8. Principle overview

	K1	K2	Disco synchro
Neutral	Open	Open	Open
Gear 1 engaged	Close	Open	Close
Gear 2 engaged	Open	Close	Open

Fig. 10. Transmission TwinSpeed DCT2

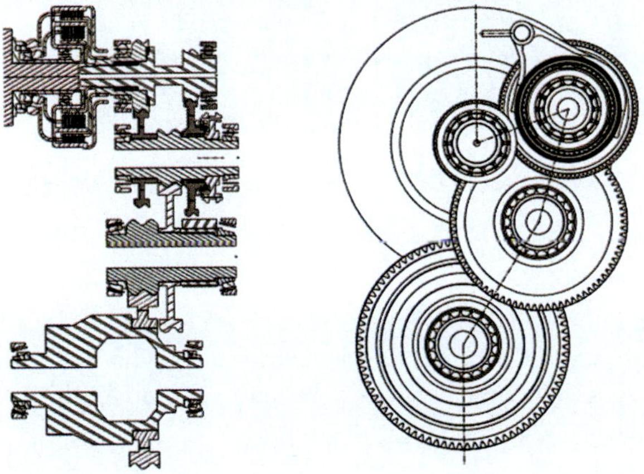

Fig. 11. Transmission architecture

- Full powershift function
- Parklock function => hydraulic technology
- Lubrication and cooling pump, since the oil is shared with the electrical machine

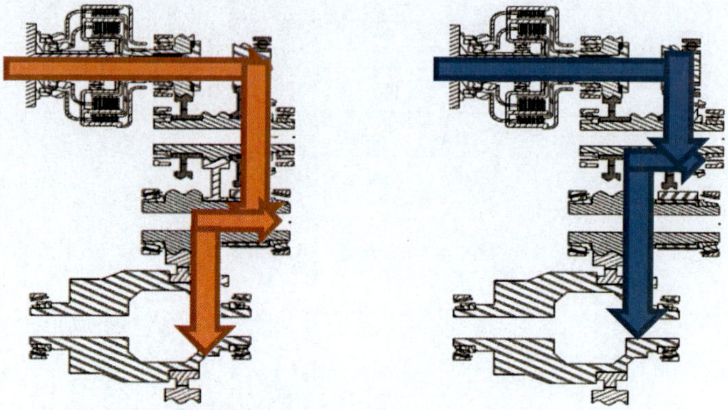

Fig. 12. Torque path in 2nd gear (in orange) and 1 st (in blue)

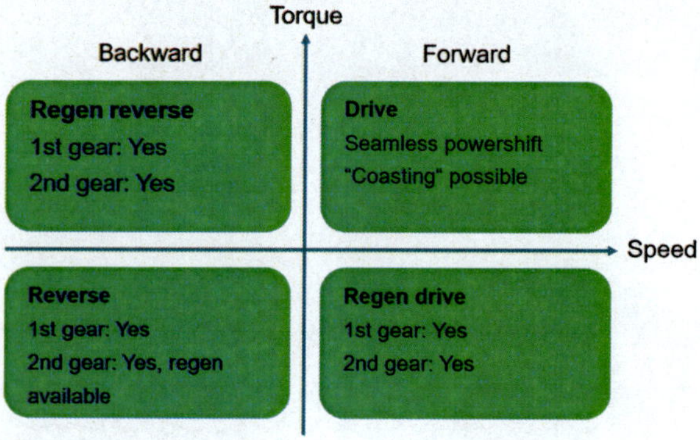

Fig. 13. Advantages of DCT 2 – reverse in 1st and 2nd gear possible

4.3.1 Full Powershift

In order to attain a full powershift function, a dual wet clutch is designed and in order to reduce the drag torque, a dCSC (double concentric slave cylinder) actuation is considered. A central HDCA (hydraulic dual clutch actuator) is used for this system. Figure 14 illustrates the clutch and the powershift function.

4.3.2 Clutch Focus

Valeo Powertrain Transmissions has a long experience and expertise in the domain of clutches. In order to meet requirements for this eAxle application, Valeo designed a optimized dual wet clutch. Main drivelines of design choices are highlighted in the following

- The new clutch combines compactness and robustness to both support efforts and to ease gearbox integration.
- The friction packs have been designed to face high slip speed levels and hot spots.

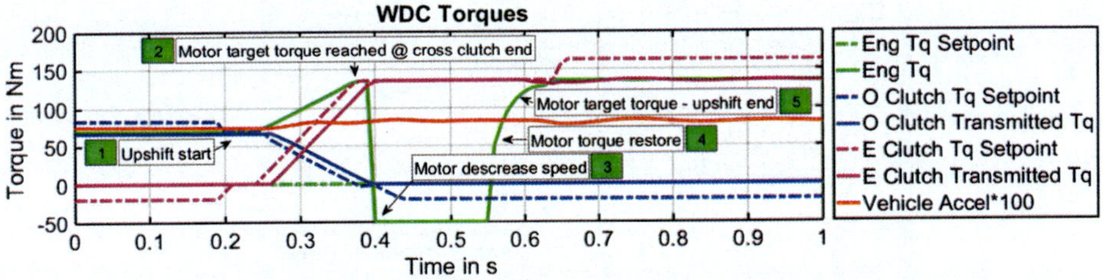

Fig. 14. Exemplary powershift function from the WLTC simulation depicted over 1s. Vehicle acceleration (red line) remains constant during shifting, illustrating reached target of 0 g/s jerk value at 40% pedal throttle.

- A particular focus is laid on lubrication and the thermal behavior. The clutch is specifically designed to meet even most challenging requirements in terms of durability and thermal duration.
- Drag torques and tumbling effects are reduced to meet best in class actual DWC and lower energy consumption and NVH issues (refer to Fig. 16). In order to reduce the overall drag torque, the dCSC actuation is specifically developed by Valeo.
- Guiding and balancing is improved further to limit NVH effects and bearings sizes.

The design was chosen based on best lowest drag torques values, in order to meet efficiency expectations. Figure 15 illustrates a dual wet clutch prototype, while Fig. 16 depicts the difference in drag torque between dCSC and piston type actuation.

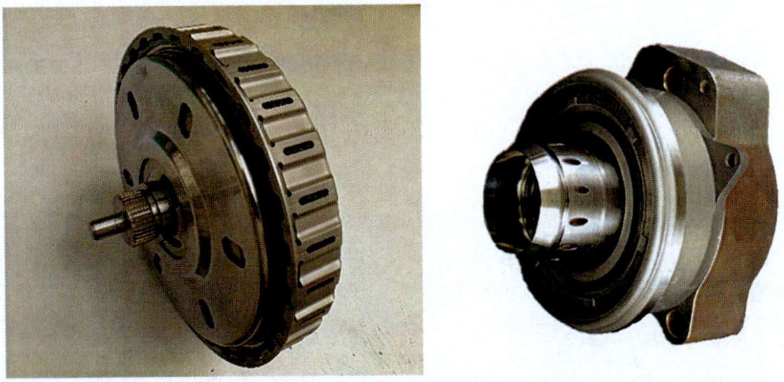

Fig. 15. Dual wet clutch prototype and dCSC (double concentric slave cylinder) [5]

By crossing the clutches and sizing the actuators in order to have the required time response, the actual designed system is able to shift without a deceleration loss at vehicle wheels. A specific control law is developed, in order to achieve a full powershift. In sport mode, a complete shifting event is made in 300 ms and in comfort mode 500 ms.

An HDCA system is designed, in order to meet the targeted requirements of the entire system. Additionally a unique actuation pump is used for the following functions: double clutch actuation, park lock and synchronization. Figure 17 depicts the working

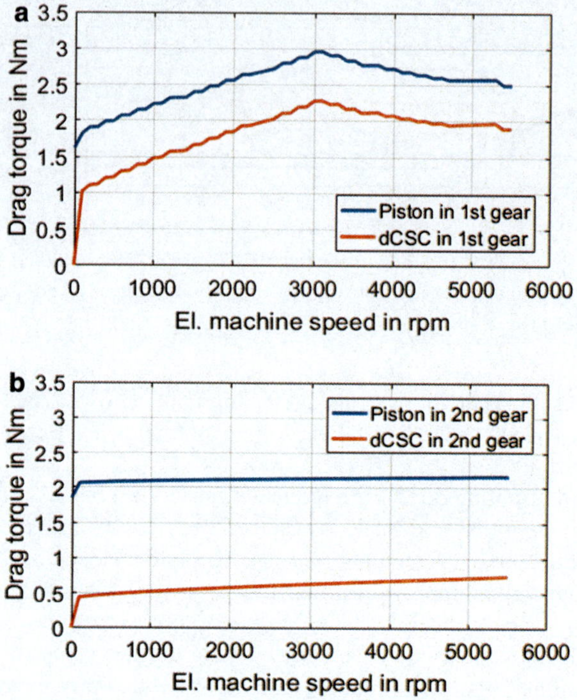

Fig. 16. Drag torque over electrical machine speed of piston type (blue) and dCSC (red) actuation for a) 1st gear and b) 2nd gear at 80 °C

principle of HDCA actuation pump, while Table 9 highlights the operation principle of the PLA (ParkLock Actuator).

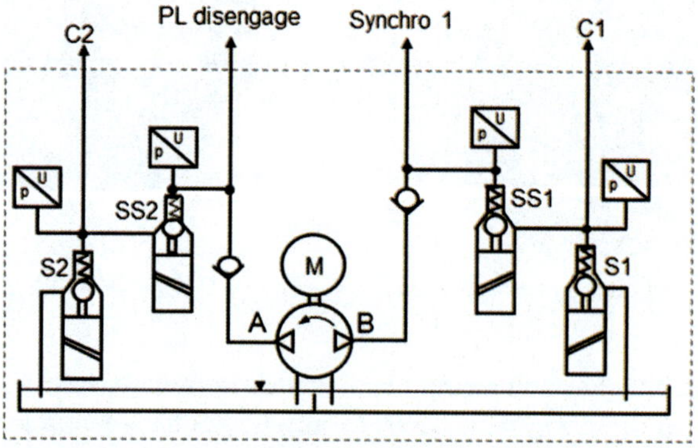

Fig. 17. HDCA working principle

Table. 9. Principle overview including PLA

	K1	K2	Disco synchro	PLA
Neutral	Open	Open	Open	Close
Gear 1 engaged	Close	Open	Close	Open
Gear 2 engaged	Open	Close	Open	Open

4.3.3 Parklock Function

Optionally a parklock function can be added. The parklock system consists of a parklock wheel, a pawl and a hydraulic parklock actuator. The system is illustrated in Figs. 18 and 19.

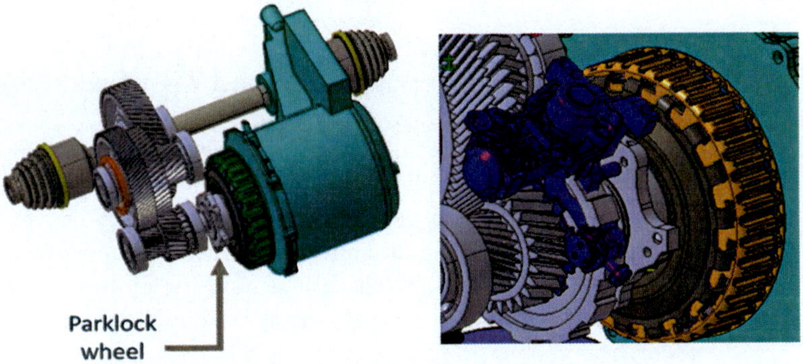

Fig. 18. Parklock wheel position and parklock actuator

The parklock wheel is situated in the output shaft of the electrical machine and the hydraulic is also integrated in a compact way, as can be seen in Fig. 19. The drivetrain is designed for shock loads coming from the vehicle wheel when parklock events are engaged. As this parklock is developed for premium vehicles, the system is able to lock >10kNm at vehicle wheel.

4.3.4 Lubrication and Cooling System

As can be seen in Fig. 19, Valeo designed a lubrication pump having one motor only. The lubrication pump is used for the cooling of the electrical machine and for lubrication of the double clutch. Based on different use cases, e.g. shifting, high power a specific optimal control strategy is used to assure adequate cooling and lubrication of all components.

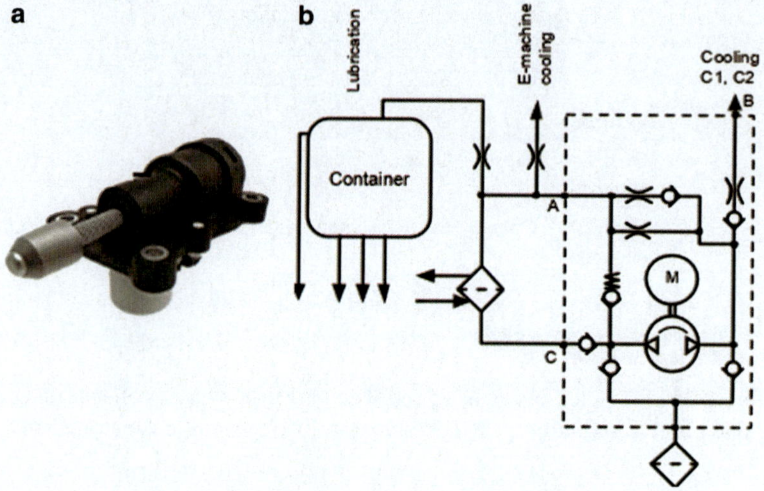

Fig. 19. a) Parklock actuator and b) detailed illustration of lubrication pump system

5 Performance Advantage of Newly Developed TwinSpeed eAxle

To design the new eAxle simulations of the full system considering several driving cycles and requirements are used. This section shows the comparison of the baseline single speed system and the final optimum multispeed design. For the drive-cycle simulations most relevant system losses are included in the simulation software, some of which are: Inverter conduction and switching losses, electrical machine Joule, iron and windage losses, gear box losses, clutch drag torque, synchronization drag torque and actuation losses.

Figure 20 shows the simulation domain, used to assess both systems. The simulations include the electrical machine, the transmission system and vehicle dynamics (aerodynamic, friction and mass). The battery system is not included. Furthermore, the following assumptions are made:

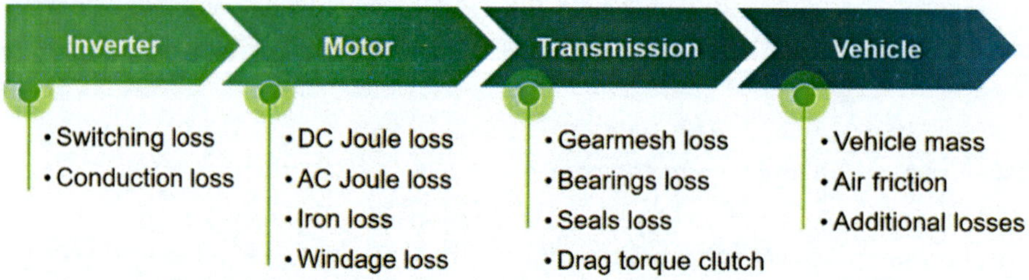

Fig. 20. Simulation domain used to assess the system performance including main losses

- Full regenerative strategy

- No boardnet consumption considered
- No heating or AC (air conditioning) consumption considered
- No detailed battery model considered, losses only

Table 10 shows the simulation results for the baseline system and the new eAxle. As the golden vehicle is used for both systems, net vehicle consumption is equivalent for both systems. With the new eAxle developed by Valeo and Valeo Siemens eAutomotive Germany relative vehicle range increases by up to 8%.

Table. 10. WLTC efficiency comparison for baseline single speed and new TwinSpeed eAxle

Consumption on WLTC	Baseline single speed eAxle	New TwinSpeed eAxle with hairpin and SiC
Total consumption in kWh/100 km*	22	20.2
Relative range increase	Ref.	**+8%**

*Full regenerative braking strategy considered.

6 Summary and Conclusions

Valeo and Valeo Siemens eAutomotive Germany developed electrical drives from the beginning as a fully integrated system providing advanced solutions for future electric mobility. Considering new developments on the SUV premium segment, this article presents the premium 250 kW eAxle including a seamless TwinSpeed transmission and an 800 V permanent magnet hairpin electrical machine. Through the combination of electrical machine and transmission, the BEV range on the WLTC increases by up to 8% for the specific premium vehicle application. For other applications, different combinations could be assessed and proposed, using different technological bricks. In the near future Valeo and Valeo Siemens eAutomotive Germany introduce new advanced solutions for the B/C vehicle class further shaping future electric mobility.

References

1. Blumenroeder K et al (2019) Volkwagen's new modular e-drive kit. In: 40th 40. Internat. Wiener Motorensymposium
2. Popescu M et al (2018) Electrical vehicles – practical solutions for power traction motor systems. IEEE Trans Ind Appl 54:2751–2762
3. Motor design limited: performance analysis of electric motor technologies for an electric vehicle powertrain – white paper. www.motor-design.com. Accessed 15 Sept 2019
4. Berandi G, Bianchi N (2018) Design guideline of an AC Hairpin Winding, In: XIII international conference on electrical machines (ICEM), pp 2444–2450
5. Olivier S (2017) Dual wet clutches for hybrid powertrains, challenges for packaging and fuel efficiency optimization. CTI symposium automotive transmissions, HEV and EV drives. Berlin

A Comprehensive Approach of the Lubrication for the Electric Powertrain Based on an Innovative Multi-purpose Fluid

Hakim El Bahi(✉)

TOTAL M&S, Centre de Recherche de Solaize, Chemin du Canal, 69360 Solaize, France
hakim.el-bahi@total.com

Abstract. Because of the increasing power density of electric motors, it is anticipated that the oil would not only lubricate the gearbox, but it will also be used as a coolant for the electric motor. TOTAL Quartz EV-Drive MP Technology has been designed as an innovative multi-purpose fluid which meets the new requirements of the combination of electric motor cooling and transmission lubrication. A comprehensive approach which identifies all the new constraints was followed in order to design such a multi-purpose fluid. Oil needs to maximize the heat transfer. Numerical simulations of an electric motor coupled with the data generated on a Stator Cooling Test enable us to identify the electric motor hot spots and to determine the fluid cooling power. As oil will be in direct contact with electrical components, it needs to be electrically insulating to avoid any current leakage. A direct oil cooling system also implies that oil should be inert towards electric motor insulation materials and bare copper. Test rigs were used to assess the oil compatibility with copper and polymer coated wires. Another critical aspect of electric vehicles is the high-speed lubrication properties. Ball bearings could experience lubricant starvation when the electric motor runs at high speed. A High-Speed Ball Bearing Test was designed to assess our fluids in this regard. Transmission components are also running at high speed, which increases oil aeration and churning losses. A specific test rig has also been designed to characterize churning losses at high speed.

Keywords: Lubrication · E-motor cooling · Multi-purpose fluid

1 Introduction

The BEV (Battery Electric Vehicle) powertrain is composed of four elements: electric motor, power electronics, reducer and battery. Today, most of the electric motors are air cooled or water cooled.

However, the power density of electric motors will increase in the years to come and the tendency will be a complete integration of the power electronics and the reducer within the electric motor. The concept of a unique fluid for the transmission lubrication and the electric motor cooling will spread as it meets the requirements of the new thermal management needed for reducing size and improving the performance of electric machines (Fig. 1).

Euroforum Deutschland GmbH (Ed.): *CTI SYMPOSIUM 2019*, pp. 80–92, 2021.
https://doi.org/10.1007/978-3-662-61515-7_9

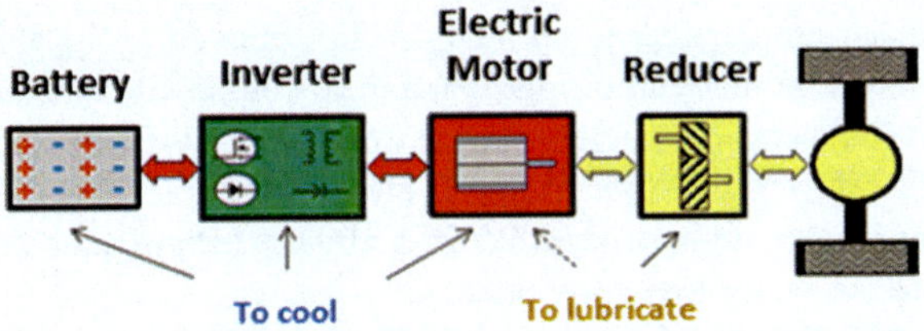

Fig. 1. Four main components of Battery Electric Vehicles (BEV)

The TOTAL Quartz EV-Drive MP Technology has been designed to be a multi-purpose fluid, i.e. an efficient coolant for the electric motor and a lubricant for the transmission and the bearing elements of the electric motor. This paper will present the new oil requirements needed for formulating such a multi-purpose fluid and TOTAL technology solutions in each regard.

2 New Requirements for the Multi-purpose Fluid

2.1 Thermal Properties

2.1.1 Numerical Simulation of the Electric Motor

Thermal properties are at the core of the development of fluids for electric vehicles. As we mentioned previously, the next generation of electric vehicles will be more powerful and smaller, which implies higher thermal stresses on the electric motor. Conventional air cooling or indirect water cooling will not be adequate for the new powertrains to come and will be replaced by a direct oil cooling thermal management system, where a multi-purpose fluid lubricates the transmission and cools down the hot spots of the electric motor.

The heat on the electric motor is created by the losses dissipated in the system during its operation. The losses in electrical machines can be separated into mechanical losses, winding losses and iron losses as shown in Fig. 2 [1].

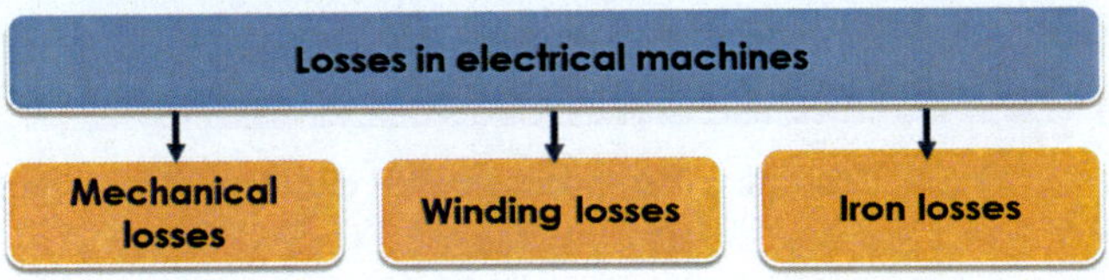

Fig. 2. General loss separation in the electric motor

The mechanical losses are due to friction losses in the bearings and to windage losses of the rotor part in the airgap. The winding losses, also referred to as Joule heating, are the losses created by the fact that current is flowing in the windings of the machine. As for the iron losses, they are created by the changing magnetic field in the stator and rotor parts.

These three losses are depending on many factors related to the electrical machine (size, geometry, magnetic material properties) and of course, are also dependent on the operating point of the electric vehicle. A study conducted by BMW [2] shows that the iron losses dominate during an urban driving cycle. At higher torque, the winding losses are more significant and at higher speed like in highway driving, mechanical losses and iron losses become the major loss component.

The heat created in the electric motor can be detrimental for the winding insulations. Motors must be designed so that the winding insulation can withstand the overall heat within the motor. The insulation materials are designed with different classes of insulation that are defined by IEC standards [3]. Copper windings may be classified in thermal classes 155 °C and 180 °C or even higher. In these cases, the maximum allowable hot spot temperatures of the winding insulation are 155 °C and 180 °C.

In the specific case of Permanent Magnet Synchronous Motors (PMSM), there is a thermal constraint regarding the issue of demagnetization. If the magnet is heated up above its maximum working temperature, it loses part of its magnetization. Many permanent magnets cannot withstand all possible operating conditions of the electric motor. This renders the machine cooling design crucial and challenging. Hence the development of the direct oil cooling solution with a multi-purpose fluid to meet the new requirements for high power density electrical machines. Modern electric motors rely on forced convection cooling that can be achieved through different configurations like jackets, spray cooling and hollow shaft [4]. In order to study thoroughly the thermal management of electric motors, TOTAL has partnered with the Institute of Power Electronics and Electrical Drives (ISEA) from the RWTH Aachen University in Germany.

Thermal Modeling was implemented on a PMSM as it is the most widespread electric motor technology in the automotive industry. The characteristics of the motor are listed in Fig. 3. Two different cooling configurations were modelled as shown in Fig. 4: the circumferential water jacket that surrounds the stator core and the direct oil cooling system, which includes both direct and indirect cooling approach.

Permanent Magnet Synchronous Motor
Stator 269 mm (diameter) x 84 mm (length)
48 stator slots
V-shape interior permanent magnet
Maximum Power = 50 kW

Fig. 3. Characteristics of the modelled electric motor

In addition to the cooling jacket, the direct oil cooling system also comprises a hollow shaft and a spray cooling arrangement for the stator end windings with four nozzles located on front and rear of the shaft. There are two types of spray: the outer spray cools down the upper part of the end winding from the oil jacket and the inner spray cools down the lower part of the end winding from the hollow shaft.

The speed-torque diagram in Fig. 5 shows the motor operating point chosen for the simulation and the corresponding heat losses. This operating point corresponds to a typical city driving speed at a high torque.

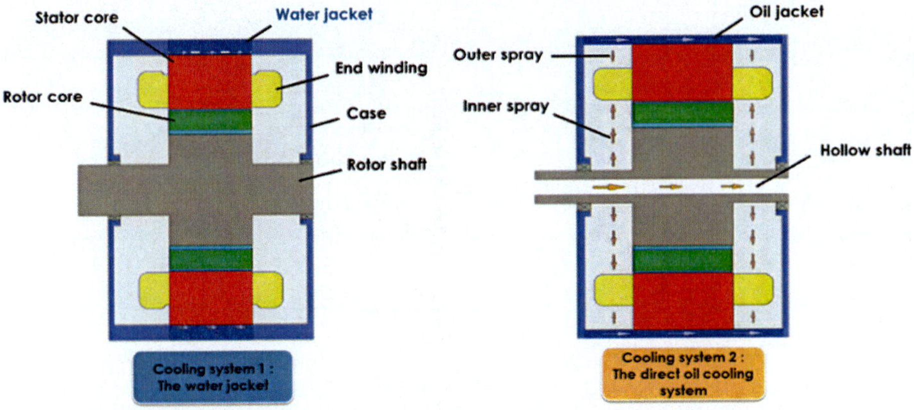

Fig. 4. Two cooling systems investigated: the water jacket and the direct oil cooling.

Winding Losses (W)	2636
Iron Losses (W)	146
Mechanical Losses (W)	790

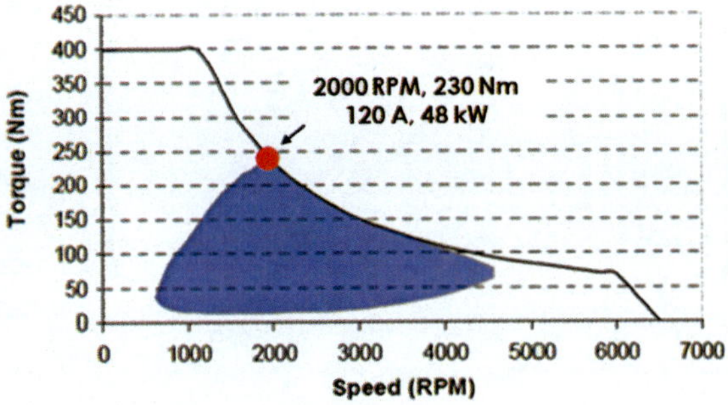

Fig. 5. Selected operating point for the simulation of losses and corresponding heat losses

The heat flux is transferred by forced convection and the cooling efficiency is approached by four main liquid properties in this context: kinematic viscosity, specific heat capacity, thermal conductivity and density. A High Reference ATF and TOTAL Quartz EV-Drive MP Technology were compared in the direct oil cooling configuration and their characteristics are shown in Fig. 6.

One of the specificities of the TOTAL Quartz EV-Drive MP Technology fluid is its very low kinematic viscosity, 3 cSt at 100 °C, whereas most of the ATFs in the electric vehicle market have a viscosity higher than 4.4 cSt at 100 °C (75 W SAE Viscosity Grade). This lowering of viscosity was accomplished in order to maximize the cooling power as it will be explained later in further details. The numerical simulation was performed with different fluid inlet temperatures and flow rates as shown in Fig. 7 and the results are presented in Fig. 8.

The maximum winding insulation and rotor magnet temperatures were determined as the heat extraction is critical for those components. Results confirm that the direct oil cooling system is more efficient than the water jacket design as it provides a greater decrease of temperature in the order of a hundred degrees. TOTAL Quartz EV-Drive MP Technology shows a better cooling performance than the reference ATF, especially

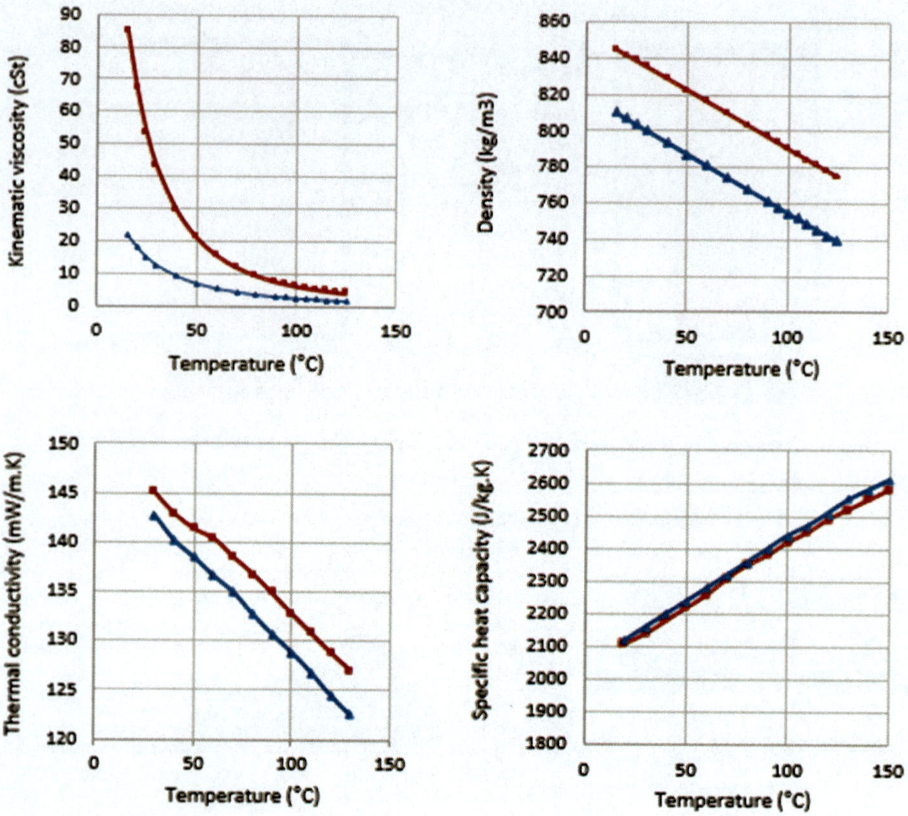

Fig. 6. Thermal properties of the investigated fluids: a High Reference ATF (red) and the TOTAL Quartz EV-Drive MP Technology (blue)

Cooling system	Water jacket	Direct oil cooling
Flow Rate	6 L/min	Oil jacket (1 L/min) Outer spray (0,5 L/min) Hollow shaft and inner spray (5 L/min)
Inlet fluid temperature	30°C - 40°C - 50°C - 60°C	30°C - 40°C - 50°C - 60°C

Fig. 7. Flow rates and Inlet temperatures implemented for each cooling system

for the winding maximum temperature, avoiding the windings to reach their maximum allowable temperature. It lowers the winding maximum temperature by an additional 14 °C and the magnet temperature by 6 °C.

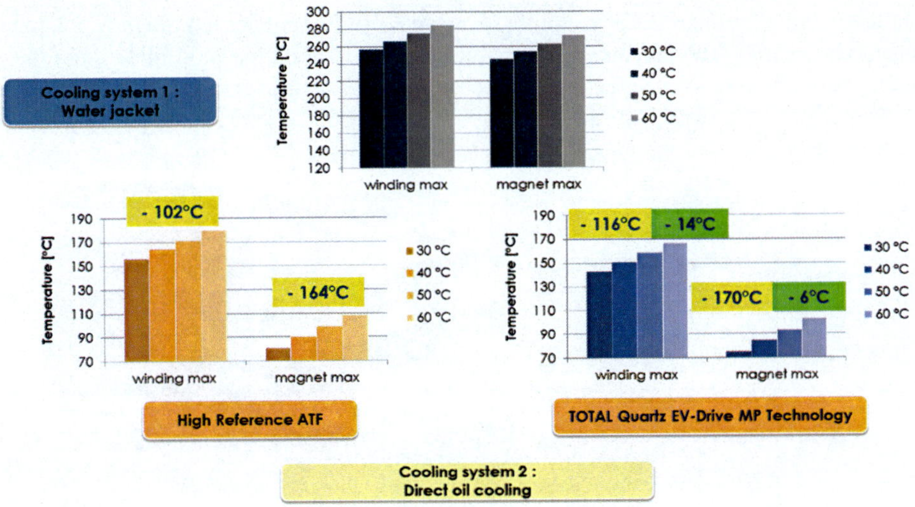

Fig. 8. Simulation results of the winding and magnet maximum temperature for the two studied cooling systems. In yellow tags are shown the reduction of temperature brought about by each oil compared to water jacket cooling. In green tags are shown the reduction of temperature brought about by Total Quartz EV-Drive MP Technology compared to the High Reference ATF.

2.2 Assessment of the Fluid Cooling Power

The simulation work shows how important the fluid cooling performance is in the thermal management of the electric motor. To evaluate the effects of combined fluid properties, researchers have suggested various Figure of Merit (FOM). For instance, Mouromtseff introduced the Mouromtseff number to evaluate the fluid cooling power on the convective heat transfer coefficient of internal turbulent flow [5]:

$$FOM = \frac{\rho^{0,5} k^{0,67} c_p^{0,33}}{\mu^{1,2}}$$

Even though the Mouromtseff number provides useful comparisons among different heat transfer fluids, it considers only a turbulent flow, which is not the case in electric motor cooling where the Reynolds number is lesser than 1000 and it does not consider the hydraulic performance characterized by the pumping power. Factors that give rise to high convection coefficients tend to give high pressure drops and high pumping costs; therefore, a new FOM is desired. In order to achieve that goal, TOTAL has partnered with the Institute of Heat and Mass Transfer (WSA) from the RWTH Aachen University in Germany.

$$NewFOM = \frac{\rho^{\alpha} k^{\beta} c_p^{\gamma}}{\mu^{\delta}}$$

A semi-empirical method was implemented. It combines both the theoretical calculations of the heat transfer coefficient and the experimental data of the Stator Cooling Test set up in order to measure the fluid cooling performance. It consists of a channel that holds three live-like stator teeth. Teeth are wrapped in copper windings which are heated electrically. The channel is fed by a closed oil circulation loop including temperature control and flow meter. The temperature of the center tooth is measured using thermocouples. The experimental setup and the test conditions are shown in Fig. 9.

Inlet temperature (°C)	50
Flow rate (mL/s)	15-30
Thermal Load (W)	50 (per tooth)

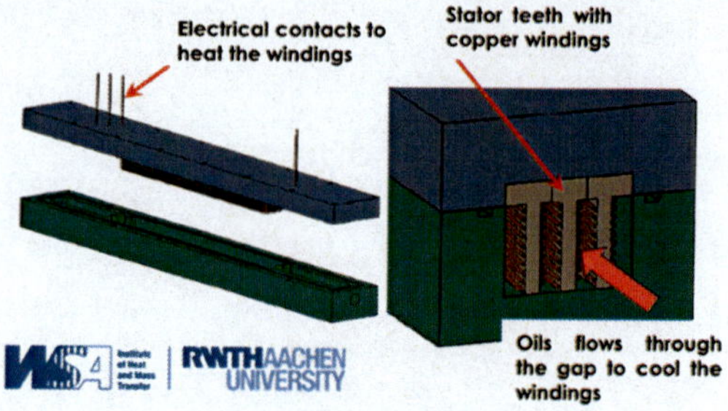

Fig. 9. The Stator Cooling Test and its experimental conditions

Numerous fluids of different thermophysical properties were tested in order to have a large statistical base from which the relevance of the new FOM can be assessed.

The new FOM determined by the semi-empirical method provides an accurate way of assessing the fluid cooling power as it shown in Fig. 10. The Root-Mean-Square of the new FOM has a value of 0.92, which indicates that there is a very good fit between the new FOM and the empirical data. Therefore, the new FOM could be used as a helpful tool for predicting fluid cooling power efficiency.

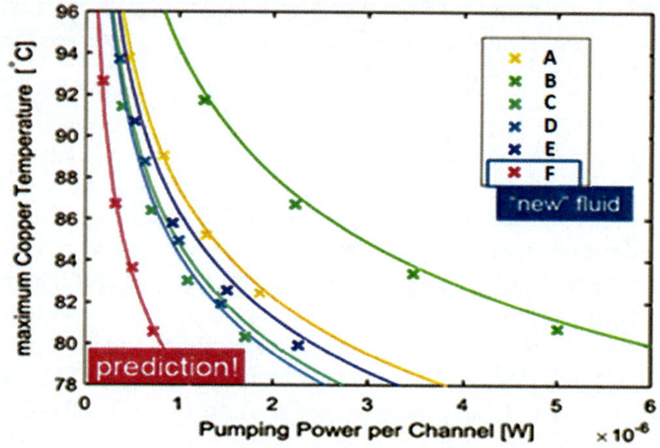

Fig. 10. Correlation between the new FOM and the results of the Stator Cooling Test

A Benchmark analysis of Automatic Transmission Fluids (ATF) currently used in Electric vehicles was performed in order to compare the TOTAL Quartz EV-Drive MP technology with the existing performances in the oil market. It comprises two high reference ATFs from the United States and one high reference ATF from Japan.

Results in Fig. 11 show that TOTAL Quartz EV-Drive MP Technology has the best cooling performance. Thanks to its higher FOM value – 284 at 50 °C – the copper temperature is lowered by 10 °C more than the other reference oils.

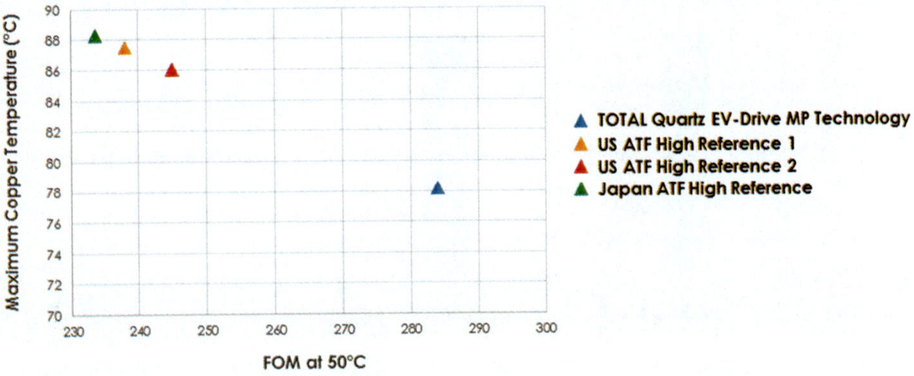

Fig. 11. Benchmark results in the Stator Cooling Test, the maximum copper temperature at a given pumping power.

2.3 Particular Issues Related to a Very Low Viscosity Oil

The multi-purpose fluid should not only be a highly efficient coolant for the electric motor, but also a good lubricant providing protection against gear and bearing wear.

A very low viscosity oil can be detrimental to the load-carrying capacity as the oil film thickness in a mechanical contact is dependent upon the oil viscosity. In this context, one might decide to formulate on oil with a high content of anti-wear/extreme-pressure (AW/EP) additives in order to form a stronger tribofilm which prevents the wear of the contacting surfaces. However, such an approach would undermine the dielectric properties and the oil compatibility with copper materials as AW/EP additives are mainly sulphur compounds. The thermal degradation of sulphur compounds generates radicals that are extremely copper-corrosive and as polar chemical species, sulphur compounds are also increasing the oil electrical conduction. Therefore, our R&D team has to wholly rethink its formulation strategy to meet all the requirements needed for a multi-purpose fluid.

2.3.1 Load-Carrying Capacity

The standard FZG oil test A/8.3/90 is widely used for the evaluation of scuffing properties of industrial gear oils. The standard procedure was severed by increasing the temperature to 150 °C in order to check that TOTAL Quartz EV-Drive MP Technology is still able to withstand such drastic conditions.

Results are presented in Fig. 12. Gears are loaded stepwise in 12 load stages between a Hertzian pressure of 150 to 1800 N/mm^2 under conditions of dip lubrication without cooling. Failure load stage is indicated when the faces of all pinion teeth show a summed total width of damaged areas which is equal or exceeds one tooth width [6]. On the basis of the scuffing tests and given its very low viscosity, TOTAL Quartz EV-Drive MP Technology shows an outstanding performance.

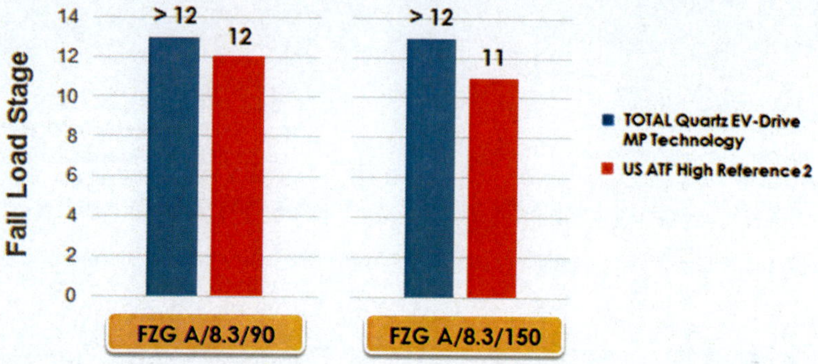

Fig. 12. Results of the two FZG Tests performed: TOTAL Quartz EV-Drive MP Technology (in blue) and US ATF High Reference 2 (in red)

2.3.2 Dielectric Property

The electrical resistivity is usually measured to characterize the oil dielectric property. The electrical resistivity represents the oil ability to oppose the electric current flow. It is expressed in ohm-meter (Ω.m). A good insulating oil is characterized by a high resistivity, in the order of Mega Ohm-meter at room temperature. An oil's electrical resistivity should not be too low to avoid short circuits, but also not be too high to avoid electrostatic discharge. Electrical resistivities of the tested fluids are shown in Fig. 13.

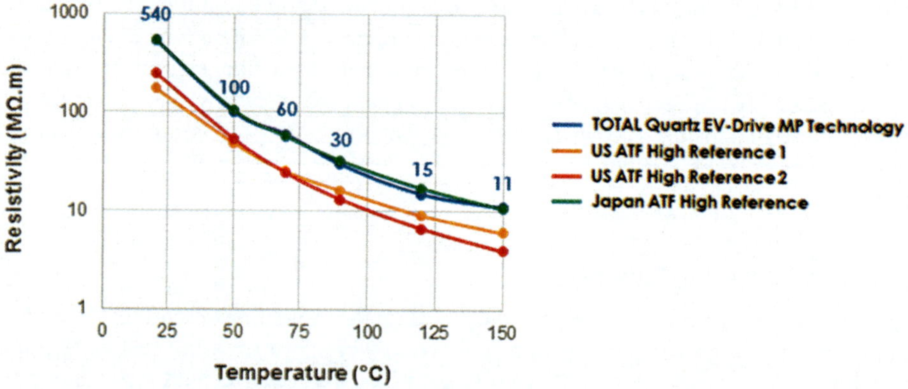

Fig. 13. Electrical resistivity versus temperature. TOTAL Quartz EV-Drive MP Technology Electrical resistivities over temperature are listed above its blue curve.

TOTAL Quartz EV-Drive MP Technology shows a good dielectric performance comparable to the Japan ATF High Reference. TOTAL Quartz EV-Drive MP Technology additives have been carefully chosen to maximize the load-carrying capacity without compromising the dielectric properties.

2.4 Material Compatibility

Formulating a lubricant with a good copper compatibility is of paramount importance as copper is widely used in the electrical machine, whether it be on windings or electronic components.

A test was developed in order to precisely monitor the corrosion kinetics of a copper wire. The test design is shown in Fig. 14 and consists of a copper wire immersed in an oil-filled tube at 150 °C with a DC current applied. A voltmeter is mounted on the electrical circuit so that the copper wire electrical resistance can be determined. As the resistance is inversely proportional to the wire diameter, it is then possible to monitor the corrosion kinetics by knowing the change in electrical resistance. A fluid with a good copper compatibility will show no reduction of the copper wire diameter over a long period.

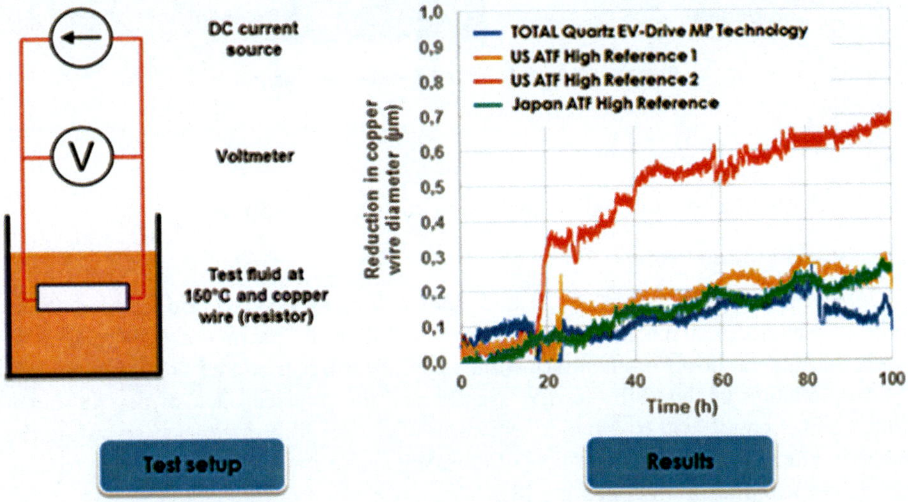

Fig. 14. Schematic view of the Copper Wire Test and experimental results

Results of the Copper Wire Test are presented in Fig. 14. Like the other high Reference ATFs, TOTAL Quartz EV-Drive MP Technology shows a good compatibility with copper as the diameter reduction is almost zero after a 100 h time period.

Copper wires are coated with a very thin layer of insulation material. Polyimide films are generally used in this application and can be affected by aggressive chemical species from the oil. The integrity of the insulation material determines the proper functioning of the electric motor. Any defects or damage to the insulation film would lead to failure. Therefore, it is important to check the oil compatibility with the insulation material.

The test setup used in our laboratory is shown in Fig. 15. The test is a modification of IEC 60851 which specifies a “peel test” for assessing the wire property to withstand twisting without showing cracks or loss of adhesion of the insulation [7]. A straight piece of wire is first immersed in the oil during 1000 h at 120 °C. Then the coating is removed on opposite sides of the wire by means of a scraper. This scraping operation should be done slightly without damaging the wire. Ultimately, the peeled wire is twisted by the test equipment shown in Fig. 15. The rotation continued until coating loss of adhesion has been reached and the corresponding number of revolutions is reported.

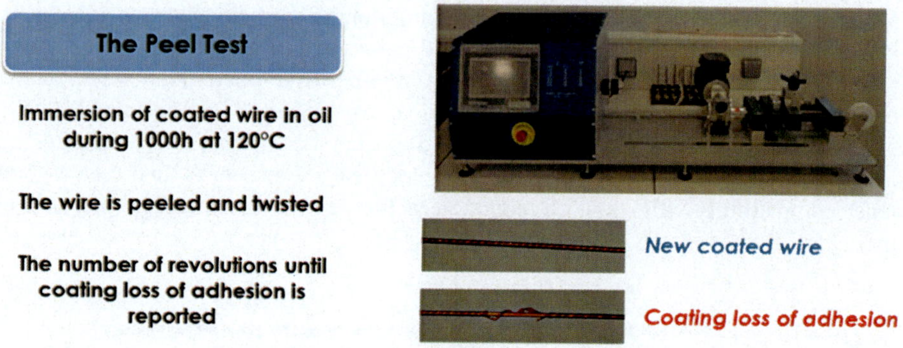

Fig. 15. The peel test setup and its procedure

2.5 Particular Issues Related to the Lubrication at High Speed

2.5.1 Bearing Lubrication

The lubrication of gears and bearings at higher rotational speed is a critical issue. Some of the current electrical machines can run up to 20,000 rpm. This increase of speed could have a detrimental effect on the lubrication of ball bearings mounted on the rotor shaft. The combination of the high thermal stresses and the physical lack of new lubricant in the contact could lead to a major reduction of the oil film thickness separating the contacting materials. The main effects of lubricant starvation are the increase of friction and the reduction of lifetime bearing [8].

Ball bearing performance tests have been designed to investigate this issue as it is shown in Fig. 16. Two operating points have been considered. The one at maximum power with maximum torque and the one at the highest speed.

2.5.2 Oil Aeration

The electric motor high speed also impacts the transmission system which is splash lubricated. The air content of the oil rises with gear rotational speed and this increasing air content dramatically increases churning losses as the oil volume expands and as the bursting of numerous bubbles by gear teeth costs energy due to their surface tension [9].

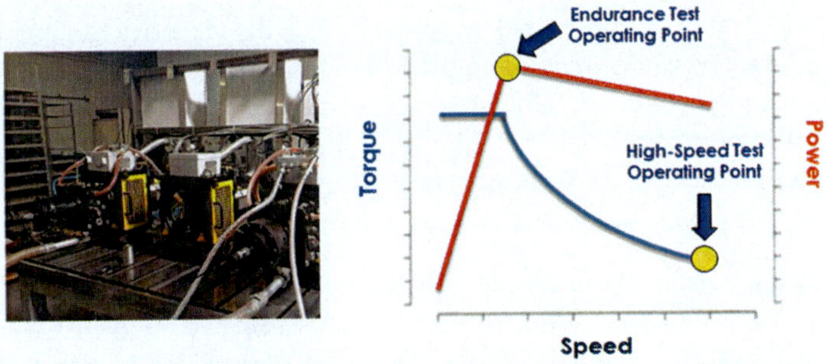

Fig. 16. High-speed ball bearing bench test picture and schematic diagram

In an attempt to have a good control over churning losses, the oil formulation has to be carefully design to limit oil aeration.

Our partner, the ECAM Lyon School of Engineering, developed a specific test rig to measure churning losses and oil aeration at very high speed as it is shown in Fig. 17. Its layout is simple: a single pinion driven by an electric motor is partially immersed in an oil bath. Churning losses are determined from direct torque measurements during the rotation of the pinion and aeration measurements are performed thanks to the "Air-X" device developed by our partner DSI. The instrument is based on low energy X-ray transmission. As X-rays absorption is dependent upon the density of the crossed material, the air content can be monitored since the oil aeration changes the oil density.

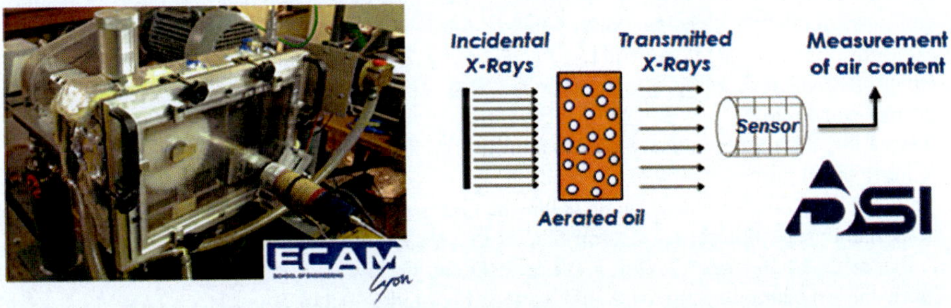

Fig. 17. High-speed churning losses test and schematic diagram of the "Air-X" device.

3 Conclusion

In the context of powertrain electrification, TOTAL has developed a new type of fluid: the TOTAL Quartz EV-Drive MP Technology. The challenge was to be able to create a multi-purpose fluid which integrates good cooling and lubrication properties. This had to be done from scratch. This led us to redefine the key principles of the fluid formulation and to develop a new base of tests to predict performance.

Formulating a multi-purpose fluid impels to reduce the viscosity in order to increase the cooling efficiency while maintaining the oil lubricating power and the other fluid requirements at a high standard, which may sound contradictory. This renders the multi-purpose fluid design inspiring and pioneering. Thanks to an intensive research from our R&D Team and partners, TOTAL has been able to achieve this goal and is still striving to formulate stepped out fluid technology for the future challenges of e-mobility.

Acknowledgements. We would like to express our gratitude to MScs. Huihui Xu and Claas Ehrenpreis from RWTH Aachen University for their valuable work on electrical machine cooling and fluid thermal properties. We are also thankful to Dr. Christophe Changenet from ECAM Lyon and DSI for their support on churning losses study. We would like to express our thanks to our colleagues Goulven Bouvier, Shimin Zhang, Mathieu Napoli, Nicolas Champagne, Grégoire Roux, Thomas Gillet and Didier Martin for the R&D work on e-mobility.

References

1. Krings A (2014) Iron losses in electrical machines – influence of material properties, manufacturing processes, and inverter operation. Doctoral Thesis. KTH School of Electrical Engineering (Sweden)
2. Merwerth J (2014) The hybrid-synchronous machine of the new BMW i3&i8. Workshop Lund University
3. International Electrotechnical Commission (IEC) (2007) Electrical insulation – thermal evaluation and designation (60085), Edition 4.0
4. Popescu M. (2015) Modern heat extraction systems for electrical machines. IEEE workshop on electrical machines design, control and diagnosis (WEMDCD). Torino
5. Mouromtseff IE (1942) Water and forced air cooling of vacuum tubes. Proc IRE 30(4):190–205
6. Hoehn BR, Oster P, Tobie T, Michaelis K (2008) Test methods for gear lubricants. Goriva i maziva, vol 47 no 2
7. International Electrotechnical Commission (IEC) (2013) Winding wires – test methods – part 3: mechanical properties (60851-3), Edition 3.1
8. Damiens B (2003) Modélisation de la lubrification sous-alimentée dans les contacts élastohydrodynamiques elliptiques. Doctoral Thesis. INSA Lyon (France)
9. LePrince G, Changenet C, Ville F, Velex P, Dufau C, Jarnias F (2011) Influence of aerated lubricants on gear churning losses–an engineering model. Tribol Trans 54(6):929–938

Assessing the Relative Endurance Capacity of Hybrid Drivetrain Components in an Early Development Stage with an Indicator Based on Preceding Drivetrain Generations

Jannick Fischer[1,2(✉)], Simon Rapp[1], Katharina Bause[1], and Albert Albers[1]

[1] IPEK – Institute of Product Engineering at Karlsruhe Institute of Technology (KIT), Sulzberg, Germany
jannick.fischer@daimler.com
[2] Daimler AG, Stuttgart, Germany

Abstract. In this paper an approach is shown to estimate the load on a component in a hybrid drivetrain. Therefore, the already existing method for conventional drivetrains is modified and expanded to fulfill the new requirements of hybrid drivetrains. The interaction between combustion and electric engine is considered to calculate a load-equivalent constant torque that is used as an input to determine a drivetrain specific, load proportional standardized value. With the help of drivetrain simulations, it is shown that the new approach leads to better estimation results in the case of hybrid drivetrains. Additionally, it is displayed that the new method is valid not only for hybrid drivetrains but also for conventional drivetrains, as it is a more general method. Furthermore, an example is given, how the knowledge about a component's usable load range and the estimated load of a new drivetrain can be used to predict roughly the necessary activities during the development of the new drivetrain.

Keywords: Load estimation of hybrid drivetrains · Prediction of development activities

1 Introduction

At the beginning of the development of a new car, there are always certain objectives regarding vehicle speed, acceleration, weight and fuel economy to achieve. While trying to fulfill these goals it is of utmost importance to survey the endurance capacity of the drivetrain. To determine whether the lifetime goal can be achieved with a certain component the load on this component has to be deduced. To avoid costs and other problems throughout later development stages this has to be done during the early stages of the product development. Usually the loads are determined either with vehicle measurements or with the help of simulations, but this is not always possible in early phases of product development.

Euroforum Deutschland GmbH (Ed.): *CTI SYMPOSIUM 2019*, pp. 93–102, 2021.
https://doi.org/10.1007/978-3-662-61515-7_10

Therefore, in [1] an analytic method was described to compare conventional drivetrains concerning their endurance capacity during the early stages of product development. In this method, the known endurance capacity of existing drivetrains is used to determine the endurance capacity of new drivetrains. However, new market requirements due to changed laws and customer needs will lead and have already led to more different and more complex drivetrains [2]. With the occurrence of hybrid cars for example, the variety of drivetrains has been increased significantly. Therefore, the method explained in [1] is not always sufficient.

2 State of the Art

2.1 Product Generation Engineering

As shown by Albers et al. [3] most of the time the development of new products cannot be distinguished by the means of the classic design methodology as for example described by Pahl and Beitz [4]. Instead, every product development is to a certain extent based on a set of existing systems or subsystems and associated documentation. Thus, every product development is in some way the development of a new product generation. This is depicted by the description model of PGE – Product Generation Engineering by Albers et al. [3]. The set of already existing systems, subsystems and documentation that is a basis and starting point for the development of a new product generation, including their interrelations is called the "reference system" [5].

Based on the reference system the structure and the subsystems of a new product generation are developed, combining carryover and new solutions to create a competitive advantage on the one hand with limited development risks on the other hand [6]. In the model of PGE [7], the different development activities are described with three types of variation which are used during the development of a new engineering generation:

- Carryover Variation: an existing solution is taken from the reference system and only the system connectors are varied.
- Embodiment Variation: an existing solution is carried over its improved in a way to ensure a competitive advantage.
- Principle Variation: here not only the embodiment but also the solution is varied.

2.2 Fatigue Calculation

Usually, to predict the failure of a component, the predicted amount of load cycles n_i is compared to the component's bearable amount of load cycles N_i at a certain load level i:

$$D = \sum_i \frac{n_i}{N_i}. \tag{1}$$

If the sum exceeds 1 a failure is highly probable. This approach is known by the name Miner's Rule [8].

2.2.1 Bearable Load Cycles

The bearable amount of load cycles is derived from the so called Wöhler curve, which describes the bearable load cycles with respect to the load:

$$N_i(M_i) = N_D \cdot \left(\frac{M_D}{M_i}\right)^{\kappa}. \tag{2}$$

In this equation M_D describes the fatigue limit and N_D the corresponding load cycles. κ (Sometimes referred to as Wöhler exponent) determines the curve's steepness and therefore the load sensitivity of the component. Normally these Wöhler curves are stressed (σ) based, but here it is replaced by torque to keep track of the drivetrain fatigue's main influence, without taking a look at specific fatigue mechanism or component behavior.

To calculate the shape of a component's Wöhler curve a lot of tests are necessary. Hence, such a curve is not available for many components and it is impossible to perform a proper lifetime or fatigue calculation. To bypass this problem a fictive Wöhler curve is used and either a virtual or relative fatigue level is calculated [9]. The calculation of a relative fatigue level only depends on the value of κ. Therefore, if this value is known or looked up in literature a relative comparison of fatigue levels can lead to a very good result, if in addition the behavior of one of the components is well known. Here, κ is set as 5 for all further calculations.

2.2.2 Predicted Load Cycles

There are two main approaches to determine the second input value of fatigue calculation, which is the expected load of a component during its lifetime. Firstly, there are measurements. In this case, a car is fitted with torque and rotation speed sensors and measured on defined routes with a certain driver behavior to mimic the lifetime loads. Secondly, simulations can be used. In this case, a detailed car model is build and a simulation is performed. To ensure comparability with the measured data the same routes and driver behavior has to be used.

2.3 Load Prediction

As already stated in the introduction, often, especially at the beginning of the product development process there is neither a car physically available that can be measured nor are there enough data to conduct a simulation. Therefore, based on the work of Ullmann [10], in [1] a method was shown to estimate the expected load on a drivetrain's component. Based on the knowledge about already existing drivetrains the applied load on a component in a new drivetrain is gauged. As shown in [1], this can be done by using exclusively the drivetrain's and car's basic parameters such as the maximum torque or the car weight.

For that a drivetrain specific, gear-dependent standardized value x_{Norm}. was derived. This value is proportional to the load on a drivetrain's component. The following two equations show the standardized value for the transmission input's load (3) and the total side shaft's load (4):

$$x_{Norm}^{TI} = (M_{max})^5 \cdot \frac{mr_{dyn}^2}{M_{max} \cdot i_{sum}^2}\left(\omega_s^2 - \omega_0^2\right) \tag{3}$$

$$x_{Norm}^{SS} = (M_{max} \cdot i_{sum})^5 \cdot \frac{mr_{dyn}^2}{(M_{max} \cdot i_{sum}) \cdot i_{sum}^3}\left(\omega_s^2 - \omega_0^2\right) \tag{4}$$

with

$$M_{max} = \min\left(M_{VM}, M_{lim}, M_{fric}\right) \tag{5}$$

where M_{VM} is the highest possible engine torque, M_{lim} the torque limit given by the transmission and M_{fric} the highest possible torque limited by wheel slip. With the help of known loads, e.g. through measurement or simulation, a regression curve as shown in Fig. 1 can be computed.

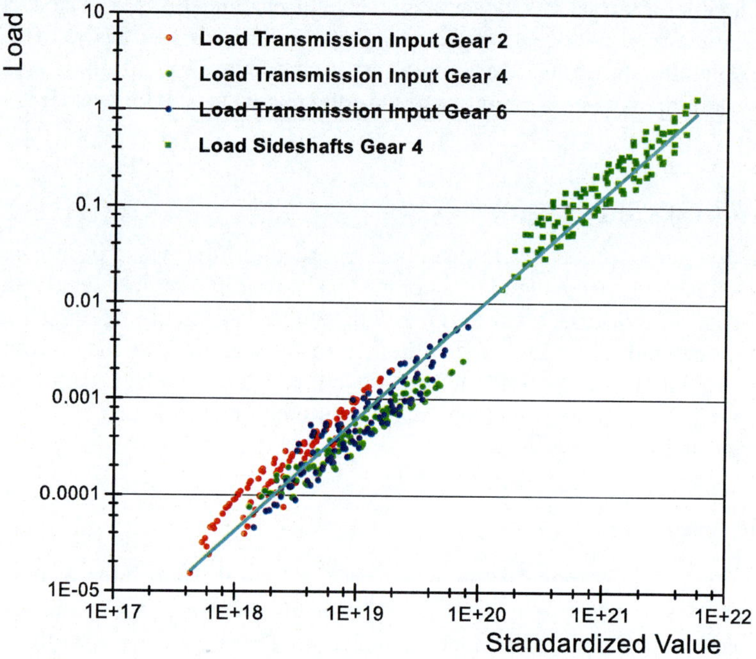

Fig. 1. Relation between standardized value and load in gear 2, 4 and 6 at the transmission entrance and gear 4 at the side shafts.

3 Motivation

As visible in Fig. 2, for a similar standardized value x_{Norm} the calculated load on a hybrid drivetrain (blue) is usually lower compared to the load on a conventional drivetrain (red).

This will cause the estimated load of a new hybrid drivetrain to be too high. The reason being that the interaction between combustion and electric engine has not been taken properly into account by just adding the respective torques. Therefore, in the following an approach is taken to expand the already existing method to hybrid drivetrains.

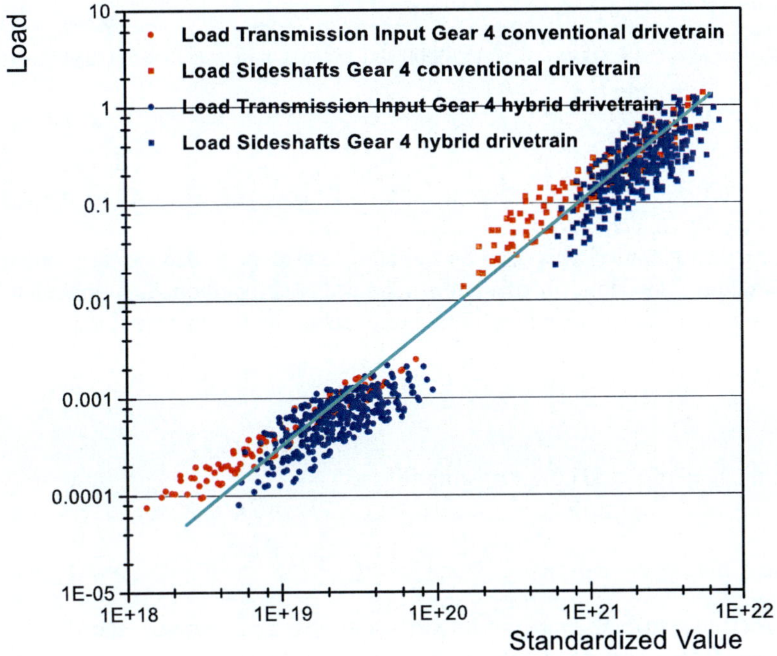

Fig. 2. Load for conventional (red) and hybrid (blue) drivetrains predicted using the standardized value derived in [1]

4 Load Prediction for Hybrid Drivetrains

To deduce a new standardized value for hybrid drivetrains, 324 different hybrid drivetrain simulations where conducted with winEVA® and compared to the 108 conventional drivetrain simulations used in [1]. In both cases, the same drive cycle and the same driver model is used. The modeled car is a rear-wheel driven car with a diesel engine and a 9-gear automatic transmission but with an added electric motor. Once again, the combustion engine, the mass, the axle ratio and the dynamic tire ratio is varied. In addition, the specifications of the electric motor are also modified. The input data is shown in Table 1.

The main idea of the following approach is to determine a constant torque M^* that is load-equivalent compared to the rotational speed depending maximum torque curve of the combustion engine – electric motor conjunction. Mathematically expressed the following equation based on Eq. (2) has to be solved:

$$M^{*5} \cdot \int \omega(t)dt = \int M(\omega(t))^5 dt \tag{6}$$

Table 1. Simulation input data

ICE M(Nm)/P(kW)	EM M(Nm)/P(kW)	Mass (kg)	Axle ratio (-)	Dynamic tire radius (m)
400/143	300/20	1800	2,24	0,307
500/170	300/60	2400	2,82	0,344
600/200	300/100	3000	3,42	0,381
	300/140			

with $\omega(t)$ the time dependent rotational speed. To solve Eq. (6) two assumptions are made to simplify this equation:

In [1] an acceleration was used as standard maneuver to deduce the standardized value. Using the knowledge about a uniformly accelerated motion the following relation between rotational speed ω and total number of rotations N can be derived:

$$N(\omega) = \frac{mr_{dyn}^2}{M \cdot i_{ges}^2} \cdot \Delta\omega^2 \tag{7}$$

With the help of Eq. (7), Eq. (6) simplifies to

$$M^{*^5} \cdot (N_s - N_0) = \int_{N_0}^{N_s} M(N)^5 dN \tag{8}$$

wherein N_0 is the current number of rotations at time t_0 and rotational speed ω_0, when a gear is entered and correspondingly N_S the rotation's amount when this gear is left. The rotational speeds at which the gear is entered and left can be calculated as described in [1] with the following to equations. These are based on the assumption that the transmission shifts, when the torque at the transmission output remains constant and a higher rotational speed will lead to a loss in propulsion force:

$$\omega_S = \frac{P \cdot 9550}{M} \cdot \frac{i_G}{i_{G+1}} \tag{9}$$

$$\omega_0 = \frac{P \cdot 9550}{M} \tag{10}$$

Here, the gear ratio of gear g is written as i_G.

The second assumption is made to simplify the rotational speed dependent sum torque curve of the combustion engine and the electric motor. Therefore, as shown in Fig. 3 the torque curve of the combustion engine is assumed constant whereas the torque curve of the electric motor is declining linearly to zero after being constant at the beginning. The rotational speed at which the change in the electric motor's curve happens can be determined with the help of the motor's power:

$$\omega_{EM} \approx \frac{P_{EM}}{M_{EM}} \cdot 9550 \tag{11}$$

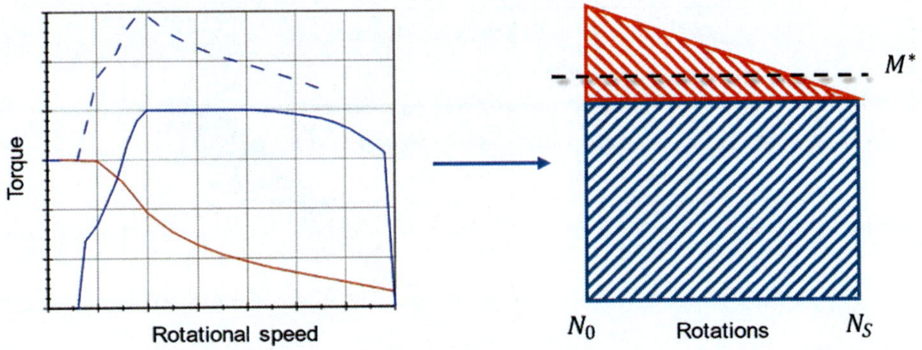

Fig. 3. Combined torque curve of an electric engine and a combustion engine (left) and proxy curve used for calculation (right).

For the purpose of calculation, the sum torque curve of both engines is written as N dependent and is in the case $n_{EM} < n_0$ as follows:

$$M(N) = M_{ICE} + M_{EM} \cdot \left(1 + \frac{N_0}{N_s - N_0}\right) - \left(\frac{M_{EM}}{N_s - N_0}\right) \cdot N \tag{12}$$

$$M(N) = c - b \cdot N \tag{13}$$

Substituting this in Eq. (8) leads to the following result:

$$M^* = \sqrt[5]{\frac{(c - b \cdot N_0)^6 - (c - b \cdot N_s)^6}{6b(N_s - N_0)}} \tag{14}$$

The results of the other cases are $n_{EM} > n_S$ and X $n_0 < n_{EM} < n_S$ are shown in Eq. (15) and (16), respectively:

$$M^* = M_{EM} + M_{ICE} \tag{15}$$

$$M^* = \sqrt[5]{\left((M_{ICE} + M_{EM}) \cdot \frac{(N_{EM} - N_0)}{N_S - N_0} + \frac{(c - b \cdot N_{EM})^6 - (c - b \cdot N_s)^6}{6b \cdot (N_S - N_0)}\right)} \tag{16}$$

Using this newly found M^* as another input value for M_{max} in Eq. (3) or (4) leads to the results shown in Fig. 4. By comparison with Fig. 3 the improvement is clearly visible, as the load of hybrid drivetrains is no longer overestimated.

In the case of a conventional drivetrain, Eq. (12) simplifies to

$$M(N) = M_{ICE} \tag{17}$$

and in this case the integral shown in Eq. (8) gives the result

$$M^* = M_{ICE} \tag{18}$$

This means by replacing M_{ICE} with M^* in Eq. (5) the standardized values given in [1] can also be used for hybrid drivetrains.

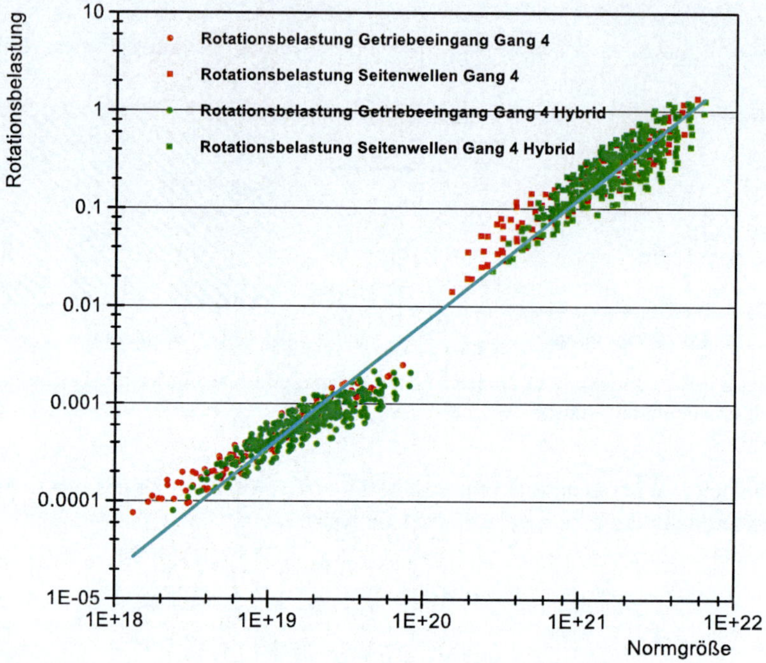

Fig. 4. Load for conventional (red) and hybrid (green) drivetrains predicted using the new standardized value.

5 Identifying Variation Types Based on the Standardized Value

As discussed before, a relation between load and drivetrain parameters can be established via regression and the help of the introduced standardized value. This is shown in Fig. 4. During the development of a new drivetrain this curve can be used to estimate the expected load on a component in the new drivetrain by calculating its standardized value. To determine whether this component can bear expected in the new drivetrain, a fatigue calculation could be made. But as described in 0, for that a Wöhler curve is required. Instead of identifying all necessary Wöhler curves one can put to use, that most of the car manufactures are using modular platform systems. With the help of former product generations, the use range of a reference system element, this means an already existing component, can be determined (Fig. 5).

Depending on the expected load on a component of a new drivetrain, there are now different cases to be looked at. The first possibility, named x' in Fig. 5, is as follows: The expected load is quite centered in the use range of a component. It has been used in a drivetrain configuration with lesser, but also with higher loads. Therefore, it can be simply carried over without further adjustments (carryover variation). The second possibility is indicated via x'''. In this case, the expected load on said component is higher than in any drivetrain ever used before. Dependent on the knowledge about the strongest component and which risk might be taken, a decision between an embodiment or a principle variation has to be made. The third possibility is, that the expected load

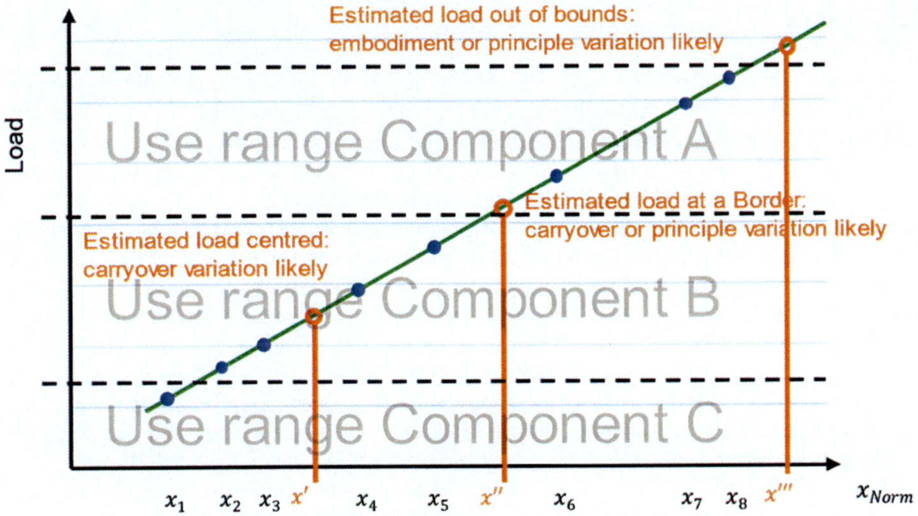

Fig. 5. Use range of three components A, B and C, load prediction (green) based on measurements (blue) and expected load on new drivetrains (orange).

lies on the border between two use ranges (x''). There are now four different options with different risks. Either, to be on the save side, the complete drivetrain using component type A can be carried over or, harboring more risks, the drivetrain using the weaker component B can be used. In both cases, it is a carryover variation for both, drivetrain and component. The other two possibilities are using either drivetrain A with component B or vice versa. In both of these cases, the component is carried over, but on the other hand, the embodiment of the whole drivetrain is altered, if this combination was never used before. With the gained knowledge about the expected variation type, the necessary development activities can be initiated in a targeted manner.

6 Summary and Outlook

With the help of the presented standardized value, it is possible to predict the load on a component in a new drivetrain. The derived load-equivalent torque M^* leads to an improvement compared to the approach shown in [1] regarding hybrid drivetrains. In addition, a method was shown to assist engineers in their decision-making process regarding the risks and benefits of choosing drivetrain components. In addition, this helps to derive the necessary development activities and these activities can be oriented targeted.

In future, the influence of the transmission and the engine type need to be further investigated, in particular the transfer of the shown approach to pure electric vehicles.

References

1. Fischer J, Albers A (2018) Charakteristische Normgröße für den relativen Schädigungsvergleich von Triebsträngen in frühen Phasen der Entwicklung. In: Proceedings, 19. VDI-Kongress SIMVEC - Simulation und Erprobung in der Fahrzeugentwicklung (2018)
2. Mosquet X, Arora A, Rodriguez-Garriga G, LaChance C, Gruskin E, Zablit H et al (2018) The electrical car tipping point. The Boston Consulting Group
3. Albers A, Bursac N, Wintergerst E (2015) Produktgenerationsentwicklung - Bedeutung und Herausforderungen aus einer entwicklungsmethodischen Perspektive. In: Binz, H. (ed.) Stuttgarter Symposium für Produktentwicklung, pp 1–10. Stuttgart
4. Pahl G, Beitz W (2007) Konstruktionslehre. Grundlagen erfolgreicher Produktentwicklung. 7th edn. Springer, Berlin
5. Albers A, Rapp S, Spadinger M, Richter T, Birk C, Marthaler F, Wessels H (2019) The reference system in PGE-product generation engineering: a generalized understanding of the role of reference products and their influence on the development process. In Proceedings of 22nd international conference on engineering design ICED
6. Albers A, Reiß N, Bursac N, Urbanec J, Lüdcke R (2014) Situation-appropriate method selection in product development process - empirical study of method application. In: Laakso M, Ekman K (eds) Proceedings of NordDesign, pp 550–559
7. Albers A, Bursac N, Rapp S (2017) PGE – Produktgenerationsentwicklung am Beispiel des Zweimassenschwungrads. Forsch Ingenieurwes 81(1):13–31
8. Miner M (1945) Cumulative damage in fatigue. J Appl Mech 12:159–164
9. Weidler A (2005) Ermittlung von Raffungsfaktoren für die Getriebeerprobung. IMA Uni Stuttgart, Stuttgart
10. Ullmann T (1986) Herleitung eines standardisierten Lastablaufs als Grundlage für die Berechnung und Erprobung von Triebstrangbauteilen, Esslingen

Drivemode – High Speed Electric Drivetrain

Mattias Flink[1](✉), Michael Burghardt[2], and Roland Bittner[3]

[1] BorgWarner, Mechanical Engineering, Landskrona, Sweden
Roman.Irlinger@euroforum.com
[2] AVL, Simulation & Projects, Regensburg, Germany
[3] Semikron, RD/SES, Nuremberg, Germany

Abstract. The paper investigates and suggests a high-speed electric drivetrain combining a SiC inverter, a PMSM electric machine (EM) and a 3-stage parallel axis gearbox. To reduce cost, a modular structure is applied allowing for installing several drive units into the vehicle according to power demand. The developed modular drivetrain concept is suitable for low and high-performance vehicles. To downsize the EM, i.e. decreasing the torque, the gear ratio and the rotor speed is increased, allowing the same power but from smaller packaging. Increasing the gear ratio gives a reduced mass and size, which primarily decreases the cost of the EM. To meet both increased speed and modular installation, different transmission concepts have been rated which is more likely to fulfill the requirements. To address cooling and rotor stresses and keep efficiency at maximum, the permanent magnet synchronous motor with buried magnets and dry rotor is developed. The peak efficiency goes above 96% with specific torque density of 4.9 Nm/kg. To supply the high-speed motor high frequency switching silicon carbide (SiC) power electronics is adopted. The converter can deliver 140 A continuous AC with a semiconductor switching frequency of 20 kHz.

Keywords: Gear ratio · PMSM · High speed · Silicon carbide

1 Introduction

1.1 Current Status

The paper describes the considered technology, the design approaches, evaluation methods and concept decision for the proposed modular drivetrain in the Drivemode project. The prototype testing phase of the subsystems is currently an ongoing process.

1.2 Background

The transition towards the electrified automobile market is steadily gaining speed, and the need for high efficiency drivetrains together with lower manufacturing cost is driving the industry forward. The trend towards reducing the active material usage in the electric machine while keeping the same power is addressed by increasing the speed of the electric machine, which reduces proportionally. To keep the same power with a smaller

Euroforum Deutschland GmbH (Ed.): *CTI SYMPOSIUM 2019*, pp. 103–115, 2021.
https://doi.org/10.1007/978-3-662-61515-7_11

package, the lower output torque from the electric machine needs to be compensated by a higher gear ratio in the reduction gearbox. The system must therefore be designed and evaluated with care, since it comes with many challenges for the designer. The EU-funded project Drivemode is investigating and suggesting such a drivetrain. The drivetrain is designed based on modular thinking allowing for installing several drive units in the vehicle according to power demand. In the project, the goal is to push the input speed to 20 000 rpm, which also calls for a high frequency current supply from the inverter to the electric machine. For demonstrating the drivetrain, a NEVS 9-3 eV was selected, and the boundary conditions were therefore set by its motor compartment space.

1.3 Specification

Already at an early stage, the Drivemode project was focusing on a drivetrain configuration with two separate drive modules positioned in the front of the car. The decision was based on the existing demonstration vehicle NEVS 9-3, which has plenty of space in the engine compartment. It was therefore natural to aim for a front wheel drive solution. The choice of excluding a differential for the demo vehicle, and instead focusing on two separate drive modules, was based on the modular thinking where one, two, or several drive modules can be used for future vehicles depending on the specific demand. It is however clear that a single drive module cannot be used for the same axle drive, unless equipped with a mechanical differential gear.

The requirements on a system level for the vehicle were derived by Chalmers University, comparing the current C-segment market for EV and ICE vehicles (2017). The target was to cover as many vehicle segments as possible by specifying an accurate torque and power level in combination with the number of drive units in the car. The total wheel level specification of the vehicle is found in Fig. 1.

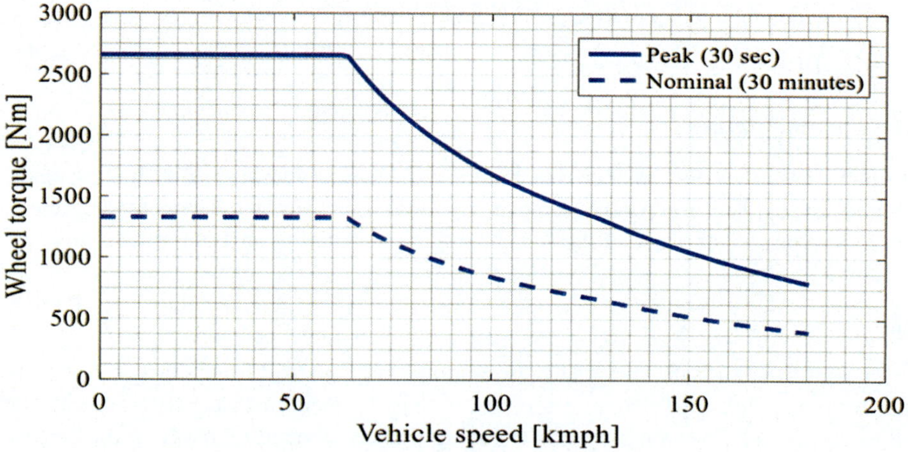

Fig. 1. Demo vehicle system requirements.

For the demonstration vehicle a total of two drive units is selected, where each of them needs to deliver according to Table 1 below.

Table 1. The system specification for one drive unit in Drivemode.

Parameter	Value
Peak power	70 kW
Continuous power	35 kW
Maximum output torque	1 350 Nm
Maximum input speed	20 000 rpm
Driveline configuration	F2
Transmission layout	Off-axis
Total gear ratio	14.1:1

2 Drivetrain Layout

One of the challenges with the design of the drivetrain is to adapt it for a modular design where one, two, or several units can be installed in the vehicle. This requires flexibility in relation to the packaging space, meaning that the electric machine might be mounted on both the right-hand side and the left-hand side in the vehicle, with the output shaft on either one or two sides. From a driving performance point of view, the possibility to have two separated units in the front (one for each wheel), could be an advantage if implementing torque vectoring in the electric machine control software.

Ultimately, a decision within the project was taken to proceed with a drivetrain layout with an identical transmission for both units, but with the possibility to connect the electric machine and drive shafts to both sides of the transmission housing. This allows for very few differences within each subsystem, the main one being two different input shafts for the transmission. It also allows for better packaging utilization, since two completely identical units would require more space. The main idea of the drivetrain layout is shown in Fig. 2 below. The modular layout also calls for a possibility to implement only one rotor shaft bearing on the non-drive end side, or only one transmission input shaft bearing. As a consequence, one of the subsystems would rely on the bearings from the interfacing subsystem. For high speed, this configuration may be both better or worse depending on the run-out and concentricity accuracy of the components.

3 Transmission

3.1 Concepts

With a selected gear ratio of 14.1:1, the input speed in the transmission reaches 20 000 rpm to meet the requirements of a top speed of 180 km/h. With such a reduction in

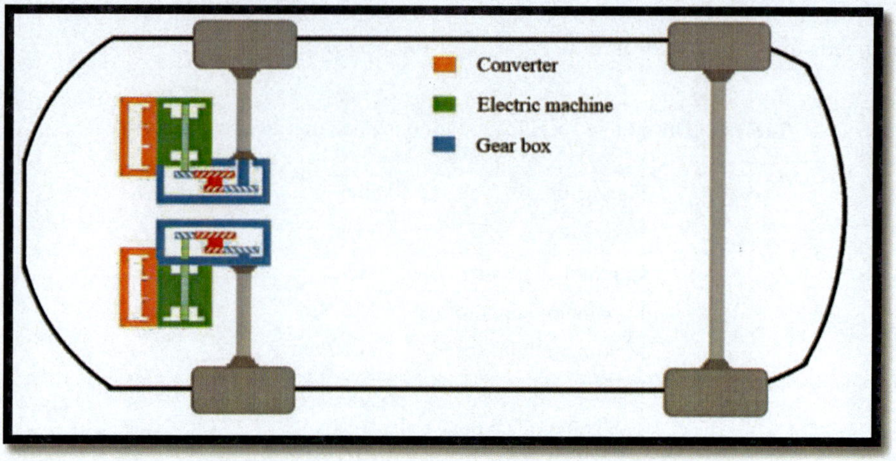

Fig. 2. The chosen drivetrain layout for the demonstration vehicle.

the gearbox, the input and output gears have large differences between the conditions of speed, loading and lubrication. These parameters, in combination with others such as size, Noise Vibration Harshness (NVH) and cost have been considered when comparing different transmission concepts. The considered concepts are:

- Concept 1–2-stage parallel axis
- Concept 2 – Input planetary gearset followed by a parallel axis helical gearset
- Concept 3 – Input parallel axis gearset followed by a planetary gearset
- Concept 4 – 3-stage parallel axis

To better compare the different concepts, a system model was built in MASTA 9 [1] for each of them. Mass, cost, packaging, and number of gears and bearings can be derived from these models which is used as a basis for decision making of the final concept. The transmission concepts were scored based on the parameters mentioned, which in turn were weighted with different factors depending on their importance. For instance, durability, NVH and cost were the most important parameters. In order to verify the scoring and findings, a second opinion was sought from a external independent party, highly experienced within the automotive industry. As a result, Concept 4 was chosen to be the most appropriate alternative for this system, closely followed by Concept 1. The result can be seen in Fig. 3.

3.2 3-Stage Parallel Axis

With such a high input speed, the first transmission shaft benefits from having as small diameter as possible. In particular radial shaft seals and bearings are limited by their tangential sliding speed, which becomes higher as the diameter grows. Throughout the transmission, only ball bearings were adopted in order to keep the drag losses as low as

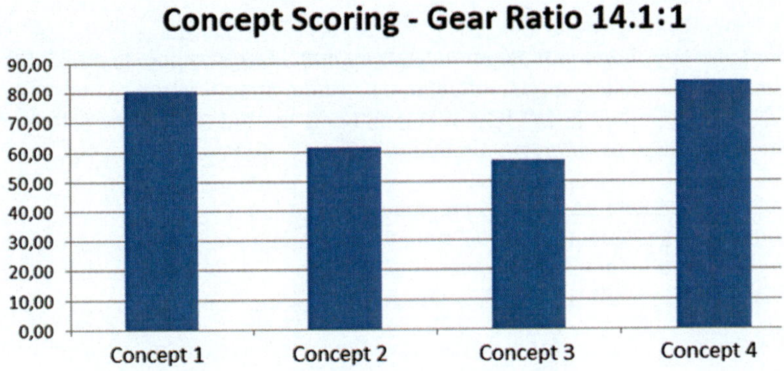

Fig. 3. The result from the scoring of the transmission concepts.

possible. The scope of the project also covered ceramic ball bearings for the input shaft, but their cost in large scale volumes did not make them feasible.

The lubrication system in the transmission unit is crucial to ensure that the bearings, gears and seals can work properly and endure a full life. Friction is reduced by the lubricant partly by a separating oil film and partly by additives transported to the tribologically loaded interfaces. In particular, the first gear mesh is crucial for sufficient lubrication as the speed and centrifugal forces are high. To develop the lubrication system, the CAE software ParticleWorks [2] was used in the design phase of the transmission. Figure 4 below shows the simulation of oil distribution at 4500 rpm input speed. The simulation tool has also been used in order to assure that a sufficient oil volume is used in the transmission unit. This is crucial for driving scenarios such as uphill driving over a long period of time.

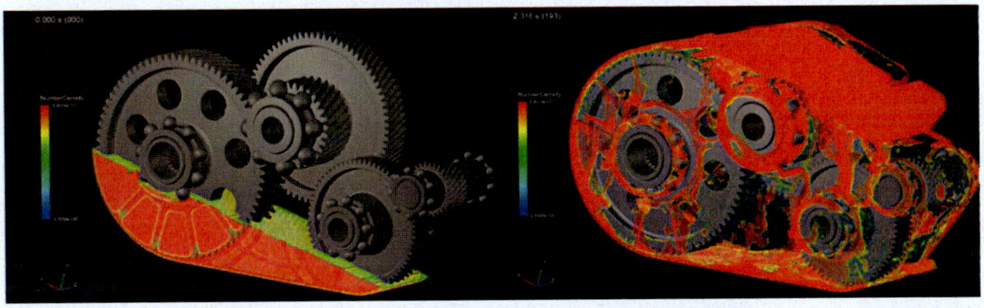

Fig. 4. The oil distribution in the transmission unit.

The number of gear teeth for each gearset has been chosen to minimize order interference with the electric machine. This means that the number of poles and slots for the electric machine must be known. Furthermore, the involute gear design is approached by prioritizing minimized peak to peak transmission error rather than creating the lowest possible gear losses. Therefore, the transversal and axial contact lines in the gear mesh are designed and iterated for low NVH in the simulation model.

When reaching a certain tip sliding speed in the gears, the risk of scuffing significantly increases, especially in the case of absent lubrication. As the input speed for this drive unit suggests that the system might suffer from scuffing, an evaluation of the safety factor to scuffing was performed in MASTA 9 [1], and was carried out for the following load cases:

- Maximum speed and corresponding maximum torque.
- Low speed and maximum torque.
- Corner speed and maximum torque.

The lowest safety factor for scuffing from these simulations resulted in a value of 2.2 for the input gear stage. This indicates that a problem with scuffing is not likely to occur in the transmission given the condition that enough lubrication is present. If this in fact turns out to be a problem, the transmission group has undertaken a twin-disc test investigation to predict the added performance from a super-finished input gear contact, as well as from a DLC-coated gear. These tests indicate that super-finished gears decrease the wear and generated heat in the gear contact, which would minimize the risk for scuffing.

4 Electric Machine

4.1 General Functional Requirements of the Traction E-Motor

In this chapter the E-machine development basis is summarized where the Work package (WP3) evaluated two E-Machine technologies in detail:

- Permanent magnet synchronous machine (PMSM)
- Induction motor (IM), squirrel cage asynchronous motor

Efficiency
For traction E-Motors with the wide speed range capability and the high efficiency aspects, many representative load points must be investigated simultaneously. The best efficiency contour of the torque speed characteristic should coincide with its most frequent operating points.

Material
Improved steel grades and thinner laminations shall be used

Windings
Proper winding configuration and technology shall be selected.

Windings and winding material shall sustain problems with high switching frequencies and high dV/dt due to the high frequency SiC switch technology.

Driving aspects
Increased iron and copper losses due to high rotational speeds and high driving frequencies shall be considered.

Sinusoidal space vector pulse width modulation (SVPWM) will be used to drive E-Machine and applying field-oriented control (FOC) with implemented maximum torque per ampere (MTPA) motor control algorithm.

Mechanical aspects

Designed high speed E-Machine shall sustain high mechanical stresses. And, used materials must withstand the high mechanical forces and vibrations.

Thermal aspects

Improved cooling shall be achieved to increase the E-Machine over-load-ability and to avoid permanent magnet demagnetization.

The E-machine should fulfil the performance requirements and still have the reasonable feasibility from the financial aspect of mass production. Therefore, the cost reduction was also considered. The following process within WP3 evaluated extensive knowledge with simulations data to clear the decision for the PMSM E-Machine. Nonetheless, it must be pointed out that the calculated results for both machines were very close.

4.2 Concept Results

The PMSM V2a machine is a three-phase machine with 36 slots stator and 8 poles rotor. The achievable output electrical properties are shown in Fig. 5. The electrical steel material properties for stator and rotor used in the design and iron loss calculation is M270-35A, but the prototype will be produced from electrical steel NO20-1200 with significantly improved performance, especially in terms of reduced iron losses. Hence the iron losses in the prototype are expected to be lower resulting in an overall improved efficiency.

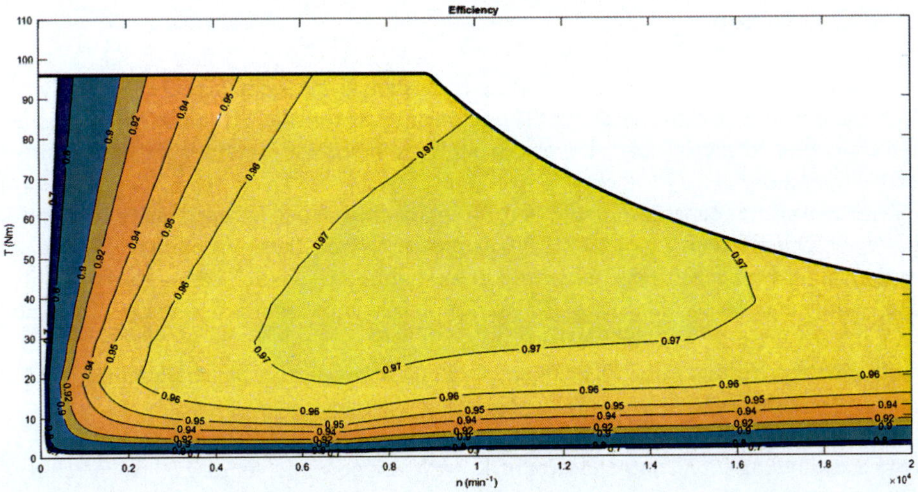

Fig. 5. Achievable output electrical properties from the electric machine.

Working points for PMSM performance evaluation and comparison are presented in Table 2 below. Phase voltage amplitude of 410 V corresponds to 720 V DC battery, which is the lowest allowable battery voltage (nominal battery voltage is 800 V).

Table 2. Electromechanical performance at peak operational conditions for PMSM V2a

Peak performance					
n (rpm)	50	7000	10000	15000	20000
T (Nm)	106.3	106.4	72.4	43	31
Iph, rms (A)	140	140	115	95	85
alpha (deg)	45	45	61	70	75
Uph, max (V)	10.9	391.2	413.4	410.2	413.2
P, mech (kW)	0.6	78.0	75.8	67.5	64.9
T, ripple (Nm), peak-to-peak	6.50	6.52	6.32	4.37	3.80

Thermal behaviour of PMSM V2a is studied in transient state at continuous operational mode. The temperature distribution is calculated for different machine parts. The temperature changes at overload (peak power) operation for 60 s with starting temperature 100 °C are within and below the specified requirements.

The PMSM design is completed and is currently in the state of manufacturing. Soon, the test phase will be initiated with the main goal to substantiate the calculations and simulation results and keeping the total performance at the requirements level.

5 SIC-Inverter

To supply the high-speed motor, high frequency switching silicon carbide (SiC) power electronics is attached to the motor. The characteristics of the SIC-MOSFETs enable high efficiency at partial load, i.e. at operating points highly relevant for the electric vehicle operation.

The converter is capable of delivering 140 Arms continuous AC current at a semiconductor switching frequency of 20 kHz. This switching frequency represents a challenge for standard power electronics, is nevertheless required to control the motor within its speed range and up to the maximum AC frequency with appropriate resolution. The inverter SOA includes DC-link voltages up to 1000 V to fit to vehicle architectures with nominal battery voltages of 800 V. The maximum temperature of the coolant is assumed to be 65 °C at the inlet.

The design of the Drivemode inverter is based on 3 parallel SiC MOSFET (1200 V) per phase allowing for an output power of 117 kW at a power factor of 0.85 and a battery voltage of 800 V. This output power capability is realized in a total volume of 2.6 L for a power density of 45 kW/l. This value corresponds to an increase by approximately 200% compared to standard inverters employing silicon-based power semiconductors. The motor control supports torque control and speed control including comprehensive protection features. In Fig. 6 below, the SiC-inverter prototype unit designed for the project is shown.

Fig. 6. The DRIVEMODE SiC-inverter

5.1 High Component Utilization

The key to achieve high power density and to minimize expensive SiC MOSFET die area and other materials, is the high utilization of the power dissipating components.

This is achieved mainly by means of thermal optimization of the cooling system and forced cooling of all power dissipating components. In practice, this means provide coolant flow not only below the power substrate and the semiconductors, but also enable cooling of the DC-link and the busbars in between. The DC-link active material and volume could be reduced significantly by this measure.

The coolant flow, power substrate and busbar design were optimized by iterating flow and pressure drop simulation and FEM simulation of the complete power path, Fig. 7. In addition, alternative materials for the heat sink and the thermal interface were investigated. To support excellent thermal performance, a Si3N4 ceramic substrate is used for electrical isolation of the heat sink and the power electronics.

To save expensive SiC MOSFET die area, a detailed FEM simulation study was performed. The aim of the different optimization steps was to improve the thermal performance of the single dies resulting in higher output current per die. In particular, the optimization of the distance between the semiconductor dies has proven to allow significant reduction of the overall temperature as well as of the temperature difference between the dies. To select the optimal chip spacing, various test structures have been analyzed. Figure 8 shows one of the simulation runs where the power semiconductor die spacing was investigated and optimized for the requirements of the inverter.

The maximum safe operating voltage is also an important parameter for high SIC MOSFET utilization. To achieve this, the commutation circuit was optimized for low total stray inductance. The dynamic behavior of the SIC MOSFET switching cell at the turn-off of the semiconductor, shows overvoltages of less than 10% of the semiconductor

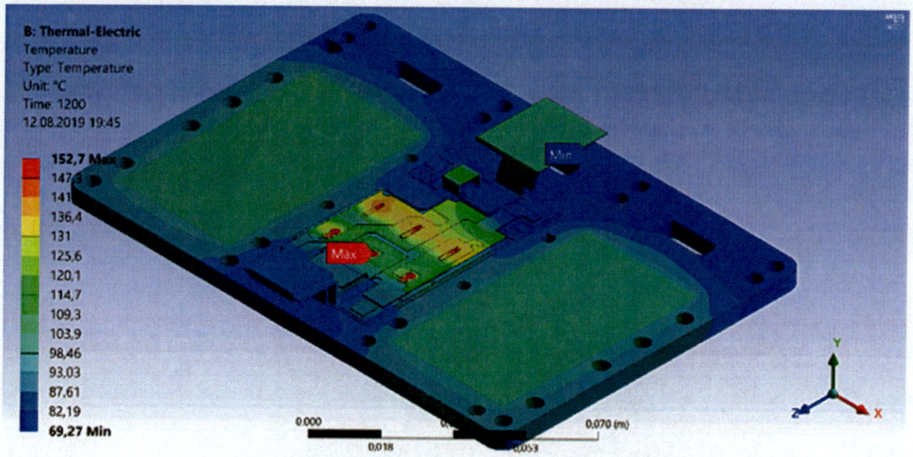

Fig. 7. FEM simulation of the complete power path.

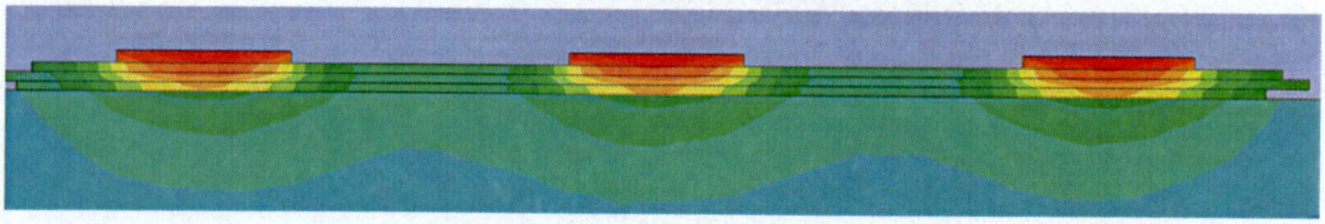

Fig. 8. Cross-section of paralleled SIC-MOSFETS (FEM simulation).

blocking voltage (Fig. 9). Also, no ringing during the transient or after the overshoot is visible. Due to the low overvoltages at turn-off, the maximum safe operating dc-link voltage for the inverter can be set above 1000 V.

5.2 Size Reduction of Auxiliary Functions

With every optimization of the main components, the volume share of components used for auxiliary functions increases. Current sensing and discharge functionalities become therefore relevant in the volume count due to the needed power dissipation.

Compared to commonly used hall sensors with core, current sensing by shunt resistors represents a good alternative in terms of footprint and accuracy. Thanks to the integration on the power substrate and shunt value of less than 1 mΩ, effective power dissipation can be achieved. Looking at the total volume i.e. including the required circuitry to provide an isolated signal to the control electronics, the volume saving is still above 50%. The low level of the sense signal can lead to acceptable noise immunity if 4 wire design together with short and paralleled connections is employed.

For safe operation and handling, the inverter must be equipped with passive and active discharge functionality to secure dc-link discharge within 5 and 5 min respectively. Due to their high power dissipation, volume consuming discrete PTCs or thick film components are commonly employed. In the Drivemode inverter, Si-chip resistors are employed to implement the discharge functions as well as the high voltage sensing (Fig. 9). The

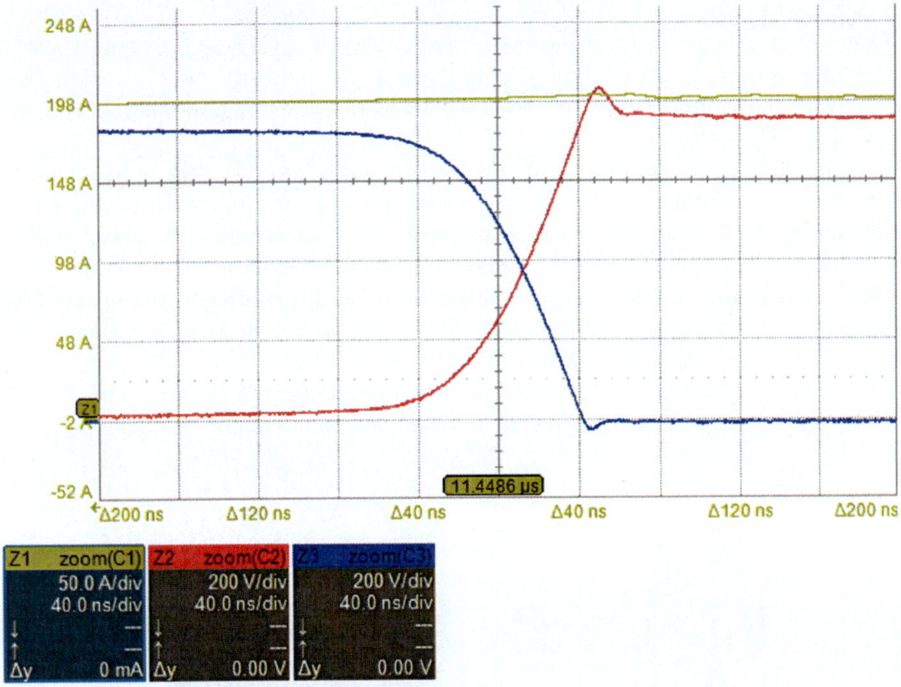

Fig. 9. Dynamic measurement of inverter output.

devices are designed specifically for this purpose and are placed close to the DC-link directly on the power substrate by means of the same production processes employed for the attachment of the power semiconductors. The forced cooling at this position enables high power density and reduced volume which results in up to 90% volume reduction compared to commonly used PTCs or thick-film components (Fig. 10).

Fig. 10. Power substrate mounting of the Si-chip resistor.

5.3 Motor Mount

The reduced volume of the inverter enables its direct mount on the motor or integration in an extended motor housing. The common goal in this case is to eliminate the cost and effort of the AC-cabling.

For the surface mount version of the inverter, an additional goal is to achieve a robust IP67 design of the inverter without the need to open the motor or inverter during its mounting. This requirement is realized by a high current pressure contact system consisting of a busbar with a molded sealing and a spring providing the required preload for the pressure contact. This design is illustrated in Fig. 11. During the mounting of the inverter the contact is pushed back into the inverter after touch-down on the contact area of the motor, increasing the contact force to the desired values.

The motor requires a simply shaped contact area for each AC phase on its surface. The Drivemode application uses end sleeves mounted on the motor windings.

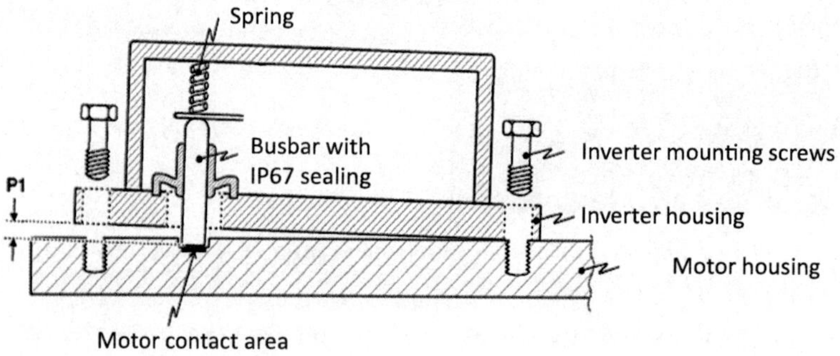

Fig. 11. Illustrating the principle of the high current pressure contact.

6 Conclusions

- The project team has designed a modular high-speed drivetrain with high efficiency based on normal operating points. The drivetrain is shown in Fig. 12.
- For the given vehicle specification, a 3-stage parallel axis gearbox and a PMSM electric machine is suggested.
- The inverter size and material utilization has been significantly improved.
- The power density of the electric machine is reaching 4.27 kW/kg
- For a single drivetrain unit, a mechanical differential is required in the transmission
- Subsystem and total system testing are required to evaluate and validate simulations done by each partner
- The mechanical interface between the rotor shaft and the transmission input shaft is crucial when testing at 20 000 rpm.
- The 3-phase AC supply interface does not require phase lead cables.

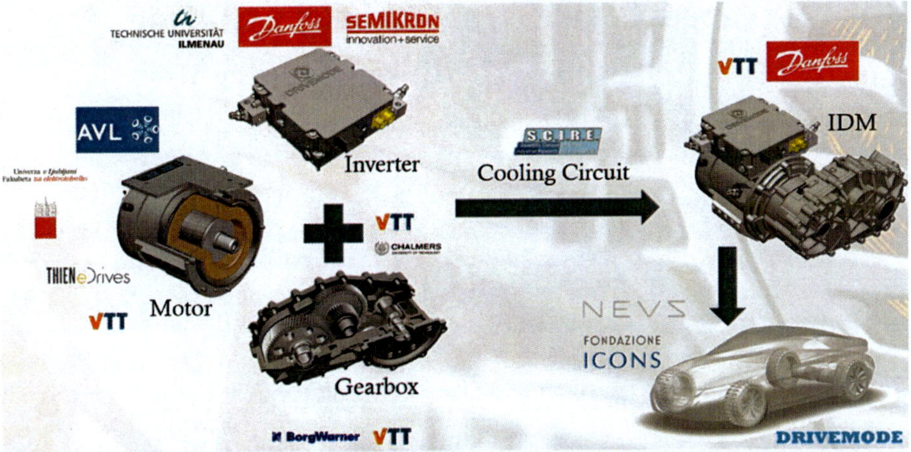

Fig. 12. Showing all partners involved in the project with the suggested 3D drivetrain assembled.

References

1. Simulation software, SMT Masta 9.0, 64x
2. Simulation software, EnginSoft, Particleworks v6.2.0

Advances in Drivetrain Lubricating Fluid Technology for Hybrid Electric Vehicles

Michael Gahagan(✉)

Lubrizol Ltd, The Knowle, Nether Lane, Hazelwood, Derbyshire DE56 4AN, UK
michael.gahagan@lubrizol.com

Abstract. The automotive transmission market has seen an increase in the number of hybrid electric vehicles (HEV), and forecasts predict additional growth. As the design of HEV drivetrain hardware evolves; so must the lubrication technology to anticipate this pace. The reason is that in HEVs the hybrid drivetrain hardware may increasingly combine electric motor, friction devices, gearbox, electro-hydraulics and the control unit all of which will be expected to be lubricated by a suitable drivetrain fluid. There is an extra challenge posed by incorporating an e-motor within the transmission housing while being constrained to ever smaller packaging dimensions, since the e-motor provides an extra source of heat which would be expected to be managed by the lubricating fluid. This leads to the need for an understanding and optimization of corrosion protection and thermal management characteristics of the fluid, and the development of new high-performing lubricating fluids suitable to the challenges is therefore paramount. There are further new challenges to lubricants. In addition to ensuring the lubricant will be durable in the harsher thermal environment, it must also be compatible with new plastics and able to protect them from degradation. There are ramifications that the hotter transmission design and environment has had on shaping the latest lubricant technology for electrified transmissions. The heat transfer characteristics of lubricants and the fundamental factors that impact them and fluid and transmission design are reviewed. The critical factors of protection against corrosion, including the ability of the fluid to allow both fluid and vapour-phase corrosion protection of copper-based componentry is addressed by the fluid technology. A secondary impact on the design of other fluid attributes, such as minimizing noise vibration and harshness (NVH), is addressed and the new fluid technology formulation approach to deal with such challenges is highlighted.

Keywords: Hybrid · Electric · Transmission

1 Introduction

The idea of employing electric motors to drive a vehicle emerged with the innovation of the motor car itself. From 1897 to 1900, EVs became 28% of the total vehicle population and were preferred over vehicles with an internal combustion engine (ICE) [1]. In 1898, Ferdinand Porsche presented the 'Egger-Lohner electric vehicle, C.2 Phaeton model'. This is known as the world's first Porsche design, or Porsche no. 1 and shown in Fig. 1.

Euroforum Deutschland GmbH (Ed.): *CTI SYMPOSIUM 2019*, pp. 116–132, 2021.
https://doi.org/10.1007/978-3-662-61515-7_12

With the advent of very low oil prices and fuel availability the ICE types soon dominated the market however and became increasingly efficient, reliable and more advanced. Concerns over vehicle emissions led to a resurgence in some electrified vehicles in the 1990s, which are themselves now obsolete apart from the Toyota Prius in an evolved form. The market has now expanded, electrified vehicles have gained in popularity and many different designs of vehicles, whether full or hybrid electric, with associated transmissions co-exist.

Fig. 1. Porsche Egger-Lohner electric vehicle from 1898 (Porsche Museum, Stuttgart)

As a vehicle type becoming more commercially viable and available, their attraction is their quietness, ease of operation particularly due to compatible automatic transmissions plus they do not have the fuel costs associated with the often highly taxed hydrocarbon fuel. They allow for energy recuperation through regenerative braking. The instant torque availability is making them desirable for motor sports, and their quietness and low infrared signature for EVs is making them useful for military use as well.

1.1 Transmission types

Of the hybridisable architectures available, stepped AT (automatic transmission), belt CVT (continuously variable transmission), twin motor DHT (dedicated hybrid transmission), the hybridised DCT (dual clutch transmission) is finding favour for various reasons. These are elimination of torque lag due to e-motor launch, compatibility with existing gearbox hardware architectures and ability to have a modular (e-DCT or non-e-DCT) design with only one e-motor if required which helps OEMs minimise design complexity. The efficiency of DCT operation makes them an appealing choice since drain on the battery would be subsequently reduced. The e-DCT design can be described analogously to the now standard DCT units except for the inclusion of an e-motor commonly inline in the so-called P2 position or "P2.5" where the motor sits to the side of the transmission parallel to the transmission axis. One attractive feature of the e-DCT is the ability to modularise the construction for electrified or non-electrified versions thus making the design a highly adaptable one. Figure 2 shows a schematic with the e-motor linked to the 2,4,6, R unit, through which it can launch the vehicle in pure EV or hybrid mode [2]. The e-motor can be immersed in the common lubricating fluid used for the rest of the gearbox.

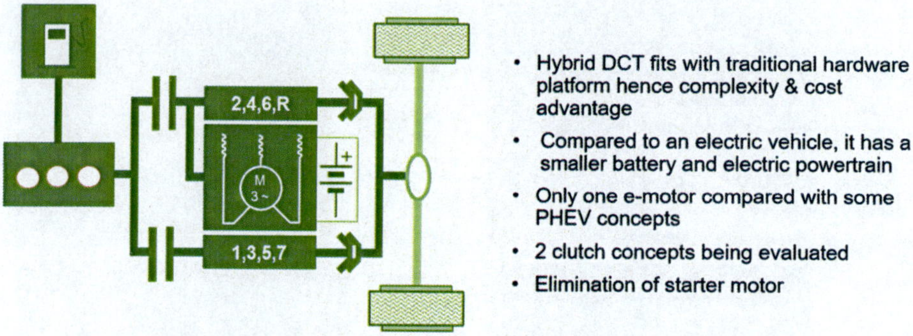

Fig. 2. Schematic of e-DCT

This however leads to questions around the suitability of lubricating fluids in such an electrified environment where the higher-voltage electrical units are exposed to the lubricating fluid and vice-versa.

1.2 Lubricants for electrified transmissions

The most important features for such a transmission fluid are corrosion protection, oxidation and deposit control and fluid electrical and thermal properties suitable for the application. Whilst attending to these properties in common with fluids for the other electrified transmission types, the ability of the fluid to deliver the correct frictional properties within the particular gearbox architecture should not be overlooked, and hence NVH becomes more important as driver expectations of an improved driving experience increase. Some important factors are highlighted in the following sections, where the development of a suitable e-DCT fluid is described.

2 Lubricant electrical properties

The most commonly measured electrical properties for a lubricating fluid are electrical conductivity, relative permittivity and dielectric breakdown strength [3]. The results are summarised in Fig. 3 and the e-DCT fluid shows the appropriate electrical conductivity properties currently desirable for such a fluid. The fluid is a static dissipative insulator [3], the relative permittivity and therefore polarizability of the fluid is at the lower end of the range for common materials (range can be 1 to several thousand), and the dielectric breakdown strength being in the tens of MV/m and is in the expected range for such a fluid. The needle-on-ball configuration being perhaps the more severe condition due to the intensity of electrical field lines in the vicinity of the needle as a point-source.

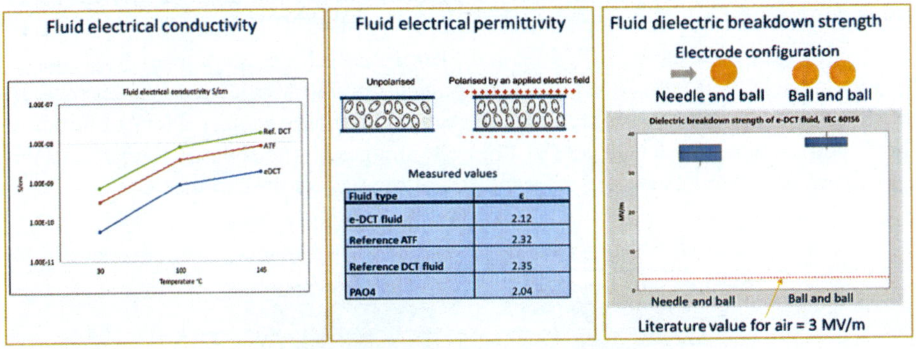

Fig. 3. Electrical properties of e-DCT fluid

3 Corrosion performance

One role of the lubricating oil is to protect metal surfaces from corrosion, a function it performs remarkably well by acting to exclude oxygen and moisture from attacking metallurgy. It provides a mobile physical barrier which can carry away contaminant matter due to its liquid nature and the inclusion of corrosion inhibitor chemistry from the oil additive system enhances its protective capability.

Even if the environment is a benign one, the presence of some additive species, particularly active sulfur from extreme-pressure (EP) additive chemistry intended to protect iron surface from wear, can at elevated temperatures lead to corrosion of copper-based metallurgy. This situation can occur if the lubricant is inappropriate for the use, perhaps with an overly large amount of EP and so the lubricant should be carefully screened for the appropriate corrosion protection characteristics. Copper-based metallurgy for example can be present in a gearbox as friction surfaces, bearing supports or washers, but it is its role as an electrical conductor in wiring and circuits that has led to increased focus due to its relevance for electrified vehicle transmissions.

Whereas much of the wiring present is insulated, were this to become compromised or where the metal surfaces are already uninsulated, the effect of corrosion can be inimical to the operation of the transmission, and material failures can be rapid at elevated temperatures.

3.1 Oil and vapour-phase corrosion protection by e-DCT fluid

Although much focus is on corrosion caused by lubricant in direct contact with copper surfaces and is well-known, it is also possible to generate vapors from the lubricant which can corrode copper surfaces in confined environments at temperatures as low as 80 °C in the vapour-space. This vapour-phase corrosion has serious implications for circuity above the fluid not normally in direct contact with the oil such as in leadframes.

3.1.1 Copper strip testing

This can be demonstrated by means of a simple copper-strip placed in a closed jar with the strip half-in and half-out of the oil, and the pieces rated after test. Any vapours are allowed to accumulate and the upper dry section and the wetted lower section are rated at the end of a selected period, in this case after 168 h at 80 °C. The results are shown in Fig. 4, where for both the e-DCT fluid and the reference fluid, there is no corrosion in the liquid phase; the fluids have protected against this. The vapour-phase shows where the e-DCT is beneficial and effectively protects the exposed copper against corrosion where the reference fluid is poorer at this. The images to the right of the jars are the copper strips removed from the jars at the end of the test for ease of inspection.

Fig. 4. Vapour-phase corrosion test

3.1.2 Copper wire testing

A more objective method of monitoring the corrosion of copper is my means of a copper wire test [3] whereby corrosion is monitored by a resistance increase in a non-energised fine bare copper wire exposed to the lubricant. This Lubrizol designed test method allows the monitoring and accumulation of data in real-time and the data can also be processed to allow further research into the reaction kinetics taking place [4].

Any corrosion taking place leads to an increase in electrical resistance due to loss of conducting metal in the process. Figure 5 shows the results from an enhancement of this method whereby an independent circuit is placed in the vapour-phase, but in the same test-cell allowing both sets of data to be acquired at the same time. The results show a stable resistance response even at the elevated temperature selected.

3.1.3 Conductive deposit testing – protection by e-DCT fluid

Corrosion processes can also be monitored by means of specially designed printed circuit-boards and held for long periods energised by electricity, in the oil at elevated temperatures.

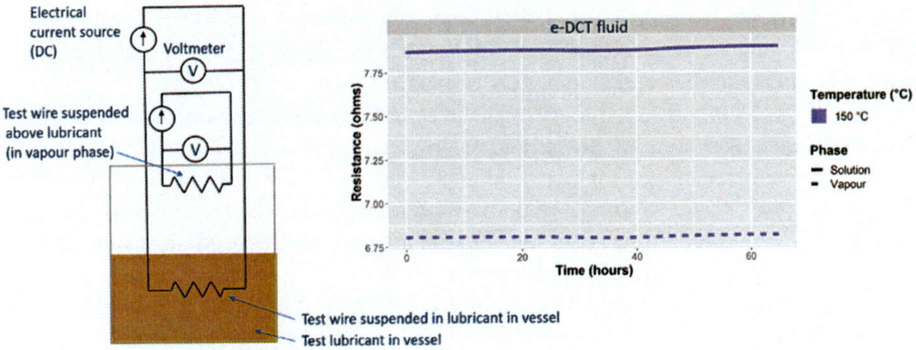

Fig. 5. Tandem vapour-phase wire resistance corrosion test

Fluid which generates deposits due to decomposition, formation of copper sulfides or both can interfere with the effective operation of circuits and the test is designed as a way to test for this.

Figure 6 shows the results of a severe-operation study where the test circuit-board has remained well-preserved by the e-fluid in contrast with a poor reference fluid.

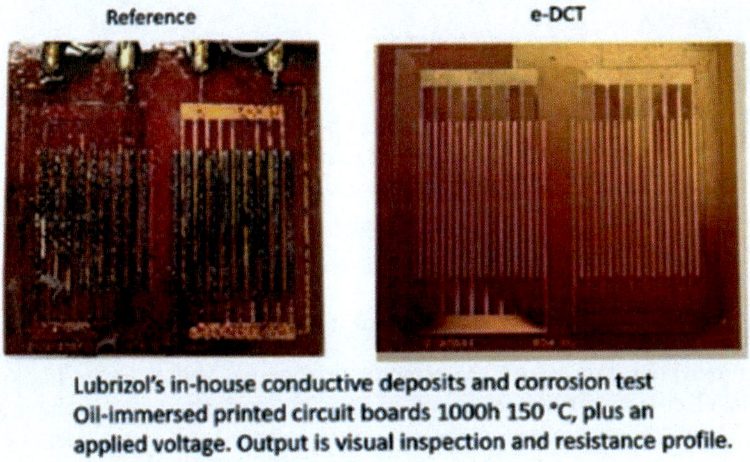

Fig. 6. Conductive deposits circuit-board test

4 Contributors to NVH

NVH problems can be very complex ones. In a vehicle solely powered by an ICE, acceptable noise and vibration limits are determined primarily by what is experienced over the ICE. In hybrid or purely electrically powered vehicles, the engine noise may be intermittent or completely removed, making drivetrain NVH more prominent. There is no singular primary contributor to NVH issues encountered in drivetrain systems and there is no singular solution either. Each system design is unique and may originate as a result of friction surfaces of a clutch, gear or bearing micro or macro geometry, or even the transmission supporting structures and it would therefore be a mistake to suggest that NVH issues can be resolved by addressing one individual source. One approach is through hardware engineering [5] which can lead to measurable improvements. The lubricant however also has a part to play in minimizing NVH whether that be to dampen the noise and vibration from gears and bearings or to ensure good shift quality and control in lubricated clutches and synchronisers.

Figure 7 has an outline of where the lubricant can impact the NVH in electrified vehicles. The focus in the following sections will be on clutch and synchroniser friction control enabled by the lubricant.

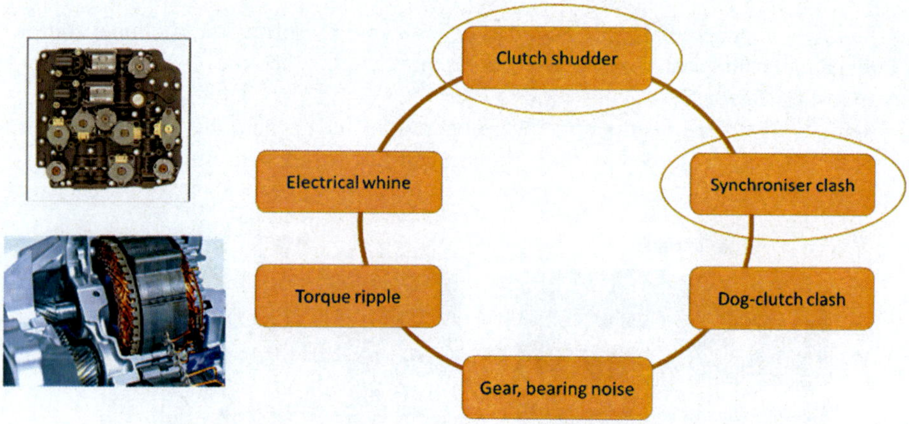

Fig. 7. Possible lubricant impacts on NVH

4.1 Gear and clutch shifting strategies and fluid influence

The synchronization process for the e-motor driven gear shifts can be implemented through the electric motor torque to control its speed during the inertia phase [1]. For this reason, a mechanical synchroniser may not be needed and in the design of some e-DCTs dog clutches may suffice.

However, where present and required for engine or hybrid mode shifting, the synchroniser can be subject to high speeds and inertias. This requires good quality synchromesh friction surfaces to be preserved either to allow the use of engine or hybrid drive modes or to avoid harsh, low differential speed engagements in EV mode due to damaged, degraded or poor friction/fluid interface properties.

This is especially important since in pure EV mode the gear engagements for example $2 \rightarrow 4$ cannot go through $2 \rightarrow 3 \rightarrow 4$ by means of pre-engagement on the odd-shaft but has to undergo active synchronization on the even shaft [2]. Any difficulty shifting in this mode may lead to a poor shifting experience, increased battery drain and longer time lags due to the nature of the consequent torque-interrupt mode thus employed. Figure 8 illustrates the various torque flows for each mode. For this reason it was important to ensure the lubricant and friction material were fully compatible in terms of friction response and friction and material durability.

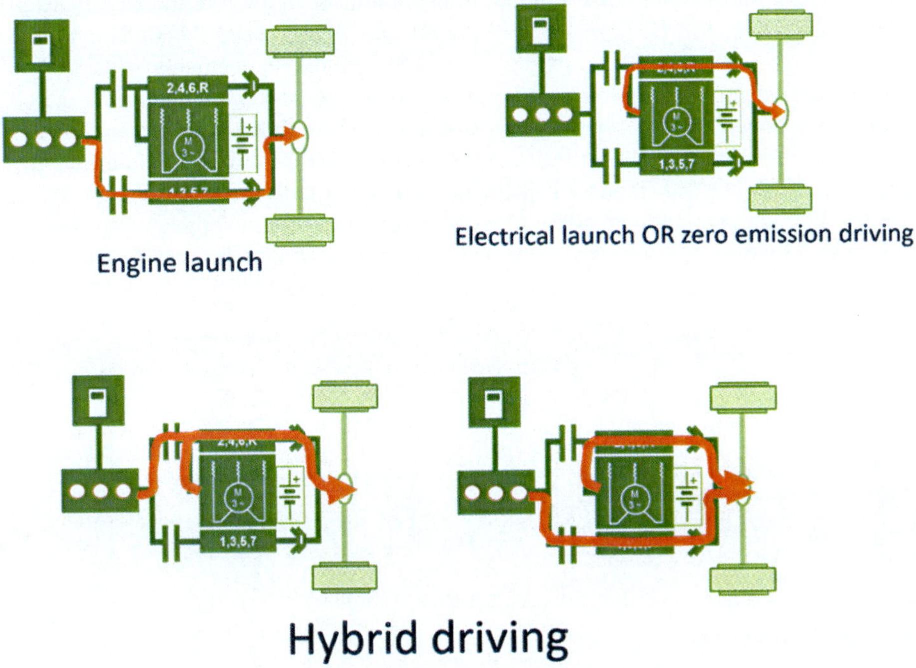

Fig. 8. Torque flow in HEV

One example of a frequently used friction material in modern synchroniser arrangements is pyrolytic carbon weave. A severe load-collective test was selected to test this synchroniser assembly and e-DCT fluid combination. The results are displayed in Fig. 9 and show in the left figure a beneficial single shift engagement (a) with a rising but stable friction coefficient (highlighted in bold) and a similarly stable friction trends graph (b) highlighted.

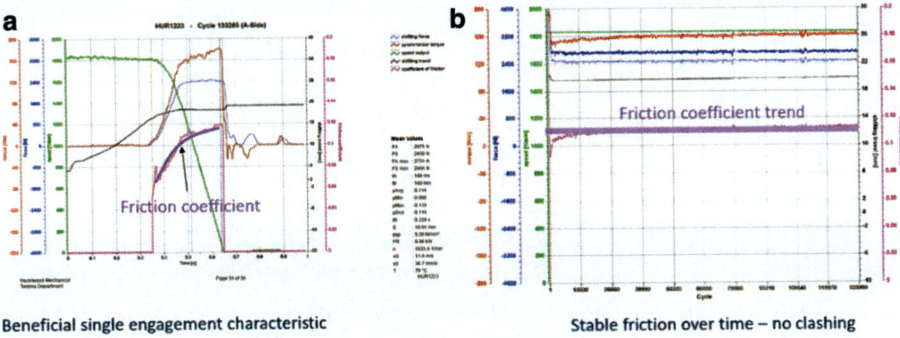

Fig. 9 Synchroniser friction

4.2 Clutch

A good clutch motion control strategy can ensure the vehicle has an excellent shift quality. Such control can be aided by the transmission fluid since in the transmission an integrated DCT will be lubricated by the oil which will assist the shifting of the clutches. The benefits the fluid provides are smooth clutch engagements, reduction in hot spots, and wear meaning less force required to engage and minimization in clutch torque-oscillation. This is a phenomenon that can lead to clutch shudder and clearly audible noise in some cases. Figure 10(a) shows the result for a fluid in a high-energy launch clutch test performed on a ZF GK friction rig using full-scale DCT clutch packs. A selected torque amplitude response, in red, shows a sharp initial engagement around 1 s. followed by two undesirable oscillation events maximizing at around 10 and 45 s.

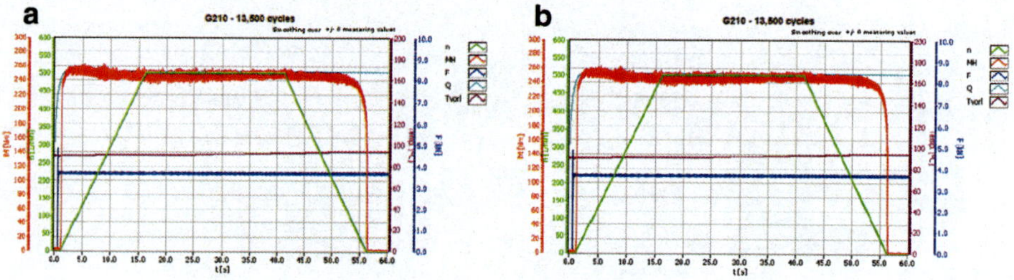

Fig. 10. GK test friction results

The results for the e-DCT fluid is shown in (b) and has the smoother and more stable torque response. The friction traces can be plotted as friction slope versus cycle number, where positive friction slope trend seen for the e-DCT fluid is desirable. The slope is determined by subtracting a lower-speed friction value form a higher-speed one. The slope plots are shown in Fig. 11.

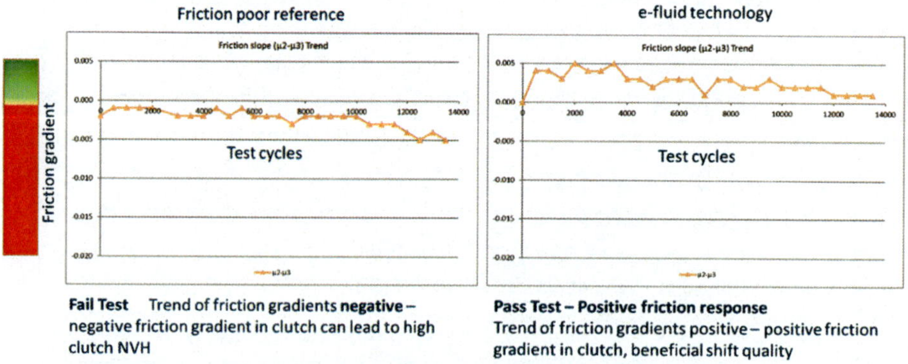

Fig. 11. Friction trends, gradient versus test cycle

5 Engineering plastics and insulation compatibility

The temperature capability of a given material is influenced strongly by the environment to which it is exposed. Therefore the temperature capability of a material should be defined in terms of the conditions to which it is exposed and this includes the lubricating oil. Given the vital role insulation and engineering plastics play in preserving the function and support of the electrics and other components in the EV powertrain, the lifetime of the device is only as good as the lifetime of the engineering plastics and insulation [6]. Three materials were selected of relevance to the powertrain and exposed to heat and the e-fluid over a prolonged duration for this reason. The materials investigated were polyamide and polyphenylene sulfide typically used for insulating mechanical or electrical support as well as polyetheretherketone - a material used in rotor slot-liners but also finding consideration in hardware as a lightweight but durable material to possibly replace metal components such as in bearings.

Three authors that have contributed significantly to understanding the problems of insulation aging are Steinmetz, Montsinger and Dakin. Steinmetz held that electrical insulation suffered insignificant deterioration below 90 °C, but that above 100 °C the rate of deterioration was rapid. In contrast Montsinger held that deterioration was a continuous reaction to temperature akin to the "10 degree rule". It was Dakin's use of the Arrhenius description of chemical rate phenomena and his use of experimental data that became the backbone of the technical understanding of thermal aging processes. This work produced knowledge making e-motors smaller and lighter and the concepts are therefore still of relevance to today's investigations. The Dakin relationship also allowed for a more descriptive, condition-relevant relationship for insulation lifetime as given in Eq. 1.

$$\Upsilon = Ae^{[B/T]} \tag{1}$$

Where

Υ = lifc

A and B are constants determined by the activation energy of the particular reaction

T = temperature in K

The physical changes during thermal aging were found to be a reflection of internal chemical change. For this reason, in testing of the e-DCT fluid technology, the tensile strength measurement of the material at high temperature exposed to the lubricant was taken an selected time periods to indicate any change in the material itself caused by heat or the chemistry of the lubricant. The immersion test protocol for the polyamide, PPS (polyphenylenesulfide) and PEEK (polyetheretherketone) was described previously [3]. To be found there as well is testing information on the polyimide insulation material Kapton® - an archetype for wire and other electrical insulation already described [3]. During thermal aging, chemical reactions in the oil can lead to acids and sludge formation potentially affecting the insulating quality, hence the high temperature of 150 °C and long duration of 1000 h was selected. The results for polyamide and PPS as % tensile strength change versus reference fluid are given in Figs. 12 and 13, and the results for PEEK tensile strength measurements tested according to reference [4] on the pure material is in Fig. 14. The results show that the e-DCT fluid/material combination is a durable one in all these cases shown with no appreciable material degradation.

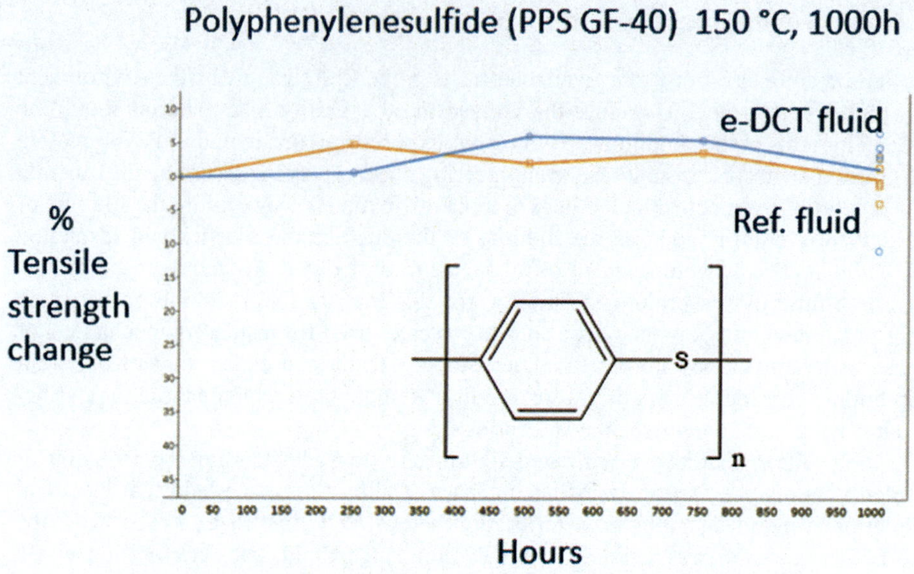

Fig. 12. Results for PPS, with material molecular structure illustrated

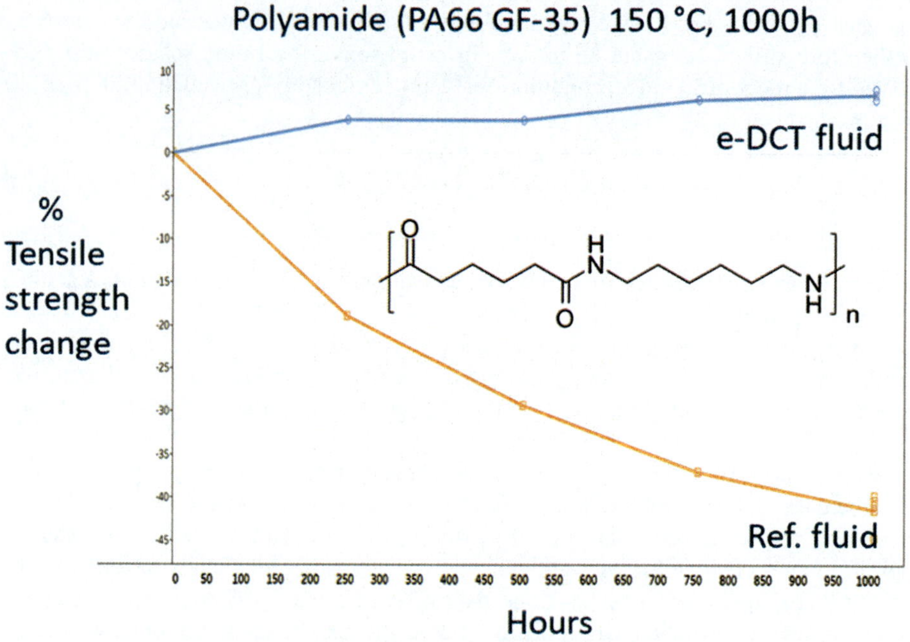

Fig. 13. Results for polyamide, with material molecular structure illustrated

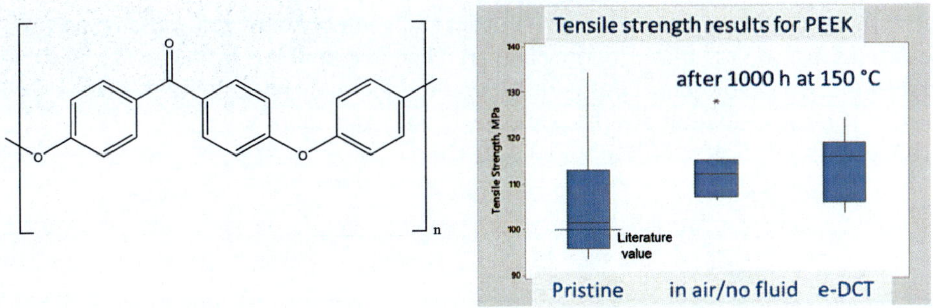

Fig. 14. Results for PEEK, with material molecular structure illustrated

6 Thermal properties

One of the critical functions of the lubricant is to cool the components it comes into contact with. It achieves this by absorbing heat from working surfaces and transferring the heat to cooler surfaces away from the sources in question. The fact that the lubricant is a fluid over a wide temperature range is therefore an appealing feature for having lubricant technology in contact with operational parts. There is more emphasis now on fluids as thermal management systems where many electrical and electronic systems are designed to be more efficiently operational within specific temperature ranges as distinct from running permanently cold. This applies mostly to battery operation currently however.

6.1 Lubricant contribution

As lubricants become cooler they become significantly more viscous and this has implications for heat transfer as subsequent fluid mobility becomes more difficult. Conversely the lower viscosity and higher fluidity whilst not evaporating at high temperatures is of significant benefit. Fluid viscosity stability across temperature ranges is therefore extremely important. This has led to the selection of API Group III and IV base oils since they can have the advantageous viscosity indices required. The four main factors for thermal transfer in order of importance are:

1. Viscosity (μ) *cP*
   - *How easily a substance can flow*
   - *Can be lowered as long as hardware protected*
2. Density (D) *g/ml*
   - *How much substance in a given volume*
   - *Some ability to increase*
3. Specific Heat Capacity (C) *J/(g*K)*
   - *How much heat a substance can store*
   - *Similar for hydrocarbon oils*
4. Thermal Conductivity (λ) *W/(m*K)*
   - *How easily heat travels through a substance*
   - *Similar for hydrocarbon oils*

Such properties are dominated by the lubricant and non-polymeric transmission performance additives do not play a significant part where the fluid is a fresh one [4]. As can be seen in Fig. 15, the impact of additive on fluid thermal conductivity whether it be additive content, oxidation or vehicle service is not significant.

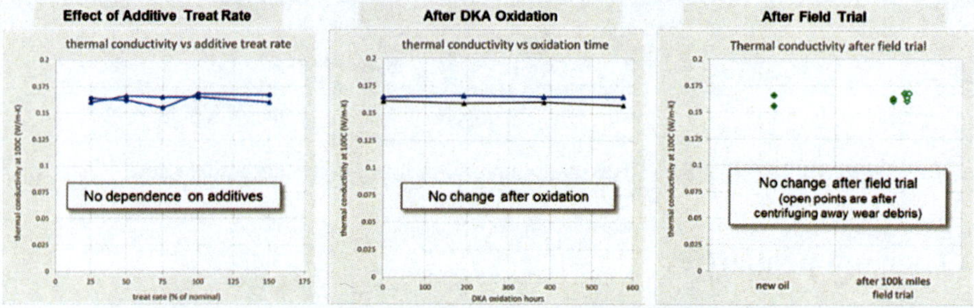

Fig. 15. Thermal conductivities of selected lubricants in W/mK

Figure 16 shows graphs for properties of different lubricating oils at varying temperatures. It is clear that viscosity is the only property significantly affected by temperature or between fluid type, so for example, relying on the thermal conductivity data of a lubricant alone is not sufficient to understanding its overall cooling ability.

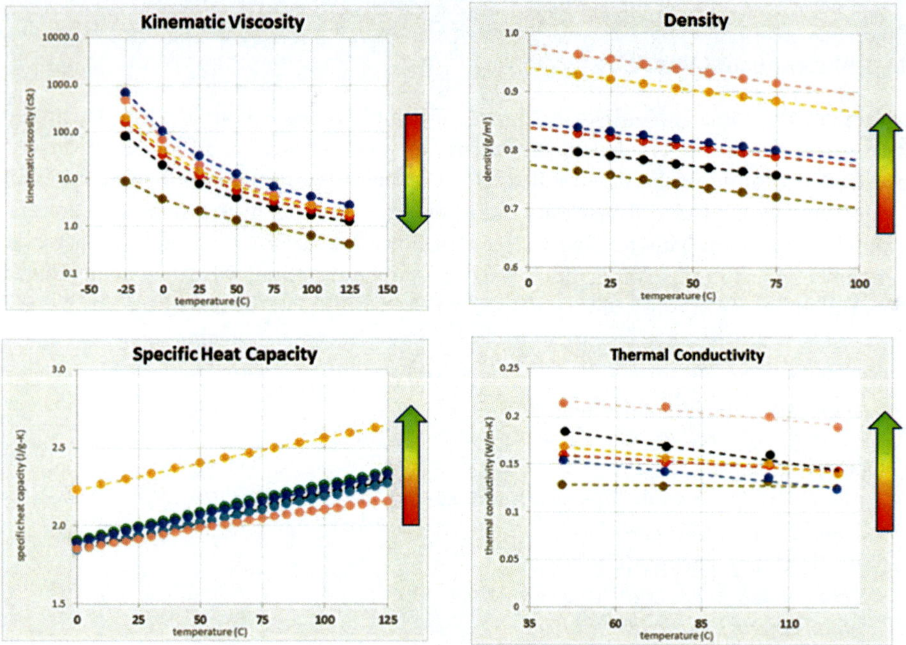

Fig. 16. Thermal management properties and beneficial qualities indicated by green in arrow

6.2 Engineering aspects

The inclusion of e-motor into the transmission bathed in oil has become a common feature in many designs; however a more recent innovation is the inclusion of integrated inverters using the lubricant as the heat transfer medium for the electronics. This has led to even more intense interest in thermal properties of lubricants where the oil would become part of the hardware, e-motor, and power electronics thermal circuit. While the lubricant provides a benefit in terms of thermal management, it is limited by the physics and chemistry of its hydrocarbon nature required for a currently viable transmission fluid.

Particular benefits are where stable liquid range and particular fluid electrical insulating properties are required and delivered by the oil. In Fig. 16 for example the viscosity range for lubricating oils is shown as from −25 °C where water, with better thermal conductivity than oil (0.6 W/mK versus 0.1–0.2 W/mK for an oil) would normally be solid and therefore immobile. Alternatively, water at 100 °C would normally be a gas, with significant evaporation occurring at temperatures below this point. Additionally water vapour is a known antagonist to normally stable electrical insulation [7], hence limiting its suitability for not only for uninsulated electrics due to higher electrical conductivity but even in some insulated environments. Suitable transmission lubricating oils can be stable at temperatures comfortably above the 100 °C threshold, and be compatible with materials typically present in the drivetrain.

Solutions are being found to overcome any thermal transfer limitations of lubricating oils by making use of fluid flow control management techniques. Figure 17 illustrates differing flow regimes and significant factors and the fact that for particular flow regimes, thermal conductivity of a fluid is no longer the most impactful especially with the more desirable turbulent flow from a heat transfer perspective. Integration of power electronics in the transmission system appears to be a trend and requires consideration, where laminar flow could ordinarily dominate.

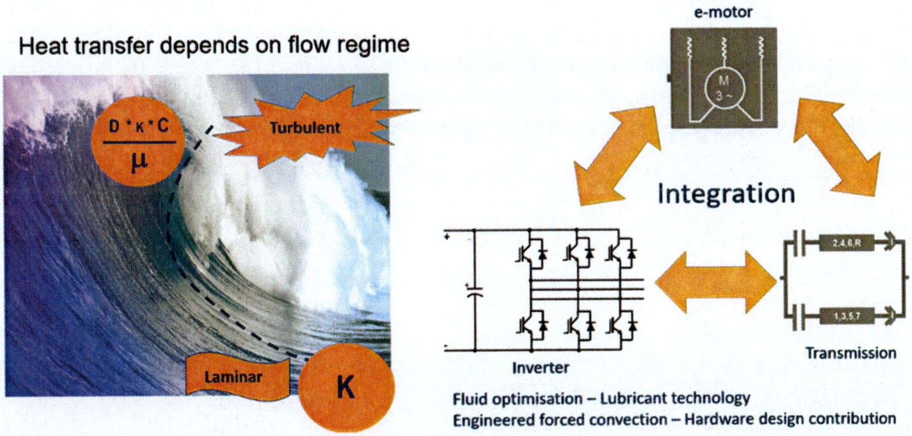

Fig. 17. Flow regimes and integration of power electronics in the transmission system.

Engineers aware of the physical limitations of the lubricating oils are designing-in to the integrated inverter the use of forced convection and turbulent flow regimes to optimise what can be provided by the oil. As a background, inverters – the essential power electronics devices for electrified vehicles deliver high currents within compact spaces, and much has been done hardware-wise to reduce the known temperature sensitivity of the inverter, for example in the use of silicon carbide metal–oxide–semiconductor field-effect transistors (MOSFETs). Even then effective heat dissipation is still important and one example of such is the "Ecochamps" initiative [8]. Here a transmission-integrated inverter is cooled by lubricant jet impingement as well as oil-cooled capacitor banks leading to improved electric current stability. Their use of computational fluid dynamics (CFD) and heat transfer analysis was key the delivering the solution to the oil-cooled power electronics challenges. The solution to the thermal challenges posed by increasingly integrated hardware and electrics is neither a purely lubricant physical chemistry nor an engineering one therefore, but an effective combination of both.

7 Oxidation and thermal stability

Whilst contribution to the thermal conductivity or specific heat capacity of an oil by means of additive chemistry is limited, there is still much the additive package can do in terms of affecting the thermal behavior of the lubricant longer term. This is because the formulated transmission fluid itself will contain performance additives which influence how the fluid responds to heat, and oxygen from the air. A lubricant without suitable antioxidants or dispersant technology would rapidly increase in viscosity due to oxidation processes. Therefore any benefit gained in optimizing the viscosity to improve the thermal transfer characteristics of an oil when freshly used, could be countered by its transformation possibly into a more viscous, oxidised fluid. The formation of sludge and intractable deposits on heat-radiating surfaces such as electromagnet coils and electronics would be harmful.

This would interfere with cooling oil flow around such components and consequently impact the lifetime of the components because of the Dakin relationship. The e-DCT fluid makes use of advanced antioxidant and dispersants to achieve the beneficial properties required. The results from an industry standard transmission oxidation test (CEC L-48-A-00 (B), 192 h) are presented in Fig. 18 compared with some oils for reference.

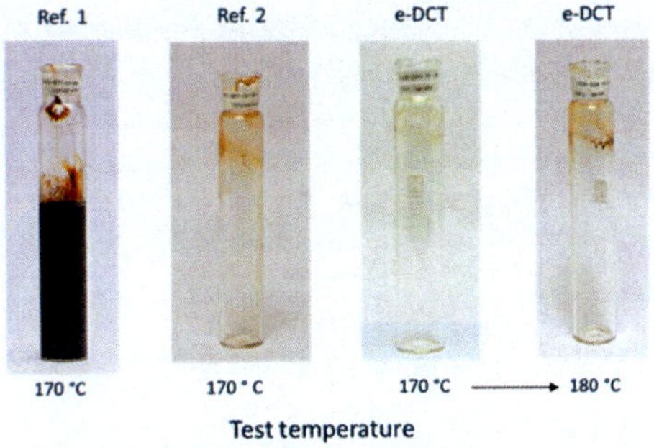

CEC L-48-A-00(B)	Ref. Fluid 1	Ref. Fluid 2	e-DCT	e-DCT
Test temperature	170 °C	170 °C	170 °C	180 °C
Kinematic viscosity 100 °C, % change	15	10	2	3
Blotter Spot, % dispersancy retained	28	100	100	100

Fig. 18. Test-tubes after oxidation testing

The results show that for the e-DCT fluid the viscosity increase is lower than a standard transmission fluid technology [3]. Also by elevating the test temperature 10 °C beyond what would normally be required, to make the test even more severe, the viscosity increase is still well-controlled. The blotter-spot test assesses the ability of a lubricant to disperse any sludge formed, and the fluid performs perfectly well (100% dispersability). The tube is also rated for the appearance of sludge and this is most prominent in Reference 1 fluid. This oil was selected as it is a commercial transformer oil and demonstrates that an oil type suitable for one type of electrification regime does not correlate to suitability in another such as in an e-transmission with its own specific demands. Due to oxidation and thermal stability therefore the oil plays a larger than perhaps initially expected role therefore in the whole thermal management process.

8 Summary and Conclusions

The use of Hybrid & Electric vehicles is growing and transmission designs are evolving to meet market needs. An electrified DCT unit is an effective way to achieve this. New e-DCT fluid technology has consequently been developed to:

- Meet the demands of electrified vehicles utilizing a hybridised DCT unit.
- Provide the benefits of superior corrosion protection of copper in oil and vapour phase.
- Protect against oxidation and be thermally stable
- Manage friction in synchronisers and clutches contributing to good NVH characteristics therefore impacting vehicle drivability.

References

1. Un-Noor F et al (2017) A comprehensive study of key electric vehicle components, technologies, challenges, impacts, and future direction of development. Energies 10:1217. https://doi.org/10.3390/en10081217
2. Piracha MZ, Grauers A, Hellsing J (2016) CEVT "The control methods for an electrified double-clutch transmission". Energirelaterad Fordonsforskning

3. Gahagan M (2017) Lubricant technology for hybrid electric automatic transmissions. SAE technical paper 2017-01-2358. ISSN: 0148-7191. E-ISSN:2688-3267
4. Gahagan M (2018) CTI conference 2018. Lubricant concepts for future HEV transmissions. 5 Dec 2018
5. Bryant M (2018) Solving NVH issues in hybrid and EV powertrains. https://www.drivesystemdesign.com/wp-content/uploads/2018/01/Solving-NVH-issues-in-hybrid-and-EV-powertrains.pdf
6. Brancato EL (1978) Insulation aging. IEEE Trans Electr Insul EI-13(4):308–317
7. Kumagai S et al (2015) Steam pyrolysis of polyimides. Environ Sci Technol 49(22):13558–13565. https://doi.org/10.1021/acs.est.5b03253
8. Ricardo "Ecochamps-European Competitiveness in Commercial Hybrid and Automotive Powertrains" Ricardo publication document V2 18 K U (2018)

AVL High Performance 48 V Integrated Electric Axle

Inigo Garcia de Madinabeitia Merino[1](✉), Christian Schmidt[1], Julian Pohn[1], Mohamed Essam Ahmed[1], and Klaus Kronfeldner[2]

[1] AVL List GmbH, Graz, Austria
inigo.garciademadinabeitiamerino@avl.com
[2] AVL Software and Functions GmbH, Regensburg, Germany

Abstract. Compared to conventional 48 V drive systems, the AVL P4 e-axle layout enables the customer to use the vehicle in pure electric drive modes, with much higher performance compared to currently available systems on the market.

A set of use cases is defined to calculate the torque and power requirements for the e-axle. Based on this, the required torque-speed characteristic of the e-axle can be determined. To be competitive on the market, the target package space is defined by the available package of a conventional 4WD rear axle to minimize vehicle modifications. Mechanical interfaces such as e-axle bearing support and half shafts to the wheels of the 4WD system shall be reusable. Additionally, a co-axial design is used to minimize the package space.

The definition of the e-motor torque and the transmission ratio is a compromise between available package for the e-machine, maximum rotational speed of the e-machine, 10500 rpm, and the fixed value of required e-axle torque, 1280 Nm. The overall gear ratio of 9,45 is divided in two steps, equipped with a disconnect element placed on the intermediate shaft. This e-axle shall be able to support the powertrain up to 130 kph as well as to be disconnected from the wheels above 130 kph to improve the overall efficiency.

For achieving best-in-class power density with low torque ripple, the e-machine is designed with 6 phases and distributed winding. The 6-phase technology shifts the 48 V system to the next higher level of electrification and allows pure electric drive capability. The inverter consists of 2 modules, each controlling 3 phases of the e-machine and is fully integrated into the e-axle, having all connections routed internally. The cooling system of inverter and e-machine is connected in series.

Within this paper the 48 V e-axle system will be outlined, and the technical solution and its features will be explained.

Keywords: 48 V · e-axle · Hairpin · PMSM · 4WD · All-wheel drive

1 Introduction

Stringent future CO_2 limits and local driving restrictions for conventional vehicles require not only improvements of engines and transmissions but also powertrain electrification. 48 V systems seem to be an attractive compromise between today's 12 V electrical

Euroforum Deutschland GmbH (Ed.): *CTI SYMPOSIUM 2019*, pp. 133–146, 2021.
https://doi.org/10.1007/978-3-662-61515-7_13

power supply systems and high voltage hybrids. 48 V allows moderate hybridization by enabling functionalities ranging from advanced stop start via sailing, boosting and recuperation even up to limited electric driving features while keeping the system affordable. 48 V technology presents itself as a highly suitable solution to the vehicle challenges shown in Fig. 1.

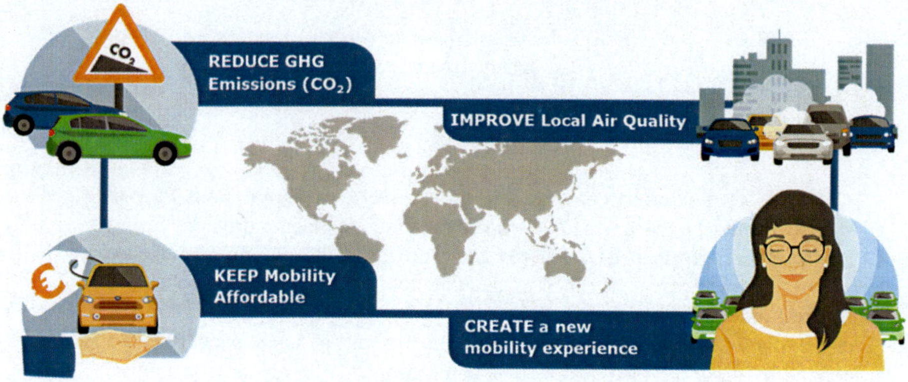

Fig. 1. Challenges for future vehicles

With today's 48 V systems, customers face the problem that pure electric drive isn´t possible due to restricted power and therefore, zero emission zones cannot be entered. In addition to that, a certain degree of driveability should be achieved to make the product attractive for the customer.

In this project, AVL's target is to create a 48 V system which allows adequate HEV functionalities, including pure electric drive mode, with a low voltage system, attractive driving performance and at moderate cost. Compared to current 48 V e-drive systems, the new AVL 48 V e-axle shall enable the customer to use the vehicle in pure electric drive mode. However, this leads to much higher performance requirements compared to currently available systems on the market, which implies that new technologies need to be considered.

To identify the performance requirements of the e-axle, a set of use cases are defined, which consider driving maneuvers out of daily life as well as RDE and WLTP driving cycles:

- Curb climbing: climbing from stop with 1 wheel at the 120 mm curb in a 45° angle (flat surface only)
- Leaving the garage: 15% incline (creeping only)
- Zero Emission Vehicle operation (inner city bans) – WLTC urban part: 22 km Range (driving through European big cities), max. 10% slope
- Zero Emission Vehicle operation in WLTC and RDE: 100% coverage of WLTC and RDE in ZEV mode (urban part only), hybrid support up to 130 kph
- 4-wheel drive operation: take off capability enhancement (HEV mode), low-μ operation at low speed

Especially pure electric driving modes become more and more important due to planned driving restrictions for conventional cars in inner city zones, displayed in Fig. 2. The new 48 V e-axle enables affordable powertrains, adequate driving performance and is a key requirement to comply with real world CO_2 emission limits.

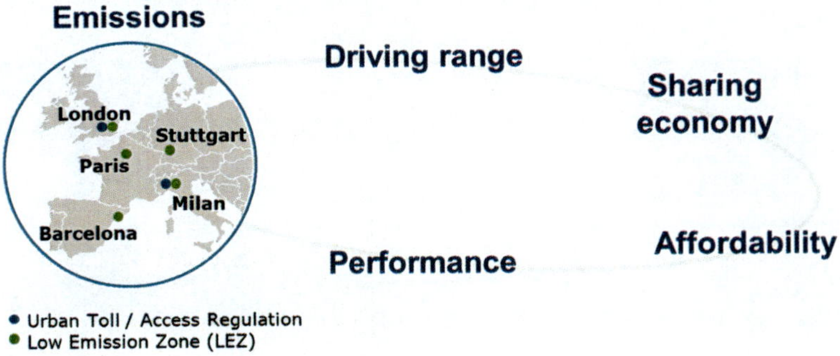

Fig. 2. Challenges for future powertrains

1.1 Hybridization and 48 V Systems

As shown in Fig. 3, 48 V is a main trend to reduce CO_2 and emissions in general. In the past, 48 V has been restricted to mild and micro hybrid applications. However, lately 48 V components have become able to support full hybrid and plug in hybrid applications with the performance increase due to new technologies.

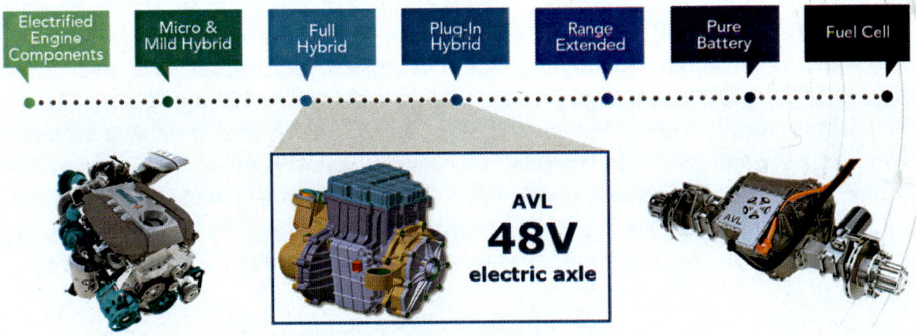

Fig. 3. Hybridization and 48 V systems

1.2 Powertrain Configuration

By increasing the performance of the 48 V system, it can be shifted from combustion engine side to the transmission side and the usage in an e-axle. The highest potential of

48 V systems can be achieved by integrating the electric drive systems in the driveline and adding the ability to decouple the engine, see Fig. 4. This enables propelling the vehicle purely electric and by that, reducing CO_2 emissions similar to a high voltage full hybrid system. The target vehicle configuration shall be a P0/P4 layout to gain maximum efficiency and to enable pure electric driving.

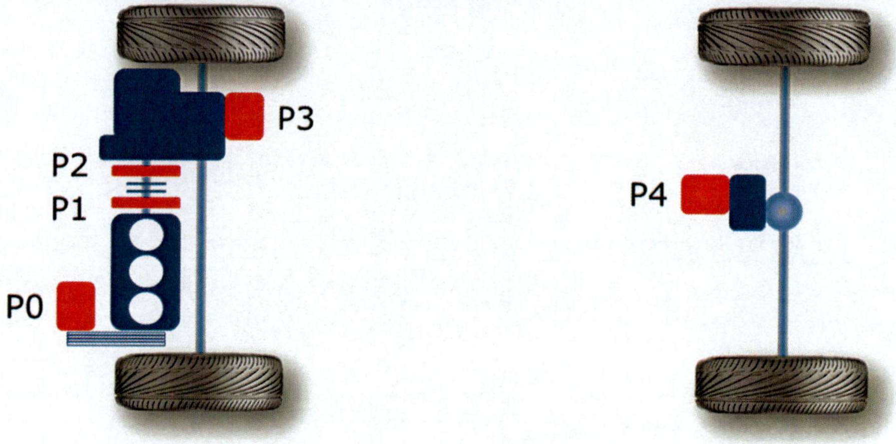

Fig. 4. 48 V powertrain configuration

2 AVL 48 V e-Axle System

As described in the introduction, a set of use cases have been used to determine the torque and power requirements for the e-axle. The target vehicle for the 48 V e-axle was a D-segment vehicle. The basic parameters of the e-axle can be found in Table 1.

The e-axle system, shown in Fig. 5, is designed to meet the target vehicle package requirements and allows the integration in a conventional 4WD rear axle. To gain the maximum advantage from the support of the electric system in lower speed area and avoid drag losses at high speed, a dedicated disconnecting system has been integrated. The e-drive can be decoupled from the powertrain above 130 kph to improve the overall efficiency.

Inverter and e-motor are cooled by water. The cooling systems are connected in series whereas the inverter cooler outlet is directly connected to the e-motor water jacket inlet to maximize integration of every element in the e-axle.

2.1 Electric Machine

The e-machine presents a 6-phase layout with distributed windings to achieve the high power requirements while maintaining low voltage and NVH levels. The e-motor itself is a permanent magnet synchronous machine (PMSM) with 6 poles and 36 slots, see

Table 1. 48 V e-axle system specification

Item	Unit	Specification
Target vehicle	–	D-segment
e-axle peak torque (1 s)	[Nm]	1260
e-axle peak power (1 s)	[kW]	31
e-axle cont. power	[kW]	15
Nom. system voltage	[V]	48
Max. Speed e-motor	[rpm]	10.500
Transmission ratio	–	9,45
Disconnect device	–	Electro-mechanic
Max. speed e-drive support	[kph]	130
Transmission architecture	–	Lay shaft/coaxial
Lubrication	–	Splash lubrication
Cooling system	l/min @ °C	8 l/min @ 60 °C
Inverter	–	Fully integrated

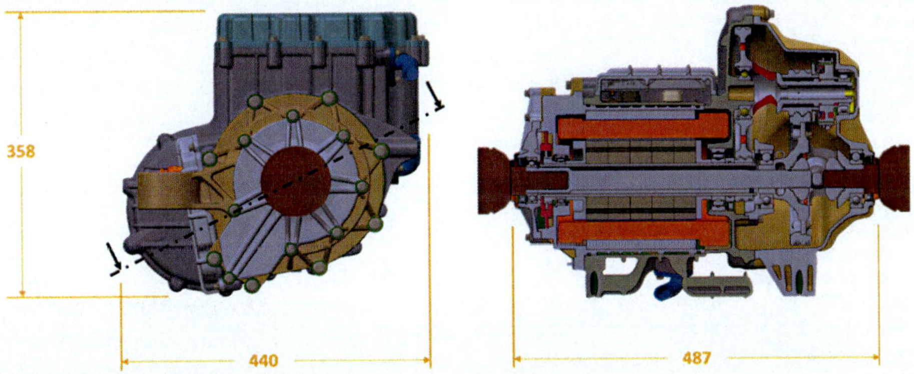

Fig. 5. 48 V e-axle main dimensions

Fig. 6. The maximum rotational speed is 10.500 rpm. For the stator and rotor lamination sheet a bonded package of 0.3 mm thin steel sheets with low iron loss density (voestalpine compacore®) has been chosen to improve the thermal and NVH behaviour of the machine. This results in a torque ripple of 5.5% (peak to peak) in the maximum torque region, up to the knee point. The outer diameter of the stator is 190 mm and the active length of the motor is 150 mm. Figure 6 shows the EM characteristics with a peak torque around 140 Nm and a peak power of more than 32 kW, limited by the system internal max. currents.

Table 2 shows the e-motor target values vs. achieved values. To utilize the whole potential of a 48 V hybrid powertrain the electric system efficiency is most relevant.

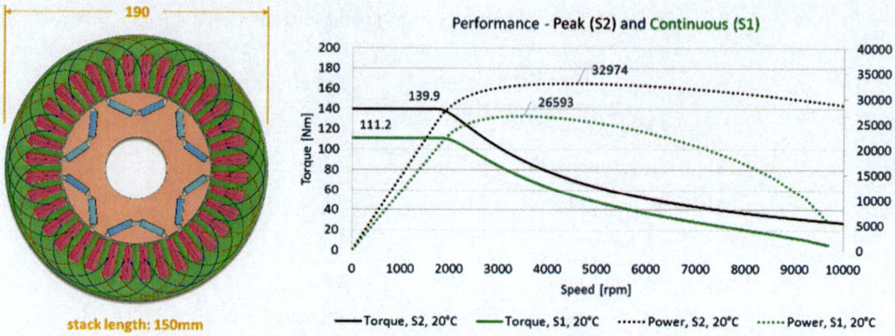

Fig. 6. e-machine characteristics

Figure 7 shows the EM efficiency maps at different temperature levels. The diagrams indicate high efficiency of >90% in operation area of 75%. Since the purpose of this machine is to propel the vehicle purely electric in city driving conditions, the highest efficiency area has been designed to match city driving requirements.

Table 2. Electric machine parameters

Item	Requirement	Final design
Active stack length	≤200 mm	150 mm
Stator outer diameter	≤200 mm	190 mm
e-machine torque (S2)	≥135 Nm	142 Nm @20 °C 136 Nm @80 °C
e-machine torque ripple at max. torque (S2)	≤15%	5.5% @20 °C 5.3% @80 °C
e-machine power (S2)	≥31 kW	32.5 kW @80 °C, 48VDC 26.6 kW @80 °C, 40VDC
e-machine power (S1)	≥5 kW	27.6 kW @80 °C, 48VDC 22.0 kW @80 °C, 40VDC
e-machine max. speed	≥10500 rpm	10500 rpm
Copper fill factor with 0.8 mm diameter copper wire	≤42%	40.1%
Inverter DC voltage	≤48 V	36–52VDC

2.2 Transmission

For the transmission layout a lay-shaft coaxial design has been chosen to meet the restricted package space in the rear axle. This layout is characterized by compact overall dimensions compared to other variants from Fig. 8, while keeping high efficiency and good NVH levels.

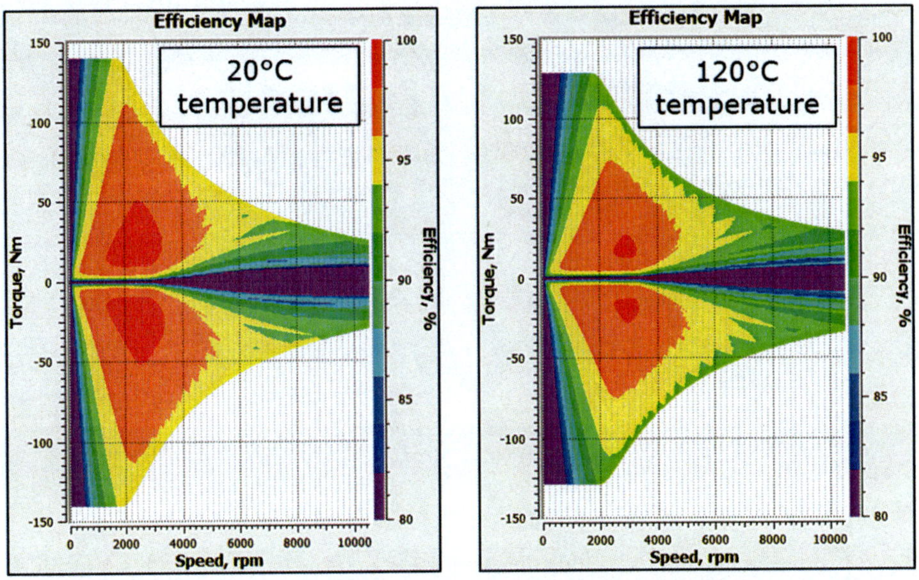

Fig. 7. e-motor efficiency map

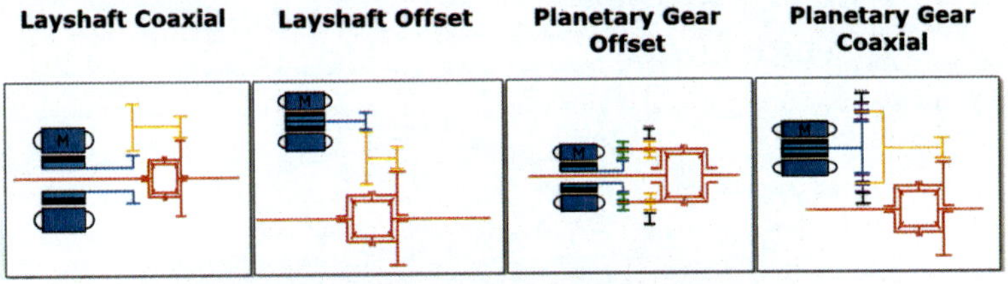

Fig. 8. e-axle transmission layouts

A hollow shaft (supported by voestalpine tubular) is used to enable a smart integration of the side shaft through the rotor. This high precision pipe provides minimum wall thickness while granting maximum strength and stiffness at minimal cost.

The definition of the e-motor torque and the transmission ratio is always a balancing compromise between available package for the system, rotational speed of the e-machine and the necessary e-axle torque output.

The required torque on the drive shafts was calculated with 1260 Nm. For the e-motor a maximum torque of 135 Nm could be achieved in the available package. As the target is to support the hybrid mode up to 130 kph, a gear ratio of i = 9,45 has been chosen. This ensures sufficient e-axle torque to achieve the performance targets and enables support up to highway speeds. Above 130 kph the e-motor can be disconnected from the powertrain with a disconnect element to prevent dragging losses. An overview of the transmission design can be seen on Fig. 9.

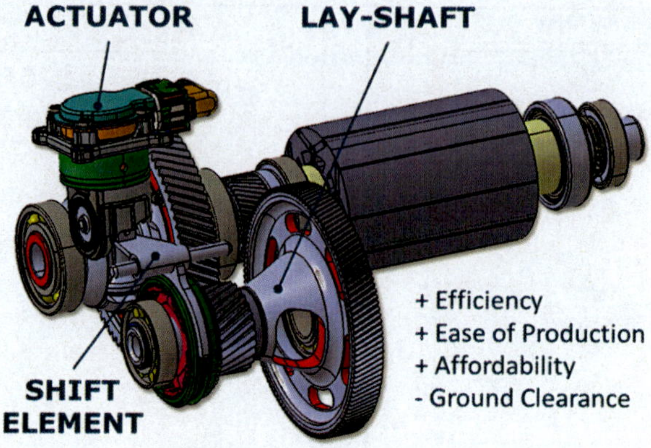

Fig. 9. Transmission overview

2.3 Inverter

The AVL 6-phase inverter, which can be seen in Fig. 10, is equipped with 4 Fairchild N-channel PowerTrench® MOSFETs in parallel per each high side and low side.

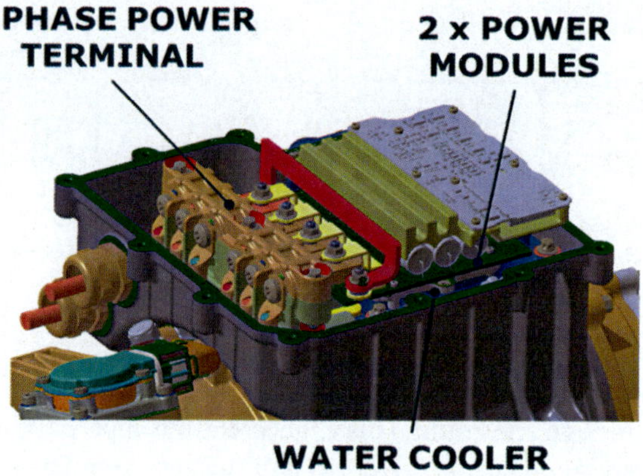

Fig. 10. 6 Phase Inverter

The field-oriented control of the PMSM with a PWM frequency of 10 kHz is taken over by an Aurix TC297 Microprocessor. Torque and speed control as well as derating functions can be applied via CAN. Service functions such as resolver angle calibration and current sensor calibration are integrated.

The inverter is integrated into the e-axle housing. This allows one joint cooling circuit for inverter and electric machine and reduces the number of interfaces and thus decreases complexity and system cost.

Thermal vias and liquid gap-filler enable efficient cooling on the rear of the PCBs. Figure 11 shows the cooling concept for the power electronics.

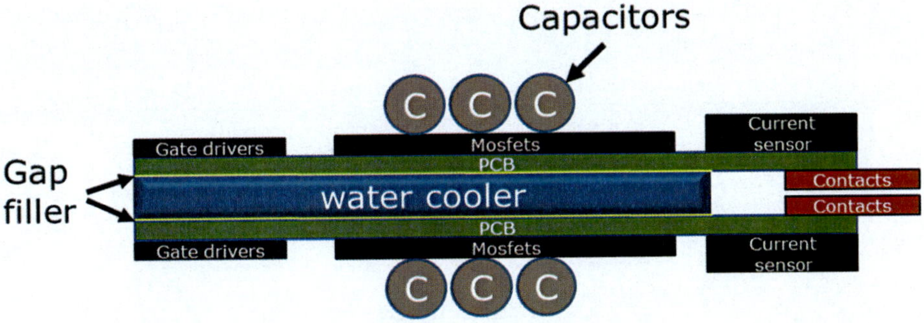

Fig. 11. Inverter Cooling Schematic

As a result of these measures, the AVL inverter can output a maximum current of 340 A (RMS) per phase and 850 A peak transient ASC (active short-circuit) current. This is crucial to realize the high power output of the 48 V electric machine.

3 Outlook

3.1 Hairpin Electromagnetic Design

Regarding the series production of the AVL 48 V P4 module, an updated electric machine design with downsized dimensions and hairpin windings, shown in Fig. 12, has been developed. This allows full industrialization of stator production and an increase in power density. Additionally, the peak torque increases by 12% and the maximum speed by 52% while volume and weight are drastically reduced, as described in Table 3.

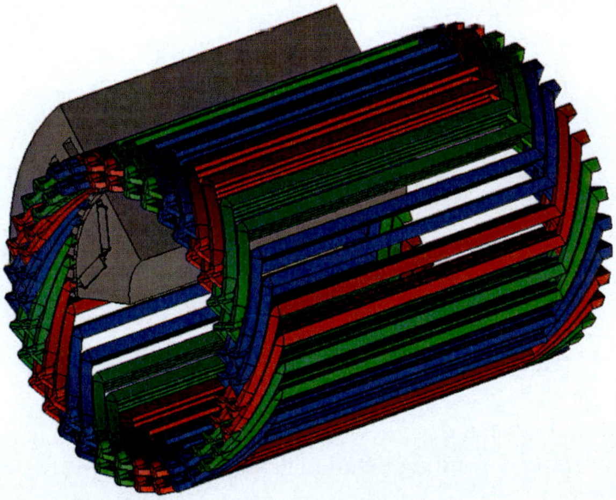

Fig. 12. 48 V Hairpin e-motor

Table 3. Hairpin e-motor performance

Item	Unit	Baseline design	Hairpin design	Benefit of hairpin
Winding	[–]	0,8 mm round wire	3 × 2,3 mm hairpin	
Copper fill factor	[%]	40	60	+20%
Active length	[mm]	150	120	−20%
Outer diameter	[mm]	190	170	−11%
Stack weight	[kg]	27,5	20	−27%
Stack volume	[L]	3,5	2,6	−26%

Special considerations have to be made regarding the AC losses inside hairpin windings, as the currents tend to accumulate close to conductor outer surfaces due to skin and proximity phenomena. These effects lead to increasing AC resistances especially at higher frequencies. Therefore, the non-uniform current distribution and losses due to skin and proximity effects must be simulated, as well as the fringing field effect on the current density. It is possible to observe that the AC copper losses for maximum torque operation in the innermost conductor of the hairpin windings for this machine can be divided as 31% due to skin effect, 8% due to proximity effect and 61% due to fringing field. By analyzing the losses of each conductor independently, detailed loss models of the winding can be calculated and more accurate thermal CFD assessments can be implemented.

To increase simulation accuracy even further and improve front-loading of the e-drive system development, the effect of inverter harmonics is integrated in the electromagnetic simulations. Table 4 shows how a PWM frequency of 8 kHz affects torque ripple and different losses of the machine. Eddy loss density plots for the permanent magnets and conductors are shown in Fig. 13.

Table 4. Effect of PWM harmonics on e-machine

Item	Change due to PWM harmonics
Torque ripple	×1.85
Iron loss	+43%
Magnet loss	+2400%
Copper loss	+8.1%

Close cooperation between designers and simulation engineers is required for multiple aspects such as the end-winding detailed design. The winding transposition configuration and connections have to consider both electromagnetic performance and manufacturing process. As shown in Fig. 14, the end-winding connections are designed showing the detailed geometry on each side of the stack as the crown side and welding side including the welding points. The authors have observed that, taking this electric

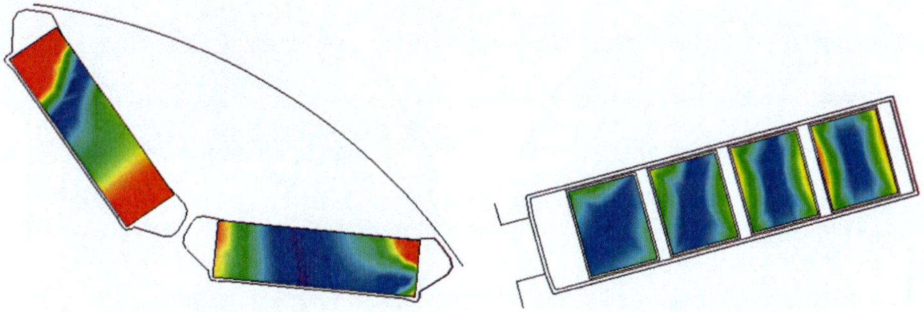

Fig. 13. Eddy current density on magnets (left) and conductors (right)

machine as reference, the Total Harmonic Distortion in the Back-EMF of single-layer winding is 46%, whereas is would be 12% for double-layer winding.

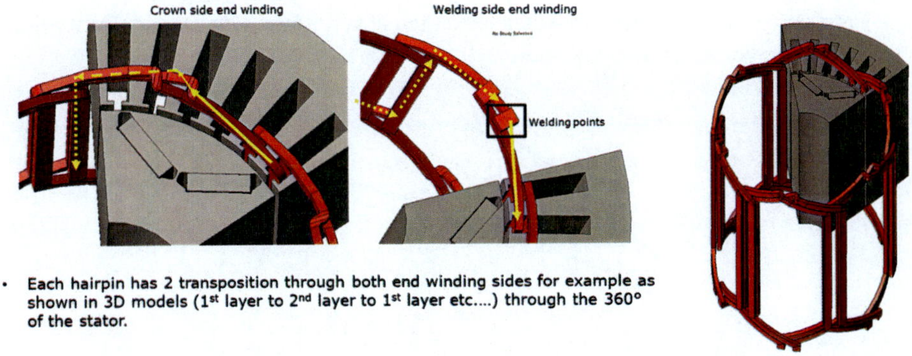

Fig. 14. 3D geometry of hairpin conductor with welding points and crown side

3.2 Hairpin Design Thermal Integration

As illustrated in Fig. 15, this machine shall be modularly implemented for different applications by using either water jacket (WJ) cooling for standard traction applications or direct slot (DS) cooling for high performance and racing purposes. This modularity could potentially reduce system weight and number of interfaces while enhancing performance and cost minimization in series production for different electric vehicles.

Figure 16 shows that direct slot cooling improves the duration of peak operation by 7.5% in comparison to axial water jacket. Apart from the improvement in this critical aspect, the temperature gradient of continuous operation is smaller in direct slot cooling as well, which contributes to lower component thermal stress and inferior steady state temperature. These improvements come from the vicinity of the direct slot cooling media to the heat source, resulting in better heat transfer. Nevertheless, such narrow slot cooling ducts are only recommended for high performance applications due to the three times

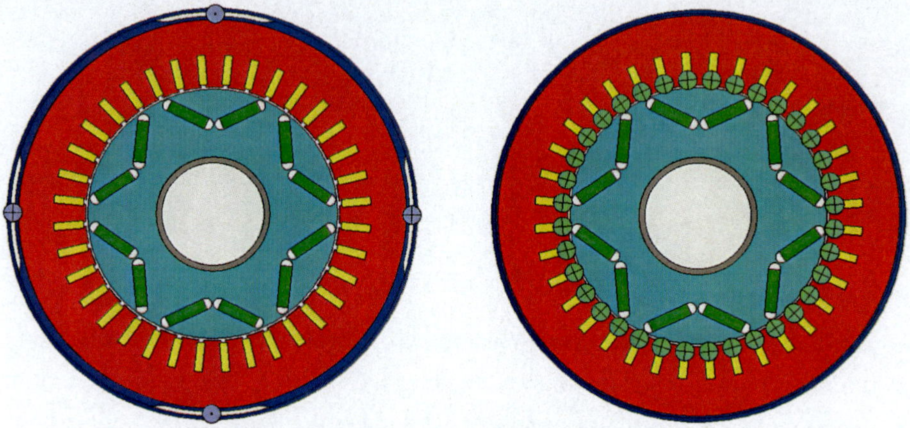

Fig. 15. Electric machine cooling concepts: water jacket (left) and direct slot cooling (right)

higher pressure drop compared to water jacket ducts and limited design flexibility due to the fixed number of cooling channels.

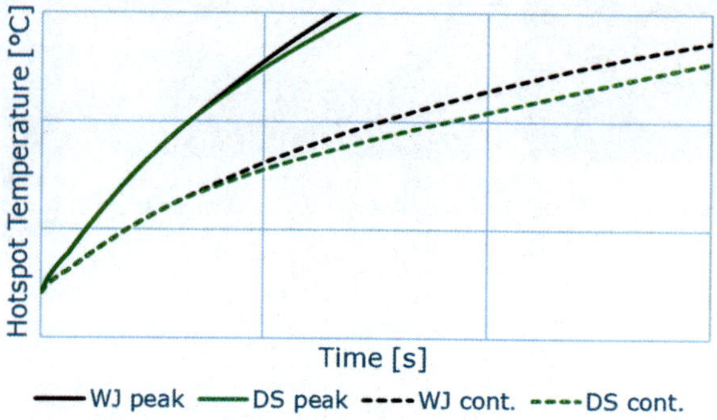

Fig. 16. Water jacket (WJ, in black) and direct slot (DS, in green) cooling effect on machine temperature for peak and continuous performance

3.3 Rotor Structure Optimization

A further improvement is the enhancement of the rotor structure by means of topology optimization. The improved design shows 27% less rotor area and almost 2 kg in weight savings, apart from a considerable inertia reduction for improved driving dynamics. Cutouts within the rotor structure allow a free thermal expansion which supports rotor air cooling and stress relieve under extreme thermal conditions. Figure 17 illustrates the relative material distribution (simulation result) [1–7].

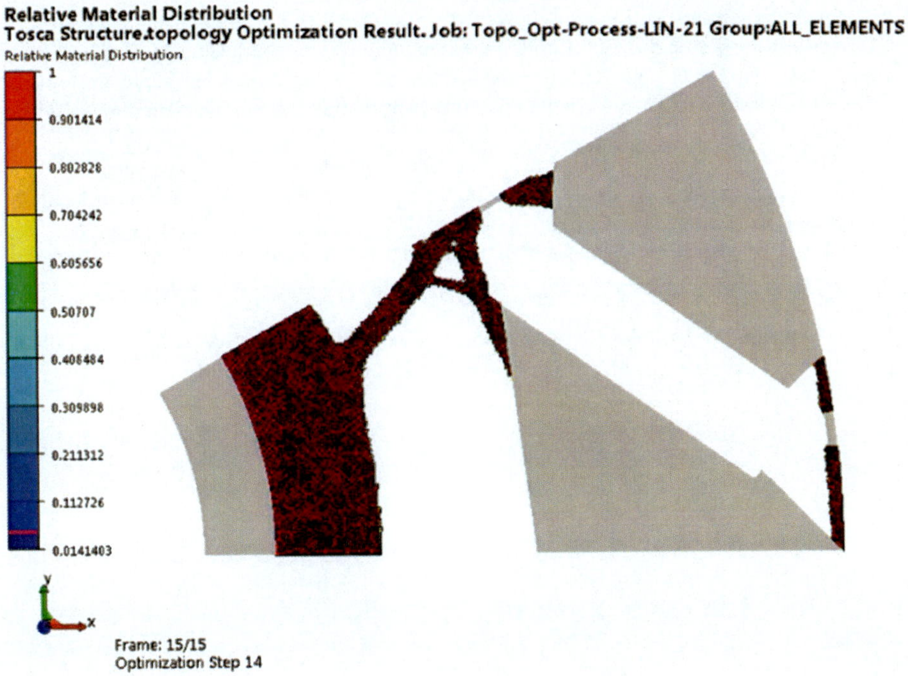

Fig. 17. EM rotor structural optimization

4 Conclusion

48 V will become a widespread standard in the automotive world as it can provide substantial benefits in CO_2 and emissions. Beyond that, the system topologies will transform towards more P2 and P4 powertrain architectures.

The future scalability of 48 V to higher power ranges enables affordable e-mobility especially within smaller cars. The trade-off between the high voltage and lower voltage systems needs to be investigated closely.

With the high performance 48 V e-axle system, AVL can provide an affordable, high efficient and smart solution for future powertrains.

Acknowledgements. This project has received funding from the European Union's Horizon 2020 research and innovation program under grant agreement No 724095.

References

1. Pels T, Davydov V, Ellinger R, Kaup C (2017) Where to place the e-machine?. In: 11th international MTZ conference on future powertrains. Frankfurt/Main (Germany)
2. Teuschl G (2018) Antriebssystementwicklung - schlüsselrolle integration. AVL electrification TechDay - affordable e-mobility. Heimsheim (Germany)

3. Teuschl G, Weissbaeck M, Kaup C, Kapus P, Sams C (2018) An affordable approach towards RDE and zero emission. In: 37th FISITA world automotive congress. Chennai (India)
4. Brendel M, Teuschl G, Pels T, Kapus P, Sams C (2018) An affordable approach towards local zero emission. In: 27th Aachen colloquium automobile and engine technology. Aachen (Germany)
5. Teuschl G, Kapus P, Weissbaeck M, Pels T (2019) Minimalemission basierend auf 48 v – ein leistbarer einstieg für grüne städte. In: 13th international MTZ conference on future powertrains. Frankfurt/Main (Germany)
6. Stenzel U (2019) Requirements of a high power 48 v battery system for future xEV applications. In: 11th International symposium "advanced battery power – kraftwerk batterie. Aachen (Germany)
7. Pels T, Neumann M, Schmidt Ch, Kronfeldner K, Teuschl G (2019) AVL's low voltage high power electric axle system. SIA Powertrain & Electronics, Paris (France)

48 V High Power: Electric Drive for Excellent CO_2 Emissions and Electric Driving Features

Friedrich Graf(✉), Martin Beiderbeck, Thomas Knorr, Dietmar Ellmer, and Mattia Perugini

Vitesco Technologies, Siemenstraße 12, 93055 Regensburg, Germany
friedrich.graf@continental-corporation.com

Abstract. Future motor vehicles regulatory requirements for CO_2 emission reduction require a significant increase in hybrid vehicles in the fleet mix of manufacturers.

The use of the internal combustion engine is increasingly replaced by the electric drive in hybrid vehicles and the challenges for exhaust aftertreatment increase, as the thermal supply of the internal combustion engine for thermal management decreases. The electric machine must have sufficient performance to enable corresponding driving patterns not only in the test cycle but also in real-life operation.

Results from Vitesco Technologies 48 V Eco-Drive System have shown that for P2 compared to a P0 hybrid, the challenge for emissions is significantly exacerbated by an increase in pure electric driving capability, especially for plug-in hybrid vehicles. At the same time, technological advances allow 48 V drives to penetrate areas of application that were previously reserved for high-voltage solutions.

This paper presents a compact design 48 V 30 kW electric machine solution, suitable for a P2/P4 configuration intended for full hybrid applications or even a plug-in hybrid and characterized by compact design with a very high-power density. At the same time, the system approach of exhaust aftertreatment is illustrated. A 48 V 30 kW "electric" means not only less CO_2, but also increased electric driving and more total driving performance with significantly less effort and complexity due to 48 V as compared to a high-voltage solution.

With the help of simulations and vehicle measurements, CO_2 potentials as well as various emission-relevant driving conditions of a 48 V 30 kW high-performance hybrid are examined. At the same time challenging environmental conditions and the potential for improvement is determined by the advanced Emicat® exhaust aftertreatment system.

CO_2 performance is also the result of intelligent operating strategies. In the future, they will have to make their contribution robustly to the lowest CO_2 in every driving cycle. For this purpose, connectivity and optimization algorithms in real time are enablers [6].

Keywords: 48 V P2 topologies · High power and torque density · Permanent magnet synchronous machine · Embedded PCB inverter · Cost efficient plug-in hybrid

Euroforum Deutschland GmbH (Ed.): *CTI SYMPOSIUM 2019*, pp. 147–163, 2021.
https://doi.org/10.1007/978-3-662-61515-7_14

1 Introduction

Since the European Parliament has agreed beginning of 2019 to reduce by 2025 CO_2 emissions by 15% and by 2030 CO_2 emissions by 37.5% from new passenger cars (31% for new vans) in comparison to the already existing target of 95 g/km of CO_2 from 2021, the European automotive industry needs to achieve a significant step towards decarbonising and modernising the European mobility sector.

In addition, the main pollutant emissions limits are drastically reduced from 2025 onward in comparison to 2020 especially for Nitrogen Oxides (NOx), Particle Mass (PM) and Particle Number (PN).

Finally, a strengthened market monitoring and surveillance system will be set up with the adoption of the new test cycles (e.g. WLTP [Worldwide harmonized Light- Vehicle-Test Procedure] and RDE [Real Driving Emissions]) to enhance the representativeness of the official test procedure for determining the emissions with respect to real-world driving (Fig. 1).

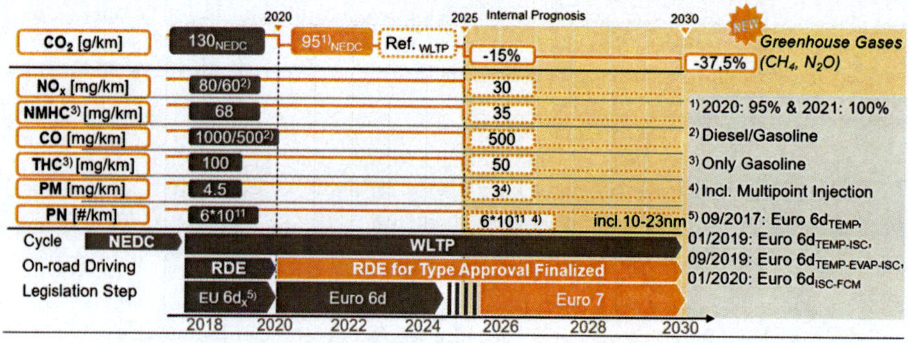

Fig. 1. Europe further consumption, emission limits & test procedures

As the combustion engine has already reached a very high level of technical sophistication and further efficiency improvement potential is more and more limited, it is widely accepted that (48 V) electrification will be necessary to comply with the future CO_2 requirements in and in respect to an OEM CO_2 reduction strategy.

As consequence, there are an important number of 48 V P0 mild HEV applications already in development or even in serial production, since the technology provide many advantages with following characteristics and contexts [2]:

- Main focus is CO_2 reduction,
- Mainstream architecture is P0 where the 48 V Belt Starter Generator (BSG) can be integrated easily as a replacement for existing alternators with minimal impact on the existing vehicle designs,
- 48 V belt starter generator are optimized to deliver requested performance for current architectures, with dedicated DC/DC, 48 V battery and belt starter generator as main ingredients
- Diesel scandal accelerated serial applications: from niche to standard

- Technical standards have been introduced and are effective

In a next step, much higher CO_2 reduction can be achieved by so-called Px hybrids, which are obtained by integrating the 48 V drive system into the gearbox or a hybrid module (P2, P2.5, P3) or by adding a second electrically driven axle (P4).

These Px-systems have in common that the drag losses of the combustion engine are avoided during braking, enabling coasting and electric driving, which all result in remarkably low CO_2 emissions and the experience of pure electric driving features.

Therefore, the 48 V electric drive system has to be consequently designed towards high power density and maximum efficiency to optimize not only the energy recuperation but also to maximize the traction performance.

2 Requirements for Future 48 V Electric Drive System

The current generation of 48 V electric drive addresses mainly the P0 architecture is designed to provide up to 60 Nm and 15 kW peak power. With such system, the CO_2 reduction on WLTC cycle is limited to a maximum of 10% (depending of the component sizing, vehicle size, transmission type and others powertrain characteristics) and the following hybrid functions can be enabled: start/stop, recuperation, torque assistance, coasting, electric creeping and reduced emission for gasoline or diesel based internal combustion engines thanks to the combination of 48 V system and electrically heated catalyst – EMICAT [3, 4].

Now with the usage of 48 V electric drive system in Px- hybrids, additional CO_2 reduction is expected up to 20% and also electric driving features on a C segment vehicle with P2–P4 architecture (Fig. 2):

	Key Requirements
SOP	2023+ incl. Commercial Vehicles
Hybrid architecture	Px
Performance	>20 kW peak
Hybrid function	Same as 2nd generation: + Enhanced recuperation + Electric driving features
CO_2 reduction (WLTC*)	Up to 20 %

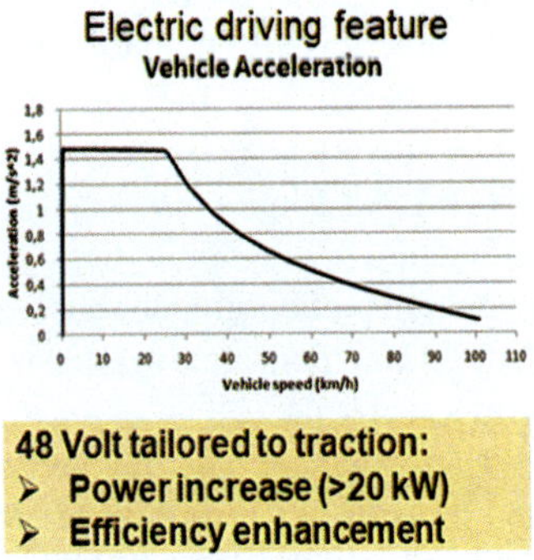

Fig. 2. Key requirements for future 48 V electric drive system

- An acceleration of 1.5 m/s^2 up to 30 kph in full electric mode with
- A continuous electric sailing up to 55 km/h

These requirements translate to an increase in power requirements for the 48 V electric drive in a power increase as well as an enhanced efficiency (>90%) to enable an extended recuperation potential:

- For engine restart, 30 kW (mechanical) for 5 s at 2200 rpm crank shaft speed
- For electric driving, 20 kW (mechanical) for 20 s
- For electric sailing, 15 kW (mechanical) continuous at 2200 rpm crank shaft speed
- For electric boost and diesel cold crank, 70 Nm for 5 s

3 '48 V High Power' Electric Drive Component

Based on these requirements, Vitesco Technologies has developed a '48 V High Power' electric drive system with a peak performance of 30 kW and an excellent efficiency. Parking, active electric sailing and following the traffic in city areas can all be achieved pure electrically with this '48 V High Power' electric drive. Its compact design with a very high-power density flexibly allows the integration into a P2-hybrid module, into a gearbox (P2.5, P3), as well as the realization of an electric axle drive (P4) and to enable 4WD capability.

Technology: It consists of two main sub components:

- An electric permanent magnet synchronous motor (PSM) with an I Pin stator winding technology (Vitesco Technologies patent) able to reach a maximum shaft speed of 20.000 rpm.
  The permanent magnet rotor provides many advantages in comparison to an AC induction machine (asynchronous machine). The volumetric torque and power density are higher, the motor's efficiency is higher, less current is used (as no magnetizing current are necessary), lower temperature is reached, and ramp-up time is shorter.
- An inverter with 6 phases based on Embedded PCB technology which includes up to 3 MOSFET's to increase the provided current (Vitesco Technologies patent).
  The inverter PCB technology ensures a high-power density of 25 kW for 0.79 L/1.51 kg.

In terms of cooling, a shared cooling concept has been designed between inverter and electric machine: the inverter is liquid cooled as the electric machine's design benefits from a water jacket.

Performance: The development of the first sample of electric machine and associated inverter has been finished since mid-2018 resulting in a compact design (diameter 175 mm) with an integrated inverter. Sample performances have been measured on an 48 V electric drive test bench with water cooling temperature below 85 °C and water-cooling flow of 3 l/min (Fig. 3) marked by solid points. Efficiency is high ~90% in motoring mode 14.5 Nm, 10 kW, 36 V at 6600 rpm.

For a hybrid vehicle, it means than we can benefit from a theorical maximum additional crankshaft torque of 70 Nm (multiplied usually by 2–3 gear ratio) provided by the '48 V High Power' electric drive to boost the already available torque from the

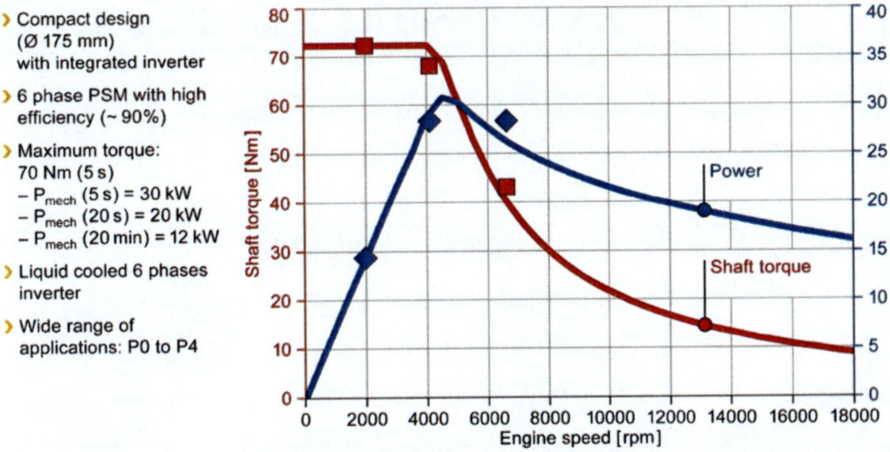

Fig. 3. '48 V High Power' electric drive sample, power and torque in function of shaft speed

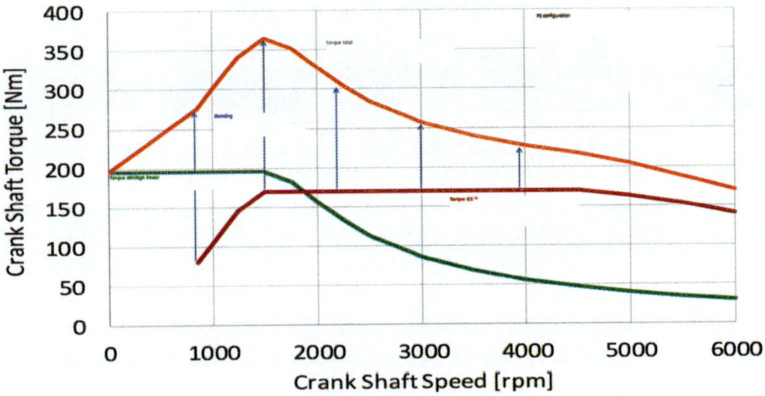

Fig. 4. Combined ICE and 48 V High Power torque in P2 configuration

Internal Combustion Engine (ICE). Therefore, the vehicle offers more dynamic acceleration which can be useful for instance when merging into traffic. Generally, the low-end engine behavior is significantly improved, which combines very well with highly charged, power-dense gasoline engines. The transmission input torque capacity has to cover this torque growth (Fig. 4).

4 Utilized Powertrain System Architecture

The next step was to integrate and to assess the performance of the '48 V High Power' electric drive in a real use case on a demo car. The choice has been made to re-use as a basis the P2 hybrid architecture already developed by Vitesco Technologies [5, 6], and then to replace the previous generation of 48 V electric drive motor by the new '48 V High Power' electric drive. The 48 V Li-Ion battery has been also replaced by a storage with more capacity at 6.5 kWh (useable) and providing more max. current for (dis)charge.

This so-called P2 hybrid architecture shows the electric traction machine and the additional K0 clutch being installed between the internal combustion engine and the transmission (Fig. 5).

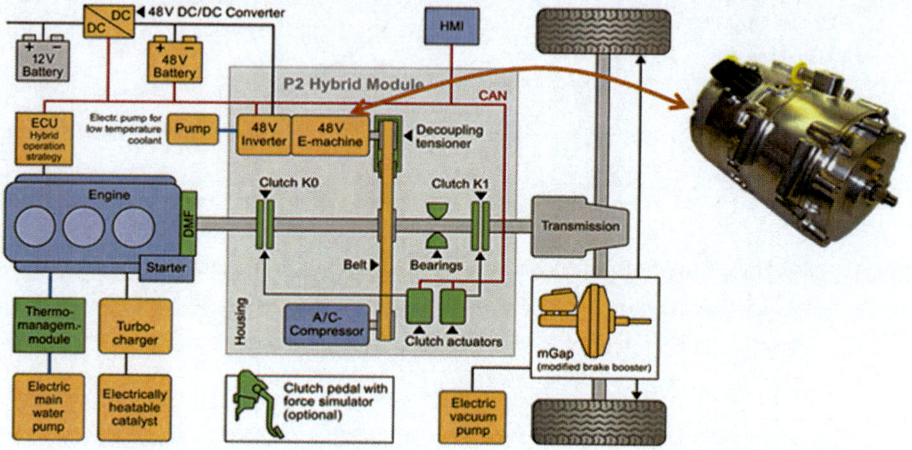

Fig. 5. System architecture of the 48 V P2 full hybrid vehicle

Hence drag losses in hybrid states such as recuperation, coasting (rolling with open powertrain), sailing (driving electrically at constant speed) and pure electric driving are eliminated. The combustion engine is the Ford 1.0L EcoBoost three-cylinder gasoline engine. It is enhanced by a matching RAAX turbocharger [1], a 200 bar fuel pressure injection system, an electrically heated catalyst (EHC) and an engine control unit coordinating the entire powertrain. A further key engine component is the Vitesco Technologies RAAX turbocharger with a low inertia radial-axial turbine, which generates high boost pressure and fast pressure build-up even at low engine speeds.

As required for a P2 hybrid vehicle a small electrical water pump enables electric machine operation when the ICE is shut down.

Another key subsystem is the high level 'hybrid operation strategy', which is executed by the engine control unit. It is implemented in a model-based manner and controls the entire hybrid powertrain.

As shown in Fig. 5 the vehicle is principally able to operate in all hybrid modes which are known from high-voltage hybrid vehicles. In particular, it is possible to drive purely electrically which can be used to perform electric creeping and launch (in combination with ECM) as well as to maintain constant vehicle speed up to 80 km/h ('sailing').

5 Vehicle Performance with '48 V High Power' Electric Drive

The vehicle hast been revealed to journalists at the Continental TechShow in July 2019. Before performing real tests on vehicle, Vitesco Technologies has used its powertrain system simulation tool chain called Vehicle Simulation Suite (VSS) to evaluate the performance of the vehicle equipped with the '48 V High Power' electric drive (Fig. 6).

Vehicle Simulation Suite is based on a model in the loop approach with a physical plant model and control functions working in co-simulation. This complete tool chain includes various powertrain architectures, libraries of components, functionalities for automated simulations and analyses. The simulation results have been correlated to real experimental measurement to ensure the required level of accuracy).

Fig. 6. System evaluation vehicle at Continental TechShow 2019

CO_2 reduction on WLTC cycle: The 48 V P2 mild hybrid demonstrator already revealed a measured CO_2 reduction of 15% on the WLTP cycle. By replacing the previous generation of 48 V electric drive motor with the new '48 V High Power' electric drive, the simulation showed an enhanced potential of CO_2 reduction up to 19% under same conditions (Fig. 7).

These results are similar to the CO_2 level reached by state-of-the-art high-voltage hybrid vehicles, demonstrating the high potential of 48 V technology in regard to high voltage solutions.

Electric driving: The driving mode simulation of the 48 V P2 mild-hybrid vehicle on a WLTP cycle shows that a first start of combustion occurs after 12.75 min which means an All-Electric Range (AER, EU type approval 2017/1151) value of 4.7 km (depletion mode, assuming no battery capacity limitation). Indeed, a very significant section (94% of time) of WLTC can be driven without the use of combustion.

Following the definition of the 'Equivalent All-Electric Range' (EAER) the driving range is 34 km with a battery capacity of 5.3 kWh net (useable) (Fig. 8).

Real driving conditions have been also a source to define power requirements. Figure 9 compares WLTC with an on-road driving cycle, the 'Regensburg City Cycle' (RCC, defined in Fig. 11).

Simulation calculation has shown the need for more peak electric power for recuperation as well as for traction in comparison with WLTC.

Other two important variables to observe are the DC peak current and the DC current distribution.

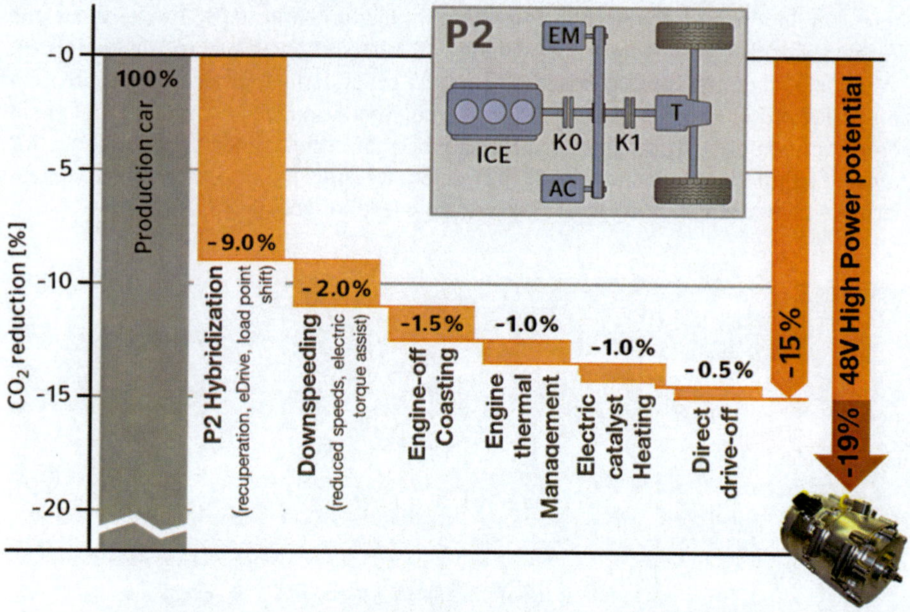

Fig. 7. CO_2 reduction in WLTP cycle for a P2 full hybrid vehicle

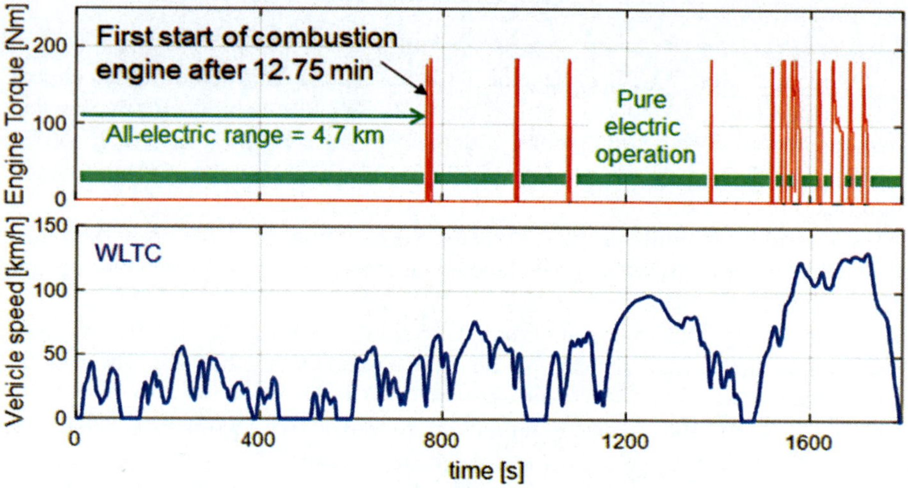

Fig. 8. Electric driving capabilities on WLTP for a '48 V High Power' P2 mild hybrid vehicle in depletion mode

Figure 10 shows a limited increase of peak current in case of battery charge sustain, however the mean value is moderate, mitigating the technical solution of wiring on the DC side.

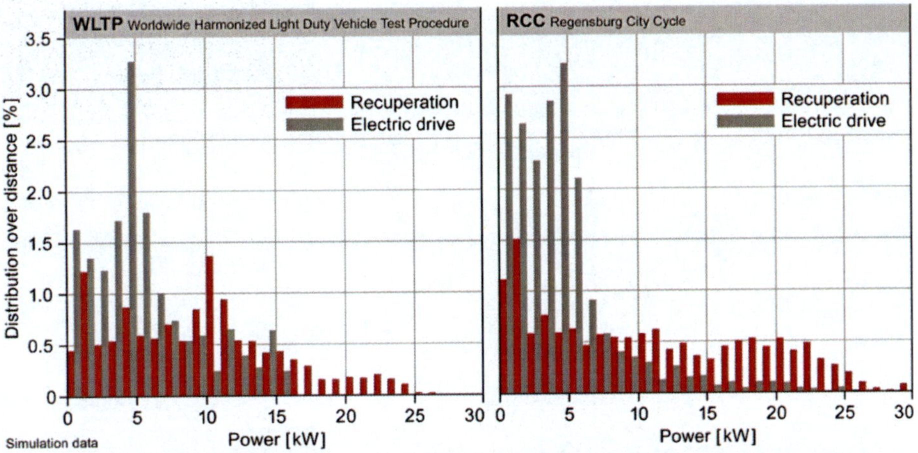

Fig. 9 and 10 (below). Power and DC current characteristics WLTC vs. RCC (charge sustain)

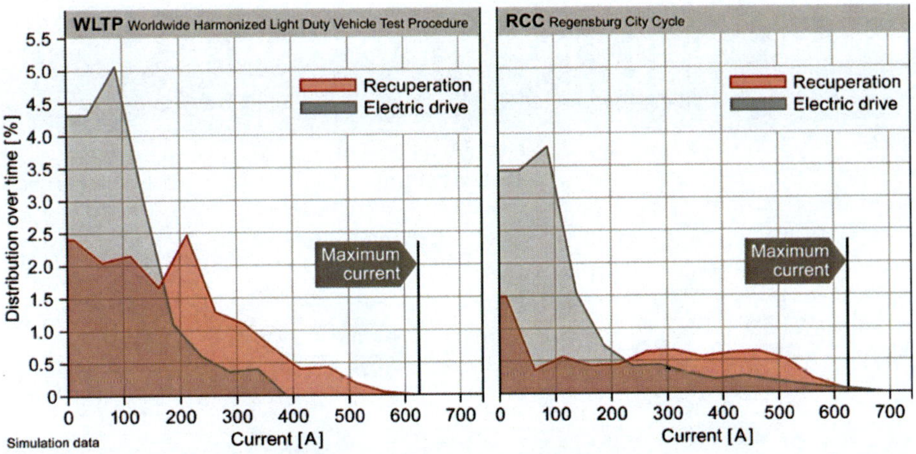

Fig. 9 and 10 (below). (*continued*)

Figure 12 shows a real driving example in RCC and the distribution among the drivetrain states in charge sustain (full hybrid). The high portion of eDrive mode even under balanced SoC is evident and underlines the electric driving capability.

The ICE on-time is reduced to 23% (distance), eDrive is at same level - ICE function is reduced to acceleration work. The practiced vehicle style is 'traffic flow'.

Figure 13 is same as from Fig. 12, however in charge depletion condition. The eDrive mode is extended to 35% and covers now nearly all acceleration manoeuvres ICE is needed only to help to accelerate, especially in higher speed condition.

Concerning driveability, e.g. elasticity is strongly improved among other manoeuvres (Fig. 14).

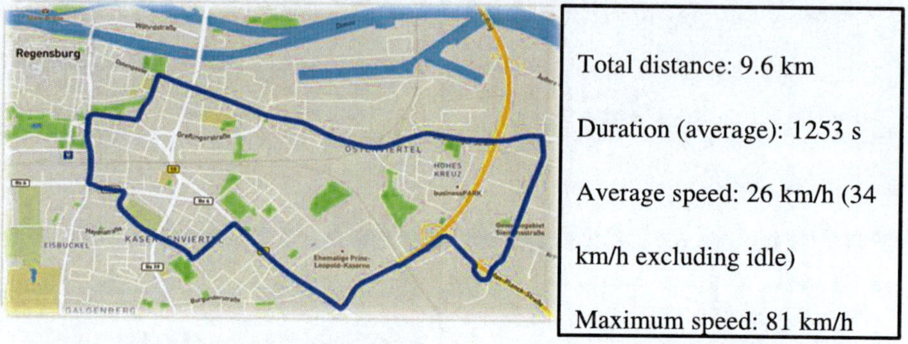

Fig. 11. Regensburg City Cycle' RCC as urban test cycle used for drivability tests

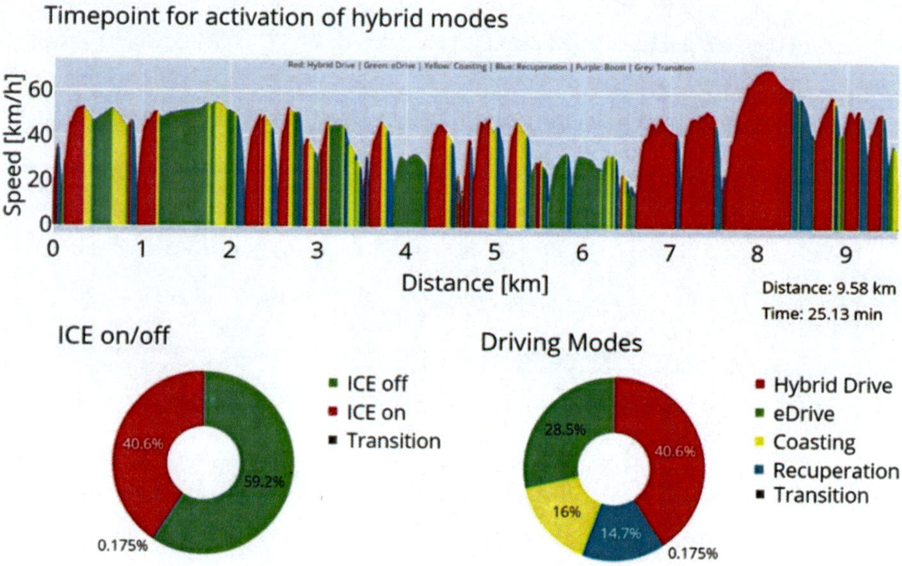

Fig. 12 and 13 (below). real on-road driving the RCC, charge sustain- and depletion mode

All these results underline that the '48 V High Power' electric drive in a P2 configuration provides extended electric driving capabilities, is highly beneficial in urban conditions and enables distinct driveability improvements in hybrid mode. In essence, the '48 V High Power' supports all full-hybrid functions and enriches the driving experience correspondingly.

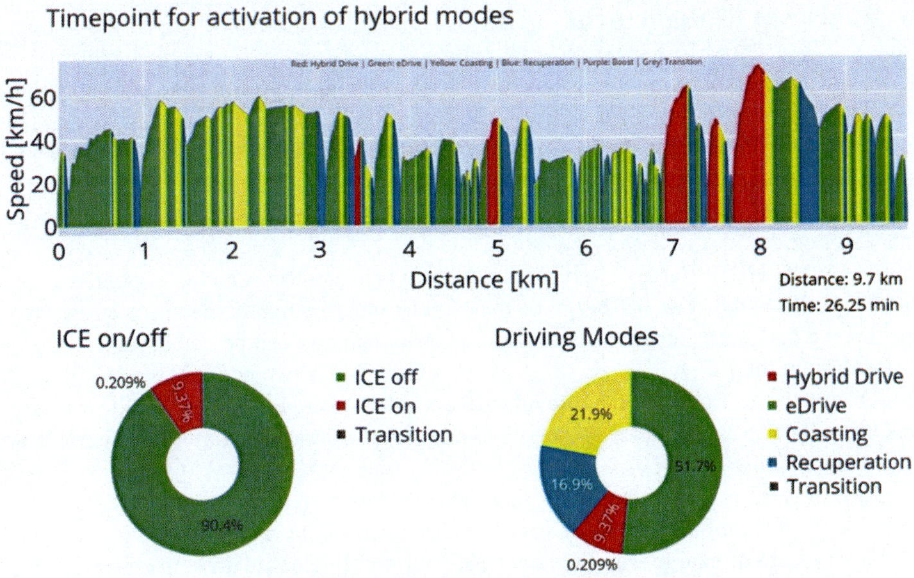

Fig. 12 and 13 (below). (*continued*)

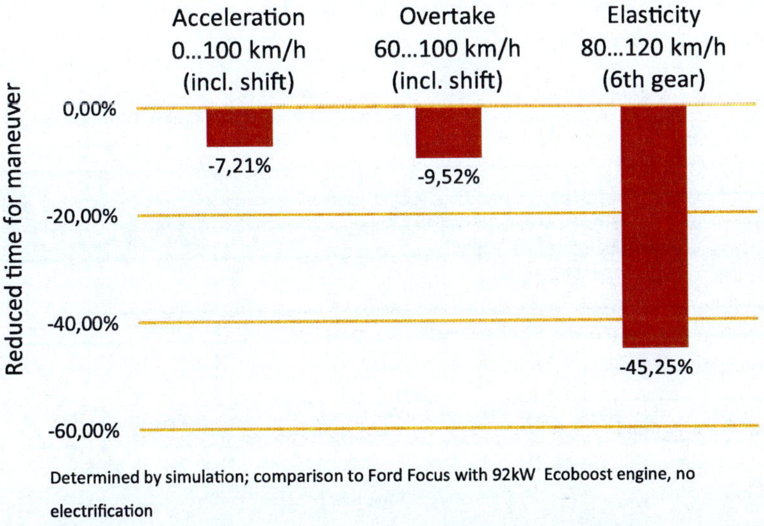

Fig. 14. Driving performance improvement by '48 V High Power'

6 Emission Management

RDE and future emission regulations lead to a more and more complex emission management. Especially at hybrid vehicles the trade-off between pollutant emissions like NO_x and HC, on one side, and CO_2 respectively energy on the other side, can only be solved by a comprehensive system management.

The two basic forms of hybrid electric systems require different emission management strategies:

At NOVC-HEVs (Not off-vehicle charging hybrid electric vehicles) – often realized as P0 hybrids – the state of charge of the battery must be maintained on average over the driving distances, because no external electrical energy can be recharged.

Although the purely electric drive of NOVC-HEVs is extended by the '48 V High Power' and thus the operation of the internal combustion engine is reduced, the resulting heat contribution is guaranteed, albeit superimposed by the uncertainty of distribution over time and distance.

At OVC-HEVs (Off-vehicle charging hybrid electric vehicles) the average heat contribution of the combustion engine is only guaranteed in the charge-sustaining mode as the battery state of charge is at the lower end and purely electric drive is constrained.

But OVC-HEVs are clearly focused on the electric drive in battery charge-depleting mode.

With the 48 V HP drive and with moderately larger batteries compared to NOVC-HEVs, it is possible to cover the often recurring short-distance operation with regular charging of the battery to a high degree with purely electric drive. Due to the resulting intermittent, stochastic heat flow, a different emission management strategy becomes necessary for the charge-depleting mode.

Figure 15 shows the challenging thermal boundary conditions of the OVC-HEVs operated in battery depleting mode.

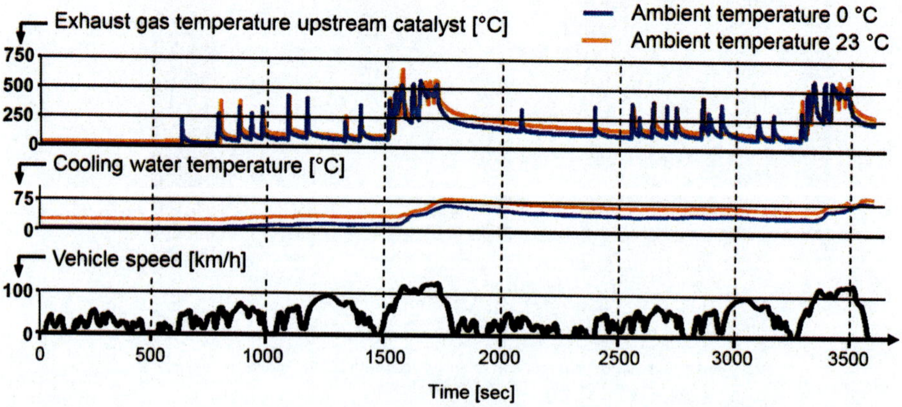

Fig. 15. Thermal exhaust behavior at P2 full hybrid vehicle (battery depletion mode)

This example of two consecutive WLTCs in charge-depletion mode operation according to the COMMISSION REGULATION (EU) 2017/1151 shows the temperature behavior at different ambient temperatures (23 °C/0 °C).

Although the cooling water circulation is not active during almost the entire test period, the normal operating temperature of the engine is not reached and decreases more than 20 °C during the urban drive after the highspeed phase.

The exhaust gas temperature before the closed coupled catalyst often barely reaches the light off temperature of the catalyst caused by the CO_2 optimal short engine activation times. Additional heating measures are required to activate the exhaust aftertreatment system.

The Vitesco Technologies EMICAT® with two electrical heating discs represents a logical step towards fast heating of the largest catalytic volume possible at this intermittent engine operation.

The local catalyst temperature increase caused by two 4 kW heating discs is exemplarily visualized in Fig. 16.

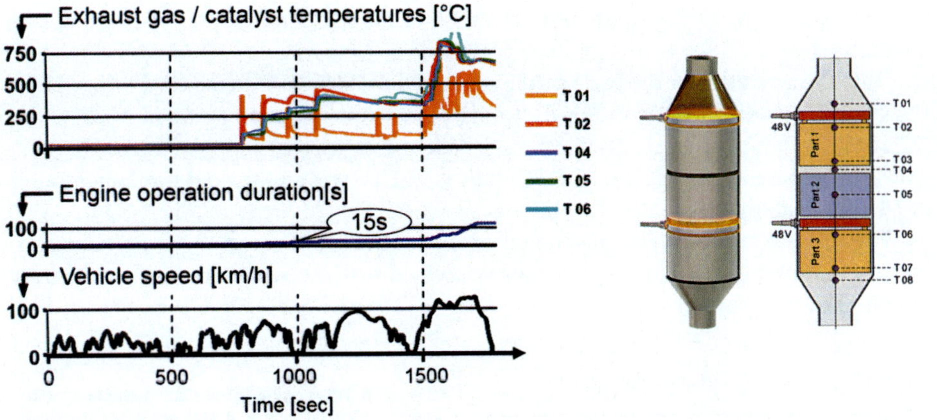

Fig. 16. Electrical heating of the EMICAT® with two heating discs, driven in battery depletion mode

All catalyst core temperatures are above 250 °C latest after 15 s of combustion engine operation. Figure 17 shows the resulting tailpipe non-methane hydrocarbon (NMHC) emission behavior at 23 °C ambient temperature. Three different hybrid operation strategies show the NMHC concentrations integrated and weighted with the actual driven distance over time.

This comparison of the emission profiles demonstrates that the NOVC 48 V hybrid vehicle with the EMICAT® (2 x 4 kW) undershoots the EURO 6d limit by more than 50%. No preheating of the catalyst is necessary before drive-off. The required electrical energy for catalyst heating is below 60 Wh without any combustion measure for heating.

Without any heating measure the OVC '48 V High Power' hybrid vehicle exceeds the emission limit more than twice (blue graph). It is obvious that a heating measure is necessary, but it should be as CO_2 neutral as possible.

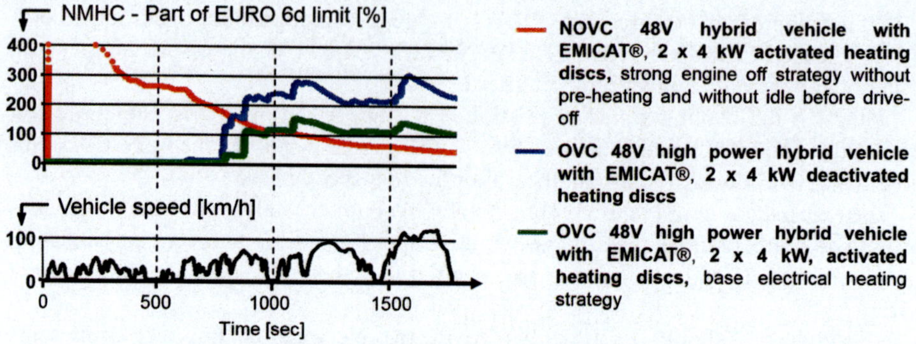

Fig. 17. NMHC-emission behavior in WLTC at different 48 V hybrid operation modes

For '48 V High Power' hybrid vehicles, heating measures from the first second of the driving cycle are often not suitable in terms of energy consumption respectively CO_2 emissions. Additional power demand beyond the electric drive, spontaneously requested by the driver, requires the immediate activation of the ICE without pre-heating steps of the catalyst.

With the activation of the two heating discs of the EMICAT® in the OVC '48 V High Power' hybrid vehicle synchronously to the combustion engine operation NMHC are significantly reduced. An electrical energy of about 150 Wh for heating of the EMICAT® is nearly sufficient to meet the NMHC limit in WLTP within this first implementation of heating strategy.

The resulting nitro-oxygen (NO_x) tailpipe emission behavior at 23 °C ambient temperature during the WLTC, integrated and weighted with the actual driven distance over time is shown in Fig. 18.

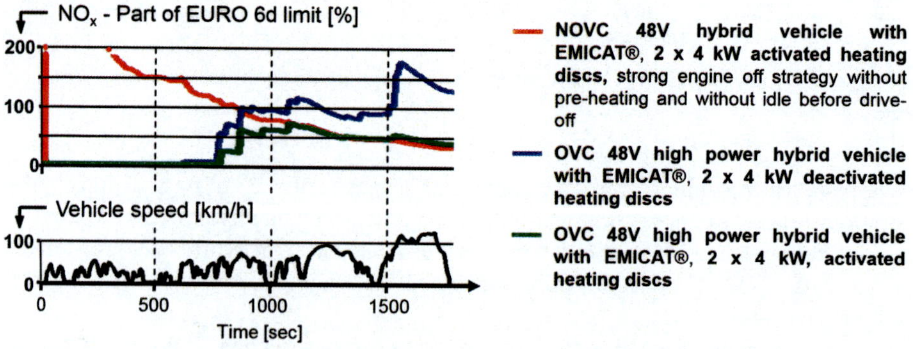

Fig. 18. NO_x-emission behavior at different 48 V hybrid operation modes during the WLTC

The diagram shows that the NOx emissions in both the NOHC- and the OHC-hybrid vehicle in depleting mode are well below EURO 6 d limit. The advantages of the EMICAT® with electrical heating discs in comparison to engine-based catalyst heating operation can be summarized as following:

- CO_2 impact-free thermal management of the exhaust aftertreatment system
- Heating operation without combustion engine activation before drive-off
- Local catalyst temperature control with disc selective heating activation
- Greater independence of the position of the catalyst to the internal combustion engine

Essentially, the new ‘48 V High Power’ electric drive in association with the electrically heatable EMICAT® shows the potential for fulfilling the challenges of the CO_2 and emission reduction.

7 Cost Comparison

Besides its very good performance, the ‘48 V High Power’ electric drive in a P2 configuration is also very well positioned cost-wise. Vitesco Technologies has compared its solutions with a power-split high voltage system using a Multi-Point Injection (MPI) as ICE and an electric Continuous Variable Transmission (eCVT). The study based on non-binding estimation shows a delta cost of (Fig. 19):

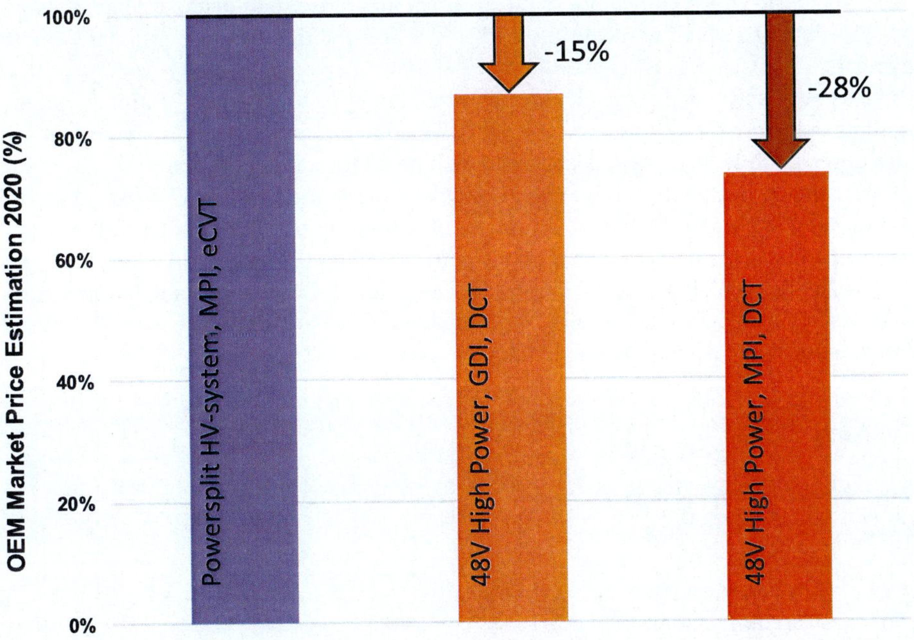

Fig. 19. Price estimation of powertrain & components related to electrification

- −15% for a ‘48 V High Power’ electric drive using a Gasoline Direct Injection (GDI) as ICE and a Dual Clutch Transmission (DCT),
- −28% for a ‘48 V High Power’ electric drive using a Multi-Point injection (MPI) as ICE and a Dual Clutch Transmission.

- Main contributors to cost saving:
  Usage of 48 V technology for the electric drive & power net (e.g. DC/DC, electrical AC compressor) instead of more expensive technology with high voltage components
- Cost efficient electric machine technology
- Low impact on engine & transmission technology

This comparison demonstrates that a '48 V High Power' electric drive technology is suitable for mass market relevant car segments and could be a strong basis for high market penetration in the future.

8 Conclusion

With the innovative '48 V High Power' electric drive, Vitesco Technologies has greatly improved 48 V electrification in different aspects. This electric motor delivers peak output of up to 30 kW, thus offering more torque to support the combustion engine electrically to a high extent, leading to a performance at low end rpm known so far only from diesel engines. Furthermore, '48 V High Power' enables quiet, purely electric inner-city journeys at significantly reduced cost compared to state-of-the-art high voltage hybrid solutions, which could bring such 48 V full hybrids to more popular vehicle segments.

The entire unit comprising electric motor and integrated inverter is extremely compact and delivers an efficiency of over 90%. This facilitates mechanical integration for different Px architecture options and ensures high CO_2 reduction benefits.

By utilizing this '48 V High Power' electric drive in the described P2 hybrid architecture, a full hybrid performs at a high level – with up to 20% CO_2 reduction - are achieved while providing enhanced pure electric driving experiences. In a next step, the Plug-In Hybrid Vehicle (PHEV) option will be explored. Simulation calculation results show less than 50 g/km (New EU Type approval 2017/1151), which would mean a classification as a Low-Emission vehicle.

Glossary

AER: All Electric Range
BSG: Belt Starter Generator
CH_4: Methane
CO: Carbon Monoxide
CO_2: Carbon Dioxide
DCT: Dual Clutch Transmission
EHC: Electrically Heated Catalyst
EVAP: Evaporative Emissions
FCM: Fuel Consumption MonitoringFuel Consumption Monitoring
eCVT: electric Continuous Variable Transmission
GDI: Gasoline Direct Injection
ICE: Internal Combustion Engine
ISC: In-Service Conformity

LIVC: Late Intake Valve Closing
mHEV: mild Hybrid Electric Vehicle
PHEV: Plug-In Hybrid Vehicle
PM: Particle Mass
PN: Particle Number
MPI: Multi-Point Injection
NEDC: New European Driving Cycle;
NMHC: Non-Methane Hydrocarbons
NO_x: Nitrogen Oxide
NOX: Nitrogen Oxides
N_2O: Dinitrogen Monoxide
RDE: Real Driving Emission Test
THC: Total Hydrocarbons
VSS: Vehicle Simulation Suite
WLTP: Worldwide Harmonized Light duty Test Procedure

References

1. Maiwald O, Schamel A, Wagner U (2016) 48 V P2 hybrid vehicle with an optimized combustion engine – fuel economy and costs at their best combined with enhanced driving behaviour. In: 37th international vienna motoren symposium
2. Schöppe D, Knorr T, Graf F, Klingeis B, Beer J, Gutzner P, Hager S, Schatz A (2014) Downsized gasoline engine and 48 V Eco Drive – an integrated approach to improve the overall propulsion system "efficiency".In: 35th international vienna motoren symposium
3. Korr T, Ellmer D, Baensch S, Schatz A (2018) Optimization of the 48 V hybrid technology to minimize local emissions in the RDE. In: 27th Aachen Colloquium automobile and engine technology. Aachen
4. Avolio G, Brück R, Grimm J, Maiwald O, Rösel G, Zhang H (2018) Super clean electrified diesel: towards real NOx emissions below 35 mg/km. In: 27th Aachen Colloquium automobile and engine technology. Aachen
5. Lauer S. Graf F, Springer M, Wechler S (2017) 48 volt hybrid with e-drive features - excellent fuel efficiency and drivability. In: Electric & electronic systems in hybrid and electrical vehicles and electrical energy management. Bamberg
6. Graf F, Lauer S, Baensch S, Knorr R, Dr Sans M (2017) 48 volt technology in the light of the connected vehicle and electrical board net advancements. In: 26th Aachen colloquium automobile and engine technology. Aachen

Efficient CFD Simulation Method for Calculation of Drag Torque in Wet Multi-plate Clutches in Comparison to Test Rig Results

Daniel Groetsch[1(✉)], Rudi Niedenthal[2], Katharina Voelkel[1], Hermann Pflaum[1], and Karsten Stahl[1]

[1] Gear Research Center, FZG, Technical University Munich, Boltzmannstraße 15, 85748 Garching, Germany
groetsch@fzg.mw.tum.de
[2] SIMERICS GmbH, Gartenstraße 82, 72108 Rottenburg, Germany
rudi.niedenthal@simerics.de

Abstract. Wet multi-plate clutches and brakes are important components of modern powershift transmissions and industrial drive trains. In the open state, drag losses occur due to fluid shear as load-independent losses. Determination of drag losses can be done by experiment or by CFD simulation. CFD simulation has many advantages over the experiment process, but often has the strong disadvantage of high computational times and in consequence, strongly simplified models. Therefore, the target is to develop an efficient and validated model to compute drag losses for any clutch design in a short computational time.

This work presents an innovative and validated CFD model to calculate drag losses of wet clutches. To ensure a convenient implementation of the model with high computational efficiency and quality at the same time, the commercial Software Simerics MP+ is used. Because of efficient modeling and solvers, the CFD model considers the geometry of a complete clearance between a steel and friction plate. This enables the model to calculate the drag losses for any possible groove design, which is a tremendous advantage in comparison to state of the art models, which often use circumferential symmetry. Furthermore, the model development is focused on a short calculation time: computational time is very short, which enables the overnight calculation of a whole drag-torque-differential speed curve. This allows variational calculations and at the same time, detailed investigations that are important both in early and advanced stages of development.

Extensive measurements on a component test rig using automobile series production parts allow a thorough validation of the model. The influences of changes in operating conditions, groove design, differential speed, oil injection temperature and volumetric flow rate are shown both in the measurements and in the simulation results.

Keywords: CFD simulation · Drag torque wet clutch · Experimental validation

Euroforum Deutschland GmbH (Ed.): *CTI SYMPOSIUM 2019*, pp. 164–176, 2021.
https://doi.org/10.1007/978-3-662-61515-7_15

1 Introduction and Motivation

Wet multi-plate clutches and brakes are important components of modern powershift transmissions and industrial drive trains. In the open state, drag losses occur due to fluid shear as load-independent losses. The trend towards greater efficiency in powertrains and the reduction of CO_2 emissions makes the reliable quantification of drag losses of clutches in an early stage of development essential [1].

A drag torque over the differential speed curve typically characterizes the drag loss behavior of wet multi-plate clutches. According to Oerlke this curve can be classified into different phases [2]. The basis of the definition of the phases in this publication are the physics of the oil flow. At low differential speed Phase I describes the linear increase of the drag torque. Phenomenologically this is because of the single-phase flow, in other words the oil flow capacity of the clutch is less or equal the delivered oil flow rate [3].

The following Phase II manifests itself through a decreasing of drag torque – often reaching a steady value. In Phase II a multi-phase flow consisting of an oil-air-mixture is present because the oil flow capacity of the clutch exceeds the feeding oil flow rate [3]. Figure 1 shows a sketch of a typical drag-torque over the differential speed curve in Phase I and II.

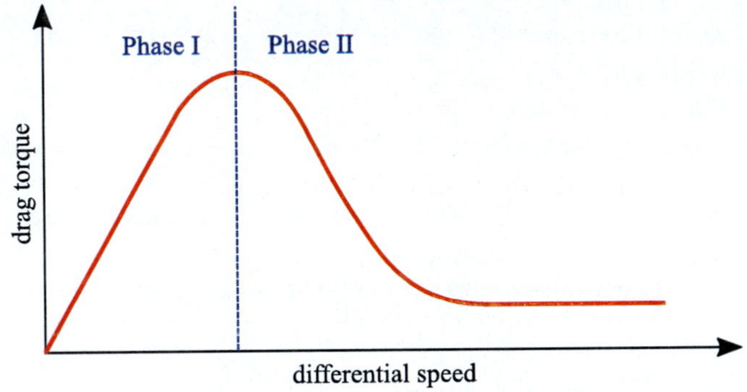

Fig. 1. Schematic sketch of drag torque over differential speed for open wet multi-plate clutch with phase boarders

Drag losses can be determined by experiment or by CFD simulation. The advantage of simulative investigations is that it is possible to do them in the early stages of development. There is no need for the availability of prototype parts [4]. This makes it possible to consider and optimize various designs. However, it is important to use an efficient and validated model to get reliable estimates in manageable time. Due to the need for considering a multi-phase flow at higher differential speeds, it is often challenging to develop simulation models, which give accurate drag torque results and at the same time, have practical relevant short calculation times [4, 5].

Literature shows modeling of the two-phase flow with the volume-of-fluid (vof) method, through mixed viscosity models or with a cavitation model [6–11].

The vof method is the most general method, because it explicitly solves the air and liquid phases. Besides the drag losses, it is also possible to specify oil distribution in the clutch. Due to the high computational costs and calculation times of this method in applications found in the literature, strong simplifications of the clutch geometry are often applied through assumption and application of circumferential symmetry [3, 12].

In this contribution, we develop a model based on the application of a cavitation approach, which combines high accuracy calculation of drag torque for any clutch for preliminary design and short calculation times. Because of efficient modeling and solvers, the CFD model considers the geometry of a complete clearance between the steel and friction plate [13]. This enables the model to calculate the drag losses for any possible groove design. We show validation of the simulation in comparison with characteristic values from test rig measurements [5].

2 Experimental Investigations

We conducted experimental investigations to determine the drag loss behavior of several wet multi-plate clutches. The data are used for validation of the CFD model presented in Chap. 3. In the experiments the following parameters have been varied:

- Mode of operation (braking mode/clutch mode)
- Feeding oil temperature (30/60/40/80 °C)
- Feeding specific oil flow rate (0.50/0.8/2.5 mm^3/mm^2s)
- Groove design (waffle/group parallel)
- Material of friction surface (paper/sinter)
- Clearance between friction plates (0.10/0.15 mm)

In braking mode only the friction plates are rotating, in clutch mode the ratio of rotational speed $n_{friction,plates}/n_{steel,plates} = 2$.

The parameters of the test procedure are summarized in Table 1.

Table 1. Summary of test procedure conditions

	Clearance/mm	T_{oil}/°C	$v_{oil}/mm^3/mm^2s$	Mode	Groove	Surface
1	0.15	40	0.80	Brake	wf	Paper
2					gp	
3				Clutch	wf	
4					gp	
5		80		Brake	wf	
6					gp	
7	0.10	30	0.50		wf	Sinter
8		60	2.50			

2.1 FZG Test Rig LK-4

We use the FZG test rig LK-4 for investigations of the drag losses of wet multi-plate clutches. Figure 2 shows a schematic sketch of the test rig setup. The test rig has two independent shafts. The outer shaft is engaged with the outer carrier, the inner shaft is engaged with the inner carrier. Electrical engines power both shafts through belt drives. The inner carrier is mounted on a torque-sensing element, which is applied to the inner shaft. The measurement of drag torque takes place directly at the inner carrier. The friction of sealing or bearings does not influence the drag torque signal.

The test rig can vary the operating conditions of the clutch in a wide range. It is possible to change the rotational speed of the inner and outer carrier independently in both rotational directions, oil injection temperature, oil flow rate and the clearance of the multi-plate clutch [1].

2.2 Parts and Lubricant

For the experimental investigations and simulations, parts from the serial production of automotive and industrial applications were used. The friction materials are a paper-based friction lining (representing automotive applications) and a sinter-based material (representing industrial applications). The paper-based plates have group parallel (gp) and waffle (wf) grooves, the sinter material has a waffle groove design. Figure 3 shows a schematic sketch of the different groove designs.

One clutch consists of five steel plates and four friction plates. This means that there are eight clearances between friction plates and steel plates. The nominal clearance between the friction surface and steel plate is 0.15 mm for paper-based friction materials and 0.10 mm for sinter.

The geometric details of the clutches under investigation are summarized in Table 2.

All experiments are conducted with serial production lubricant (ATF). The physical characteristics are extracted from the safety data sheet. The conversion of viscosity and density values to different temperature conditions is done according to DIN 51563 [14] and DIN 51757 [15]. All values are summarized in Table 3.

2.3 Evaluation Method

Figure 4 shows recorded data of an exemplary measurement in braking mode. From these continuous signals, the stationary drag torque is determined by extracting the values where the rotational speed is constant. The first 30 s of data in each stage are skipped to eliminate unsteady effects (e.g. acceleration). We then calculate the arithmetic mean of the data for each stage (constant rotational speed) in an interval of approximately 300 s to determine the steady values and compensate temperature effects on the drag torque signal. In the experimental design the rotational speed is increased in fixed steps and then kept constant for about 360 s. The control of the feeding oil temperature results in small variations in oil temperature, which affect the drag torque.

Figure 5 gives a graphical representation of the characteristic values used in this publication. The characteristic values are a good representation of the drag torque behavior of

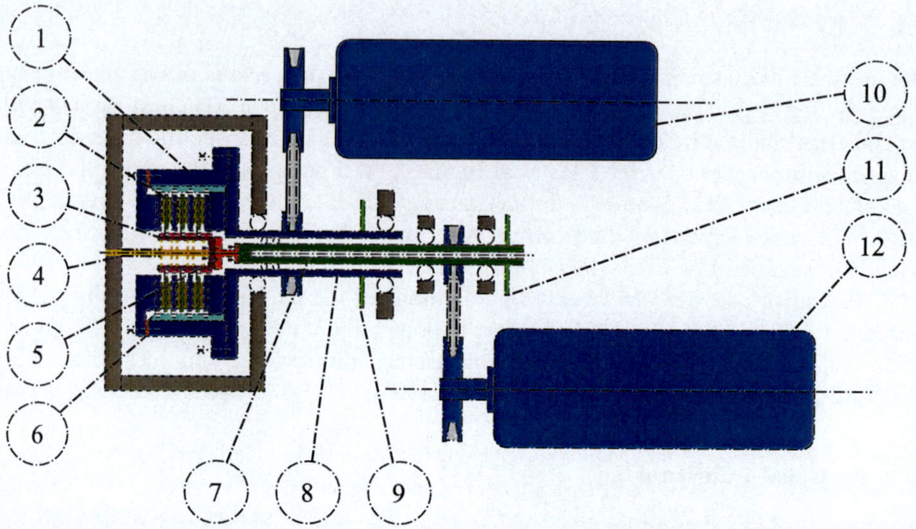

1) outer carrier
2) clutch plates
3) inner carrier
4) oil injection
5) viscous torque sensor
6) clutch clearance spacer
7) inner shaft
8) outer shaft
9) tachometer outer shaft
10) main drive outer shaft
11) tachometer inner shaft
12) main drive inner shaft

Fig. 2. Schematic sketch of wet multi-plate clutch test rig LK-4

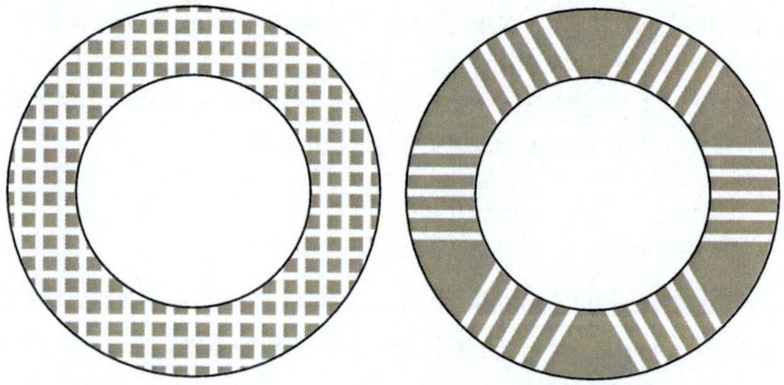

waffle (wf) group parallel (gp)

Fig. 3. Sketch of different groove designs

the multi-plate clutch (Phase I and II) and can be used, for example, as input parameters in test cycle simulations.

We calculate the linear slope $\Delta T_d/\Delta n$ of the drag torque with linear regression from the values at differential speeds in the interval 0–200 rpm in clutch mode and 0–400 rpm in brake mode. $T_{d,max}$ is defined as $\max(T_d)$ in the whole range of investigation, here

Table 2. Geometric details of multi-plate clutches

	Inner diameter	Outer diameter
Steel plates	163.8 mm	191.5 mm
Friction plates	163.1 mm	187.5 mm
Friction material	165.0 mm	187.5 mm
	Depth	Width
Waffle groove (paper)	0.15 mm	1.2 mm
Waffle groove (sinter)	0.20 mm	0.48 mm
Group parallel grooves	0.18 mm	1.0 mm
Clearance		
wf (paper)	gp (paper)	wf (sinter)
0.15 mm	0.16 mm	0,10 mm

Table 3. Physical characteristics of lubricant (ATF)

Oil temperature in °C	Density in kg/m^3	Kinematic viscosity in mm^2/s
15	818–825	–
30	808.6–815.6	26.6–27.1*
40	802–809*	18.5
60	789.5–796.5*	9.84–10.14*
80	775–782*	5.99–6.28*
100	–	4.00–4.25
*) calculated values		

0–1,000 (clutch) or 0–1,000 (brake) rpm. $T_{d,stat}$ is found at differential speed of 1,000 (clutch)/2,000 (brake) rpm.

To summarize, the following characteristic values are defined and evaluated:

1. In the linear range the slope of the drag torque over the differential speed curve $\Delta T_d/\Delta n$ is determined.
2. The value $T_{d,max}$ and Δn at $T_{d,max}$ are evaluated to determine the shift point from Phase I to Phase II.
3. The value $T_{d,stat}$ is evaluated when the drag torque reaches a nearly steady value at high rpm.

3 CFD-Simulation Model

This work presents an enhanced innovative and validated CFD model to calculate the drag losses of wet clutches. To ensure a convenient implementation of the model with

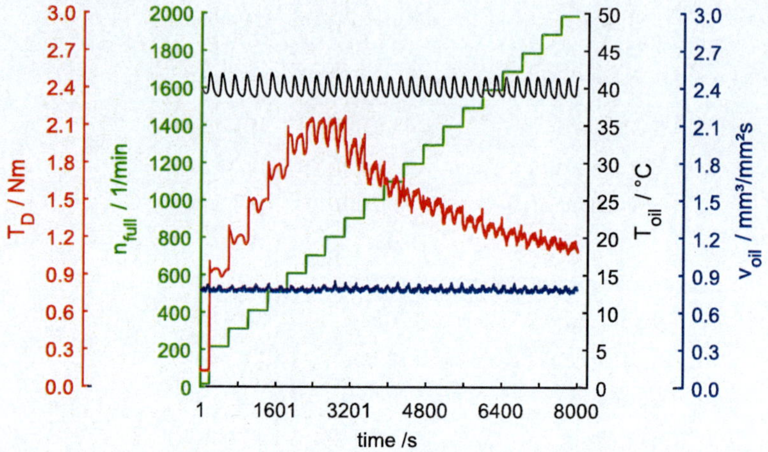

Fig. 4. Example of recorded data (drag torque T_D, rotational speed n_{full}, specific oil flow rate v_{oil}, oil injection temperature T_{oil}) over time from test rig in braking mode, gp groove (paper)

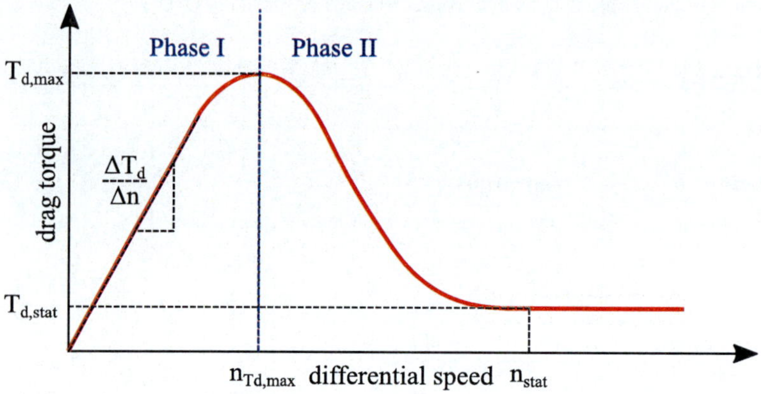

Fig. 5. Schematic sketch of drag torque over the differential speed and characteristic values

high computational efficiency and quality at the same time, the commercial Software Simerics MP+ is used.

3.1 Setup

The calculations consider one complete clearance of a wet clutch in the open state [5]. We take into account the oil supply through the grooves of the inner and outer carrier, the geometry of the steel plate and the friction plate including groove design. Due to high computational efficiency the model does not use circumferential symmetry. This makes it possible to consider any groove design and to easily adopt other designs in the calculations by only replacing the CAD-model of the grooves.

The calculation of drag torque is split in two CFD models. First, a transient single-phase model is used to calculate the oil carrying capacity of the clutch and the drag torque in Phase I. Following this, all operating points where the oil flow capacity of the clutch exceeds the feeding flow rate (Phase II) are recalculated with a transient cavitation model approach [13]. Figure 6 shows the fluid domain and the corresponding boundary conditions. For single-phase calculations, both inlets are set to ambient pressure and the flow is supposed to be laminar. For the cavitation model, the inlet is set to the constant mass flow rate and the outlets are still at ambient pressure. In Phase II the RNG-k-epsilon model is activated [16]. All rotating parts are set to rotating walls. The fluid domain is fenced by rotating walls on the one side and symmetry boundary conditions on the other side in axial direction.

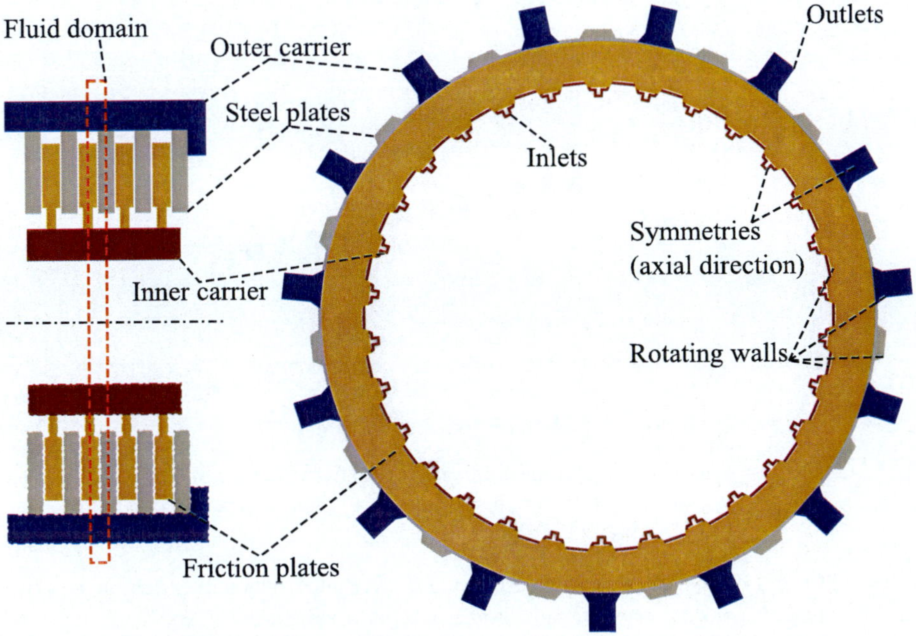

Fig. 6. Sketch of fluid domain including boundary conditions

For meshing the binary tree mesh technique of Simerics MP+ is used. Meshes with approximately 3.5 million cells are used for all calculations. Mesh and sensitivity studies have shown that this mesh quality is sufficient for drag torque calculations [5].

3.2 Results and Validation

For validation of the CFD model we compare the characteristic values (see Sect. 2.3) of the drag torque over the differential speed curve obtained from experiments with the calculations. By comparing the test rig measurements with CFD calculations we always have to keep in mind that both are models of the real clutch application and neither the value obtained from measurements nor calculation is the absolute true value. Keeping this in mind we first look if the trend of the data is qualitatively the same. Second, we consider the quantitative match.

Figure 7 exemplary depicts a whole drag torque over the differential speed curve obtained from the measurement at the test rig and from CFD calculations. The calculation time for this whole curve on a typical workstation is about 12 h, which compared to other approaches is briefly. We can see that the qualitative behavior between the graphs is good. The simulation shows linear behavior in Phase I, whereas the measurements show a nearly linear development. The maximum of the drag torque is the same quantitatively, however there is a small shift in rotational speed where the maximum value occurs. At higher rotational speeds the simulation and measurements are increasing in proximity.

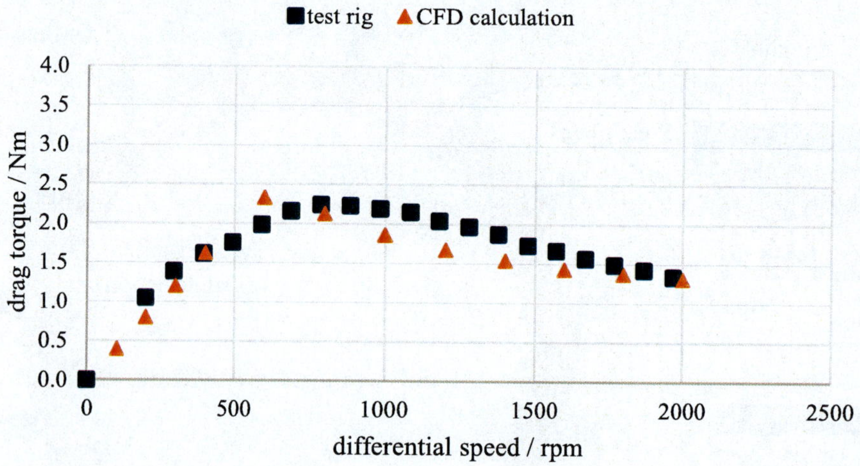

Fig. 7. Comparison of drag torque over the differential speed from test rig with CFD calculations, brake, $T_{oil} = 40\ °C$, $v_{oil} = 0.8\ mm^3/mm^2s$, $h = 0.15\ mm$, wf groove, paper friction lining,

In the further course we investigate the quantitative accordance between the test rig and CFD calculations by comparing the characteristic values $\Delta T_d/\Delta n$, $T_{d,max}$, Δn at $T_{d,max}$ and $T_{d,stat}$. Therefore, we show bar graphs of the characteristic values. In the graphs (Figs. 8–11) the indices of test procedures correspond to the lines in Table 1. For every test procedure, we contrast the measurement with the simulation for validation. Furthermore, we describe influences on the drag torque by comparing results from different operation modes e.g. 1 vs 3 is braking mode vs clutch mode for same groove design, friction material, oil flow rate and oil temperature.

Figure 8 shows the findings of the characteristic value $\Delta T_d/\Delta n$ for all operating conditions (see Table 1).

One can see that the linear slope of the drag torque curve corresponds well between CFD calculations and the test rig in all operating points.

The main influencing factors of the quantity of the linear slope of the drag torque seem to be temperature and clearance. The maximum slope is where the minimum oil feeding temperature and minimum clearance occurs (test procedure 7), the minimum is at the highest oil temperature and maximum clearance (test procedures 5, 6). The mode of operation (brake/clutch) has almost no influence on this characteristic value (test procedure 1, 2 vs. 3, 4).

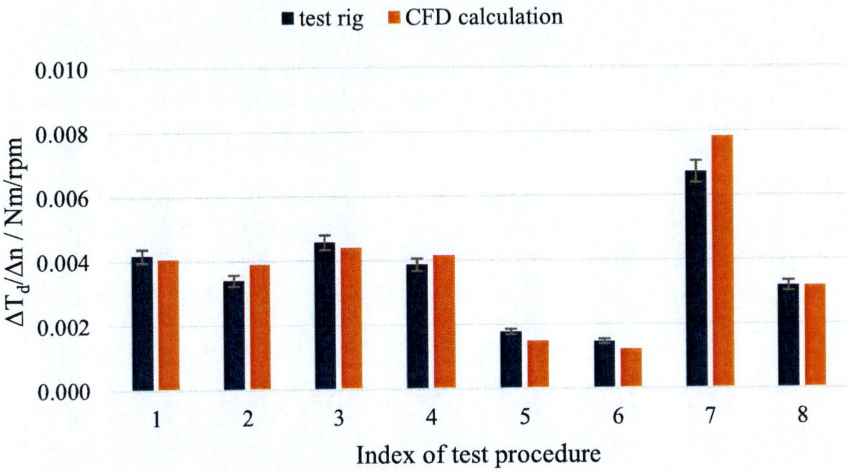

Fig. 8. Comparison of characteristic value $\Delta T_d/\Delta n$ from test rig measurements and CFD calculations for all operating points (see Table 1)

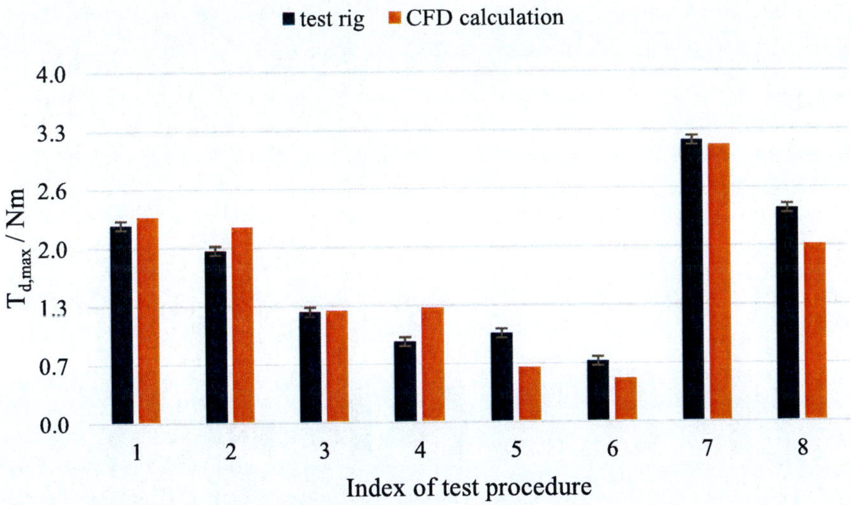

Fig. 9. Comparison of characteristic value $T_{d,max}$ from test rig measurements and CFD calculations for all operating points (see Table 1)

Figure 9 compares the characteristic value $T_{d,max}$ for all operating conditions (see Table 1). Again, there is a good correspondence between measurement and calculation. The influencing factors of $T_{d,max}$ are oil temperature, mode of operation, feeding flow rate and clearance. From theory it is known, that in Phase I there is a linear proportionality between drag torque and fluid viscosity (compare Table 3), in other words a higher oil feeding temperature reduces $T_{d,max}$ (test procedure 1, 2 vs. 5, 6). Clutch mode reduces $T_{d,max}$ in comparison to brake mode (test procedure 1, 2 vs 3, 4). This can be explained

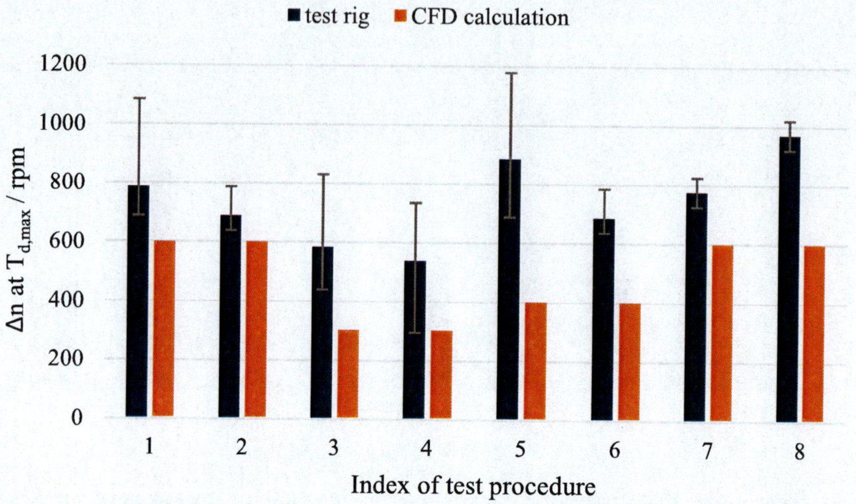

Fig. 10. Comparison of characteristic value Δn at $T_{d,max}$ from test rig measurements and CFD calculations for all operating points (see Table 1)

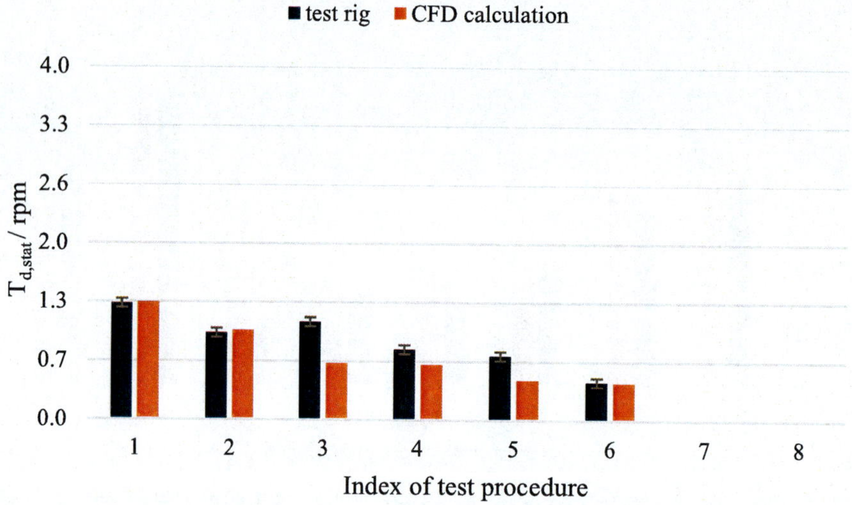

Fig. 11. Comparison of characteristic value $T_{d,stat}$ from test rig measurements and CFD calculations for all operating points (see Table 1)

through the higher oil carrying capacity of the clutch at the same differential speed between the plates. A higher feeding flow rate and smaller clearance shifts the maximum drag torque to higher differential speeds. This can be seen by comparing results from test procedure 8 vs 1 and 2.

The value Δn at $T_{d,max}$ quantifies the shifting point from Phase I to Phase II. As seen in Fig. 10 this point is not an easy fix in measurements. The errorbars indicate differential speeds Δn where the value of T_d is not more than 5% less then $T_{d,max}$. There is a qualitative match between $\Delta n(@T_{d,max})$ as one can discriminate between different designs. From a quantitative perspective, the CFD calculations seem to under predict the value of $\Delta n(@T_{d,max})$. A possible reason for this could be that the current model design neglects surface tension effects. Further studies will be done to improve the quantitative match.

Figure 11 shows the comparison of drag torque at high differential speed $T_{d,stat}$ ($\Delta n =$ 1,000/2,000 rpm) in Phase II. We can see that for braking mode at lower oil temperature (test procedures 1, 2) there is an excellent qualitative and quantitative match. For clutch operating mode (test procedures 3, 4) the stationary drag torque is under predicted. For the sinter based material there is no test data available.

4 Conclusion and Outlook

We show an evaluation method for drag torque measurements of wet clutches and define characteristic values for the drag torque over the differential speed curve. Furthermore a transient CFD calculation model, which takes into account the geometry of a full (360°) clearance between friction plates, steel plates and the grooves of inner and outer carrier for calculation of drag torque, was developed.

To ensure validity of the modeling approach we validated the model in various different operating conditions (compare Table 1).

Validation of the model by comparing a complete differential speed over the drag torque curve from measurement and CFD calculation displays an excellent match of results. The data indicate the influence of oil feeding temperature, oil flow rate, clearance, operation mode (brake, clutch), groove design and friction material on the characteristic values of the drag torque curve. The accordance of the linear slope of drag torque in Phase I, $T_{d,max}$ and $T_{d,stat}$ between the test rig and simulation model is very good for all operating conditions. The shifting point $\Delta n(@T_{d,max})$ from Phase I to Phase II is difficult to predict in measurement and calculation. Further studies have to be done to minimize the variation of the measurements, for example by using the design of the experiment methods.

To summarize, the CFD model is capable of predicting the drag torque behavior of wet multi-plate clutches in various operating conditions qualitatively and quantitatively with short calculation times (over night) and high accuracy. Enhancements should focus on improving the estimate of the shifting differential speed from Phase I to Phase II without losing the advantage of very short computation times.

References

1. Dräxl T, Pflaum H, Stahl K (2015) Neue Erkenntnisse über Schleppverluste an Lamellenkupplungen. VDI-Fachtagung "Kupplungen und Kupplungssysteme in Antrieben". VDI-Berichte Nr. 2245:149–161
2. Oerleke C (2000) Einflußgrößen auf die Schleppmomente schnellaufender Lamellenkupplungen in Automatgetrieben. Diss., Bundeswehr-Univ

3. Neupert T, Bartel D (2015) Schleppmomentuntersuchungen an nasslaufenden Kupplungslamellen mithilfe von Prüfstandsmessung und CFD-Simulation: analysis of the drag torque in a disengaged wet clutch disc by test rig measurements and CFD-simulations. VDI-Fachtagung "Kupplungen und Kupplungssysteme in Antrieben", VDI-Berichte Nr. 2245
4. Albers A, Ott S, Basiewicz M et al (2017) Variation von Nutbildern mittels generativer Verfahren zur Untersuchung von Schleppverlusten in Lamellenkupplungen. In: VDI-Fachtagung Kupplungen und Kupplungssysteme in Antrieben. VDI Verlag GmbH, Düsseldorf
5. Grötsch D, Niedenthal R, Völkel K et al (2019) Effiziente CFD-Simulationen zur Berechnung des Schleppmoments nasslaufender Lamellenkupplungen im Abgleich mit Prüfstandmessungen. Forschung im Ingenieurwesen (83):227–237. https://doi.org/10.1007/s10010-019-00302-3
6. Neupert T, Bartel D (2019) High-resolution 3D CFD multiphase simulation of the flow and the drag torque of wet clutch discs considering free surfaces. Tribol Int (129):283–296. https://doi.org/10.1016/j.triboint.2018.08.031
7. Yashwanth BL, Ngo D, Schroeder D et al (2018) Drag and cooling characteristics of circular pin-fin groove pattern of a multi-plate clutch pack using CFD. In: SAE international400 commonwealth drive, Warrendale, PA, United States
8. Behzad M, Saxena V, Schaefer M (2018) Thermal-hydrodynamic optimization of grooves in a wet clutch. In: Dritev - drivetrain for vehicles, EDrive, transmissions in mobile machines: international VDI Congress, June 27 and 28, Bonn, Nichtredigierter Manuskriptdruck. VDI Verlag GmbH, Düsseldorf, pp 101–117
9. Mahmud SF, Pahlovy SA, Kubota M et al (2016) Multi-Phase simulation for studying the effect of different groove profiles on the drag torque characteristics of transmission wet clutch. In: SAE 2016 world congress and exhibition. SAE international400 commonwealth drive, Warrendale, PA, United States
10. Pahlovy SA, Mahmud S, Ogawa M (2018) Development of new groove design for reduction of drag torque or spin loss of disengaged wet clutches in the high speed region. SAE Technical Paper Nr. 2018-01-1300. https://doi.org/10.4271/2018-01-1300
11. Rudloff M, Bartel D, Deters L (2011) Simulation der Strömung in nasslaufenden Lamellenkupplungen. In: Kupplungen und Kupplungssysteme in Antrieben 2011: VDI-Fachtagung mit Fachausstellung, Wiesloch 2011, Nichtred. Ms-Dr. VDI-Verl., Düsseldorf
12. Neupert T, Bartel D (2017) Einfluss des Nutdesigns von nasslaufenden Kupplungslamellen auf das Strömungsverhalten im Lüftspalt. In: VDI Fachtagung - Kupplungen und Kupplungssysteme in Antrieben 2017. VDI Verlag GmbH, Düsseldorf
13. Singhal AK, Athavale MM, Li H et al (2002) Mathematical basis and validation of the full cavitation model. J Fluids Eng 124(3):617. https://doi.org/10.1115/1.1486223
14. Prüfung von Mineralölen und verwandten Stoffen - Bestimmung des Viskosität-Temperatur-Verhaltens – Richtungskonstante m: Prüfung von Mineralölen und verwandten Stoffen DIN 51563(51563)
15. Prüfung von Mineralölen und verwandten Stoffen - Bestimmung der Dichte DIN 51757(51757)
16. Orszag SA, Yakhot V, Flannery WS, Boysan F, Choudhury D, Maruzewski J, Patel B (1993) Renormalization group modeling and turbulence simulations. In: International conference on near-wall turbulent, Tempe, Arizona

Significant Drag Torque Reduction and Improved Clutch Dynamics by Innovative, Very Compact Separating Springs for Wet Clutches

Hüseyin Gürbüz[1,3](✉), Jörgen Schulz[1], Ferit Kücükay[2], Fatim Scheikh Elard[2], and Sascha Ott[3]

[1] Mubea Tellerfedern GmbH, Daaden, Germany
hueseyin.guerbuez@mubea.com
[2] Institute of Automotive Engineering (IAE), TU Braunschweig, Brunswick, Germany
[3] Institute of Product Engineering (IPEK), Karlsruhe Institute of Technology (KIT), Karlsruhe, Germany

Abstract. Due to the CO_2 limits, which are becoming even harsher, losses in automatic transmissions must be drastically reduced. Furthermore, electrification of the drivetrain requires better controllability in order to improve interaction between combustion engine and electric motor. Mubea separating springs offer a good solution to achieve these goals. In order to minimize clutch losses, separating springs are accepted as an innovative component of current transmission programs in which discs in the clutch system are actively separated from each other. Extensive investigations have shown that drag torque losses can be reduced by 40% and more and that CO_2 savings of 0.5 g/km (in WLTP) can be achieved. First of all, this lecture presents the validation of earlier simulation results regarding a positive effect of separating springs on shifting comfort through measurement of real starting-up with a dual clutch transmission. Here shift comfort and NVH behaviour could be greatly improved by the use of separating springs. In addition to well-known wave-shaped separating springs, a new type of so-called "torsional separating spring" is presented for the first time, which offers the same technological advantages by improving the installation space requirement. Furthermore, excerpt from test results will be presented with the new torsional separating spring. In addition to already known advantages in reducing drag torque, it offer further important advantages with regard to clutch dynamics. Tests were carried out at the IPEK and IAE on a clutch test bench with near-series hardware. All results show a better controllability and a more stable control behaviour with separating springs, when engaging the clutch due to a better linearity of the torque/pressure curve. When clutch is disengaged, a higher dynamic is achieved by a faster separation of the discs.

Keywords: Wet multidisc clutch · Drag torque reduction · Improved clutch dynamics

Euroforum Deutschland GmbH (Ed.): *CTI SYMPOSIUM 2019*, pp. 177–190, 2021.
https://doi.org/10.1007/978-3-662-61515-7_16

1 Introduction

1.1 Separating Spring in Wet Multi-disc Clutches

Wet multi-disc clutches and brakes are used in many areas of drive technology to synchronize differential speeds of shafts, to decelerate masses or to prevent torque overloads. They are used in automatic transmissions, CVT's and dual clutch transmissions due to their load shifting capability, high power density and good control and comfort characteristics. In most cases they operate according to the "normally open" principle, whereby a hydraulic pressure is built up to engage clutch and a spring takes over the disengagement. These springs can be realized as coil spring packages or as space-saving disc springs. In addition, cushion springs are used as damping elements in many clutches and are placed in front of the clutch package to dampen the impact of the piston when clutch is engaged. This has a positive effect on shifting quality. Cushion springs can be designed as wave springs or disc springs. With increasing optimization of transmission losses, attention is focused on drag torques generated by differential speeds. In connection with the CO_2 topic, reduction of these drag torques is being worked on in addition to reduction of the space requirement and resulting weight reduction.

For some years now, separating springs have further been used to minimize or eliminate the drag torques occurring between the individual discs when the clutch is open. In new designs, separating springs also have the potential to take over piston return function, which can lead to a further reduction in installation space and weight, The idea of using spring elements in disc pack to evenly adjust the clearance dates back to the 1970s and 1980s. Although the idea is not new, this option has been used very little in recent years. Only a new industrialization concept on Mubea's side made it possible to produce separating springs at an acceptable market price (Fig. 1)

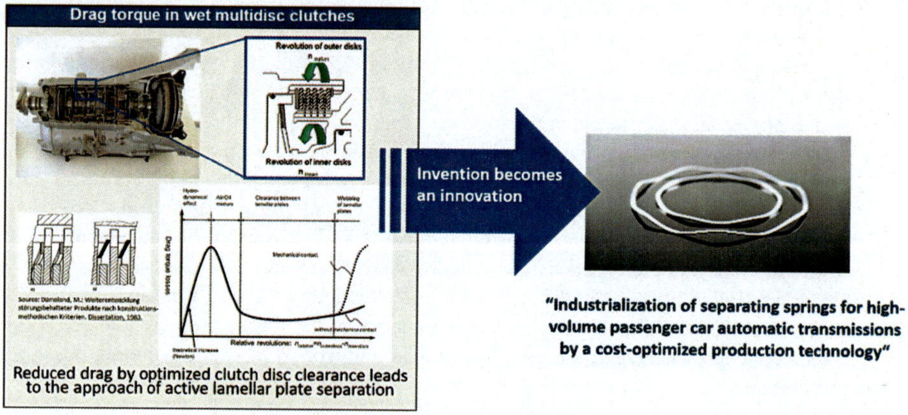

Fig. 1. Innovative Mubea separating springs

On the one hand, the Mubea separating springs are designed to withstand such high force and tension that the function of the piston return spring is performed at the same time and this can therefore be eliminated. On the other hand, the production costs of springs are noticeably reduced through the use of winded and then welded wire. This has led to a breakthrough of the invention separating spring.

1.2 Estimation of CO_2 Reduction by Separating Springs in the WLTP

In a collaborative project with the Institute of Automotive Engineering (IAE) and Mubea Tellerfedern GmbH, a calculation tool was developed to estimate influence of separating springs on fuel consumption and CO_2 emissions in the WLTP via variable drag torque curves. The calculation tool is based on a backward calculation of the required energy at wheels towards the energy to be provided in the form of fuel (Fig. 2).

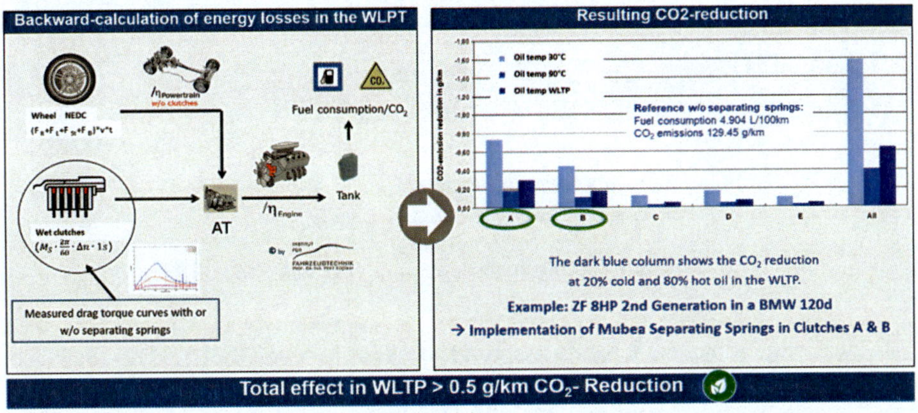

Fig. 2. Backward calculation of energy losses in the vehicle with and without separating springs in the WLTP

First, the energy required at wheels is determined from the driving resistances to be overcome. This energy is then used to calculate the energy required at the transmission input by dividing it by the efficiency of the drive train. The model offers the possibility to vary the drag torque curves (with and without separating springs) and thus to consider the influence of them at this point. Furthermore, the energy to be supplied from the fuel tank is calculated by specifying an engine efficiency. By means of the known reactions of the combustion of petrol or diesel and the resulting amounts of carbon dioxide, CO_2 emissions can be calculated from the consumption of the vehicle and differences due to variable drag torque losses can be identified.

Figure 2 shows calculated CO_2 reduction when using separating springs in a BMW 120d with an 8-speed automatic transmission (8HP 2nd Generation) from ZF. By integrating separating springs in clutches A and B, a potential of 0.5 g/km CO_2 reductions is achieved.

1.3 Separating Spring Applications

In effort to reduce fuel consumption in order to achieve the CO_2 emission targets, Mubea separating springs has been accepted as an innovative component by many transmission manufacturers to minimize energy losses of automatic transmissions. They are used in large-scale transmission programs such as the ZF 8HP and Ford 10R80. Through their own publications, our customers have confirmed the effectiveness of our springs in reducing drag losses and thus CO_2 emissions (Fig. 3).

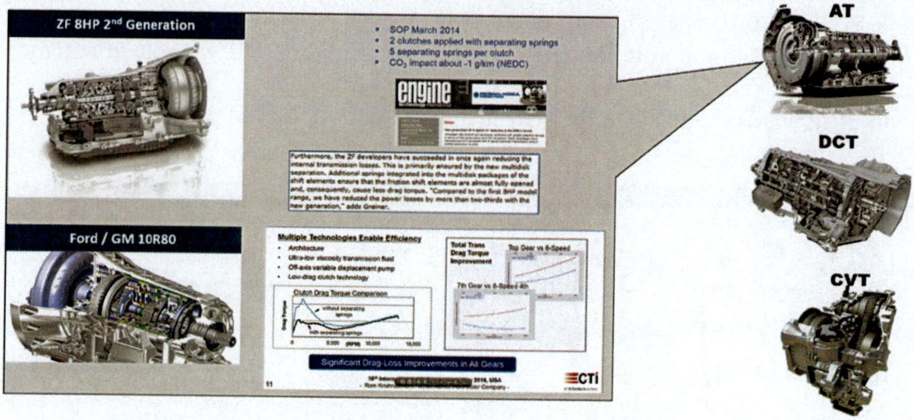

Fig. 3. Separating spring application in ZF 8HP and Ford 10R80

In dual clutch market, separating springs are used in VW DL382. Here they also take over the piston return function and enable a more compact design. The reduction in drag torques was also confirmed by VW (Fig. 4).

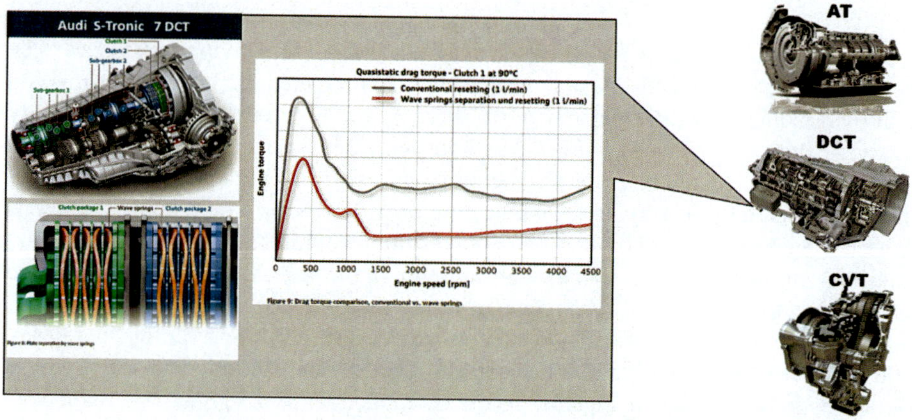

Fig. 4. Application of separating springs with piston return function in VW DL382

A more recent application is the use of separating springs in CVT field. The revers brake system in the Hyundai gamma CVT, which is open most of the time, has been designed with separating springs to reduce drag losses (Fig. 5).

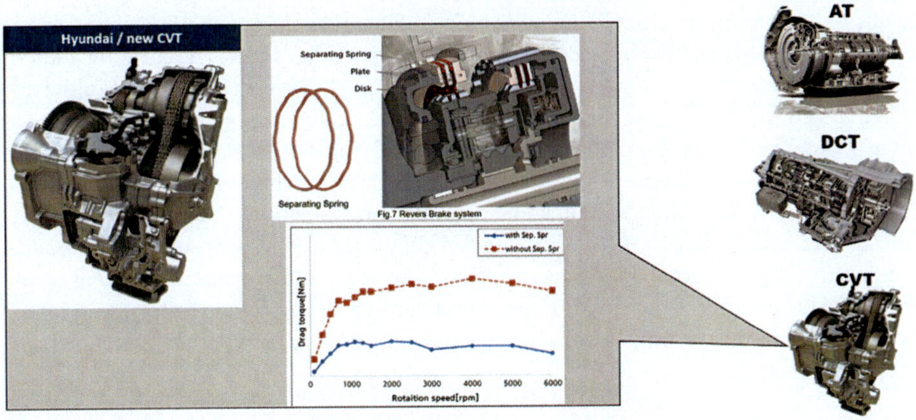

Fig. 5. Separating springs in revers brake system of Hyundai gamma CVT

2 Positive Effect of Separating Springs on Shift Comfort and NVH Behavior

The controllability of wet multi-disc clutches is influenced by a two-phase characteristic of the multi-disc package. The piston first closes clutch clearance against little resistance and then presses the disc pack together to ensure torque capacity. Since clearance is 1 to 3 mm, while the stroke for the clutch compression is only 0.1 to 0.3 mm long, piston is pressed together against little resistance to ensure torque capacity. Main oil volume is required for piston movement in the clearance. However, piston area must be designed according to required contact force in order to generate the required torque. This means that closing the clutch clearance is a relatively slow process. Since transmission control cannot distinguish between these two processes, errors or fluctuations in pressure can lead to torque shocks [1].

In 2017, a simulation model was developed in cooperation with IAE and results published in order to assess the impulses when the piston hits the disc pack and the influence of disc springs on comfort behaviour.

The model essentially represents equations of motion of individual masses and, according to the impact theory, also takes into account spontaneous changes in velocity upon impact of the masses. The model described was used to simulate 6 different spring arrangement scenarios. A clutch is simulated in scenario 1 which contains only a piston return spring. In scenario 2, the clutch is extended by a cushion spring.

In Scenario 1 and 2, 1000 variants with different stochastic initial positions of discs were simulated. Starting from scenario 3, initial positions of clutch discs are evenly distributed, since in scenarios 3 to 6 separating springs are used.

In scenario 3, a separating spring is used in addition to the return spring.

The clutch in scenario 4 is extended by a cushion spring, so that in this scenario all three spring types are installed. Piston return in scenarios 5 and 6 is no longer performed by the classic return spring, but by separating springs, which for this reason are designed to be much more rigid, so that same piston return load results as in the other scenarios. In scenario 5 separating springs are the only spring elements in the clutch, while in scenario 6 also a cushion spring is simulated. In the simulation model, shock effects in the clutch are calculated and can be displayed in a travel-over-time diagram. Individual impulses are calculated for all impacts between the various masses (especially pistons and plates) and described in an impulse-over-time diagram.

Result of the simulation is that an even distribution of the clutch clearance leads to significantly reduced torque shocks. Separating springs actively ensure such an even positioning and therefore improve shift comfort (Fig. 6 left). If drag torque is not an issue, combination of piston return spring and cushion spring results in optimum shifting comfort. When drag torque reduction is required, separating springs with return spring function allow optimum shifting comfort, drag torque reduction, installation space savings and system costs [2].

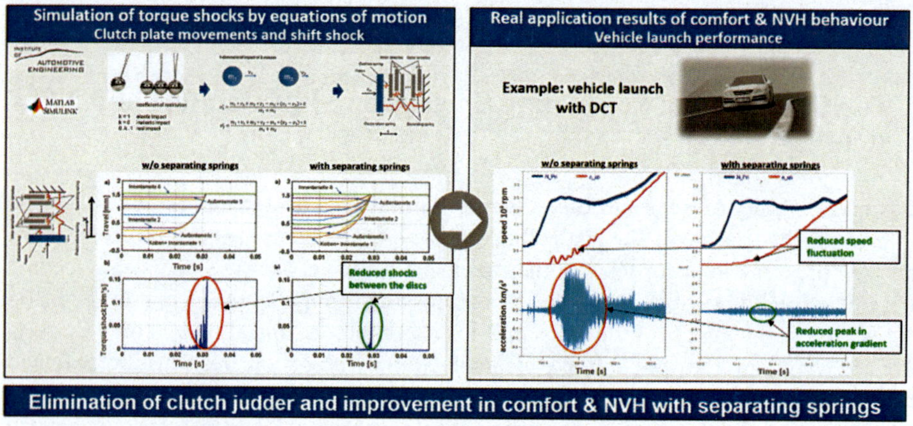

Fig. 6. Positive effect of separating springs on shift comfort and NVH behaviour Simulation results from the year 2017 (left) new measurements of real start-up processes with a dual-clutch transmission (right)

On right side of Fig. 6, vehicle test results of a start-up process with a dual-clutch transmission are shown. In order to determine the influence of separating springs on shifting comfort and NVH, tests were carried out with coil spring pack as piston return spring on the one hand and with separating springs with piston return function on the other hand. The diagrams show input and output speeds as well as the measurement of vibration accelerations at transmission output. The test setup with coil spring pack has speed fluctuations and vibrations that can occur during the slip phase due to clutch judder and other vibration phenomena. The measurements with separating springs show in comparison that speed fluctuations and oscillations in the clutch system could be significantly reduced. In addition to preventing clutch judder, it was also possible to optimize the NVH behavior. In the configuration with Coil spring pack, a high-frequency starting noise could be noticed, which was no longer subjectively detectable when separating springs were used.

3 Types of Separating Spring

Separating springs can be designed either as wave springs or as disc springs. In applications shown, they were previously used in the wave design, because the cost structure could be greatly reduced by waved and then welded raw material. Here, weld seam can be placed in a load-uncritical area (transition from wave top to wave bottom). Disc springs do not give this possibility. In addition, this design offers advantages for oil distribution in clutch system, as its wave form prevents oil flowing out to the outside less than disc springs. Mubea has been supplying wave springs for many years, initially in the stamped version. Since 2014, separating springs for the large ZF 8HP and VW DL382 transmission programs have been supplied in the winded and then welded variant. In these cases, they were used intensively to reduce drag losses and to optimize the axial installation space when taking over the piston return function. For the integration of separating spring, however, a radial installation space is required. This leads to the requirement that needed space must be taken into account at an early stage of development when calculating the friction surface and the torque capacity. In ongoing series applications, where installation space is already defined, integration is more difficult. This led to further development at Mubea in order to also develop an separating spring for space-critical applications and running series. Figure 7 shows for the first time the new Mubea torsional separating spring, which was developed especially for this purpose.

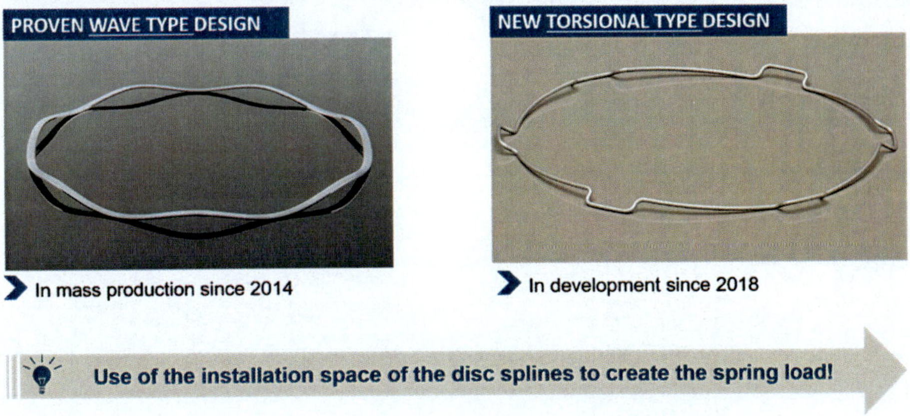

Fig. 7. New type Mubea torsional separating spring

The core idea here is to use the spline spaces in outer or inner disc carrier to generate spring load. As a result, installation space required in friction system could be significantly reduced.

This, however, changes the operating principle of the separating spring.

With wave spring, bending stress is generated by compressing top and bottom sides of the wave according to principle of the bending beam clamped on both sides (Fig. 8 top).

In the case of the torsional separating spring, twisting of angled cantilevers leads to torsional stress on the wire segment in between (Fig. 8 bottom).

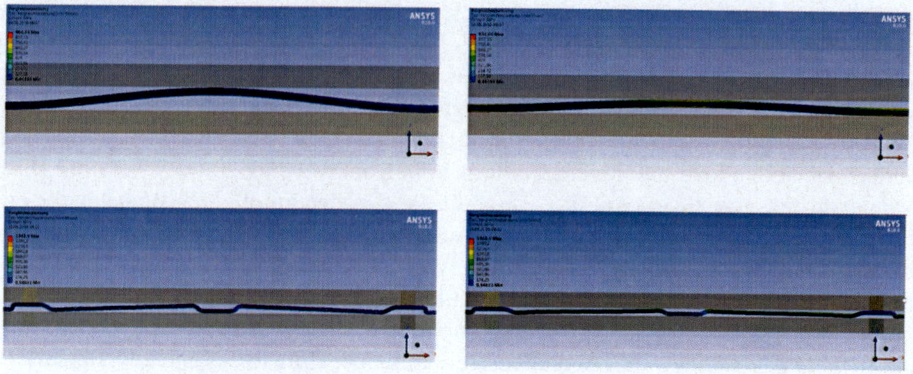

Fig. 8. Working principle wave type and torsional type separating spring

A comparison of the two operating principles also shows that the torsional stress has a higher material use.

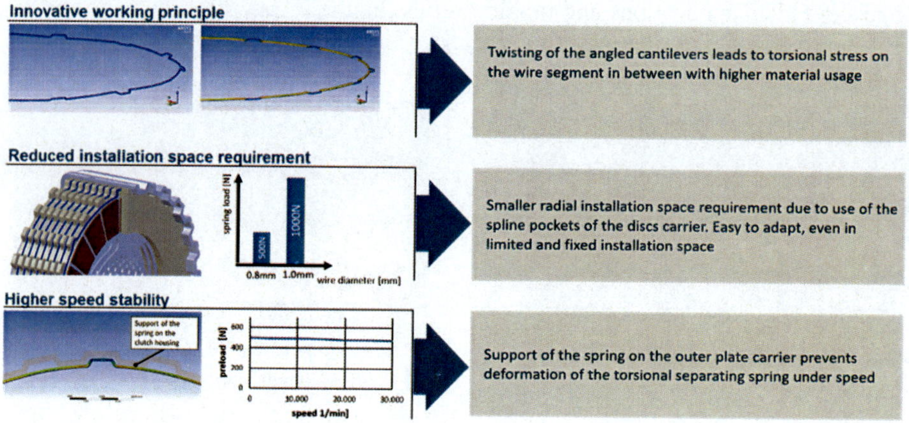

Fig. 9. Improvement of the separating spring with regard to installation space and speed stability

This also results in a reduced installation space requirement. With a wire diameter of 0.8 mm to 1.00 mm, either pure disc separating or the integration of piston return function with up to 1000 N can be carried out. Furthermore, torsional separating spring is better used at higher rotational speeds. Waved type springs have the disadvantage that they lose their wave height and thus their spring load due to the centrifugal force at high speeds. Although loss of load can be greatly reduced by a simulation-accompanying design, it cannot be completely prevented. With torsional separating spring, on the other hand, spring material is supported by the disc carrier when centrifugal force is applied and does not lose the spring load (Fig. 9). FEM calculations have shown that force loss at 30,000 rpm is less than 5% [3].

4 Real Tests by Use of the Innovative Mubea Separating Springs

4.1 Validation Environment and Test Setup

During validation of the new Mubea torsional separating spring, aim was to take customer requirements and general conditions into account at an early stage of product development. Among other activities, 2 cooperation projects were started with potential customers and institutes in the field of drive system technology. During cooperation with IPEK a brake of an 8-speed automatic transmission was investigated. In cooperation with IAE, a brake was also investigated which is, however, installed in a CVT transmission. In both cases, the reference system had a coil spring pack for piston return function and was compared with the new Mubea torsional separating springs with piston return function during the tests. In order to change the residual system as little as possible, original components (transaxle case, discs, etc.) were used. The test parameters are shown in Fig. 10.

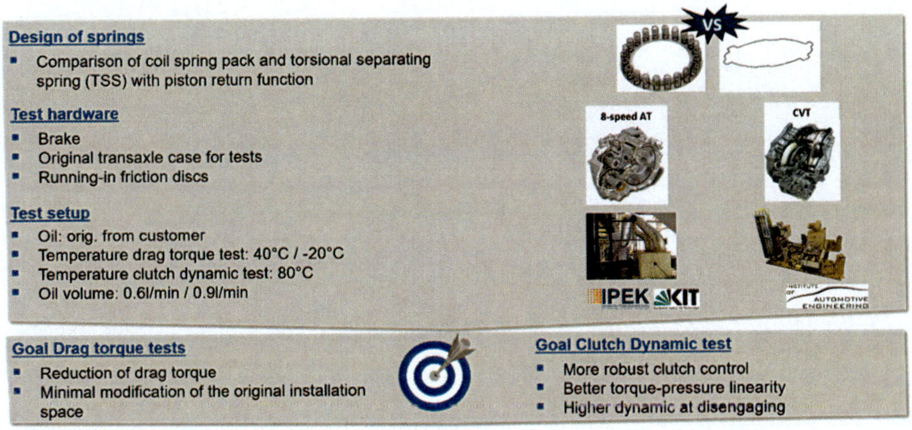

Fig. 10. Validation of the new torsional separating spring and test setup drag torque/clutch dynamic tests

In both cooperation projects, drag torque and dynamic investigations were carried out, whereby the drag torque investigations at the IPEK also included tests at −20 °C. The aim of the drag torque investigations was to reduce the losses with minimal modification of the original installation space. Here the running-change potential of the torsional separating spring should be demonstrated. In the dynamic tests, the main goal was to achieve a more robust clutch control. Furthermore, a linear behaviour of the torque-pressure curve should be achieved when engaging the clutch and a faster dynamic should be achieved when disengaging.

4.2 Test Results of Drag Torque Investigations

As presented in introduction, reduction of drag losses due to separating springs is known and confirmed by many applications. Results of the IPEK tests are presented here as an example for new drag torque investigations, as drag torque tests at minus temperatures were carried out here for the first time. The temperature selected was −20 °C, as the oil used is still pumpable at this temperature. Furthermore, necessary modification of installation space was to be determined in these tests in order not to negatively influence drag torque characteristic by reducing flow cross-sections and to estimate the running-change potential.

In these tests, outer steel discs were separated. Results have shown that radial distance between friction disc and separating spring must be adjusted in order to reduce drag losses over the entire speed range. The diameter of the friction discs was reduced without reducing friction surface. Only diameter of carrier plate was reduced. A modification of friction disc diameter by 0.8 mm was necessary. The results of drag torques show that losses at 40 °C could be reduced by up to 40%. In tests at -20 °C, drag losses could also be reduced over the entire speed range between 25% and 40% (see Fig. 11).

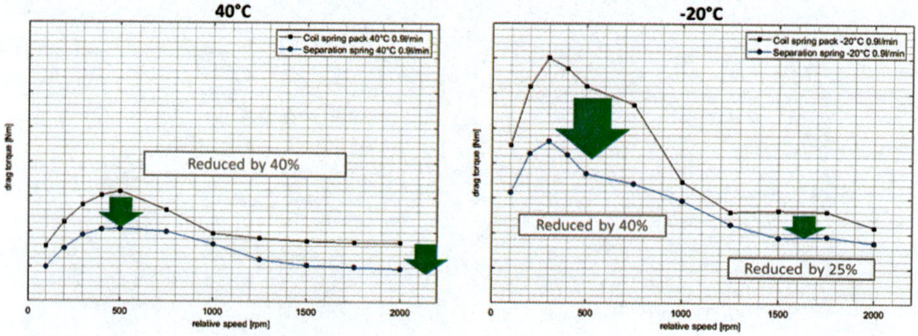

Fig. 11. Comparison of drag torque results at 40 °C and −20 °C with coil spring pack and new torsional separating springs by Mubea

4.3 Influence of Separating Springs on Clutch Dynamics

A requirement for good calibration and quick and easy adjustment of shift quality, especially in automatic transmissions, is a reproducible opening and closing of clutches. The shifting behavior is significantly influenced by the characteristics of the components involved in the shifting process. In most cases, engaging is performed hydraulically and so-called piston return springs are used for disengaging. These are located outside the friction system and only indirectly influence engaging and disengaging behaviour via the spring characteristic curve. Separating springs, on the other hand, are integrated in the friction system and can be used to actively influence clutch system control (Fig. 12). Numerous tests have shown that shifting characteristic can be significantly effected by springs located in the friction system and changing package elasticity [4]. Separating springs therefore offer potential to positively effect lubrication and alignment of discs in the clutch pack. For this reason, influence of separating springs on shift behavior and especially on clutch dynamics in comparison to conventional piston return springs will be discussed.

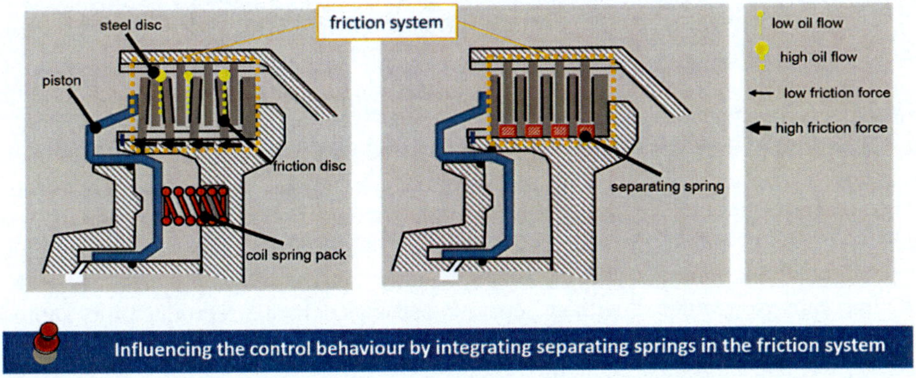

Fig. 12. Comparison of positive influence possibilities with transmission springs on system behaviour (left) Conventional piston return spring (right) separating spring with piston return function

The engaging and disengaging of a wet multi-disc clutch can be divided into different phases in which viscous friction and static friction can occur individually and also together as mixed friction (Fig. 13). In phases with viscous friction and mixed friction, a more linear behaviour and a higher reproducibility of torque/pressure characteristic, especially in the area of the Kisspoint, is expected with separating springs due to uniform distribution of clutch clearance and parallel closing of all friction pairs (Fig. 13 left). During disengaging, parallel opening of all friction pairs is expected to result in faster release of discs and thus higher clutch dynamics. Furthermore, a higher reproducibility of slip limit and torque-time characteristic is predicted due to the equilibrium of forces in the disc pack (Fig. 13 right).

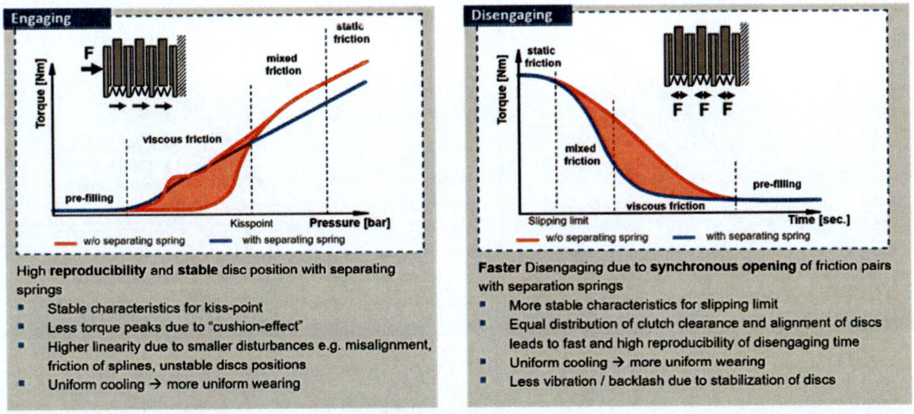

Fig. 13. Engaging and disengaging phases as well as comparison of characteristic with and without separating spring in wet multidisc clutches

In the following, the phases of the engaging and disengaging process are described in more detail.

Starting from open state of clutch with an existing drag torque, a pre-filling pressure must first be reached for closing, at which point piston begins to move. This corresponds to the preload of the piston return spring and depends on its spring characteristic curve. Pressure is then increased to move piston and overcome clutch clearance. The torque transmission in this phase takes place through viscous friction and corresponds to a drag torque with decreasing clearance. Therefore, transmitted torque is also strongly dependent on distribution of the clearance, volume flow as well as the viscosity of the oil. In comparison with and without separating springs, a higher reproducibility should be achieved here by means of separating springs, since an even distribution of clutch clearance can also be achieved here.

In the second phase of closing process, Kisspoint, also known as Touchpoint, is reached by increasing the actuating pressure. This is where the transition to contact between clutch discs begins, whereby viscous friction and static friction initially dominate simultaneously. This phase is called mixed friction. Also in this case a higher reproducibility and a more even contact is expected by use of separating springs. If pressure is further increased, oil is squeezed out of the individual air gaps and friction lining and static friction is present. In this phase, torque and pressure depend linearly on one another. Here no difference is expected.

Figure 14 shows results achieved by engaging the brake in an 8-speed automatic transmission with coil spring pack and with separating springs. In comparison, it can be seen that torque/pressure characteristics of the configuration with coil spring pack and separating spring differ greatly.

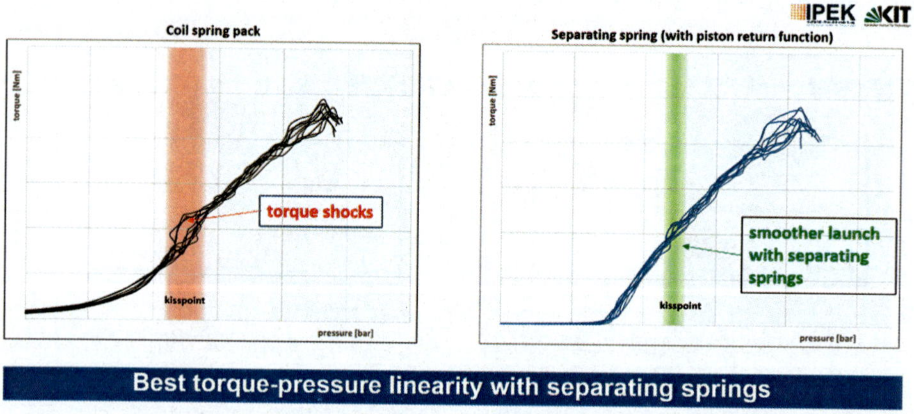

Fig. 14. Test results (Torque-pressure-curve) of engaging in comparison with coil spring pack and separating springs with piston return function

At low pressures, the configuration with coil spring pack has a higher drag torque and torque does not depend linearly on the pressure. With separating springs, by contrast, there is a linear dependency between torque and pressure and a high reproducibility can be clearly seen. In the area of the Kisspoint, results with Coil spring pack also show that torque shocks are present. With separating springs, on the opposite, torque shocks are not visible and a softer start-up is possible.

When clutch is disengaged, static friction suddenly reduces by reducing the pressure until slip limit is reached and clutch slips. Figure 15 shows results of disengaging the brake of a CVT transmission.

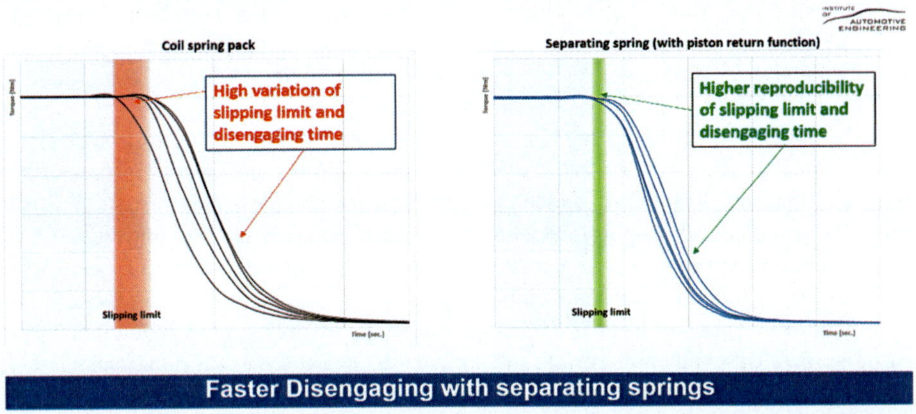

Fig. 15. Test results (Torque-time-curve) of disengaging in comparison with coil spring pack and separating springs with piston return function

Piston is returned in comparison with coil spring packs and separating springs with piston return function. It can be seen that opening behaviour of the configuration with coil spring pack varies greatly. There are very fast opening processes as well as delayed ones. The explanatory model for this includes that piston is pushed back by coil spring pack and discs closest to piston release first. All other discs are released sequentially, resulting in a high degree of torque variation. In the case of the separating springs, all friction pairs are opened in parallel by equilibrium of forces. In addition to higher reproducibility, this also results in higher dynamics at disengaging.

In summary, the test results on the dynamic behaviour of brake system show that higher reproducibility and dynamics can be achieved by integrating separating springs into the friction system. The positive effect of separating springs will be more noticeable in clutch systems with a higher oil volume (e.g. starting clutch in a DCT) and number of discs, since viscous friction has a greater influence on torque transmitting.

5 Summary and Conclusion

In the context of the necessary reduction of fuel consumption and therefore CO_2 emissions, focus is on energy losses of automatic transmissions. In order to minimize these losses, so-called separating springs are an important component in current transmission programs, in which disc discs are actively separated from each other in clutch system. This can significantly reduce drag torque when the clutch is open. These separating springs can also be designed with such high forces that they simultaneously take over the piston return.

Extensive investigations have shown that drag torque losses can be reduced by 40% and more and CO2 savings of 0.5 g/km (in WLTP) can be achieved - despite a conservative calculation approach.

Furthermore, validation of simulation results regarding a positive influence of separating springs on shifting comfort was presented by real start-up processes with a dual clutch transmission. Here, a clutch judder and a NVH phenomena could be eliminated.

In addition to well-known wave-shaped separating springs, a new type of so-called "torsional separating spring" was presented, which offers the same technological advantages by reduced installation space requirement.

If the use of separating springs is taken into account in the early pre-development phase, the installation spaces for waved separating springs are available or can be designed. However, installation space and potential for change are extremely limited, especially in the case of ongoing transmission programs. This is why innovative "torsional separating springs" have been developed that can avoid major changes to the clutch design by retaining the same technological advantages as waved springs.

In addition to already known advantages in terms of drag torque reductions, separating springs offer other important advantages in regard to clutch dynamics. These advantages could be fundamentally demonstrated at IPEK and IAE by tests on a clutch test bench with near-series hardware. The results show better controllability and more stable control behaviour with separating springs when engaging due to better linearity of the torque/pressure curve in the lower torque area. When the clutch is disengaged, higher clutch dynamics are achieved by faster separation of discs.

References

1. Hemphill J, George P (2014). https://www.schaeffler.com/remotemedien/media/_shared_media/08_media_library/01_publications/schaeffler_2/symposia_1/downloads_11/Schaeffler_Kolloquium_2014_19_de.pdf
2. Schulz J, Seekamp C-P (2017) Optimization of wet clutches in automatic transmissions by use of different types of springs. In: 16th international CTI symposium automotive transmissions, HEV and EV drives. Berlin
3. Gürbüz H, Ott S (2019) Albers, A.: Potentials for improving the system behaviour of highspeed clutches for BEV using optimal transmission springs. In: 19th international VDI congress dritev. Bonn
4. Wimmer Th (2004) Zu- und Abschaltverhalten von nasslaufenden Lamellenkupplungen, FVA, Forschungsvorhaben Nr. 335 Heft 726

48 V AWD Demonstrator with P0 + P4 Close to Wheel Concept

Matthias Werra[1], Matthias Ristau[1], Arno Ringleb[1], Sven Oliver Hartmann[2](✉), Julian Kumle[2], and Daniele Rosato[2]

[1] Institut für Fahrzeugtechnik, Technische Universität Braunschweig, 38106 Brunswick, Germany
[2] SEG Automotive Germany GmbH, 70499 Stuttgart, Germany
contact@seg-automotive.com

Abstract. This paper presents a 48 V Hybrid demonstrator with an electric rear axle which is setup by SEG Automotive Germany GmbH in cooperation with the Institute of Automotive Engineering of the Technical University Braunschweig.

First, the concept of the 48 V Hybrid demonstrator is presented. It consists of a P0 electric machine which replaces the alternator of the conventional vehicle. The conventional front wheel driven vehicle is extended to an all wheel driven vehicle. Two identical new 48 V electric machines - developed by SEG Automotive Germany GmbH - are integrated into the rear axle. Therefore, a rear axle with a new gear box is developed. For this reason, the dimensioning of the gear box ratio is analysed. Afterwards, the design and construction of the gear box and all necessary changes in the vehicle to realise the electric rear axle are shown. Finally, the concept of the operating strategy is presented.

Keywords: 48 V · HEV · E-axle · Electric powertrain · P0 + P4 topology · eAWD

1 48 V HEV Concept

1.1 CO_2 Reduction by Electrification

This chapter summarizes the key messages of the SEG Whitepaper Transformation of the Powertrain – A balanced view of technologies [1]. E-mobility is undergoing an exponential growth – sales in 2018 rose by 74% year-over-year to 1,261,000 electric vehicles (EVs). Nevertheless, the market share of new registrations in 2018 only equaled 1.5% [2]. So far, electric cars only play a subordinate role in the global vehicle population of 1 billion cars with only around 3 million units. In total, over 86 million passenger cars (PCs) were sold in 2018, 0.5% less than in the previous year [3]. At the same time, there has been a sharp increase in heavy and energy-hungry Sport Utility Vehicles (SUVs).

In addition, purely from a CO_2 perspective, currently Electric Vehicles (EVs) are not by default a better choice compared to other powertrain technologies. In many regions, India for example, even in 2040 the energy mix will still be dominated by fossil fuels to

Euroforum Deutschland GmbH (Ed.): *CTI SYMPOSIUM 2019*, pp. 191–202, 2021.
https://doi.org/10.1007/978-3-662-61515-7_17

such an extent that EVs will offer no significant advantages over efficient combustion engines. Conversely, in regions with a good energy mix – most notably Norway and France – EVs can already save more than 50% of CO_2 emissions over the lifetime of the car [4].

There are a number of obstacles to be removed before true global prevalence can be achieved. Most importantly, further technological breakthroughs and price reductions in battery technology are essential – as well as the elimination of possible bottlenecks in production and scarcity of key raw materials such as cobalt and lithium. Presently, higher costs, limited range and doubts regarding the charging infrastructure still deter buyers from purchasing an EV.

Looking ahead, EV market share will continue to grow – but even in 2025 it is expected to reach no more than 7 to 14%. According to current forecasts, at least 85% of passenger cars will still be produced with an internal combustion engine in 2025. CO_2 emissions for these vehicles must be kept as low as possible. In addition to established technologies such as start/stop, the 48 V hybrid technology appears to be a promising solution available in the short term.

Unlike a PHEV, no high-voltage safety architecture or charging devices are required for this mild hybridization. As a result, a large part of the performance spectrum of a PHEV can be reached at a much lower cost. In markets with a poor energy mix the 48 V hybridization even leads to a better CO_2 balance than a plug-in.

1.2 48 V Suitable for MHEVs (P0) and HEVs (P > 0)

Figure 1 shows basic configurations of 48 V hybrid systems. SEG's Boost Recuperation Machine (BRM) in P0 configuration was first presented in 2013 and went into series production for certain cars in 2017. Currently, the technology is expanding, especially in the upper mid-class, and will subsequently penetrate other vehicle segments. By 2025, it is forecast to reach 20% global market share.

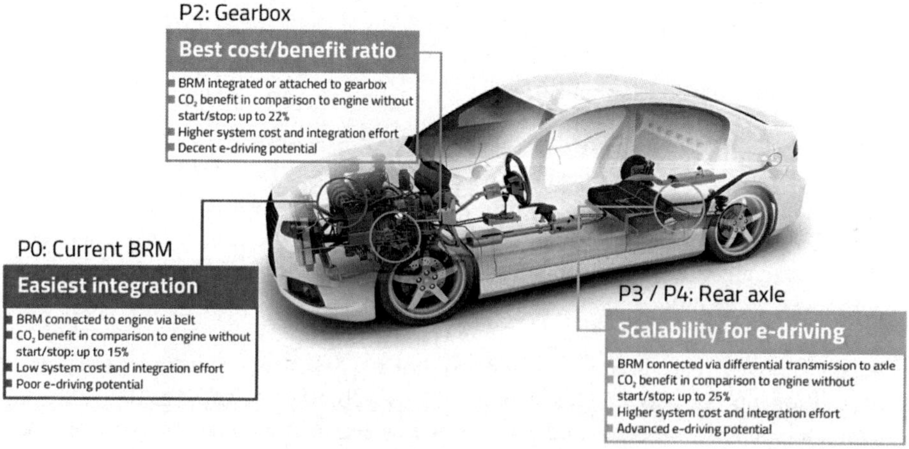

Fig. 1. 48 V powertrain topologies [1]

For the 48 V technology in P0 configuration, the conventional alternator is replaced by a (BRM). It employs the braking force recovery technology ("recuperation") to charge a 48 V battery. In real world driving, up to 15% fuel and thus CO2 emissions are saved and additional functions such as "Boosting" and "Coasting" can be enabled (see info box - BRM performance spectrum). A major advantage of 48 V is a low implementation effort. The BRM replaces the alternator in the belt drive ("P0 topology") – it can therefore be integrated into the powertrain architecture of most common vehicle platforms without significant modifications.

In the meantime, 48 V has also emerged in advertising as an essential technical feature, for example in the new Mercedes E-Class. 48 V powertrain solutions are being promoted as an integral part of the vehicle to the customer with features such as Comfort Start and Boost, as shown in Fig. 2.

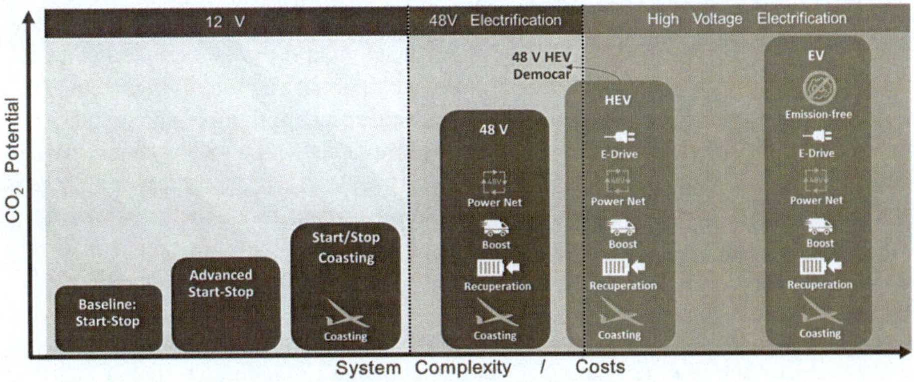

Fig. 2. Comparison of CO2 potential, system complexity/costs and applications of 12 V, 48 V and high voltage applications.

At the same time, development is already underway on future generations of 48 V powertrain systems in order to deliver even higher performance and further CO_2 savings. Those considerations are looking at the placement of the E-machine in the vicinity of the transmission (P2) or on the axle (P3/P4) instead of in the belt drive (P0).

In the medium term it is also conceivable to combine one BRM in the belt drive with one or more 48 V E-machines on the independent axle. In this scenario, the P0 positioning ensures maximum comfort during start/stop operations, while the P4 positioning provides additional recuperation benefits and sufficient power for purely electric driving at lower speeds. Overall, this results in a more cost-effective alternative to PHEVs, without the need for a high-voltage protection architecture in the vehicle.

In this paper, a P0 + P4 concept of a 48 V hybrid system, combining a 48 V BRM with a rear axle, electrically powered by two 48 V E-machines is presented. The system design and the fitting into a car with a conventional drivetrain are explained.

1.3 Targets for 48 V Demonstrator Vehicle

Since theoretical approaches and simulation on the system level are necessary for making concept decisions a validation of those methods are necessary if a concept should become a series product for the market. Therefore, a 48 V HEV demonstrator vehicle is build up in order to fulfill following targets.

First, numbers for CO_2 emissions have a huge impact on design decision on the powertrain. Therefore a total vehicle simulation in combination with a dynamic optimization approach is used for a nested system level design method, as described in [5, 6]. One main target for the demonstrator vehicle is to analyze the difference between simulation and a real vehicle on the road. As an example the application of a brake blending strategy can be implemented in a simulation but the comfort of this strategy have to be experienced and fine adjusted in the vehicle. After analyzing the difference, the goal is to reduce this difference by adapting the simulation.

Second, one characteristic of a HEV is, as described in Sect. 1.2, purely electric driving functionality. Simulation, based on real driving data, shows that purely electric acceleration towards a velocity of 20–35 km/h can be reached by a 48 V system. This would enable a 48 V HEV to drive electric on parking spaces and in a residential district. Nevertheless, the torque reserve for such driving situations with a 48 V system is rather small. The target for the demonstrator vehicle is to investigate if the driver does feel the limited torque reserve. An additional electrical driving function for further CO_2 reduction will be to keep the velocity at a certain speed constant. This functionality is known as e-coasting or sailing and speeds between 70 and 120 km/h are predicted for 48 V Systems.

Furthermore, the system has to be able to start the internal combustion engine for further acceleration. The next target for the demonstrator vehicle is to prove that this functionality can be realized by 48 V system as a combination of a P0 and P4 machine.

Next functionality to validate is a torque vectoring functionality. In [7] a method is developed to improve vehicle dynamics by a 48 V rear axle. The demonstrator vehicle will be used to implement the described method and to validate the improvements in a real environment.

Besides validating the described functionality, the process of building and setting up the demonstrator vehicle gives the opportunity to experience the difficulties for an OEM to integrate such a system into their conventional drivetrain.

Measurements on the driving demonstrator vehicle will give an insight into the NVH and reliability requirements of a customer project.

2 System Level Design

Derived from the described targets of the demonstrator vehicle in Sect. 1.3, a vehicle configuration as shown in Fig. 3. An electric rear axle (1) is installed to recuperate brake energy, support the combustion engine by boosting and enable purely electric driving functionalities. The electric axle is equipped by two wheel-individual electric machines to enable torque vectoring functionality. Those machines are mounted to one gearbox housing in which two separate transmission connect the machines separate to the wheels.

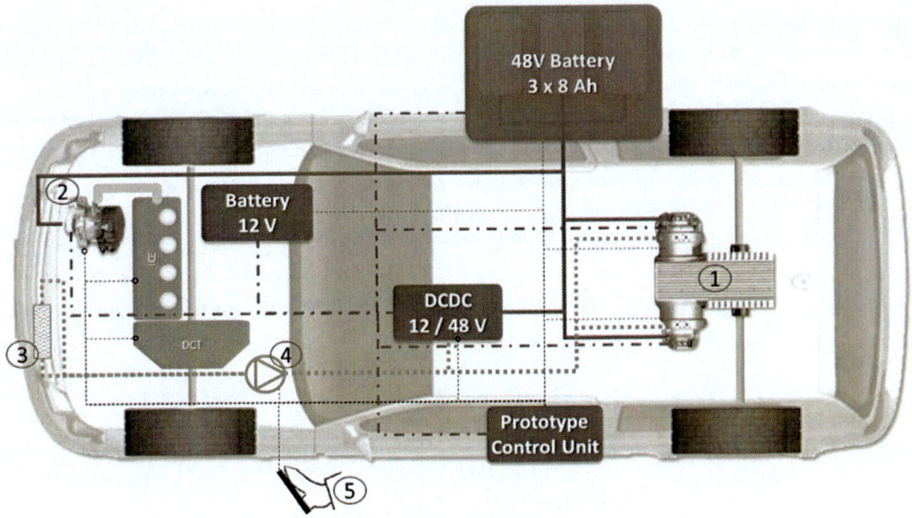

Fig. 3. Vehicle architecture of the 48 V demonstrator vehicle

The conventional alternator at the belt will be replaced by a 48 V BRM (2) for a comfort start and a load point shift of the engine. In addition, the BRM can support a maximum acceleration of the vehicle with its boost function.

All three machines are connected towards a 48 V battery. A 12/48 V DC/DC converter balances the state of charge of the 12 V and 48 V power supply net. The conventional 12 V power supply net is extended such that it supplies the control of the components.

The two electric machines on the rear axle and the DC/DC converter are equipped with a fluid cooling system which is installed separate from the existing cooling systems in the conventional vehicle. Therefore, a heat exchanger (3) is installed in the motor compartment. A water pump (4) controls the fluid flow through the cooling circuit. The BRM is air cooled.

The hybrid control strategy is running on an installed prototype control unit. This control unit is the interface between conventional powertrain, installed 48 V powertrain and sensors. According to the driver demand (5), state of the battery and temperatures in the powertrain the control unit defines the driving mode and distribute the torques to ICE, electric axle and BRM.

2.1 Component Sizing and Their Base Functionality

In order to find the machine size with the most market potential the simulation described in Sect. 1.3 is used in the WLTP-cycle. The optimization parameter of the powertrain configuration are the machine size on the P4 position (varying from 10 to 35 kW), the battery size (16, 24 and 32 Ah) and the transmission ratio (varying from 4:1 to 14:1). The P0 is always equipped with an available 48 V BRM out of the series production from SEG.

In Fig. 4a.) the result of the simulation is shown as an example of a D-class vehicle with a weight of 1590 kg. In order to gather all kinetic energy, the CO_2 benefit increase as the recuperation power of the machine increase. Nevertheless, the CO_2 benefit approaches a constant value as we reach 25 kW. This value depends on the application. Since a market potential for this power class is identified, validated by customer requirements, SEG decided to start a series development for a 25 kW electric machine. Due to the availability of 18 kW prototype machines two of them are mounted on the rear axle for the demonstrator vehicle. The energy storage has to be able to gather all the recuperation power. Therefore, it needs a specific power and energy capacity. The battery simulation is based on an 8 Ah cell and the scaling is implemented by connect-ing cells in parallel. The figure shows, that there is a difference between the CO_2 benefit for 16 and 24 Ah and a minor difference between 24 and 32 Ah. For the demonstrator vehicle a 24 Ah battery is configured by connecting three 8 Ah batteries in parallel.

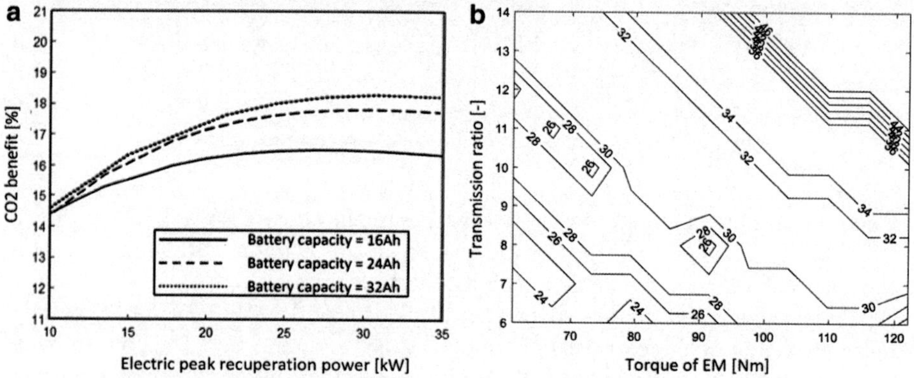

Fig. 4. a.) CO_2 Potential of a 48 V HEV in P4 topology. b.) Acceleration towards a specific velocity (z-axis) as a function of the available torque (x-axis) and the transmission ratio (y-axis).

The transmission ratio from the electric machines towards the wheel defines the available torque on the wheels for electric acceleration. In Fig. 4b.) the acceleration towards a specific velocity is shown as a function of the machine torque and the transmission ratio. In order to experience 48 V electric driving functionality a transmission ratio of 12 is chosen. An available torque greater than 1400 Nm on the axis is calculated to enable a sufficient acceleration towards 34 km/h.

2.2 Transmission Design

The transmission design is, besides the powertrain architecture and the transmission ratio, defined by the available installation space. Therefore, a precise installation space analysis is necessary at the beginning of the design process.

For selecting the basic vehicle of the demonstrator vehicle, various vehicle models were analyzed on the suitability of a drive unit integration at the rear axle. Choosing a vehicle platform including a Front-Wheel-Drive (FWD) and an All-Wheel-Drive (AWD) variant offers the opportunity to install an electric rear axle at the FWD-vehicle since the installation space and brackets are foreseen for the AWD-vehicle. Figure 5 illustrates the idea of the vehicle conversion. Further measures on the front-wheel drive basic vehicle are necessary to integrate the 4WD rear axle including subframe this is described in Sect. 3.1.

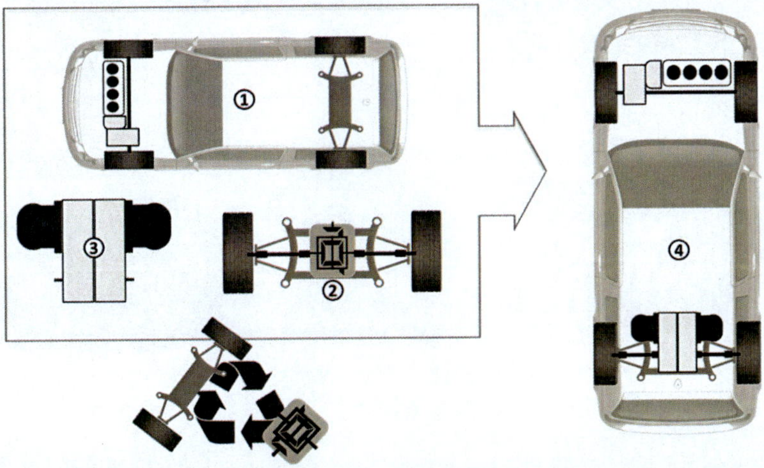

Fig. 5. Converting a FWD production vehicle to the P4 demonstrator: (1) FWD basic vehicle, (2) 4WD rear axle, (3) new drive unit, (4) P4 demonstrator

The installation space analysis is realized with a 3D measuring system on the basic vehicle and on the same 4WD model. In the process essential subassemblies were removed (exhaust system, tank, rear axle components) to fully reproduce the package. The tactile and optical measurement data are then transferred to a 3D CAD model. This CAD model is used for a deep investigation of the available installation space. Figure 6 shows an overview of the design process.

Various layouts for the arrangement of EM and gearboxes were investigated. Available transmission on the market for electric powertrains were taken into account and analyzed whether they are suitable for an integration in the demonstrator. Due to the very individual installation space a customized transmission design is necessary. After comparing different transmission layouts, a mirror-inverted configuration of EMs and gearboxes is selected, in which two independent two-stage helical gear units in a common housing provide the necessary gear ratio. The two EM are flanged to the housing in parallel to the rear axle (Fig. 6). Providing a simple transmission structure on the one hand and avoiding a collision with other components on the other hand, a big distance between the EM-shafts (input shaft) and the output shafts is chosen. This leads to other challenges within the package, which will be discussed in more detail in the next chapter. However, the choice of the drive unit layout and the design of the gearbox ensure that the concept can be implemented and integrated into the demonstrator without any changes to the vehicle structure.

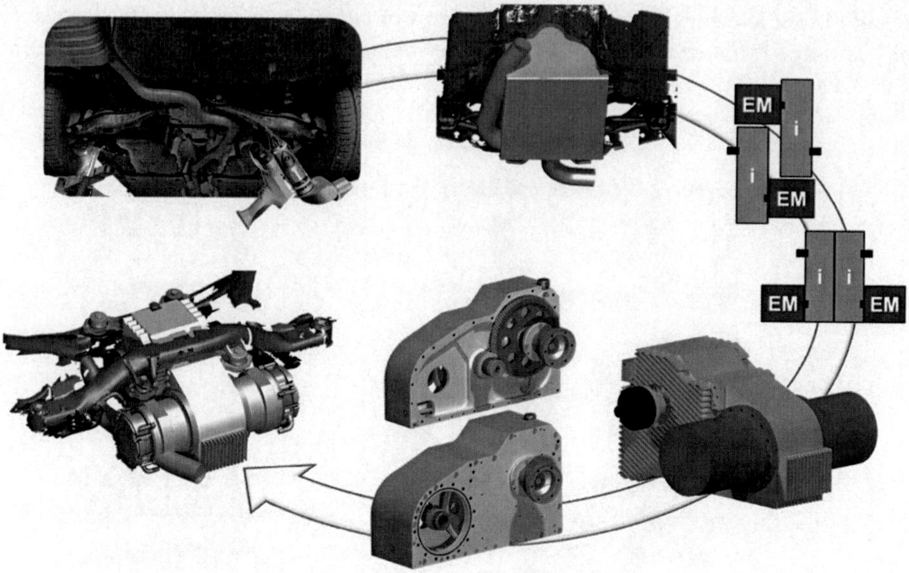

Fig. 6. Overview of the design process: Installation space analysis and identification, different layouts and transmission design

3 Mechanical Integration

3.1 P4-Integration

The basic design of the drive unit with two wheel-specific EM and two independent gear units in a common housing is described in the previous chapter. Despite the use of a different rear axle subframe and the accurate package analysis, further adaptations to vehicle assemblies are necessary in order to integrate the electric electric axle into the demonstrator. Figure 6 shows the CAD model of the drive unit mounted in the subframe at the end of the development process. The recognizable collision with the exhaust pipe is described in the next chapter.

Figure 7a shows the CAD model of the electric axle installed in the virtual installation space model of the vehicle underbody. The mirror-inverted, axially parallel configuration of the EM ensures that the width of the drive unit in the rear area (between the wheels) is quite narrow, so a free positioning of the output shaft flanges is possible. These flanges are designed in such a way that the series side shafts of the 4WD model can be used as a plug-and-play solution for the demonstrator vehicle.

The disadvantage of this layout is shown in the front area of the drive unit. A collision of the electric machines and the petrol tank is detected in the CAD model which leads to a redesign of the petrol tank. This petrol tank is based on the 4WD tank system and reduces the total volume for storing fuel. The adapted petrol tank can be seen in Fig. 7b on the right side.

The drive unit is connected to the 4WD subframe, decoupled via standard rubber-metal components, to the original positions of the final drive mounts. Furthermore, an

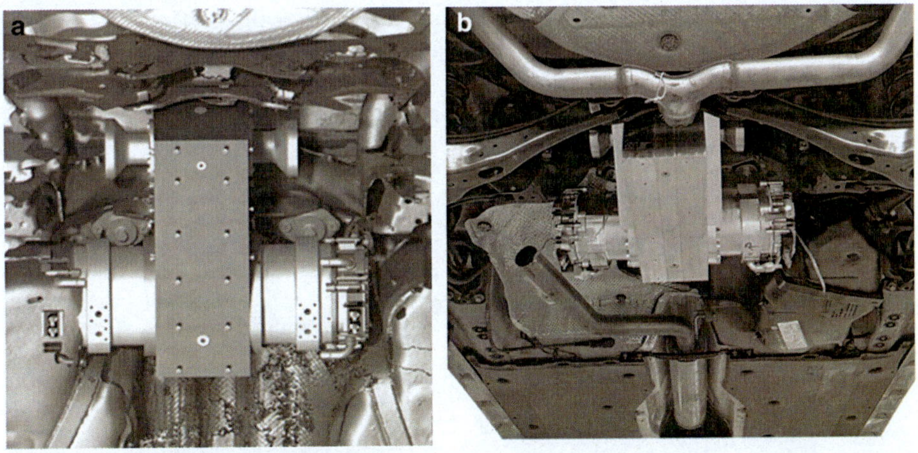

Fig. 7. a.) CAD model of the drive unit in the underbody surrounding area. b.) Drive unit in real P4 demonstrator (unfinished installation)

additional mount has been provided, which requires a mechanical modification to the subframe.

Finally, the routing of the exhaust pipe on the underbody has to be modified. For this purpose, a pipe segment is shaped individually according to the new conditions and installed in the demonstrator vehicle while re-using central and rear silencers. Due to the downsizing of the tank, the exhaust pipe could be routed around the drive unit, shown in (Fig. 7b). Figure 8 illustrates the main physical changes to the vehicle integrating the new drive unit to the rear axle.

In conclusion, the influence of the integration of the new drive unit on the vehicle structure is minimized. Additional, the goal of avoiding any changes to the car body is fullfilled. The largest mechanical intervention was made on the petrol tank system: the volume is approximately divided by two. However, the vehicle range by fuel is not a main requirement for driving the P4 demonstrator on the test track for proofing the functionality.

3.2 P0-Integration

This chapter explains the integration of the P0 Boost-Recuperation-Machine (BRM) from series production. As mentioned in Sect. 2.1, the conventional alternator is replaced by a BRM which is integrated in the belt drive in order to recuperate energy, shift the load point of the combustion engine and support the combustion engine by boosting. At first, the packaging in the demonstrator is analyzed. Therefore, the front end of the demonstrator was disassembled and the conventional alternator was removed. In order to integrate the series BRM necessary modifications are identified by using a 3D-print of the BRM. This investigation shows a collision of the 3D-print of the BRM and the cooling pipes of the air conditioning system. As a result, the layout of the cooling pipe is modified, shown in Fig. 9.

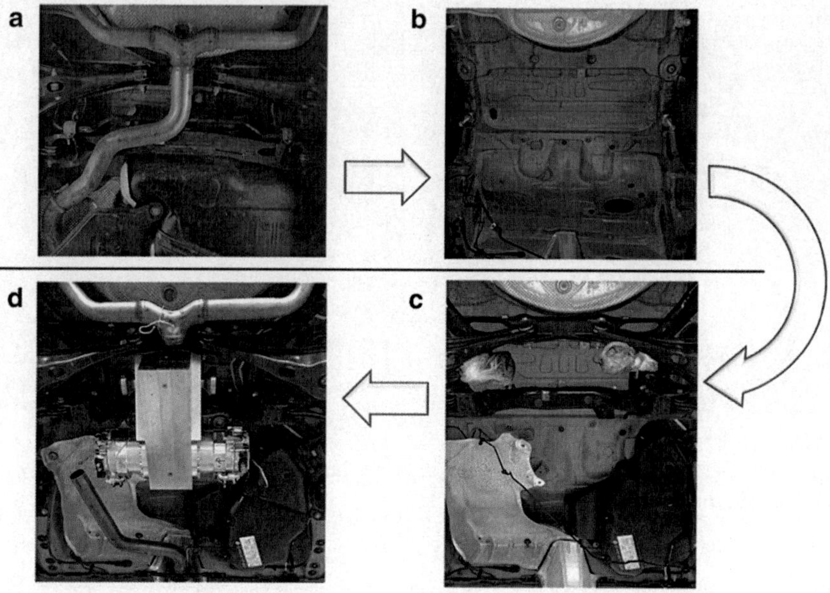

Fig. 8. Integration of the new drive unit to the demonstrator step by step

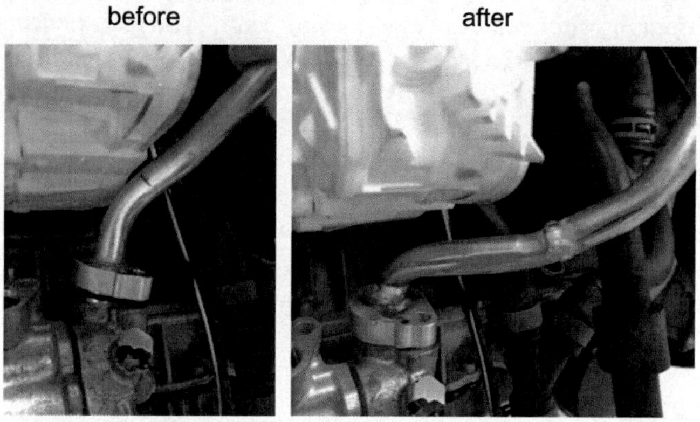

Fig. 9 On the left side, the collision of the cooling pipe and the 3D printed BRM is shown. Therefore, the layout is modified, shown on the right side such that the pipe is routed around the 3D printed BRM and the functionality of the air conditioning system is not affected.

4 Functional Integration – Base Concept Operating Strategy

This chapter presents the base concept of the operating strategy with its main functionalities, illustrated in Fig. 10.

Basically, the operating strategy can be divided into four main parts: The ICE-solo+Hybrid mode (1), the EV mode (2), the braking part (3 - divided into HEV braking and EV braking) and the ICE start/stop transition (4) between the ICE-solo+HEV mode

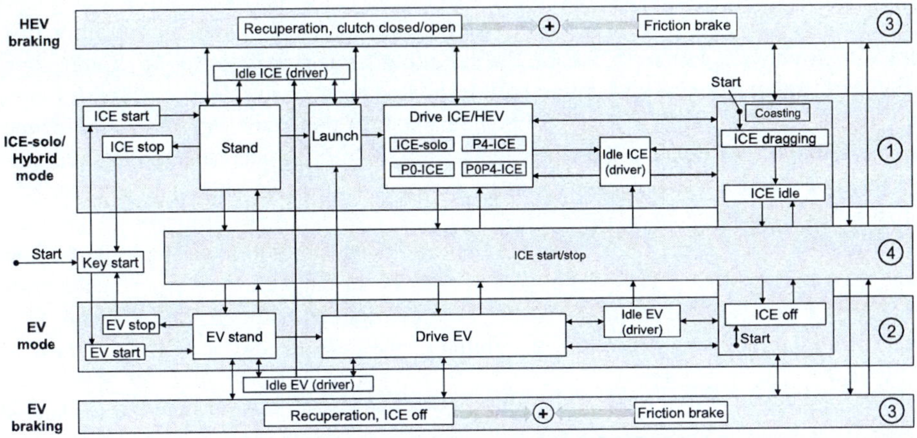

Fig. 10. Base concept of the operating strategy

and EV mode. Within these four main parts, various driving states are defined and connected with each other by transition conditions. Each driving stated defines the task of the drivetrain component. The transition conditions need to be fulfilled to get from one driving state to another.

While starting the vehicle (driving state: "Key start") the operating strategy differentiates two possibilities depending on the State-Of-Charge (SOC). If the SOC is above a certain determined threshold, the vehicle will start in EV mode. Otherwise the engine is turned on and ICE-solo mode is activated. At first, the ICE-solo+Hybrid mode is presented. The ICE start process is triggered by the driving state "ICE start". After the engine is turned on, the operating strategy turns to "Stand" mode. From that point, the engine can be turned off by passing the driving state "ICE stop". Otherwise, the launch state is performed, in case the transition condition is fulfilled. One condition for example is that the accelerator pedal is activated. After the automatically performed launch process, by the automated gearbox, is finished the operating strategy is in the driving state "Drive ICE/HEV". In that driving state, the strategy controls the torque split between the ICE, the P0 EM and the two P4 EMs. By doing so, the four stated sub driving states are defined ("ICE-solo", "P0-ICE", "P4-ICE" and P0P4-ICE"). If the driver deactivates the accelerator pedal and is not using the brake pedal, the transition condition to the driving state "Coasting" is fulfilled. In this state the ICE is dragged. After a certain time, the engine is decoupled from drivetrain by opening the clutch. Regardless of the usage of the pedals the driver can decouple the ICE from the drivetrain by opening the clutch. This is done by switching the driver gear selector lever from "D" into "N" and must be taken into account. These driving states can be reached from every driving state after the engine has been turned on.

By activating the brake pedal within the main part ICE-solo+Hybrid mode, the transition condition to the main part HEV braking is fulfilled. Within this main part, the braking energy can be realised either by electric recuperation of the electric machines, by the friction brake or by a combination of those possibilities (brake blending).

The third main part is the EV mode. After the key start and in case the condition that the SOC is above a certain threshold, the operating strategy will go to the driving state "EV start" and the engine stays turned off. From that point, if the electric machines send an "active" signal, the "EV stand" state is reached. In that state, the interaction of the driver with the accelerator and brake pedal as well as with the gear selector lever defines the various stated. The process is mostly identical with the already above described process.

The fourth part is the transition state between the upper and lower main parts. The main part ICE start/stop has to be passed in case the strategy changes from a state with ICE "on" towards a state with ICE "off" and the other way round. In this main part, the opening and closing process of the clutch as well as turning the ICE on and off is processed.

These functionalities can be realised by some hardware changes. At first, the series accelerator pedal has to be modified to modify the signal of the accelerator pedal sent to the engine control unit. Secondly, the engine start/stop button is modified to turn the ICE "on" and "off" and simultaneously open and close the clutch. The last main hardware modification is the implementation of the control unit of the electric machines. The machines can be controlled independently from the main conventional components.

5 Outlook

After finalizing the hardware set up of the demonstrator vehicle the next step will be the initial operation of the powertrain system. Afterwards the operation strategy and the torque vectoring module can be implemented and tested in the vehicle. As already presented in Sect. 1.3 the main target for the vehicle is to validate the simulation models for the CO_2 reduction, electrical driving functions and the torque-vectoring module on the dynamometer and the test track.

References

1. Hölzl C, Uhl M, Sokol P (2019)Transformation of the powertrain. A balanced view of technologies. Whitepaper, Stuttgart
2. IEA (2019) Global EV Outlook 2019
3. JATO (2018) Global car sales up by 2.4% in 2017 due to soaring demand in Europe. Asia-Pacific and Latin America
4. Vollmer P (2018) Diese Emissionen haben Batterien von Elektroautos. Bei der Herstellung von Akkus und der Förderung der dafür benötigten Rohstoffe wird CO2 frei - deshalb starten Elektroautos mit einem Klima-Rucksack
5. Werra M, Kücükay F (eds) (2019) CO2-analysis and dimensioning of 48 v hybrid drivetrains in legal and customer based cycles. 28th Aachen colloquium automobile and engine technology. Aachen, 07–09 Oct 2019
6. Kumle J, Hofman T (eds) Dynamic optimization of an operation strategy for a nested powertrain system design approach. evs30 international battery, hybrid and fuel cell electric vehicle symposium. Stuttgart, 9–11 Oct 2017
7. Tigges F, Kumle J, Scheikh Elard F, Kücükay F (eds) (2019) Vehicle dynamics improvements enabled by 48 V rear axle. Automotive drivetrain, intelligent and electrified. Novi, MI, USA, 11–14 June 2019

Industry 4.0 Applications for Improved Efficiency in EOL Testing

Ralph Heckmann(✉)

teamtechnik Maschinen und Anlagen GmbH, Planckstrasse 40, 71691 Freiberg, Germany
automotive@teamtechnik.com

Abstract. There are significant more and new data generated during EOL testing of hybrid and e-Drive components. In addition new products and new requirements in testing appear. Due to that there is a strong request to optimize production and testing processes like reducing cycle times, predictive operations, lean production and an efficient high volume production. This is why modern EOL-testing needs flexible and performant data analysis. This article gives three data based Industry 4.0 applications for improved efficiency in EOL testing.

Keywords: Industry 4.0 · EOL testing · End-of-line testing · E-Drive · Electric drive testing · Transmission testing · REST API · Predictive maintenance · Big data · Trend analysis · Testing software

1 Introduction

The growing success of electrified and electric drives means challenging new tasks for end-of-line testing. This requires improved flexibility and efficiency in EOL testing as a solution for the increased complexity due to the following reasons:

- More product complexity from Transmission to E-Drive EOL
- More subsystems in products: TCU's, Inverter, Safety code, …
- More sub-functions to be tested: Kiss-Point, Electric measurements, …
- Test Bench controls periphery has to be integrated
- Huge volume of data
- High requirement in availability of production system
- Customer need of process and test data

2 Test Bench Structure

The e-axle test bench consists of three modules, shown in Fig. 1, all designed to meet the challenges of flexibility, precision and efficiency. The design of this modular structure means that it is possible to implement mature components which have already been tested by teamtechnik in 300 test bench applications. Our customers not only benefit from the outstanding availability of the test bench during normal operation, coupled

Euroforum Deutschland GmbH (Ed.): *CTI SYMPOSIUM 2019*, pp. 203–207, 2021.
https://doi.org/10.1007/978-3-662-61515-7_18

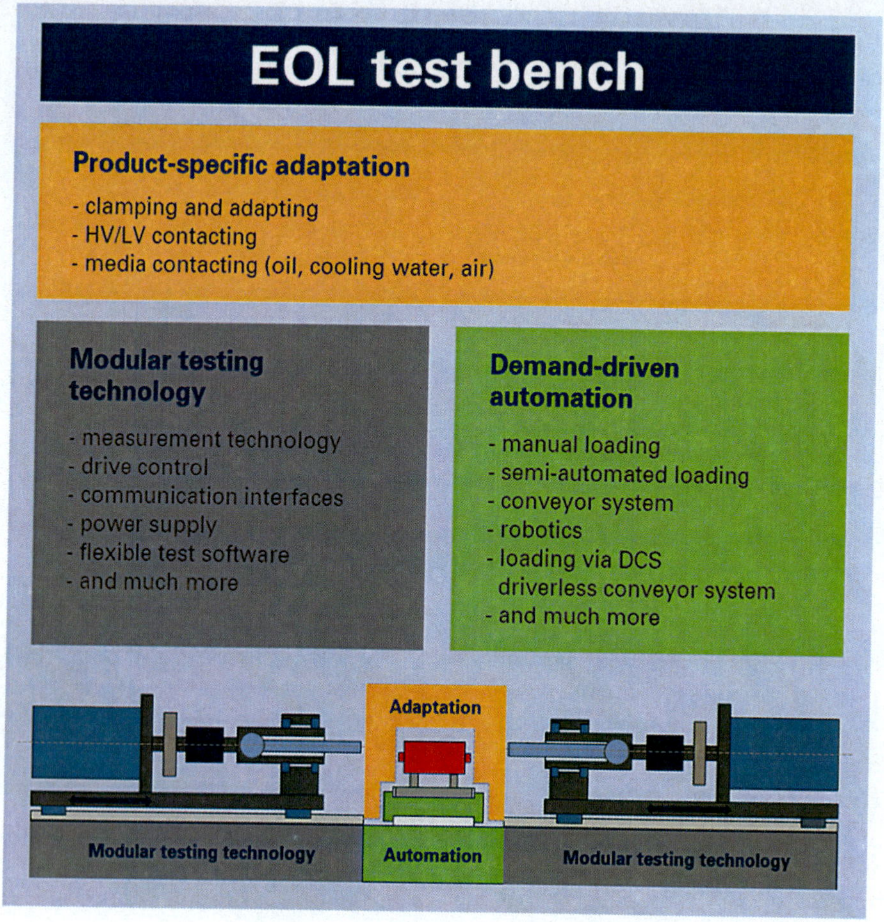

Fig. 1. E-Drive test bench structure

with our integrated maintenance and operating support strategy, but also from our low costs and short delivery times.

At its heart is the standardized testing technology (shown in gray), which is applied to the test object (shown in red). This involves a standardized module in which all the measurement technology, such as the functionality for determining electrical, mechanical and NVH characteristics, is integrated. The module also handles communications with control units and bus systems, and manages the energy supply, along with battery simulation, power flows, power recovery and power supply.

The customer- and product-specific interfaces (shown in yellow) are then installed on this module. These interfaces include the horizontal and vertical clamping devices, cooling circuits and oil supply. They are joined by flexible adaptations of power electronics, control units, sensors and the high-voltage power supply, all of which use standard, existing series equipment, as far as this is possible. Generally, the test bench must replicate the installation position in the vehicle 1:1. And, in fact, it must be even more accurate, to ensure that testing is non-reactive and that the results generated by the tests are absolutely accurate.

The third module is a system which automates the test bench in accordance with the customer's specifications (shown in green). Automation scenarios range from loading the test object manually through to fully automated conveyor systems, robotics or driverless transport systems.

3 Measurement Process

teamtechnik has also designed the measurement technology in the e-drive test benches with a flexible open architecture. The most important parameters concern the properties of the e-machine and the transmission. The test bench provides highly dynamic, reproducible load profiles, with excellent control quality, for analyzing noise emissions from electric drives. Only in this way can short test intervals with excellent reproducibility of measurement results be achieved.

A new test requirement for e-drives is the ability to measure the cogging torque in rotating electrical machines, as this effect has a major impact on the smooth running of electric drives and increases fluctuation in the e-drive torque, especially at low speeds. The quality of the e-machine winding and excitation form are determined from the progressions of the phase voltages induced by the generator. Electromagnetic compatibility requirements are also taken into account here.

Another important test determines an electric drive's correct torque characteristic. The teamtechnik software uses the curve characteristic of the torque, which is recorded in real time, to analyze the efficiency of the electric drive, and uses that to calculate the potential range of the vehicle.

4 Three Examples for Industry 4.0 Applications

4.1 Big Data

The big data volume generated in EOL-testing can be used for the optimization of testing. Powerful software tools are needed to fulfill this task. One example is the testing software teamsoft. TEST offered by teamtechnik, which has the following features:

- Presentation of test results at a glance
- Dashboard: Quick overview of production
- Powerful analysis of test result possible
- Flexible filter options of result data (incl. share function)
- Different statistics function (e.g. OK/NOK statistics or error statistics)
- Versatile analysis functions of curves
- Versatile statistic functions (histogram, control charts)
- Export functions (XLSX, CSV, TDMS)
- PDF reports

4.2 Automatic Trend Analysis

Early detection of any kind of deviations in the production process is one advantage of the automatic trend analysis based on the wide variety of data generated during EOL testing processes. Software tool are used to monitor all process values according to statistical rules. If a rule applies, this is displayed in the evaluation. Further analyzes are possible by a drill down functions. Professional software can also support to localization the source of error. Examples are the prediction of component failures to avoid system downtime or quality control by detecting deviations of production tolerances (Fig. 2).

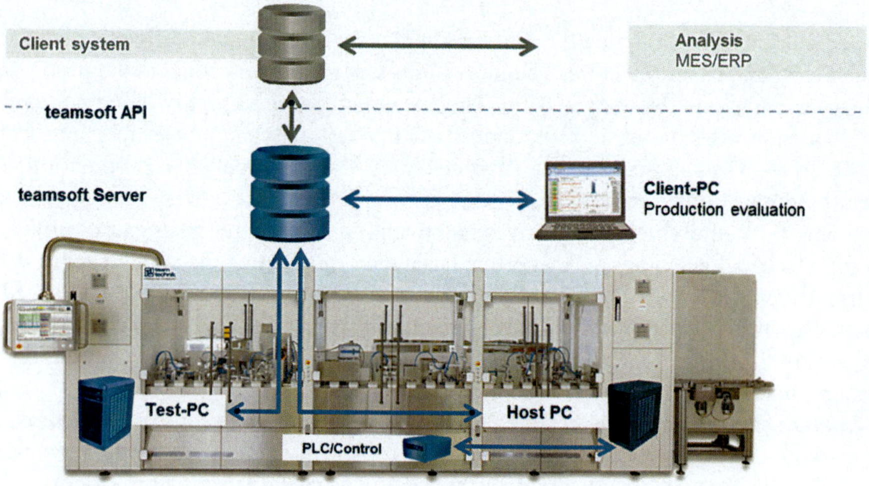

Fig. 2. REST API structure

4.3 REST API

REST API is an abbreviation for REpresentational State Transfer Application Programming Interface. It is used to access and integrate data easily. This is possible based on a clearly defined interface to extract data from the production database. One advantage is the Protection of the production server against external intervention. The REST API can be API can be used to:

- make data available to other systems (e.g. ERP)
- deliver data to special customer tools
- make special evaluations
- create special exports

5 Conclusion

There are significant more and new data generated during EOL testing of hybrid and e-Drive components. In addition new products and new requirements in testing appear.

Due to that there is a strong request in optimization of production and testing process like reducing cycle times, predictive operations, lean production and an efficient high volume production. This is why modern EOL-testing needs flexible and performant data analysis

Triple Wet Clutch e-Module for P2 Hybridization

Architecture and Technical Description of a Triple Wet Clutch Unit and the Related Active and Passive Clutch and Shift Actuation System

Olivier Simon[1](✉) and Wilhelm Heubner[2]

[1] Valeo Transmissions, 79 Rue Poulainville, 80009 Amiens, France
olivier.simon@valeo.com

[2] Valeo Actuation Systems Ebern, Andreas-Humann-Straße 2, 96106 Ebern, Germany
wilhelm.heubner@valeo.com

Abstract. Bei P2-Hybridarchitekturen haben integrierte Lösungen von Triple Wet Clutches mit effizienten Aktuierungssystemen bei mehreren OEM-Projekten einen hohen Stellenwert. Insbesondere bei DHT- oder Hybrid-FWD-Architekturen, bei denen es um die Abmessungen in axialer Richtung geht, bietet die Integration der Trennkupplung zwischen ICE und Getriebe (C0) in ein Triple Wet Clutch Modul (TWC) eine intelligente Designlösung.

Das Papier beschreibt verschiedene Architekturen von TWC mit Zentraleinrückern (CSC) und integrierten Betätigungskolben mit Schwerpunkt auf P2-Offline-Hybridarchitekturen und den damit verbundenen Vor- und Nachteilen.

Der Beitrag beschreibt auch ein innovatives Betätigungssystem des TWC Moduls, das auch weitere Getriebefunktionen wie Parksperre und Schaltzylinder betätigt. Das Betätigungssystem ist eine Innovation von Valeo, die auf die Bedürfnisse der DHTs zugeschnitten wurde. Es kann drei Kupplungen betätigen und die Schaltung der Gänge und Ansteuerung des Parksperrzylinders durchführen.

Moderne DCTs und DHTs verwenden so genannte "On-Demand"-Aktuierungsysteme, um die ständigen Verluste der zentralen Powerpacks zu vermeiden, die durch ihr konstant hohes Druckniveau und den Leckagen in den Steuerventilen entstehen.

Die hydraulische Betätigung der Kupplungen und Zylinder bringt eine hohe Leistungsdichte und damit Vorteile im gesamten Aufbau des Getriebes. Der Schwerpunkt der Entwicklungsarbeit liegt auf niedrigem Stromverbrauch, geringem Gewicht, kleinen Einbauräumen, die frei positioniert werden können, und attraktiven Preisen. Ein geschlossener Ölkreislauf mit eigenem Ölbehälter ermöglicht eine lebenslange Befüllung und freie Wahl des Schmieröls.

Am Ende des Papiers steht die Beschreibung der Systemarchitektur einschließlich der Software.

In case of P2 hybrid architectures, integrated solutions of Triple Wet Clutches with efficient actuation systems are having a big attention on several OEM projects. Efficient ways for hybridization can be achieved either using existing dual clutches gearboxes, either using dedicated hybrid transmissions (DHT). In those cases the packaging constraints in axial direction are at stake, particularly on FWD

Euroforum Deutschland GmbH (Ed.): *CTI SYMPOSIUM 2019*, pp. 208–218, 2021.
https://doi.org/10.1007/978-3-662-61515-7_19

applications. The integration of the disconnecting clutch between ICE and gearbox (C0) inside a the Triple Wet Clutch (TWC) is providing smart design solution. The paper will describe different architectures of TWC with Clutch Slave Cylinders (CSC) and piston type actuation technologies focusing on P2 Off-line hybrid architectures, and their associated advantages and drawbacks.

The paper will also describe an innovative actuation system of the TWC, managing also some gearbox functions as park lock and gear shift cylinders. The actuation system is a Valeo innovation which considers the needs of DHTs. It can actuate three clutches and operate the shifting of the gears and park lock cylinders. Modern DCTs and DHTs use so called "on-demand" actuation systems to avoid the constant losses of central power packs, which are generated by their constant high pressure level and leakages in the spool valves. The hydraulically actuation of the clutches and shifters brings a high power density and therefore advantages in the gearbox packaging. The focus of the development work is on low power consumption, low weight, small packaging free in positioning, and attractive pricing. A closed oil circle with an own oil reservoir allows lifetime filling and free choice of the lubrication oil. A description of global system architecture including software will be described at the end of the paper.

Keywords: Triple wet clutch · Hybrid P2 module · Actuation system · First keyword · Second keyword · Third keyword

1 Introduction: Several Clutches Arrangement Solutions

In case of P2 off-line gearbox hybridization, OEM can chose different architecture options. The first option is a "P2 off-line module" [1]. The P2 off-line module (Fig. 1) is a very interesting solution for FWD cars with several transmission types to be hybridized (DCT, AMT or AT gearboxes), it includes the C0 clutch and the damper, and is connected with a chain to a 48 V E-motor. The P2 off-line module as being is versatile solution, allows for example to use existing standard Double Wet Clutches, in case of hybridization of a DCT, either using a CSC actuation or a rotary feeding piston actuation technology (Figs. 2 and 3).

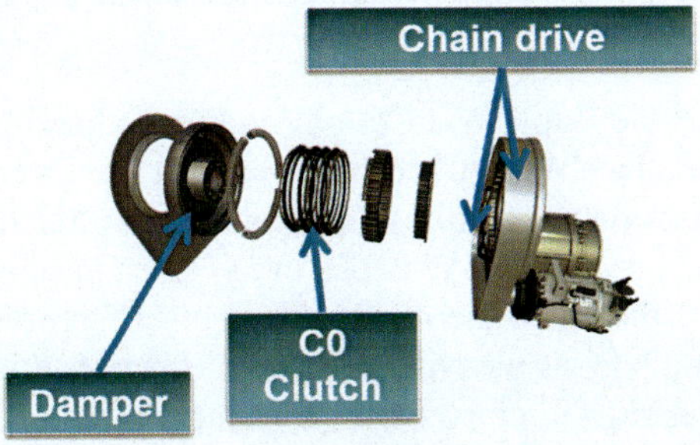

Fig. 1. P2 off-line module

Fig. 2. DWC with dCSC actuation

Fig. 3. DWC with piston actuation

The second option is the Triple Wet Clutch Module. In case there is only one transmission to be hybridized, the TWC will probably present the advantage of good packaging optimization and function integration. The packaging optimization will be achieved with a smart arrangement of the three clutches in the given packaging, and the function integration will be done thanks to the combination of different actuation components in the same actuation module, as we will see below. However this solution will remain specific to a given packaging, therefore somewhere more difficult to standardize to other applications.

2 Triple Wet Clutches off Line Unit - CSC Actuation

The TWC with CSC actuation presents the advantage of having an optimized CO_2 efficiency, thanks to CSC actuation and removal the internal seals inside the clutches, usually used with rotary feeding and piston actuated clutches. CO_2 efficiency simulations have been presented in previous papers from Valeo [2]. TWC with actuation of the three clutches by CSC is probably the most optimized architecture of wet clutches relative to CO_2 efficiency, and presents an efficiency gap compared to less optimized architectures which can reach 2%, which is very significant.

The Fig. 4 represents an example of TWC with CSC, having a connection to an off-line 48 V or high voltage E-motor realized by a gear set. The C0 clutch is actuated from the left by a CSC, and the C1 & C2 clutches from the right by a double CSC. The TWC is supported by two support bearings.

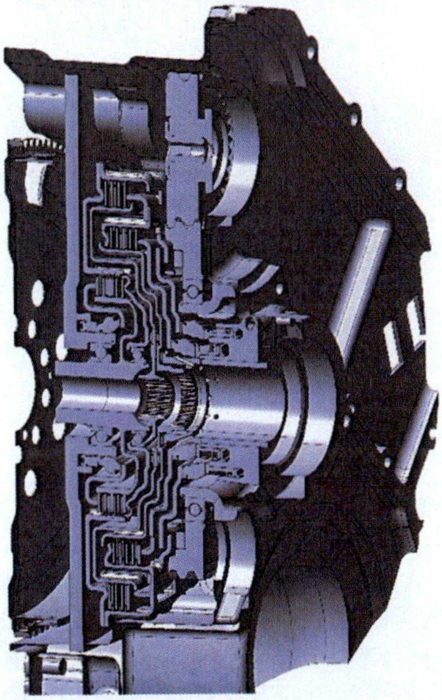

Fig. 4. Triple wet clutch Module with CSC passive actuation

The gear connected to the clutch is supported by a bearing directly linked to the gearbox. This allows to control very precisely the distance tolerance between the rotation axis of gear connected to the clutch and the rotation axis of the gear connected to the E-motor, in order to obtain good NVH performance. The misalignment between the triple clutch unit and the gear is managed by an innovative misalignment compensation device, capable to transmit torque and compensate the off axis. The design of the TWC represented on Fig. 4 is optimized to cope with very tight axial packaging, typical for DHT with FWD vehicle applications, with the three clutches nested on the same vertical axis.

3 Triple Wet Clutch off Line Module – Rotary Feeding and Piston Actuation

Other architectures are possible with rotary feeding and piston actuation, depending on customer choices relative to actuation type. The triple clutch module is still connected to a 48 V or high voltage E-motor with a gear set. The appropriate support of the TWC is still ensured with two support bearings as shown on Fig. 5. There is a similar misalignment device as described in previous paragraph, between the triple clutch unit and the gear.

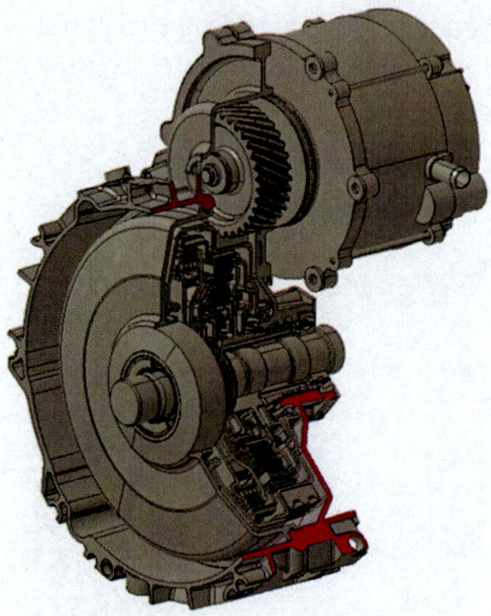

Fig. 5. Triple wet clutches unit with rotary feeding and C1 & C2 piston actuation

Depending on packaging possibilities, a possible smart arrangement is to actuate the C0 clutch by a CSC, allowing better CO_2 performances compared with C0 piston actuation.

In this configuration, actuation maximum pressure of C1 and C2 would be around 18 bars, as the actuation pressure of C0 with CSC would de around 40 bars. This difference of actuation pressure range can be easily managed with the Valeo actuation system described below with two pumps, the first one managing the C1 & C2, and the second one managing the C0, shift actuators and park lock.

4 Actuation System

New hybrid transmissions and DHTs have a big focus on efficiency. Conventional power packs generate losses by constant high line pressures, leakages in valves, rotating seals for piston type wet clutch actuation, and actuators, and in case of mechanical driven pumps by far too high oil flow at higher engine speed. Even power packs with electrically driven pumps and accumulators cannot be considered as “on-demand-actuation systems”, as they still work with high line pressures and leaky valves.

The Valeo powertrain actuation system works with real on-demand-actuation, the pressure for the actuation of the clutches is only created when it is needed, in the level which is actually needed, and can be kept without leakages and the need of running pumps for a longer period. The same philosophy is used at the actuation cylinders. Additionally, it is very important for an efficient system to separate high pressure supply for actuation from low pressure for lubrication and clutch cooling.

The clutch actuation is done by an electrically driven dual flow pump with two inlets and two outlets, see Fig. 6. The roller vane pump which is used here can provide the two independent flows with only one set of pump stator and rotor, is robust against contaminations and cost-effective. The flow in both outlets is proportional to the pump speed and is guided over check valves to the CSCs. The pressure on both outlets can be regulated independently by two proportional valves.

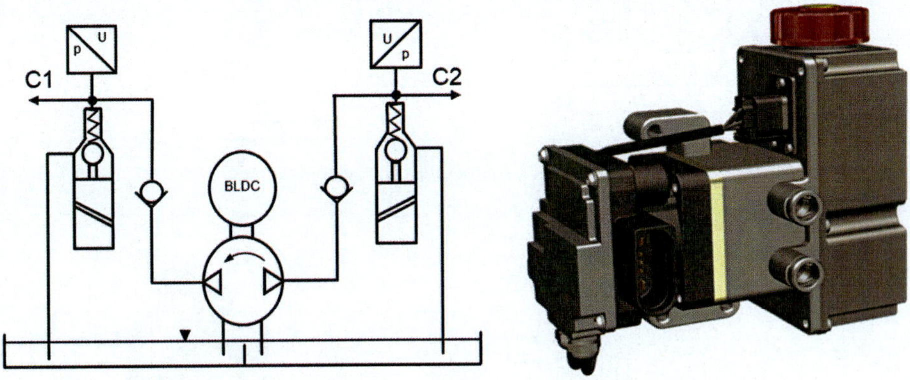

Fig. 6. Hydraulic double clutch actuator

The proportional valves are designed in ball seat technology, which is unusual in gearbox hydraulics. The reasons for this design are robustness against contamination, cost-effective components and leak-tightness in closed position. This enables, in combination with CSC-type actuation of the wet clutches, that the pressure can be kept on the desired level only by actuating the proportional valve with relatively low electrical current consumption, and the motor-pump can be stopped (Figs. 7 and 8).

The BLDC motor and the proportional valves are driven by an integrated electronic control unit (ECU), which is regulating the pressure in closed loop with the help of integrated pressure sensors. The ECU is connected to the transmission control unit (TCU) via CAN interface, receives commands for pressures on both channels and operation modes, and sends actual data and diagnostic values. It starts and stops the motor, and decides the motor speed according to the operation modes.

The actuator has its own reservoir, which is divided below a certain level into two chambers in order to maintain the function in one circle even when the other circle would be leaky. In this case the DCT would be operated similar to an AMT (automated manual transmission). The closed oil circle allows, together with the robust valves and pump, to omit a suction filter, which improves the function at low temperature. Filtering is done

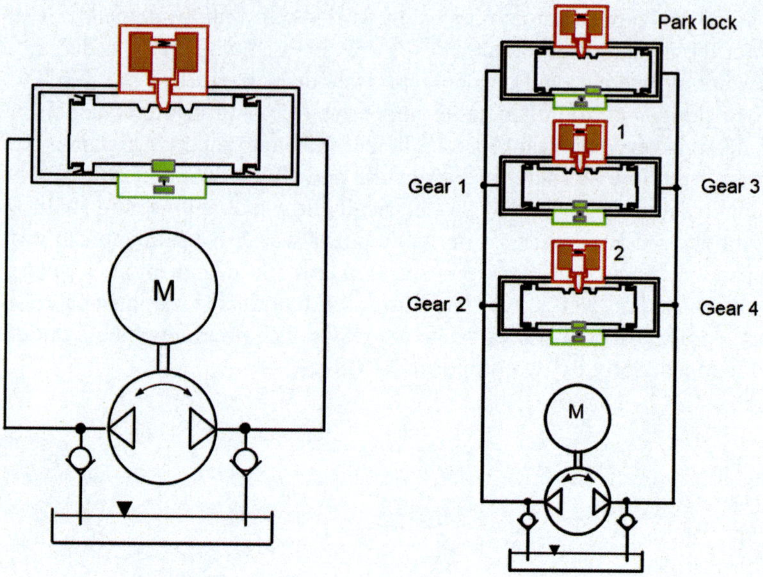

Fig. 7. Schematic shifting system

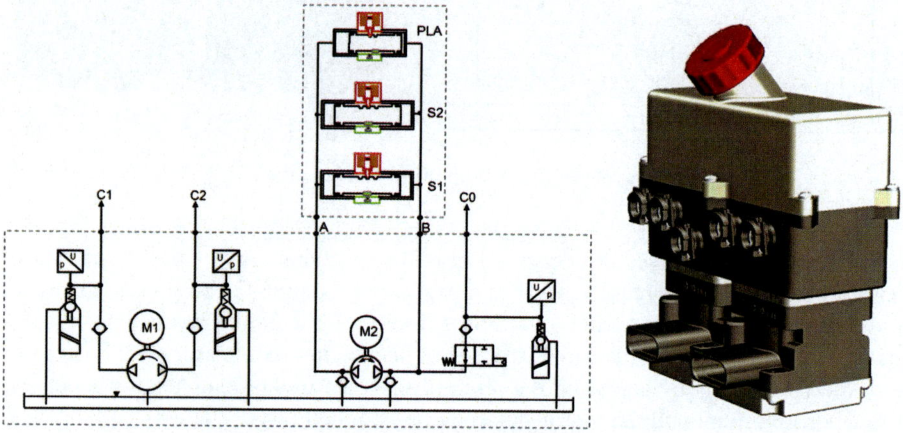

Fig. 8. Schematic triple clutch and shift actuator

on pressure side and on the connectors, to avoid contamination from outside (Figs. 9 and 10).

The shifting and the park lock actuation are made by a second electrically driven pump in combination with cylinders with integrated locking magnets. The single flow pump has two ports, which change the function from inlet to outlet or vice versa when the sense of rotation is changed.

All actuation cylinders can be connected in parallel without any valve in between. The pistons have as many grooves as desired positions, in case of a shift cylinder normally

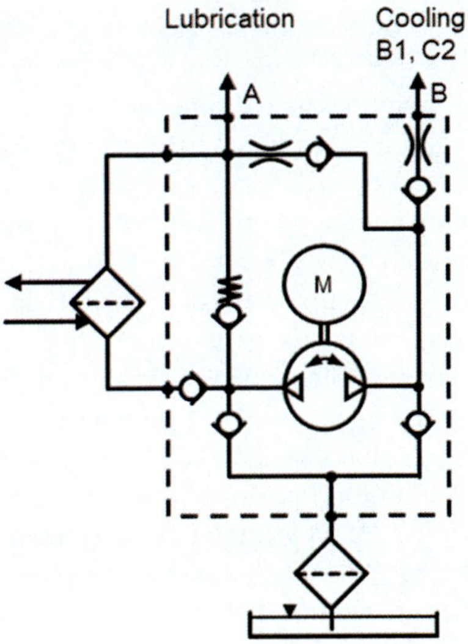

Fig. 9. Lubrication and cooling pump with heat exchanger

Fig. 10. Actuation system integration on a DHT gearbox

three, e.g. for gear 1, neutral and gear 3; in case of the park lock cylinder there are two grooves for engaged and disengaged position. The pawl magnets are moving the locking pins, which are spring loaded in the locked position. The geometry of groove and pin is chosen in a way that the locked piston does not move even under maximum pressure. The cylinders have integrated position sensors.

The reversible pump moves the released piston by pumping the oil from the one side to the other side of the double-acting cylinder; the synchronizing is done by controlling the torque of the pump motor in closed loop with the signals from the position sensor.

The pressure acts on one side of each of the cylinders, but only the released one is moving.

Both ports of the pump have a connection to the reservoir via a check-valve. This enables that volume losses caused for example by bleeding orifices can be compensated on the actual inlet port.

The motor is also driven by an integrated ECU, which is regulating the motor speed and limits dynamically the motor torque according to the commands of the TCU. The ECU is connected to the transmission control unit (TCU) via CAN interface, receives commands for sense of rotation, motor speed and torque limit, and sends actual data and diagnostic values. The pawl magnets and position sensors are directly wired to the TCU, in order to divide the tasks and to gain a higher safety level.

The actuation of the engine clutch (C0) is also made by the shift pump. When all pistons are locked in neutral or shifted position, a valve to the clutch actuation branch can be opened and the clutch can be pressurized in the same way as the other clutches.

When hydraulically shifting will not be used in the transmission, the third clutch can be operated by a modified version of the dual flow pump. The motor will be made reversible, and the 4 ports of the dual flow pump change their function from inlet to outlet or vice-versa with the sense of rotation. For example C1 and C2 are pressurized in clockwise rotation, and the C0 clutch in anti-clockwise sense of rotation. The circles are separated by check-valves.

Active lubrication and dedicated cooling of the clutch pack by a separate low-pressure optimized motor pump reduces losses in the transmission by reduced friction in the gears and bearings and reduced drag losses in the open clutches. An additional gain can be achieved with dry sump lubrication. This low pressure systems have to be adapted to the transmission, there a lot of different solutions.

The chosen example shows a schematic with a heat exchanger, a lubrication and a cooling circle. When the pump is running anticlockwise, it supplies oil for lubrication. Due to the higher pressure drop in the heat exchanger, caused by the higher viscosity, the cold oil will partly bypass the heat exchanger through the preloaded check valve. This helps for a faster warm-up and will reduce losses in the gear meshes and lubrication. Warm oil will only pass the heat exchanger, because the pressure drop in the heat exchanger will be reduced. When the pump is running clockwise, it supplies oil to lubrication and cooling, divided up by the orifices and the flow resistance of the following

5 System Architecture and Software

On below Fig. 11. is the global system lay-out showing the connections between the TCU and the actuation system. The related software architecture is described on Fig. 12. The software is composed of several layers.

- The TSM transmission supervision module usually embedded inside the TCU and under OEM responsibility will provide the clutch torque set point, the cooling demand the shift actuations and park lock positions.
- The lower layer also usually embedded inside TCU will manage the Triple Clutch, cooling and gear actuators positions. The Clutch Torque Model will calculate the

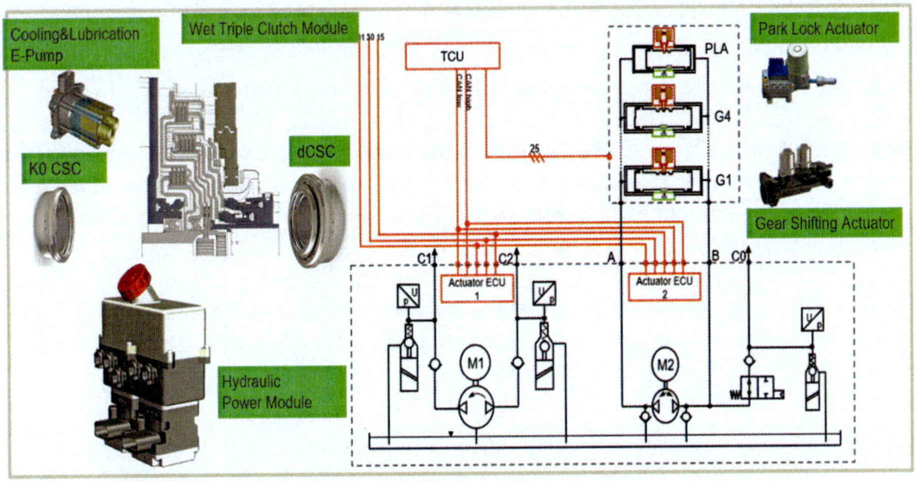

Fig. 11. System architecture

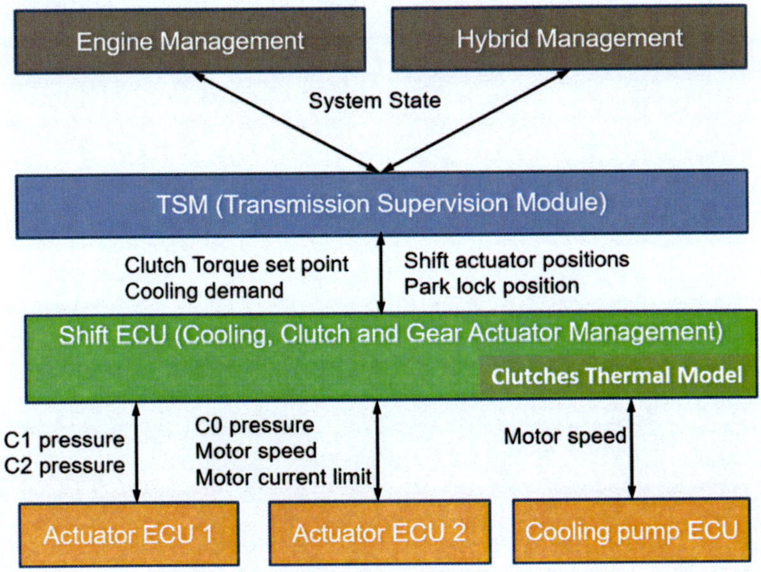

Fig. 12. Software architecture

pressure on C0, C1, C2, and will have a specific thermal model embedded, to protect the Triple Clutch from thermal abusive operations. This layer can be either under OEM or Valeo scope of supply.

- The actuators ECU and cooling pump ECU will provide the necessary current to actuators and cooling pump to obtain the requested pressure, cooling flow and gear actuators positions. This layer is under Valeo scope of supply.

6 Conclusion

The triple wet module is an attractive solution for Hybrid Transmissions. In case of P2 off-line for DHT or Hybrid FWD transmissions, where the axial packaging is at stake, an innovative arrangement of three clutches nested on same vertical axis has been proposed. In order to obtain a good NVH performance on the gear set connection between E-motor and clutches, a robust design of the gear set associated with a new misalignment compensation system has been described.

The actuation system plays an important role in the efficiency of the gearbox. Real "on-demand-actuation" avoids losses due to constantly high pressures and leakages, and reduces the stress on the components. Lubrication and oil cooling with separate oil pump systems can also contribute to an efficient system.

References

1. P2 Off Line module compared with other architectures. CTI symposium automotive transmissions, Valeo Powertrain systems. Berlin, December 2018
2. Dual Wet Clutches for Hybrid power trains, challenges for packaging and fuel efficiency optimization CTI symposium automotive transmissions, Valeo Powertrain systems. Berlin, December 2017

Method for Thermal Modeling of Electric Traction Machines for Hybrid Vehicle-Applications

Holger Hinrich(✉)

Porsche AG, Porschestraße 911, 71287 Weissach, Germany
holger.hinrich@porsche.de
http://www.porsche.com

Abstract. The method aims for maximum simulation quality on the one side and on the other side minimal effort in creating thermal models by combining test bench and vehicle tests as well as the use of thermal networks and 3D CFD- /CHT-simulation.

A self developed concept for semi-automatic creation of thermal networks for electric machines is presented and applied to the electric traction motor of the current Porsche Panamera Turbo S PHEV in parallel hybrid configuration. With the use of an automatic optimization process for the data of the thermal network, the behavior of the model can be fitted to realistic measurements. The basis is the so called calibration-cycle that was measured with a prototype car and was transferred to a test bench. The configuration of the test procedure was improved to have a maximum accordance between the two measurement methods and consequently also compared to the two simulation methods. A noticeable high improvement of the simulation quality is presented. A relevant influence of the dynamic variable thermic environment on the monitored temperature in rotor and stator is shown. The resulting models are used to predict the component behavior in the vehicle environment and to evaluate sensitivities in optimizing the durational performance of the machine for dynamic load cycles. The method can be extended to modelling machines in a wide variety of electric machines, also for purely electric vehicles.

Keywords: Electric motor · Thermal modelling · Hybrid vehicle · Validation · Test bench · Simulation · Optimization

1 Introduction

Due to the non-ideal energy transformation processes of an electric machine from electric power to mechanical power there is also a heat source in the active Material of the motor heat. This heat induction leads to a temperature rise that can eventually result in a thermal damage of the motor of the power is not reduced.

This decrease of available mechanical performance of an hypothetical electric machine is plotted in Fig. 1.

Euroforum Deutschland GmbH (Ed.): *CTI SYMPOSIUM 2019*, pp. 219–225, 2021.
https://doi.org/10.1007/978-3-662-61515-7_20

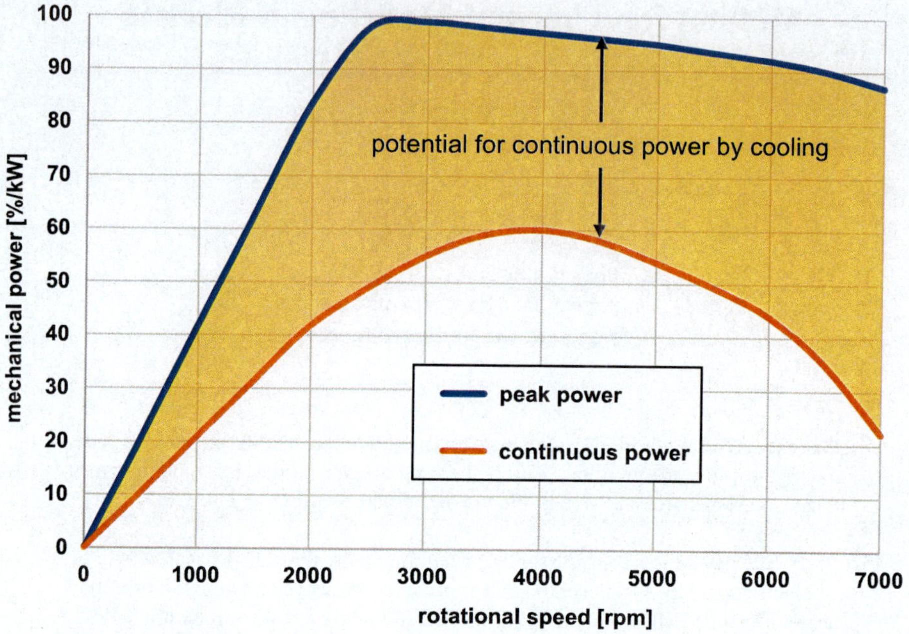

Fig. 1. Potential of continuous performance in relation to peak performance over rotational speed of an electric machine [3]

To receive the obvious potential of continuous performance, the cooling of the machine must be improved or the thermal limitations must be fully exhausted. Overheating will lead to an irreversible damage of the machines isolation in the stator, the magnetization of the rotor magnets or the bearings of the machine. An ideal cooling right at the heat source of the system will lead to a full exploration of the potential.

Thermal modeling of the electric machine helps to explore the full potential by identifying the relevant sensitivities

2 Thermal Modelling and Discretization

The basis for the model is a reduction of substructures of the machine to a number of masspoints. To determine the temperature change for each timestep a energy balance $\sum PV_{MP}$ is formed. The heat induced by losses ϕ_x, heat going to external potential sources like cooling water or ambient air $\phi_\in$ and heat going to neighboring mass points with their respective temperature $G_N \cdot \in$ is summed up. This leads to a simplified approach to calculate the temperature change during each timestep $\dot{\in}$ according to [4]:

$$\dot{\in} = \frac{\sum PV_{MP}}{C}; \quad \dot{\in} = \frac{G_N . \in + \phi_x + \phi_\in}{C} \tag{1}$$

By solving the energy equation for each mass point and each time step a continuous monitoring of temperature is created. Figure 2 shows the transfer from the energy balance in one point mass to a continuous temperature signal.

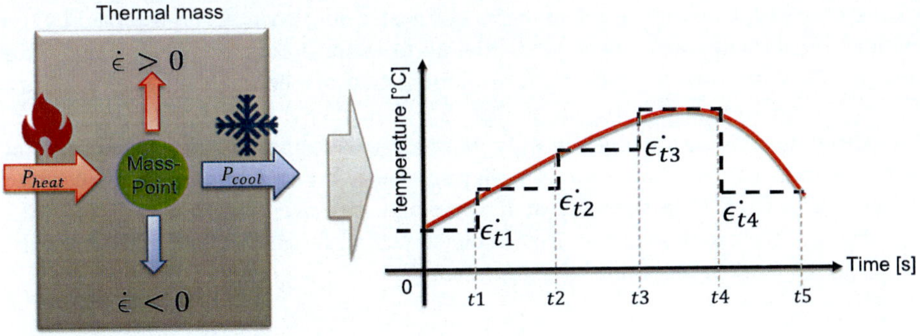

Fig. 2. Reduction of a thermal mass to a mass point and solving the energy equation for each time step for continuous temperature monitoring

The behavior of the thermal network can be influenced by changing the data for thermal mass of each mass point and the thermal resistances connecting the mass points among themselves and with their environment. By writing a automation that creates the thermal network, those basic input parameters can be changed dynamically. The algorithm needs information about the geometry of each substructure, the discretization, the specific material data for heat capacity and conductivity and information about the connection to attached substructures.

The resulting thermal model is used to predict transient thermal cycles in the monitored system. In an example of two different substructures divided in six and four point masses contacting each other heat of 30 W is homogenously induced in the substructure to the left. The substructure to the right is connected to a fixed potential source auf 20 °C at the bottom. Figure 3 shows the simple model and the resulting temperature curves.

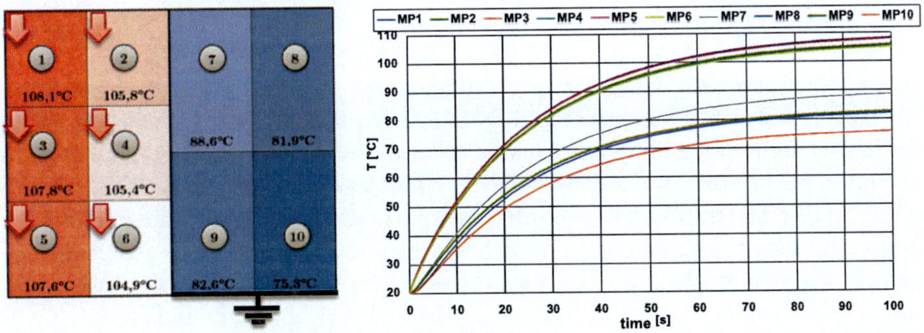

Fig. 3. Two substructures divided in ten point masses and the resulting temperature curves during heat up

The discretization of the model has a big influence on the model behavior. This concerns the resolution of time and geometry. The geometrical resolution has to be small enough to represent all relevant substructures including hot spots. When choosing a discretization that is too rough, the hotspots cannot be represented accurately, especially in structures with heat induced. Homogeneous heated substructures represented by too few load points will lead to over prediction up to factor 2 compared to the analytical solution [2]. With a fixed geometrical resolution, a corresponding time step has to be chosen. For high geometric resolution there also has to be a high temporal resolution. The energy in each mass point can only be transferred to the neighboring mass point within on time step. Too few time step during energy transfer leads to a delay in the transport. This leads to less steep heat up curve and no convergence after 100 s. With a fixed step size the example of Fig. 3 is simulated with different numbers of mass points. It leads to the final temperatures after 100 s shown in Fig. 4

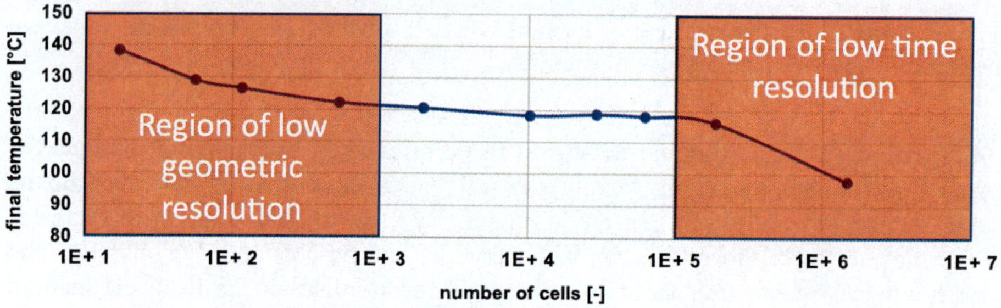

Fig. 4. Influence of number of cells on final temperature after 100 s

The effects of low geometric resolutions do not apply for non-heated structures. The discretization needs to represent solely the relevant substructures without further mass points.

3 Coupled Simulation 3D CFD and 1D Thermal Network in Calibration Cycle

This simplified approach is transferred to more complex structures with higher discretization, like an electric machine in a 3D CFD-/CHT-environment. To reduce the numerical effort to a minimum the periodic construction of the machine is used to simulate only a section. The symmetry of the geometry is used. The heat flux through the cutting planes can be transferred to the opposing side of the section for a periodic interface. In addition, the electromagnetic simulation is accomplished previously to generate characteristic maps for all load points to approximate the induced heat losses. Those loss maps and the dynamic data of the load cycle is treated in a simple Simulink model. Only the minimal necessary parts are transferred to the 3D environment. Figure 5 shows the coupling mechanism of the two models with coupling software in-between.

On the left a reduced view on the thermal network of the electric machine with different subgroups colored differently, like the coils, the rotor- and stator iron, the isolation or the magnets.

The exchanged data between the fast running Matlab model and the complex CFD-model in StarCCM+ is the cycle data and induced heat load to the substructures on the one hand and the resulting temperatures of the relevant components at specified point

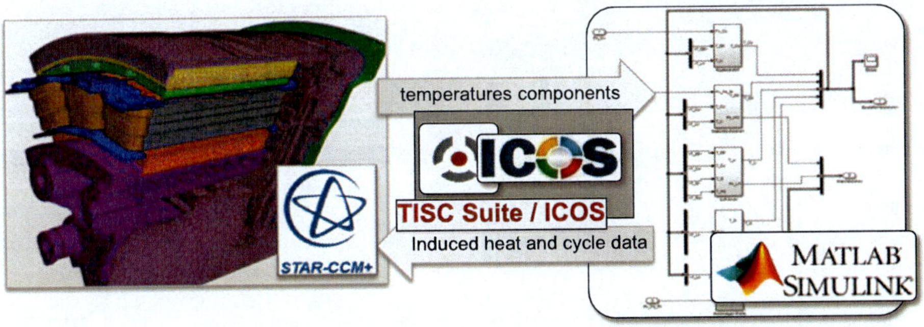

Fig. 5. Coupling mechanism 3D CFD/CHT and 1D-Matlab model

probes on the other side. A coupling software secures the synchronization of the different time steps and the minimization of numerical errors in the balance of energy. With this method a high resolution of the temperature field can be achieved. The heat load can be evaluated temperature dependent within the Simulink model. Even Derating mechanisms that may influence the performance of the machine can be taken into consideration. The heat fluxes within the structures and in the environment can be monitored in any position. By reducing the model, a moderate computing time on a CPU cluster of about 6.5 days for a real-time cycle of about 10 h can be achieved. A further parallelization of the computing problem with higher CPU performance is possible for future applications.

The load cycle is based on a real driving cycle of a prototype. The aim of the cycle to include many different load cases including high speed, high power, low power and cool down phases. This way high dynamic and long term load can be taken into account for the subsequent model calibration. This test is also performed in a test bench measurement with dynamic environment including rotational speed, torque, DC-Voltage, cooling water flow and temperature and conditioning of the bench flange. The same cycle is used for the simulation in the above-mentioned complex coupled 3D model and a much-simplified 1D model solely in Matlab. An extract of the so-called calibration cycle and the resulting temperature curves are visualized in Fig. 6

In all cases a representative point probe or physical sensor for temperature at the same copper coil in the stator winding is monitored. A very good accordance of all four methods to evaluate the transient temperature curves can be seen. There are still some slight differences in the absolute temperatures but also in the dynamics of the system. The causes are amongst others: insufficient temporal and geometric discretization, the correct assignment of heat losses to the relevant substructures in the simulations. There are also causes for differences on the measurement side. This includes accuracy of sensors for temperature, currents and torques and the ability of the test bench automation to follow the required load profile. There is also a difference in the measurements methods between the prototype car and the test bench. The scattering of the results due to measurement and simulation are in an equal range. Therefore, to increase the accuracy of the simulation it is not only necessary to control the simulation parameters but also the underlying measurement basis for model calibration.

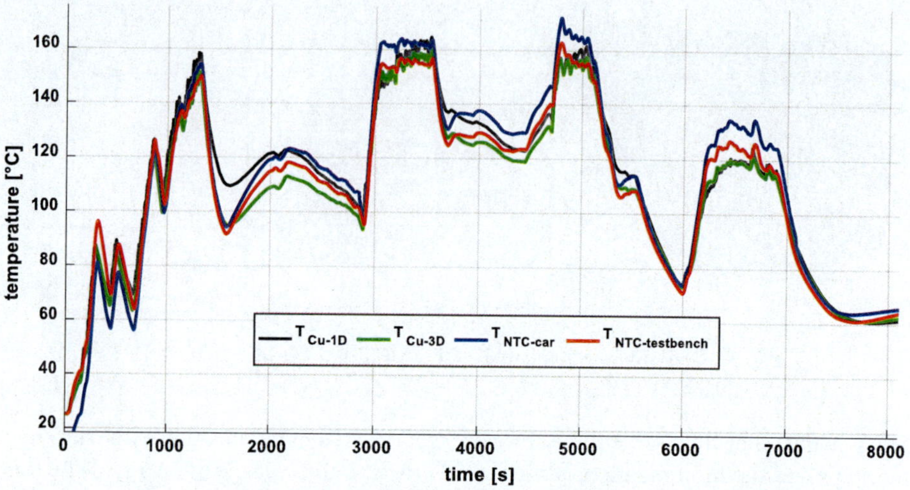

Fig. 6. Temperature curves in calibration cycle for all four disciplines

A detailed view on the underlying assumptions for the results of the calibration cycle in all four disciplines as well as a detailed discussion of the results and an interpretation can be found in [1].

4 Conclusion

A simple method for creating thermal networks for electric machines within a dynamic environment of a hybrid car is presented. A short insight in the effects of geometric and temporal discretization on the resulting transient temperature curves is given. A method for coupling fast 1D thermal networks with slower but more complex 3D CFD simulation with manageable computing resources is a further extension. By comparing the temperature curves of the two simulation methods with measurement from a prototype car and the test bench the four methods are valued. For further increase of the simulation accuracy not only the numerical error but furthermore also the measurement errors have to be considered.

The method aims to be used within the development process for hybrid vehicles. By including the resulting models into a total vehicle simulation environment, cooling concepts and their influence on the vehicles performance can be compared directly. This way the development of product substance within the drivetrain is directly aligned with the goal of maximum performance for the final customer-car.

References

1. Hinrich H (2019) Methode zur thermischen Modellierung elektrischer Traktionsmaschinen für Hybridfahrzeuganwendungen. Cuvillier Verlag, Göttingen
2. Wrobel R, Mellor PH (2010) A general cuboidal element for three-dimensional thermal modelling. IEEE Trans Magn 46(8):3197–3200

3. Reif K, Noreikat KE, Borgeest K (Hrsg.) (2012) Kraftfahrzeug-Hybridantriebe. Vieweg+Teubner Verlag Wiesbaden. ISBN: 978-3-8348-0722-9
4. Incropera FP, DeWitt DP (2002) Fundamentals of heat and mass transfer. 5. Aufl. Wiley, New York

System Cost Reduction by Electric Powertrain Design Optimization

Martin Hofstetter(✉), Dominik Lechleitner, and Mario Hirz

Graz University of Technology, Graz, Austria
martin.hofstetter@tugraz.at

Abstract. The presented optimization method supports early system design of electric axle drives (eDrives) for battery and hybrid electric vehicles. The multi-criteria design problem of finding the best suitable solutions regarding performance, efficiency, package and costs is addressed with a sophisticated multi-objective optimization method. The overall system costs are reduced by including common-part considerations for purchased parts, demonstrated on the example of bearings. Two approaches are compared: the implicit approach relies on an extended cost model that covers a quantity-dependent cost reduction effect. The explicit approach directly minimizes the number of unique parts in the system. Both approaches are compared and the effect on the optimal system solution is discussed for a case study, which shows significant cost reduction potential.

Keywords: System design optimization · Cost optimization · Electric powertrain · Gearbox design · Common parts · Carry-over parts

1 Problem

The design of electric axle drives (Fig. 1), consisting of electric machine, power electronics and gearbox, is highly complex as multiple components are involved and multiple objectives and requirements (performance, efficiency, package and costs) must be considered simultaneously. This work is focusing on the gearbox as the linking element between electric machine and drive shafts of the wheels. The gearbox subcomponents of shafts, gears, bearings and housing must be designed based on input requirements on gearbox level, whereas a large variety of possible solutions is eligible (Fig. 2).

To reduce costs, engineers try to reduce the number of unique parts (equivalent to raising the number of common parts) as several cost reduction phenomena arise (as discussed in Sect. 2). In this work, the number of unique bearings is subject to minimization and the resulting system cost reduction of this common-part consideration is investigated. It is important to consider that any change of a subcomponent (i.e. a bearing) may have a strong influence on the whole system design, as other subcomponents generally need to be adapted to the different load or package. For example, there is a strong influence from the gear design (evaluated according to ISO 6336 [1]) on the resulting bearing reaction forces and thus feasible bearing selection (load capacity evaluated according to [2], which is based on DIN ISO 76 and ISO 281). Therefore, not only the reduced

Euroforum Deutschland GmbH (Ed.): *CTI SYMPOSIUM 2019*, pp. 226–235, 2021.
https://doi.org/10.1007/978-3-662-61515-7_21

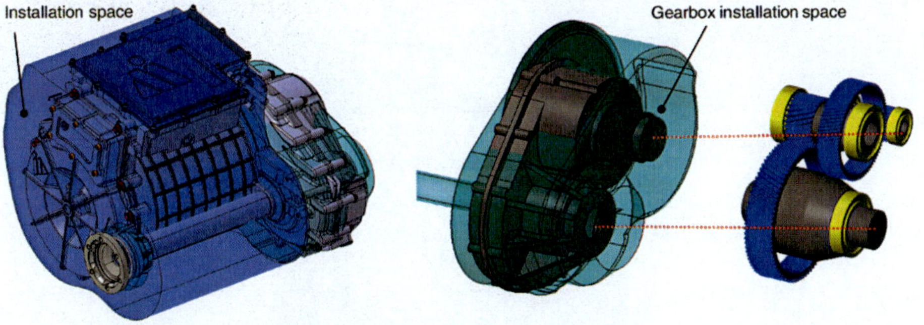

Fig. 1. Example of eDrive solution (left) and gearbox interior (right) in given installation space (transparent blue and turquoise) [3].

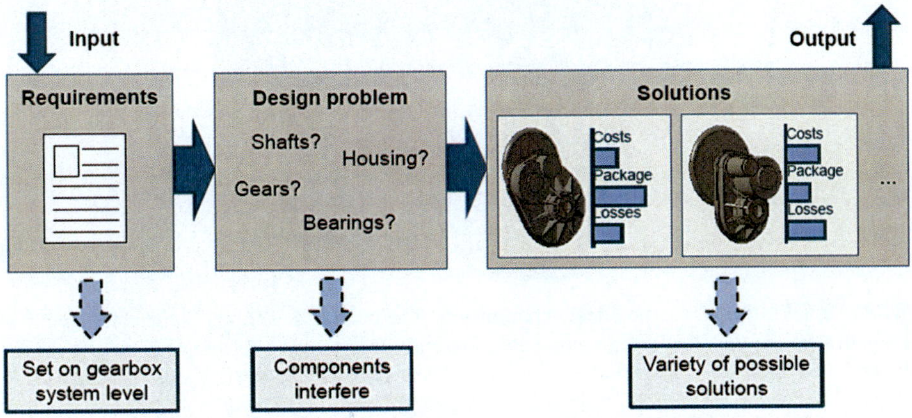

Fig. 2. Schematic illustration of the gearbox design problem [3].

purchasing prices must be considered, but also the effect on the whole system design with its multiple design objectives and requirements regarding performance, efficiency, package and system costs. The presented approach is capable of employing common parts to an extend, which is most favorable for the optimality of the gearbox system.

2 Methodology

The present contribution is based on a computer-aided gearbox design optimization process [3] supporting the early development phase of gearboxes in eDrives. For given load, lifetime and desired installation space, the differential evolution-based process (Fig. 4) delivers design parameters (Fig. 3) for shafts, gears, bearings and their arrangement to handle efficiency, package and costs in a multi-objective manner. In general, any objective derivable from the gearbox properties can be used as an optimization objective. The package integration is quantified using the so-called package metric, which is the violating volume of the gearbox with the desired installation space, as depicted in red

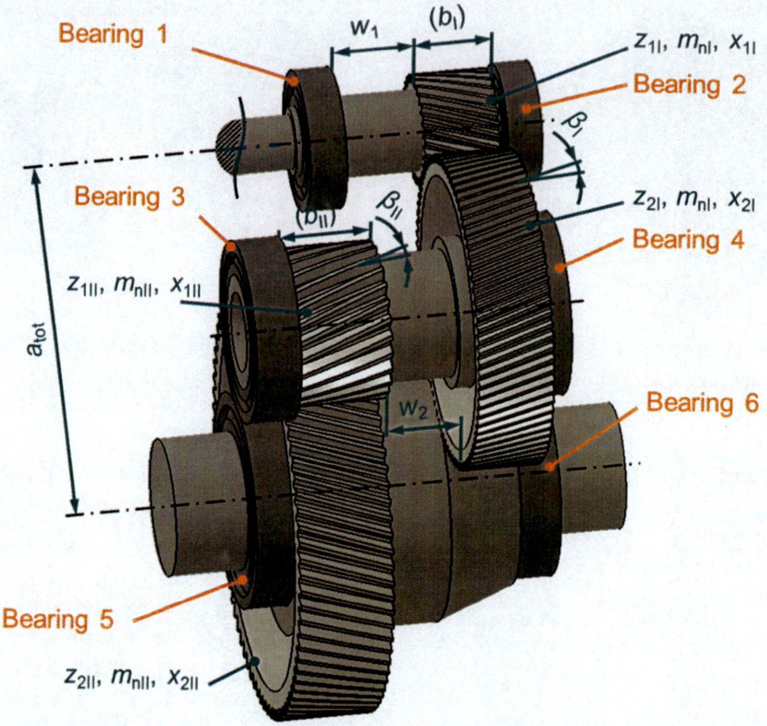

Fig. 3. Visualization of gearbox design parameters. The bearing selection is subject to common-part optimization; the other design parameters are optimized to meet objectives optimally (based on [3]).

color in Fig. 7. Altogether, six bearings must be selected for a single-speed helical gearbox architecture commonly found in eDrives (Fig. 3). For each bearing type (grooved ball bearing, cylindrical roller bearing or tapered roller bearing), a large catalogue of different sizes is available for selection with the corresponding prices. In the present approach, the costs for shafts, gears and housing are estimated with a mass-specific cost factor. The integrated differential drive is selected off-the-shelf depending on the required transmittable torque.

The design methodology [3, 4] is extended by an advanced cost model to include quantity-dependent cost reduction caused by common parts (bearings in this case) to achieve an overall system cost reduction, as described in the following section.

2.1 Cost Degression by Applying a Common-Part Approach

It is a well-known fact that a reduction of the number of unique components inside a system in general lowers the costs. A common-part approach follows this logic to decrease the total costs. The number of common parts is increased, resulting in a larger quantity demand for the common parts. Ehrlenspiel et al. [5] describe five main influences on the cost degression by increasing the quantity, which are:

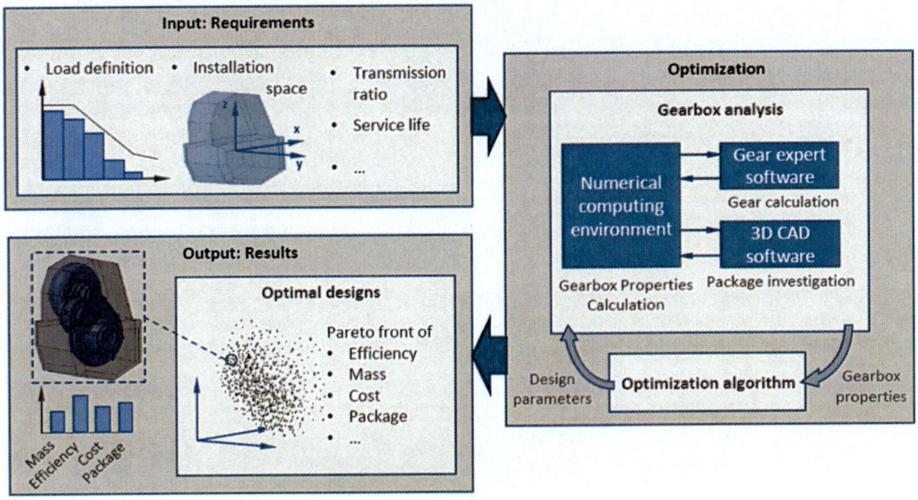

Fig. 4. Illustration of the design optimization process [3].

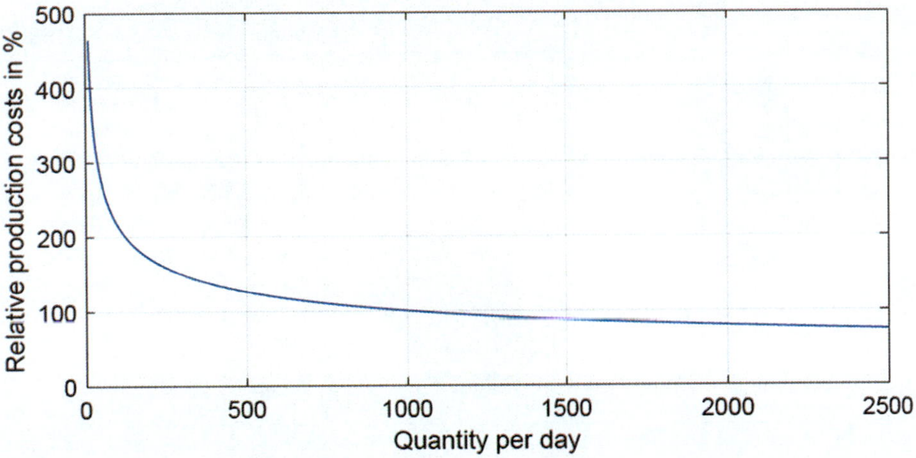

Fig. 5. Exemplary cost degression by increasing the number of produced ICEs [5], curve according to Eq. (1)

a. degression of the one-off costs,
   Each unique part has to be developed, requires a production planning and a corresponding production line and logistics strategy set up.
b. degression due to training effect,
   A new working step requires training and becomes more familiar when repeated many times. For a smaller number of unique components, training shows less effort and is more efficient.
c. degression due to optimized designs,

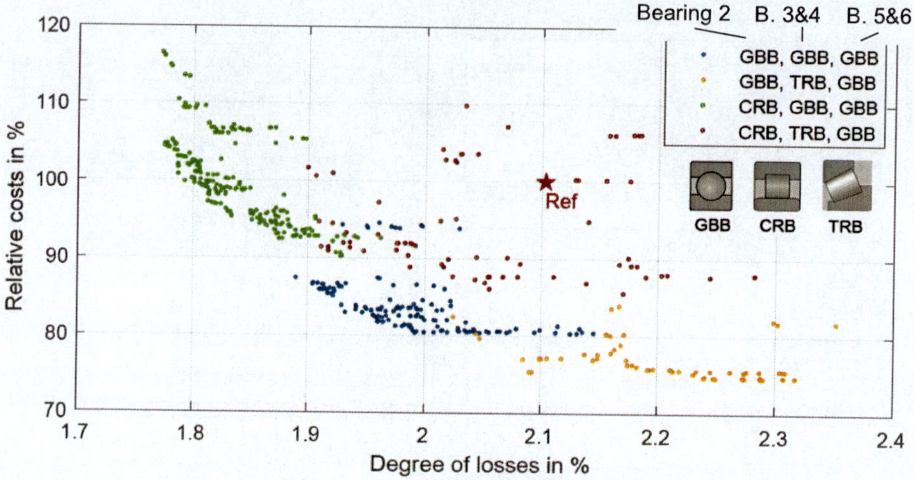

Fig. 6. Projection of the Pareto front in the dimensions "relative costs" and "degree of losses". The bearing selection has a strong influence on the costs/losses trade-off.

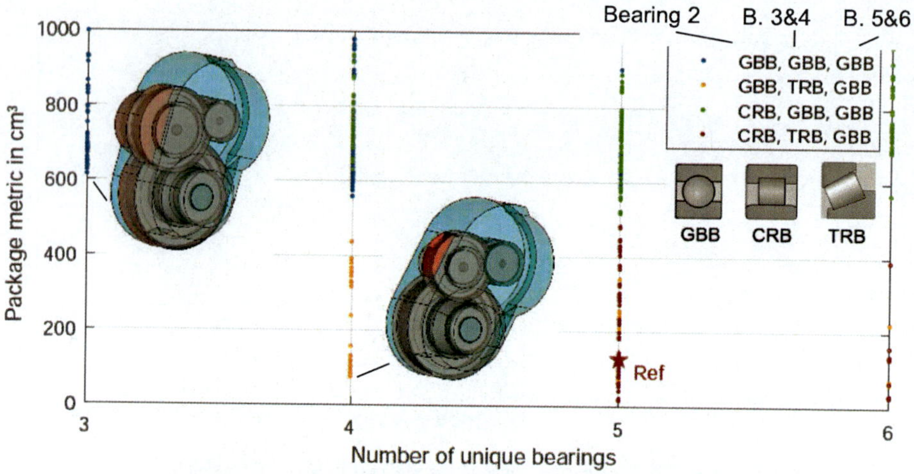

Fig. 7. Projection of the Pareto front in the dimensions "package metric" and "number of unique bearings". Enforcing a high number of common bearings leads to a package problem.

As common parts reduce the number of unique components, more time can be spent designing these parts and optimizing them for minimum production and logistics costs.

d. degression due to more efficient production processes,
The quantity required of a single part directly influences the amount of acceptable investment costs for the production line. More efficient processes that reduce the production costs of a single part in general show higher one-off costs. However, as

these costs are divided among all produced parts, large quantities justify efficient processes and are thus cheaper per part.

e. degression due to quantity discount.
A larger quantity of purchased parts and resources in general reduces the achievable purchasing prices at the market.

For purchased parts, the cost degression effects from a. to d. are passed to the customer in form of a price advantage and lead to a quantity discount (e.) for purchased parts, in this case the bearings.

The creation of a quantity-dependent cost model is a challenging task that depends on the company's and supplier's conditions. A simple example is given in [5] for production costs of an automotive internal combustion engine (ICE), illustrated in Fig. 5. With known reference costs C_0 and reference quantity n_{P0} the costs C for each part decrease with the quantity n_P according to Eq. 1.

$$C = C_0 \cdot \frac{1}{\sqrt[3]{n_p / n_{p0}}} \tag{1}$$

This means the costs are decreased by around 20% when the quantity is doubled. As the cost degression due to quantity discounts can also be described by Eq. (1) according to [5], this can exemplary be applied to the bearings inside a gearbox – they are typically purchased parts. For a two-stage helical gearbox, as shown in Fig. 3, a maximum of six unique bearings and a minimum of only one unique bearing is eligible in principle. The cost degression according to Eq. (1) between these two extreme cases is around 45%. However, this only affects the bearing costs. The shafts, gears and housing might become more expensive when such a common-part approach is enforced for the bearings. Moreover, the package demand, the mass and the gearbox efficiency might take unacceptable values. A holistic design method has to consider all influences and only choose a common-part approach to an extent, which is favorable for the optimality of the entire gearbox system.

To include common-part aspects, two possible approaches arise:

a) minimization of the number of unique components (explicit) or
b) consideration of the cost degression (implicit).

The former is (as described above) a restrictive approach as the actual cost benefit is not directly apparent, but it also eliminates the necessity of a sophisticated cost model – the number of unique components is explicitly formulated as an optimization objective. The latter indeed requires a direct description of the cost benefits by the cost model – common-part aspects are implicitly considered by the cost model and do not require an additional optimization objective.

This approach can be extended to not only consider a single design problem but to also consider multiple concurrent design problems and already existing designs.

3 Results

In the following section, the presented design method is applied for a case study. The results are discussed and the found solution set is compared to a conventionally engineered reference solution to verify the method.

3.1 Case Study Problem

The presented case study (see Fig. 1 and Table 1) is based on an enhanced analysis model and synthesis strategy compared to [3]. In addition, a different oil is applied, which affects the design of bearings and gears as well as the efficiency ratings. Therefore, the results are not identical compared to [3].

Table 1. Main requirements of the case study

Total transmission ratio	$9.32 \pm 1\%$
Total center distance	165 mm
Nominal input torque	366 Nm
Maximum input speed	15 000 rpm
Required service life	944 h
Material of shaft & gears	20MnCr5
Oil	ISO VG 46

The optimization is configured to minimize the objectives

a. costs,
b. losses,
c. package metric and the
d. number of unique bearings.

The losses are determined by application of a longitudinal vehicle dynamics model, which is used to simulate the WLTP class 3 cycle [6]. Furthermore, a degressive cost model for the bearings according to Eq. (1) is applied. This means both an explicit and an implicit common-part approach are employed, to allow a comparison of both approaches. Concerning the applicable bearing types for this case study, bearing 1 is fixed to be a grooved ball bearing (GBB), bearing 2 can be a GBB or a cylindrical roller bearing (CRB), while all other bearings can be GBBs or tapered roller bearings (TRBs).

In this case study, only publicly available costs data [7, 8] is applied to demonstrate the methodology. For use in a specific development project, company-specific cost data should be used.

3.2 Case Study Results

The final set of designs contains 534 Pareto-optimal solutions. As a four-dimensional objective space is present that cannot be directly visualized, single two-dimensional projections are used to visualize the results in the following. Furthermore, a fully engineered reference solution for this case study is known (depicted in Fig. 1). This reference solution is compared to the solution set found by the optimization algorithm and used to verify the method.

3.2.1 Efficiency vs. Costs

Figure 6 shows a strong influence of the bearing selection on the trade-off between losses and costs. Solutions with a cylindrical roller bearing for bearing 2 and otherwise grooved ball bearings (CRB, GBB, GBB) show the lowest losses but are rather expensive. Solutions with only grooved ball bearings (GBB, GBB, GBB) represent a balanced trade-off between both objectives and solutions with tapered roller bearings at the intermediate shaft and otherwise grooved ball bearings (GBB, TRB, GBB) show the lowest costs but also high losses. General engineering knowledge implies that grooved ball bearings are favorable for low losses [2]. Frankly, this is not entirely true in this case study. Bearing 2 has to support high radial loads, which requires seriously larger grooved ball bearings compared to cylindrical roller bearings due to their lower load capacity. This also means that frictional forces have a larger lever arm and thus induce a higher frictional torque than cylindrical roller bearings.

3.2.2 Common-Part Approach

As explained earlier, both an explicit and implicit common-part approach are chosen for the optimization. Thus, the number of unique bearings is explicitly minimized and additionally a degressive cost model is applied. That way, the actual influence of the number of unique bearings on the costs and the other objectives can be better visualized - designs with the lowest possible number of unique bearings are present in the solution set regardless of their actual optimality regarding the other objectives.

As can be seen in Fig. 7, a minimum of three unique bearings is found for this design problem. It can be seen that solutions with six (the highest possible number) and five unique bearings show the best package metrics. In fact, changing from six to five different bearings is a design decision that almost has no influence on the package integration in the extreme case. However, changing to four unique bearings has a small impact and changing to three unique bearings has a large impact on the package integration. This outlines the consequences of enforcing a common-part approach by explicitly minimizing the number of unique bearings – other objectives, in this case the package metric, might take unacceptable values.

To visualize the actual relation between the number of unique bearings and the costs in the system context, the projection in the dimensions "costs" and "number of unique bearings" is depicted in Fig. 8. Through use of the holistic gearbox optimization method, more than 25 percent cost reduction can be achieved in the extreme case compared to the reference solution. The global cost optimum is achieved by only four unique bearings. Further reducing this number actually increases the costs again and potentially has unfavorable consequences concerning the other objectives (e.g. the package metric as shown in Fig. 7). In fact, some solutions with five unique bearings are even cheaper than designs with three unique bearings.

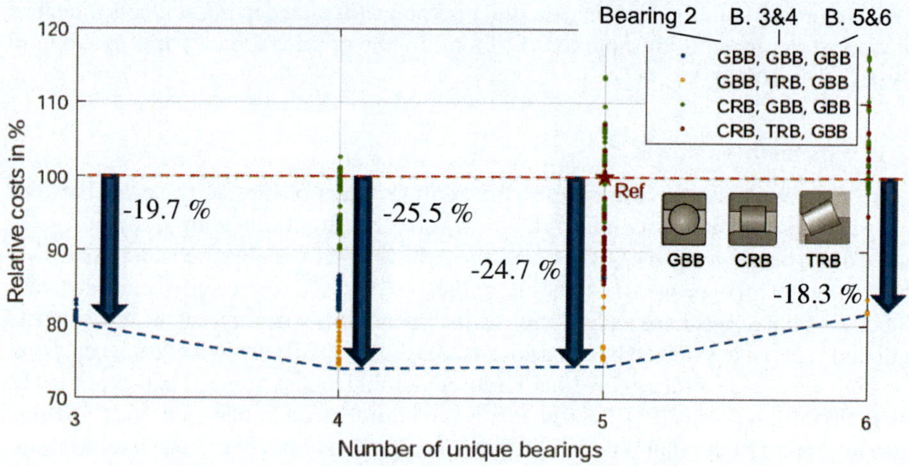

Fig. 8. Projection of the Pareto front in the dimensions "relative costs" and "number of unique bearings". Minimizing the number of unique bearings does not necessarily minimize costs.

Accordingly, the relation between the number of unique parts and the actual system costs is not monotonic. This implies that simply minimizing the number of unique bearings by an explicit approach is not equivalent to an implicit approach, which directly considers the cost degression. Only the latter is capable of effectively minimizing the total system costs by applying common parts.

4 Conclusion

The complex problem of designing gearboxes for eDrives is addressed with a multi-objective optimization method to find Pareto-optimal solutions regarding efficiency, package and costs. An advanced cost model is applied that considers quantity-depended cost degression of purchased parts, demonstrated on the example of bearings. The quantity of each purchased part can be increased by using common parts, with the possible downside of not having a tailor-fit selection and adaptations on system level may be required. In a use case, the resulting system cost savings caused by common bearings in the gearbox are investigated for different levels of communality. Additionally, also the number of common bearings is explicitly optimized to show the effect of both presented strategies.

Through use of the holistic optimization approach, more than 25 percent of cost savings can be achieved compared to the reference solution in the extreme case. Concerning the common part approach, it can be concluded that no direct correlation between a simple explicit common-part approach (minimization of the number of unique bearings) and the actual cost benefit is present. Although an explicit approach does not require a sophisticated cost model, the resulting solution set might lead to wrong implications for the system costs and has to be handled with care. Therefore, it is recommended to apply the implicit approach that directly reflects the quantity-dependent cost degression of using common parts in a holistic system optimization method, to consider all relevant factors simultaneously.

Acknowledgements. This work is supported by Magna International and Magna Powertrain. To reduce development time and effort in future, Magna Powertrain is investigating a holistic optimal design approach for electric drive systems in co-operation with Graz University of Technology. This method supports early development stages by solving conflicts between efficiency, performance, package, weight and costs. It offers engineers and decision makers a quantitative basis to process specific product characteristics with respect to individual requirements and priorities. The presented gearbox design methodology [3, 4] and common-part considerations are part of this project focusing on eDrive system perspective [9, 10], also including the electric machine and the power electronics.

References

1. ISO 6336 (2006) Calculation of load capacity of spur and helical gears. Standard, International Organization for Standardization
2. Technische Grundlagen und Produktdaten zur Gestaltung von Wälzlagerungen (2014). https://www.at.schaeffler.com/remotemedien/media/_shared_media/08_media_library/01_publications/schaeffler_2/catalogue_1/downloads_6/hr1_de_de.pdf. Accessed 04 Oct 2019
3. Hofstetter M, Lechleitner D, Hirz M, et al (2018) Multi-objective gearbox design optimization for xEV-axle drives under consideration of package restrictions. Forschung im Ingenieurswesen 82(4):361370. Distributed under the Terms of the Creative Commons Attribution 4.0 International License (http://creativecommons.org/licenses/by/4.0/)
4. Lechleitner D (2019) Design method for optimized gearboxes in electric vehicles. Master Thesis, Institute of Automotive Engineering, Graz University of Technology [unpublished]
5. Ehrlenspiel K, Kiewert A, Lindemann U et al (2014) Kostengünstig Entwickeln und Konstruieren. Springer-Verlag, Berlin Heidelberg
6. Tutuianu M, Marotta A, Steven H, et al (2013) Development of a worldwide harmonized light duty driving test cycle (WLTC). Technical report, UNECE
7. Kugellager – Direkt Ltd. & Co KG (2019) kugellager-direkt.de. http://www.kugellager-direkt.de. Accessed 04 Oct 2019
8. Kugellager-online GmbH & Co. KG (2019) ekugellager.de. https://www.ekugellager.de. Accessed 04 Oct 2019
9. Hofstetter M, Hirz M, Ackerl M (2016) System design optimization of xEV-axle drives with package restrictions. In: Proceedings of the FISITA 2016 world automotive congress
10. Hofstetter M, Hirz M, Gintzel M, Schmidhofer A (2018) Multi objective system design synthesis for electric powertrain development. In: 2018 IEEE Transportation Electrification Conference and Expo (ITEC). Long Beach. https://doi.org/10.1109/ITEC.2018.8450113

The Ideal Future Hybrid Powertrain

How to Conduct a Comprehensive Simulation and Design Analysis, Identifying Ideal Hybrid Powertrain Architectures and Operating Strategies

Y. Jokmin, F. Holldorf(✉), E. Montefrancesco, K. Loock, and N. Moeller

hofer forschungs-&entwicklungs GmbH, Rebenring 31, 38106 Brunswick, Germany
Fabian.Holldorf@hofer.de

Abstract. Studies in the automotive area agree that the number of electric vehicles has increased over the past years and will increase further in the future. The high number of electric vehicles will not be high enough to fully replace the conventional powertrains utilizing and internal combustion engine (ICE). Instead, conventional and electric powertrains are expected to co-exist in the future [1]. The closest co-existence of these two powertrain systems is the hybrid powertrain. Hybrid vehicles use two different sources of power, most frequently petrol and electricity. There are several ways to arrange these power sources in the powertrain. Solutions range from both sources providing propulsion in parallel to one feeding energy into the other in a serial setup.

None of the hybrid systems proofed to be the most favourable one so far. Car manufacturers prefer certain setups to others while competitors have contrary preferences. One setup is becoming more and more popular: The so-called dedicated hybrid transmission (DHT). The DHT attaches the electric motor (or multiple) to a central point in the transmission while common hybrids tend to connect their electric machine either before or after the transmission. The DHT is flexible in its attachments and also concerning drive modes as well as e-motor and battery size. It can thus cover a wide range of applications. Is the DHT therefore the ideal future hybrid powertrain?

This paper investigates hofer powertrain's approach to hybrid transmission development from the empty sheet to the final hardware product. The focus lies on the first steps to find the concept most suitable for the respective application. Mechanical components are considered rather than other key components like e-motors, software or system integration. The concept development follows automated analytic approaches looking at all hybrid layouts imaginable and narrowing down the solution pool step by step. Each step dismisses hybrid layouts from the pool when violating key parameters. The key parameters originate from various engineering fields e.g. manufacturing, NVH or efficiency.

hofer powertrain's development process shows that most evaluation criteria are tied to the respective application. Every application has different key parameters that automatically lead to a different ideal solution. Hybrid transmissions, common ones as well as DHTs, only live to their full potential whenever tailored to their respective application. There is – so far – no hybrid transmission setup that proofs to be the best for all hybrid applications and thus no overall ideal future hybrid powertrain.

Euroforum Deutschland GmbH (Ed.): *CTI SYMPOSIUM 2019*, pp. 236–248, 2021.
https://doi.org/10.1007/978-3-662-61515-7_22

Keywords: Hybrid transmission · Transmission layout · System analysis

1 Introduction

Hybrid cars are selling better than ever. Kraftfahrt-Bundesamt [2] shows the development of hybrid car sales from 2013 until today. The trend is clear: More and more customers purchase hybrid cars. This trend is driven by a greener mind-set of the customers as well as by legislative guidelines in many countries. Car manufacturers meet this increased demand with a wider range of hybrids in their portfolios. hofer powertrain witnesses this trend as well and is currently working on multiple hybrid powertrains for various customers all over the world.

The existing hybrid transmissions on the market show very different levels of hybridization. They vary from add-on solutions for existing powertrains to new P2-developments and go all the way up to newly developed so-called **D**edicated **H**ybrid **T**ransmission (DHTs). DHTs have one or more electric motor required for the transmission function. A DHT needs its electric machine to provide propulsion not only as a start-up element. Table 1 gives an overview of modern hybrid structures of current hofer projects and of the drive modes they provide. It outlines how wide the range of hybrid structures is. This diversity is required to meet the customer requirements that can be substantially different depending on their applications.

Table 1. Overview of current hofer powertrain's hybrid transmission projects.

Project	Structure	Start-stop	Generator	E-drive	Sailing	Boost	Reverse	eCVT
1	DHT	+	+	+	+	+	el.	+
2	P2	+	+	+	+	+	el.	–
3	DHT	+	+	+	+	+	el.	+
4	P4	+	+	+	+	+	el./mech.	–
5	P 2.5	+	+	+	+	+	mech.	–
6	DHT	+	+	+	–	+	el.	+

The first hybrids on the market were add-on solutions integrating an electric motor into an existing ICE powertrain – this is a common application for a P2 hybrid. This setup is easy to install and rather cheap but cannot make use of all the benefits of the hybrid powertrain to full extent especially facing the critical package of transversal powertrains. A more feasible solution for transversal applications is the P2.5 hybrid, where an electric motor is combined with a dual clutch transmission. The EM is attached to one of the drive shafts which enables to keep similar package dimensions regarding wheel track. Additionally, no clutch between ICE and EM (K0) is required. When requiring additional hybrid functionalities, a DHT is especially for transversal powertrains the solution of choice for max. efficiency and optimum hybrid performance.

hofer powertrain conducted all of the above projects. The projects #1 and #2 will be considered in this paper in detail. Beside many other requirements, the basic evaluation of a powertrain is done on the vehicle traction performance. Figure 1 shows the traction diagram of these two projects. It shows the traction force at the wheels over the vehicle speed. The dots represent customer requirements of the projects #1 in red and of project #2 in green. The lines represent the traction force required for driving level (full line) and on a slope (dotted line). The coloured boxes cover the operation points provided by the respective powertrain. Project #1 is an SUV application with mediocre max. speed and high traction force. Project #2 is a super sports application with low climbing ability but very high speed. The ideal future hybrid powertrain must be able to fulfil the traction requirements of both projects.

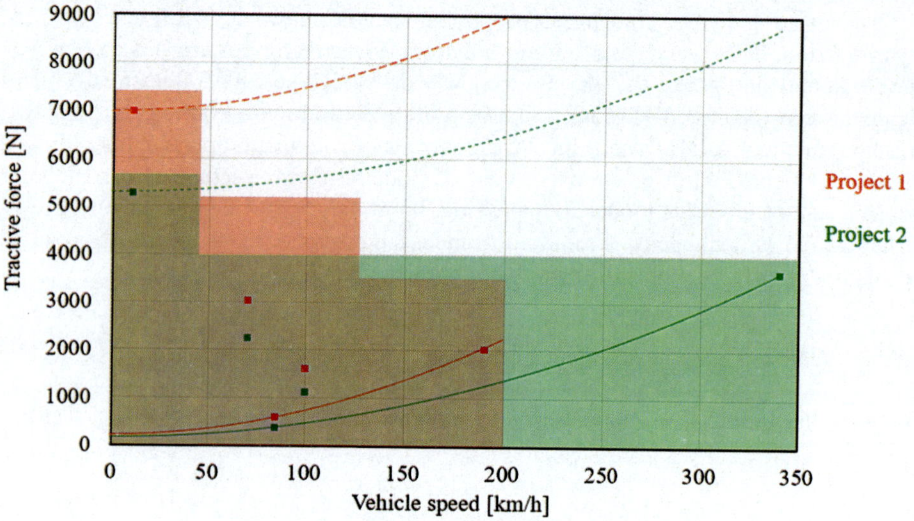

Fig. 1. Traction requirements for project #1 and #2.

hofer powertrain chose very different setups for both projects. The traction requirements as per Fig. 1 were too different to be met with one transmission and at the same time provide the desired drive modes, efficiency and NVH behaviour. Project #1 is a dedicated hybrid where two electric machines provide propulsion in addition to an ICE. This DHT combines a planetary gearset with two spur gear stages to provide the optimum combination of traction force and efficiency with two gears. Project #2 is a P2 hybrid for a super sports application. The electric motor provides electric drive for reverse and inner city driving. The ICE coupled with a DCT provides the required high power (>800 Nm input) and the e-motor boosts whenever necessary to enhance the sportive performance.

Figure 2 shows the traction diagrams of the project #2. The super sports car identifies by the number of gears, the high vehicle speeds and the enormous traction forces for quick accelerations. This is a niche hybrid application but it shows the vast range hybrids can be reasonably applied to. The P2 is the most suitable setup for this type of application. The combination of e-motor and battery are big enough to benefit the driving performance but at the same time light enough to not inhibit the car from reaching its maximum speed even at steep slopes. A DHT e.g. would be rather complex and integrating it into the longitudinal package would be challenging. The benefits of the DHT like a possible eCVT-mode would not be of much use in this super sports application that is designed for high speeds rather than comfortable take-off.

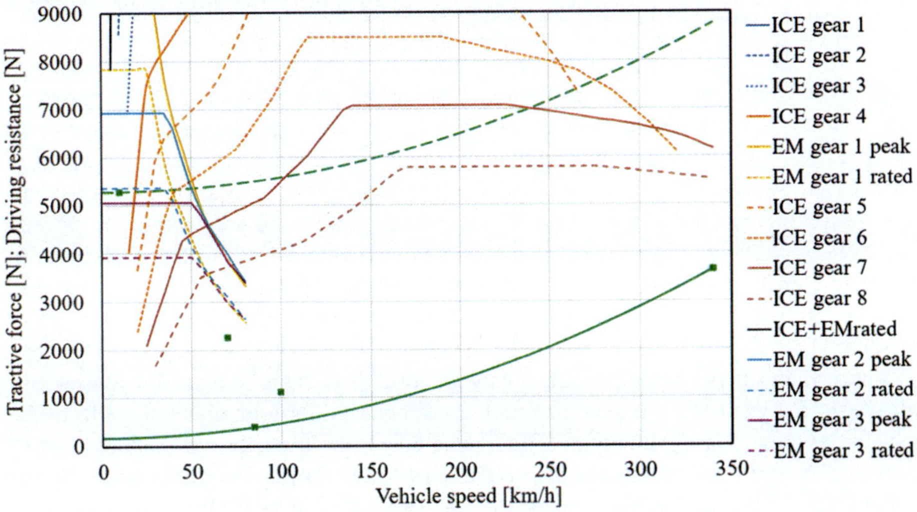

Fig. 2. Traction diagram of project #2.

The following chapters outline hofer powertrain's approach to hybrid transmission design from scratch to the first hardware product. It aims to find the ideal future hybrid powertrain and outlines why this attempt is ambitious.

2 Methodology

The challenge of every development process is to find the solution that suits our customer's application best. This chapter aims to introduce hofer powertrain's approach to this challenge. Starting point is usually a customer request with designated boundary conditions e.g. vehicle class, desired performance, type of powertrain etc.

The field of DHTs is vast and the number of solutions almost endless. The first step must therefore be to narrow down the solution set. hofer powertrain uses self-developed software code to run through the following steps to eliminate undesired solutions:

1. Combination of gears. This step considers both planetary and spur gearsets. It eliminates all solutions that do not provide the required number of gears, drive modes or include too many shifting elements or free-wheel clutches. Also, boundary conditions like max. speed and climbing ability are considered here and unsuitable solutions are dismissed.

2. Feasibility. This step takes the solutions resulting from step 1 and checks if they are feasible to assemble and manufacture. This step considers not only penetration of shafts but also package criteria (i.e. rough diameters and possible axial widths) as well as torque transfer capabilities of the gears and the shifting elements

The remaining pool of solutions is still vast, and it requires algorithms to check the solutions for further criteria. The next criteria hofer powertrain investigates for are NVH and efficiency.

Powertrain acoustics become more important in alternative drive systems. Conventional powertrains utilized a noisy combustion engine drowning most of the noises of other powertrain components. Alternative drive systems lack this property making powertrain noises more relevant. The end customer must not hear any noise in any operation point: Be it rattling, whining or howling, be it caused by the wheels on the pavement, the whimpering motor or the rattling of gears. hofer powertrain as an expert on powertrain NVH focuses on transmission- and motor-based noise. They developed methods to predict the noise emission of the transmission. These methods are algorithm based and simulate multiple aspects of the transmission's NVH behaviour from excitation to noise emission.

The excitation is considered on a general level first. This general level stretches from basic design criteria (e.g. no common divisor for meshing numbers of teeth) to the frequencies these gears excite with. The frequencies result from the numbers of teeth of every gear in combination with the respective shaft speed. Further frequencies introduced by the e-motor(s) and perhaps auxiliary drives like an oil pump. All frequencies must not be too close to each other to prevent resonance. hofer powertrain's code runs through all reasonable combinations of numbers of teeth at all possible shaft speeds to find those numbers of teeth that meet the frequency safety margin.

The efficiency of a DHT is mainly driven by the number of shifting elements, the number of gears meshing and lubrication aspects (e.g. the lubrication concept and the amount of oil inside the transmission). hofer conducts a full efficiency calculation to enable a general evaluation of the potentials based on key parameters. This investigation considers the potential for every drive mode and evaluates if these drive modes can be used in a reasonable way. The evaluation if using a drive mode is desired or not is either based on customer criteria (e.g. full electric reverse gear) or based on other boundary figures e.g. a minimum required efficiency. DHT concepts with a higher potential to reach max. efficiency values are favoured to those with less potential and the most favourable ones are investigated concerning their NVH behaviour in greater detail. A more detailed investigation of the gearsets efficiency is conducted at a later stage of the development process.

The solutions remaining in the solution set all lead to functional transmissions. They will meet the requirements concerning driving performance, drive modes, NVH, efficiency and complexity on a basic level. Further aspects may be considered to further break down the solutions. Reviewing the results with stricter boundary conditions (e.g. higher efficiency values) will show differences between them. The selection of a concept must be conducted together with the customer with all above calculation and simulation results.

The concept is selected now and needs to be designed. The concept design includes the crucial components of the gearset: E-motor(s), inverter, gears, shafts, bearings, housing(s), park lock and lubrication system as well as small parts (screws etc.) to keep them all in place. These parts are important to look at individually but their combination to a system is just as important. The software hofer powertrain uses KISSsys to consider both the overall system and the individual component. The basic geometry of the gears e.g. together with the centre distances is designated by the concept investigation. These pieces of information result in a quick 3D model in KISSsys already at a very early development stage. This model is crucial e.g. to kick-off the housing design that can be conducted parallel to gears' macro geometry design.

The next step is the detailed design of the gears: The micro geometry. The micro geometry deals with minimum changes on the gear geometry to improve the rolling motion of the gears. The micro geometry influences the teeth's entry into and exit from the mesh, it affects the transmission error as an indicator for the mesh's noise behaviour, and it can tackle manufacturing irregularities e.g. tooth flank twist related to manufacturing procedures. The transmission error is the difference between the gear's ideal and actual radial position on the pitch circle. Small deviations in the transmission error – even a few microns – affect the transmission's noise emission. The transmission error is load dependent and thus usually displayed over the transmission torque range. The design target is to keep the transmission error to a minimum in frequently used operation points. hofer powertrain's automatized transmission error analysis investigates the transmission error for countless combinations of micro geometries. The analysis also considers the sensitivity of the transmission error to housing deformation, tolerances of the bearing seats and shaft deflection. The transmission error is particularly sensitive to these three types of deviations due to the usually short shafts in alternative drive systems. Deviations in the bearing area have a greater impact on the gear mesh on short shafts compared to long shafts due to the lever principle.

The last main step of the concept design process is the housing. The housing plays a central role not only as it takes up the forces resulting from the torque transmission but also because it emits vibration and thus noise to the environment. The housing must sustain all loads resulting from drive, coast, park lock and shock events and must therefore be designed stiff yet light. It is usually made from aluminium alloy in passenger car applications. These alloys are a light base material that deforms under the loads from the gear set. The deformation of the bearing seat has direct impact on the shaft and together with the shaft deformation, the position of the gear and thus the transmission error are influenced. Every change to the housing design changes the stiffness of the bearing seats and thus the transmission error. This relation is the main driver for hofer powertrain to kick-off the housing development as early as possible in the design process. The whole system needs to be considered to find the optimum solution for the application e.g. due to the high impact of the housing deformation on the transmission error. This requires multiple loops of running through the algorithms described above which is only possible due to the high level of automatization.

3 Project Application

This chapter describes the development process of a recent hofer powertrain project. The development process followed the methodology described above and led to a DHT development from scratch to a hardware product. This project was referred to as "Project #1" in Chap. 2.

The customer in this project was a Chinese car manufacturer aiming to establish themselves in a unique position on the market with this powertrain. The main boundaries were typical of an SUV application. Optimized fuel efficiency was the primary sales criterion and the combination of driving performance with optimized NVH behaviour also had a high priority. The transmission was supposed to offer powerful EV-drive with two e-motors but also hybrid and eCVT modes as well as a pure ICE drive. The ICE and both electric motors were provided by the customer.

hofer powertrain's algorithms quickly found combinations of feasible planetary gearsets that fulfilled all the drive mode and performance criteria. They also passed the feasibility and assembly check. Figure 3 shows the resulting traction diagram of this vehicle application. It displays the traction force over the vehicle speed considering lines of driving resistances at constant slopes (between 0 and 33%) as well as significant combinations of force and speed resulting from customer requirements:

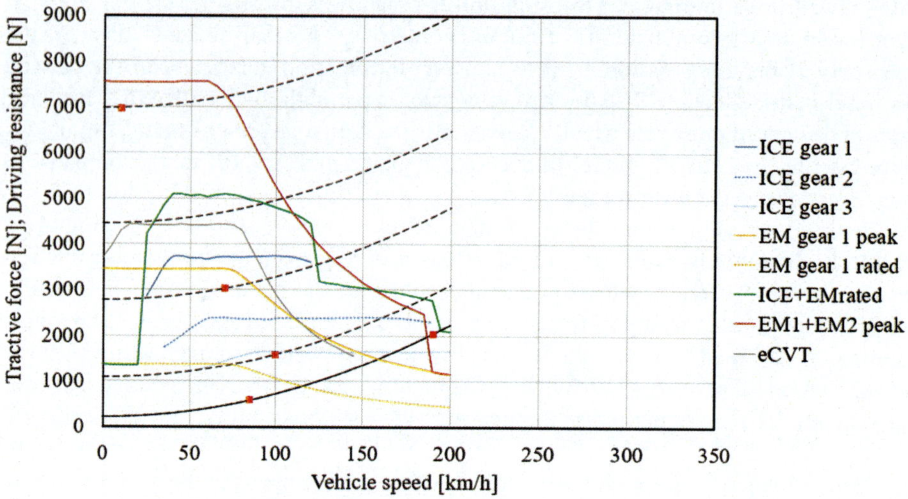

Fig. 3. Traction diagram of project application.

- T1: Max. vehicle speed in the plain
- T2-4: Partial load at certain slopes/speeds
- T5: Climbing ability

The coloured lines show the potential of every drive mode of the gearset. Every line represents a drive mode e.g. the green line stands for propulsion provided by the ICE together with the e-motor in continuous drive mode. All operation points below this green line can be provided by these two drive units together. Most operation points can be provided by multiple drive modes and one of the challenges of the DHT development is to find the most efficient drive mode for every operation point. This will be investigated at a later stage of the development process.

The initial NVH check guided the selection for numbers of teeth. Their excitation frequencies had the required distance to neighbouring frequencies over the whole speed range. The gearset related frequencies must fit around the e-motor frequencies with a certain distance ΔHz to prevent resonance. Overlapping frequencies would be subject to resonance and thus prone to loud noise emission and premature part failure.

The electric motors have fixed excitation frequencies rooted in their design. Figure 4 shows these frequencies in red and blue and the gearset frequencies of a possible combination of numbers of teeth in yellow and green over the vehicle's speed range. The min. ΔHz between all frequencies, usually a single-digit percentage value, is fulfilled in this scenario. This combination of numbers of teeth remained in the solution pool and became subject to further investigation.

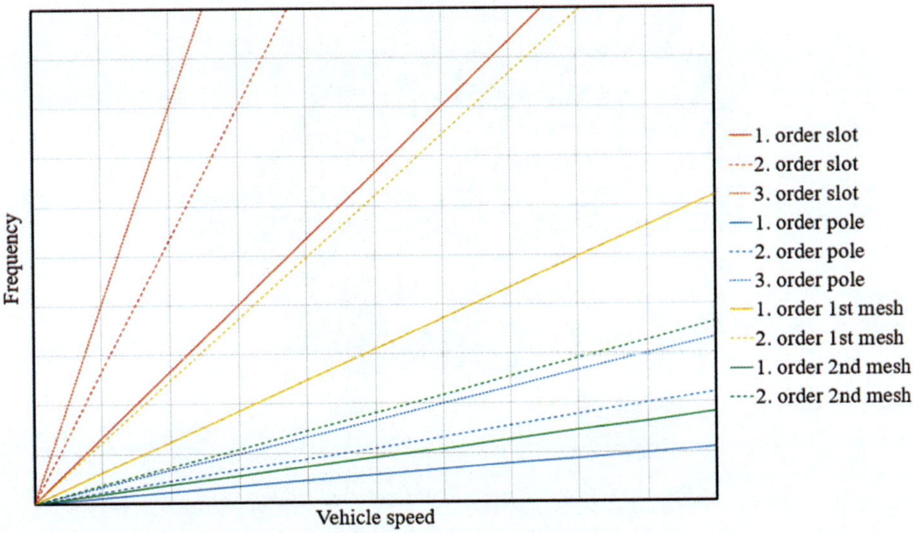

Fig. 4. Frequency check of the project application.

All gearsets fulfilling the above criteria were now checked for their efficiency. The planetary gearset had its efficiency dependent on the rotation directions. Other aspects, e.g. a rough estimation of the oil level was considered and the efficiencies at different operation points were calculated. Gearset with a good efficiency in frequently used stages were favoured to those with less efficiency in the same areas. A more detailed efficiency analysis also beneficial for the hybrid strategy was conducted at a later stage of the development process.

The remaining solutions fulfilled many criteria e.g. feasibility, performance, basic NVH and efficiency. These solutions were looked at together with the customer. Both parties agreed on a reasonable concept to move on and to go through the subsequent steps of development with. The selected concept ran through the standard macro geometry design process. Gears were defined, bearings were chosen and shafts designed. This process is standard to hofer powertrain and in transmission development in general and not investigated in detail here. The detail design of e-motor(s) and power electronics are also key knowhow of hofer powertrain. The e-motors in this project were provided by the customer so the focus of this paper lies on the mechanical gearset components.

Figure 5 shows the concept design of this project application. The two electric motors feed their torque and speed into the transmission on different shafts one through a spur gear the other one together with the ICE through the planetary gearset. The clutches and brakes in the gearset allow switching between different gears and drive modes. All motors are crucial for the application. Without either one of them, the system would be statically undetermined and not functional making this transmission a DHT.

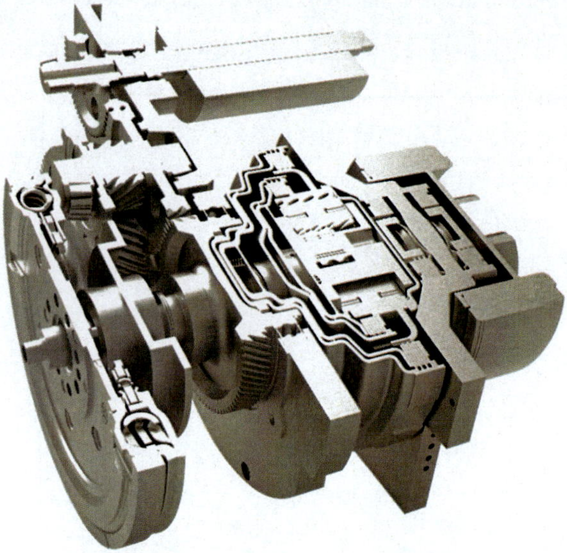

Fig. 5. Concept design of project application.

The design displayed in Fig. 5 is the first draft. The subsequent detail design considers the part applies more detail and takes NVH behaviour and efficiency into account. The NVH investigation aims to find designs with an NVH behaviour that meets the customer requirements and fulfils hofer powertrain's design criteria. The customer requirements were rather vivid stating a max. audible noise during different load stages. This is hard to relate to at this early level of development which was the main driver for hofer powertrain to develop an automatized tool aiming to predict the noise of the transmission. The tool is based on MATLAB "Simscape". It considers the modes of shaft movement resulting from the calculated excitation frequencies via the vibration transfer function. Every excitation frequency leads to a certain behaviour of the shafts, the modes. The amplitude of the excitation defines the extent of that behaviour. The combination of type and extent of behaviour is evaluated and if identified as critical it is tackled e.g. by stiffer shaft designs. hofer powertrain's method to evaluate the severity of said combinations is proofed by comparing calculation results with acoustic measurements during testing as shown in Fig. 6.

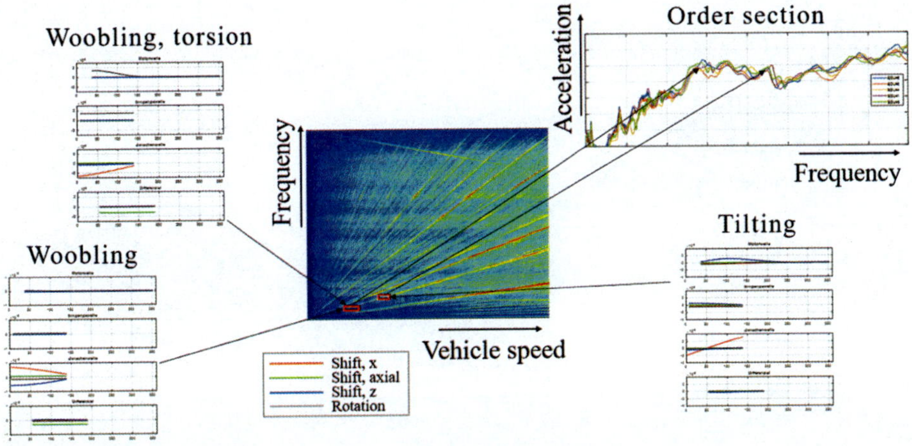

Fig. 6. hofer powertrain's NVH prediction method.

The Campbell diagram shows sound levels from low (blue) to high (red). The excitation frequencies are visibly as beams eradiating from the origin. The two small boxes highlight high sound levels related to shaft modes. The top graph shows measurements at this frequency with different electric machines conducted in parallel. All electric machines show similar behaviour leaving only the shafts as a root cause of the high sound level. The graphs to the left and right of the Campbell diagram show the relevant modes of the shafts. Wobbling and tilting shafts result in inaccurate gear meshing and thus noise emission. These modes and the resulting shaft motions can be identified in an early development stage. They can be changed e.g. by design modifications to detect and prevent noise emission even before hardware testing. This investigation runs through multiple loops to consider changes on gears, shafts and housings.

Another detail design criterion affecting the transmission to a significant extent is the micro geometry of the gears. The micro geometry mainly affects the transmission error and the contact pattern. The contact pattern is a design criterion with great impact on the durability of the teeth while the transmission error is responsible for noise development and influences the end customer directly. It is the engineer's task to find a suitable balance between them. The transmission error was evaluated more important by the customer for this project. The contact pattern was prone to more compromises therefore. hofer powertrain analytically investigated all combinations of transmission error and looked into the contact patterns of the most promising ones. The results of the analytical approach are shown in Fig. 7. These results are from the final loop – there were no changes done to the gearset afterwards. It shows the peak-to-peak transmission error in microns over the transmission torque range from 0 to 100%. Every line represents a different set of micro geometries. The sets are combined analytically with strategic boundary values.

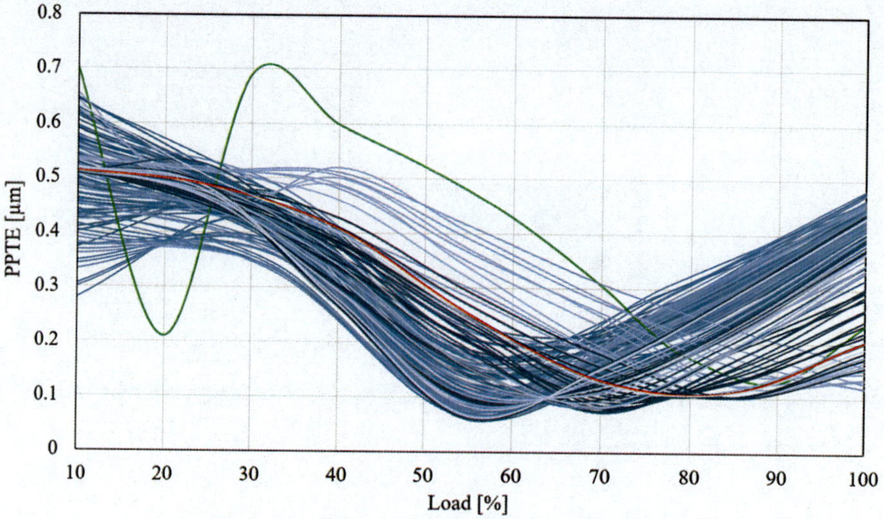

Fig. 7. Results of the automatic PPTE calculation.

hofer powertrain's engineers selected the red graph for this project application. The graph is above many others in areas of mediocre loads but far below them at high loads. The green graph is based on the same micro geometry as the red one but applied to the gearset of project #2 (c.f. Chap. 2). It shows that the micro geometry needs to be individually designed and cannot be transferred from one transmission to the next and lead to the same results. The analytic creation and subsequent selection are required for every mesh anew. Another factor that led to the decision in favour of the red graph was the tooth contact pattern which is strongly affected by the gear's micro geometry. It usually travels over the tooth width due to flattening effects, the deflection of gears, shafts and housings and manufacturing tolerances. The contact pattern should be radially central at all times and should not be close to the tooth's edges to avoid edge wear.

For a low fuel consumption, the system efficiency is evaluated in all possible modes of operation. This is done by an automated calculation, which can consider the engine/electric motor maps and simulated transmission efficiency. For DHTs the number of driving and hybrid modes can vary over a wide range (Fig. 8).

The above steps are only a brief glance at hofer powertrain's transmission development process. The whole development process contains more aspects and those aspects outlined in greater detail than shown above. This extract only shows crucial aspects of the development and provides guidance on the journey to find the ideal future hybrid powertrain.

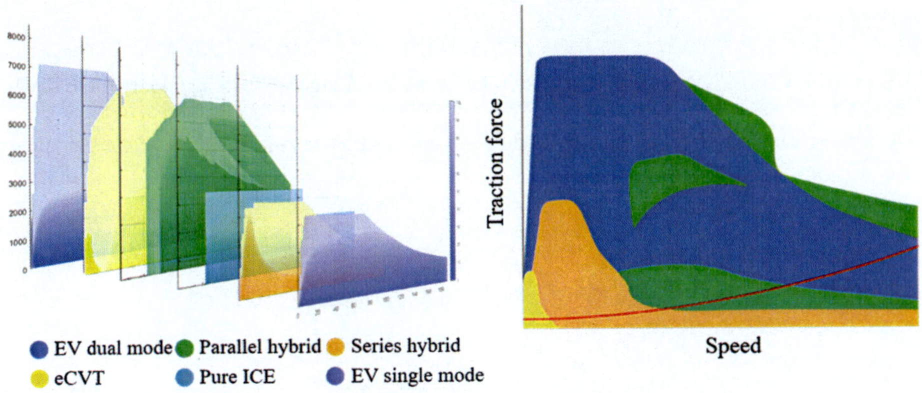

Fig. 8. Efficiency calculation.

4 Summary, Conclusion and Outlook

This paper shows the efficient hofer strategy to find best transmission architecture for requested powertrain concepts. Based on the high diversity of requested powertrains the toolbox should be very flexible.

Starting with a semi-automatically finding of best gear set structure with "hofer transmission layout creator", hofer uses the toolbox "TE automatic screening" to investigate from huge amount of micro geometry combination the best solution with small PPTE, with low sensitivity over tolerances and sufficient gear contact pattern. This strategy gives a very good chance to find a solution with good NVH behaviour and reduces the number of necessary testing loops during validation.

The result of this systematic approach shows that there is not only one ideal hybrid concepts available for all applications. The requirements to the powertrain differ very strong and so the development toolbox need to be very flexible in usage. From hofer point of view to keep this flexibility the best solution is to have high automatization grade (to investigate huge parameter fields) combined with high experienced manual decisions (set boundaries very carefully). This combination gives the best balance between effort and development time and high level of NVH prediction.

Abbreviations

ICE	Internal combustion engine
DCT	Double clutch transmission
DHT	Dedicated hybrid transmission
NVH	Noise vibration harshness
SOC	State of charge
SUV	Sport utility vehicle

References

1. IEA (2019) Global EV Outlook 2019. IEA, Paris. www.iea.org/publications/reports/globalevoutlook2019. Accessed 11 Nov 2019
2. Kraftfahrt-Bundesamt (2017) Neuzulassungen von Kraftfahrzeugen nach Umweltmerkmalen (FZ14) Accessed 11 Nov 2019

E-FDU: An Innovative Double Motor, Disconnectable Front Electric Drive Unit for Ferrari Sport Car Application

Fabio Irato[1](✉), Carlo Cavallino[1], Gianluca Quattromani[2], Giulio Lapini[3], and Giuseppe Manici[4]

[1] Dana Graziano, Turin, Italy
fabio.irato@dana.com
[2] Know-how and Simulation Department, Ferrari, Maranello, Italy
gianluca.quattromani@ferrari.com
[3] Application Department, Ferrari, Maranello, Italy
[4] Transmission Design and Development, Ferrari, Maranello, Italy

Abstract. Hybrid technology offers key opportunities for fuel economy and performance enhancement also in the sport cars application. The main driver of the technology is the vehicle requirements definition and the development of the components focused on efficiency, performances and weight reduction.

Ferrari and Dana Graziano, together with sister company Vocis and an electric motor partner, has developed and taken into production in less than 24 months a novel P4 hybrid powertrain in the form of a dual motor front torque vectoring axle. The unit is capable to drive the front wheels of the vehicle independently, enhancing traction and improving vehicle dynamics, providing at the same time full electric drive capabilities and fuel consumption reduction. The system comprises a twin independent disconnectable transmissions driven by two high revolution (25000 rpm) and high-power density electric motors (up to 162 kW in total).

The front axle is also responsible for EV mode and electric reverse.

Both left and right transmissions integrate an innovative electromagnetic disconnecting device with integrated electronic HW and advanced controls developed by Vocis and Dana Graziano. Advanced control techniques enable transmission disengagement\engagement in 100 ms and replace entirely traditional actuation systems reducing overall weight, cost and vehicle integration complexity.

Ferrari developed an optimized system control in terms of e-motors torque and transmission disconnecting strategies to achieve overall best vehicle performances.

E-AWD and torque vectoring capability improve acceleration performances, vehicle dynamics, and fuel optimization.

Keywords: Disconnecting systems · Electromagnetic actuator · Twin e-motor transmission · High speed revolution · Independent front wheels drive · HV e-motors · Torque vectoring · HEV · BEV

Euroforum Deutschland GmbH (Ed.): *CTI SYMPOSIUM 2019*, pp. 249–282, 2021.
https://doi.org/10.1007/978-3-662-61515-7_23

1 Introduction

Nowadays, electrification represents a fundamental requirement for vehicle integration in order to guarantee upcoming legislation targets and at the same time maximise driving pleasure and reduce fuel consumption.

The new Dana Graziano P4 hybrid powertrain layout (e-FDU) developed for Ferrari application, integrates high revolution e-motors in a twin disconnectable transmission (see Fig. 1)

Fig. 1. e-FDU overview.

Specific vehicle control strategies have been developed to meet hybrid functionalities and vehicle dynamic performances.

The EMA consist of and Electro Magnetic smart actuator able to connect and disconnect two rotating parts acting on a mechanical shifting element (see Fig. 2). The smart adjective is intended to define an integrated system that includes both mechanical and electronic subcomponents in the same package without any need to link a dedicated external ECU. EMA is able to directly communicate with vehicle control unit trough an electrical connector. The shifting movement is obtained by electromagnetic force.

2 Axle Layout

E-FDU architecture has been developed based on vehicle performances, weight and space optimization.

Fig. 2. EMA section installed in the e-FDU.

The main system requirements are:

- Longitudinal acceleration in e-drive condition $\geq$0.4 g
- Max e-drive speed 135 km\h
- Torque vectoring
- Brake regeneration up to 210 km\h
- Max E-FDU weight 60 kg
- Vehicle installation constraints (i.e. steering rack; air conditioning circuit)

In order to meet targets, e-axle with twin independent e-machines and gear mesh layout has been chosen (see Fig. 3):

2.1 e-Machines and Gear Ratio Sizing

In order to meet vehicle max speed and longitudinal acceleration requirements, e-machines have been designed to provide a total amount of 162 kW input power (continuous value), 170 Nm input torque (85 Nm each side) and a max speed of 25000 rpm. The gear mesh consists of a first cylindrical reduction and a secondary stage made by a planetary gear set: the ring is fixed to the chassis, so the torque coming from the last cylindrical gear is transferred to the sun shaft, satellites and the planet carrier. A connection between clutch body ring, welded on the planet carrier and a sliding sleeve (EMA disconnecting system subcomponent) provides torque transmission continuity to output flange and vehicle wheel. This particular layout, with a total gear ratio of 14.576:1 allows to fully optimize weight and space.

Boundary vehicle subsystems as air conditioning circuit, steering rack and chassis, strongly influenced e-axle layout, that assumed a "8" shape to guarantee minimum

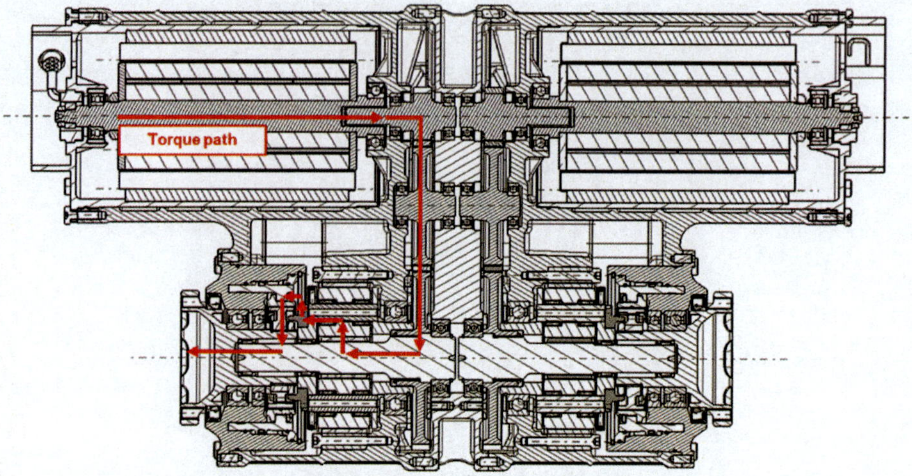

Fig. 3. Initial e-axle layout section.

clearances with other parts during every vehicle working condition (see Fig. 4). To further optimize a compact design, a dedicated external oil lubrication circuit (driven by an electric pump) has been developed.

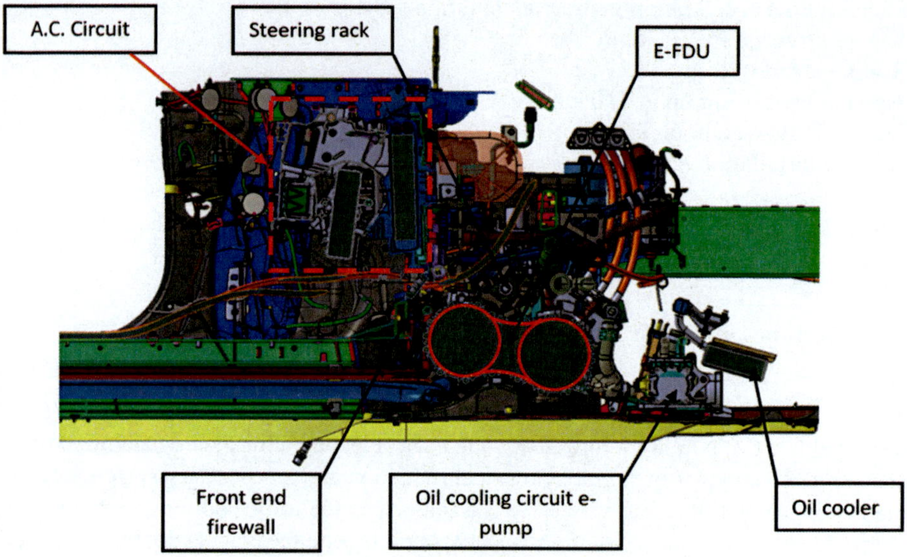

Fig. 4. e-FDU axle installation in vehicle.

3 Disconnecting System

Max e-machine speed and e-axle chosen gear ratio are not compatible with max required vehicle speed, hence, a dedicated disconnecting system able to exclude e-drive motion from ICE has been developed. The concept uses an indexing between Hub and Sliding sleeve from a synchronizer layout to optimize engagement phase, reduce angular clearance and improve NVH behaviour:

The initial actuation system was based on a "classic" hydraulic concept (see Fig. 5) made by a pump, accumulator, control valves and a piston (with position sensor feedback). Engagements and disengagements were made by synchronizing e-machines and wheel speed.

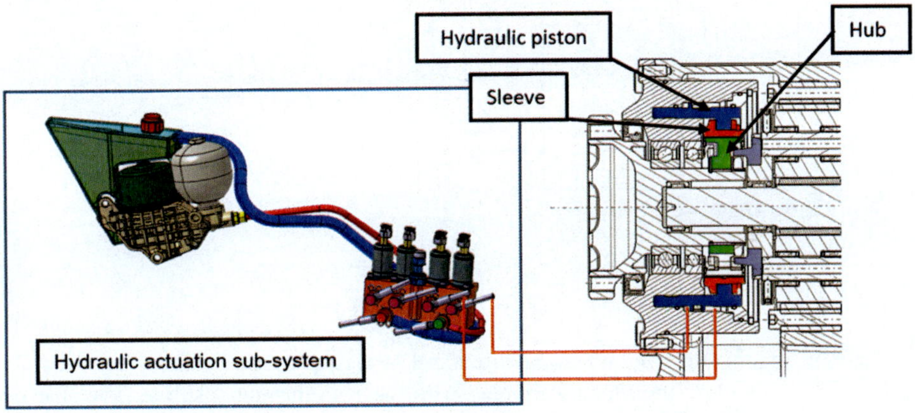

Fig. 5. e-FDU hydraulic actuation system layout.

The hydraulic system has been replaced by EMA, with the following benefits:

1 Less actuation time ➔lt; 100 ms
2 Space saving ➔no external actuation system and electronic control unit required
3 No dedicated hydraulic fluid for actuation ➔no oil management need
4 Vehicle packaging simplified thanks to a compact design
5 Weight saving ➔up to 5 kg

3.1 EMA (Electro Magnetic Actuator)

In the e-FDU, EMA acts as disconnecting system by coupling and decoupling torque path from e-machines to vehicle wheels. It is located at the output shaft of the transmission (see Fig. 6).

Since the FDU is made by twin independent e-motors, two different EMA are installed into the transmission, one for the left wheel and the other for the right one (Fig. 7).

The following section (Fig. 8) shows the EMA subcomponents layout:

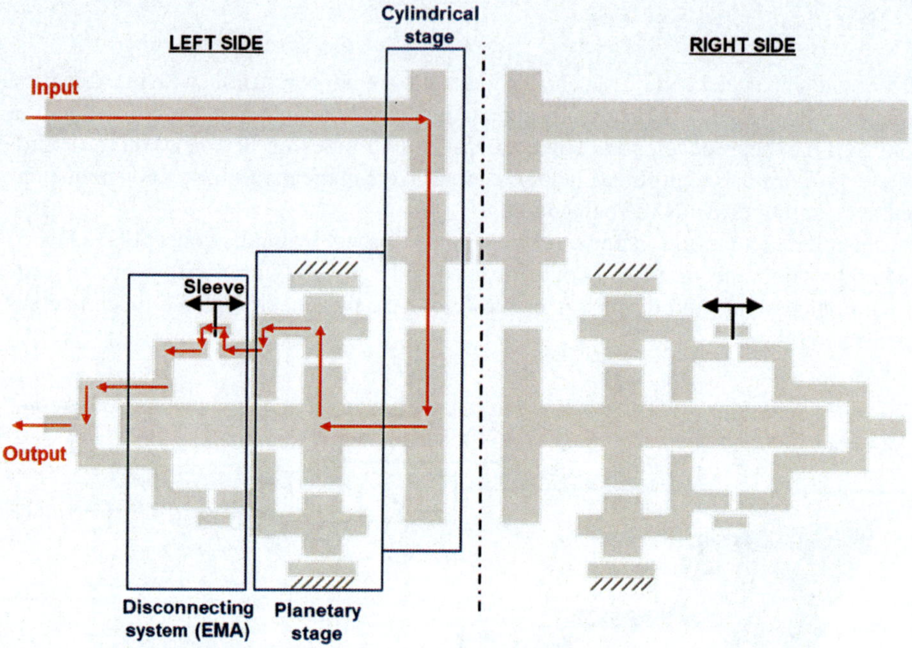

Fig. 6. e-FDU layout.

The mechanical component responsible to allow torque transmissibility between e-machines and wheels is the Sleeve, coupled with the Hub through a spline. The sleeve slides axially, in order to couple and decouple the Hub with the Clutch body ring (welded on the planet carrier). The axial stroke of the sleeve is limited by two end-stops.

Two different coils (see Fig. 9) are installed into the EMA that, properly energized by vehicle battery voltage, induce magnetomotive force that push the sleeve in the two directions. The coils are made by copper wire winded into plastic spools and fixed to the steel stators.

In order to constrain the magnetic flux to the sleeve, non-magnetic steel has been used for engagement and disengagement end-stops (see Fig. 10).

Dedicated simulation software has been used to verify coils design and magnetic flux behaviour, in order to optimize corresponding shifting force on sleeve (2D and 3D simulations) (Figs. 11, 12 and 13):

The EMA main electric system references are listed in the chart below:

EMA is electrically linked to the Vehicle Control Unit (both for power and CAN lines). For reliable operation it is connected to a bus with 2 CAN terminators. EMA is activated by communication via CAN (see Fig. 14). At the end of external communications, the EMA will perform shutdown routines and then stop drawing current from the supply (Table 1).

When an actuation is required, the EMA receives a message from vehicle ECU and starts to energize coils to enable sleeve shifting. In order to perform and engagement, clutch body ring and sleeve angular speeds must be synchronized in a range of

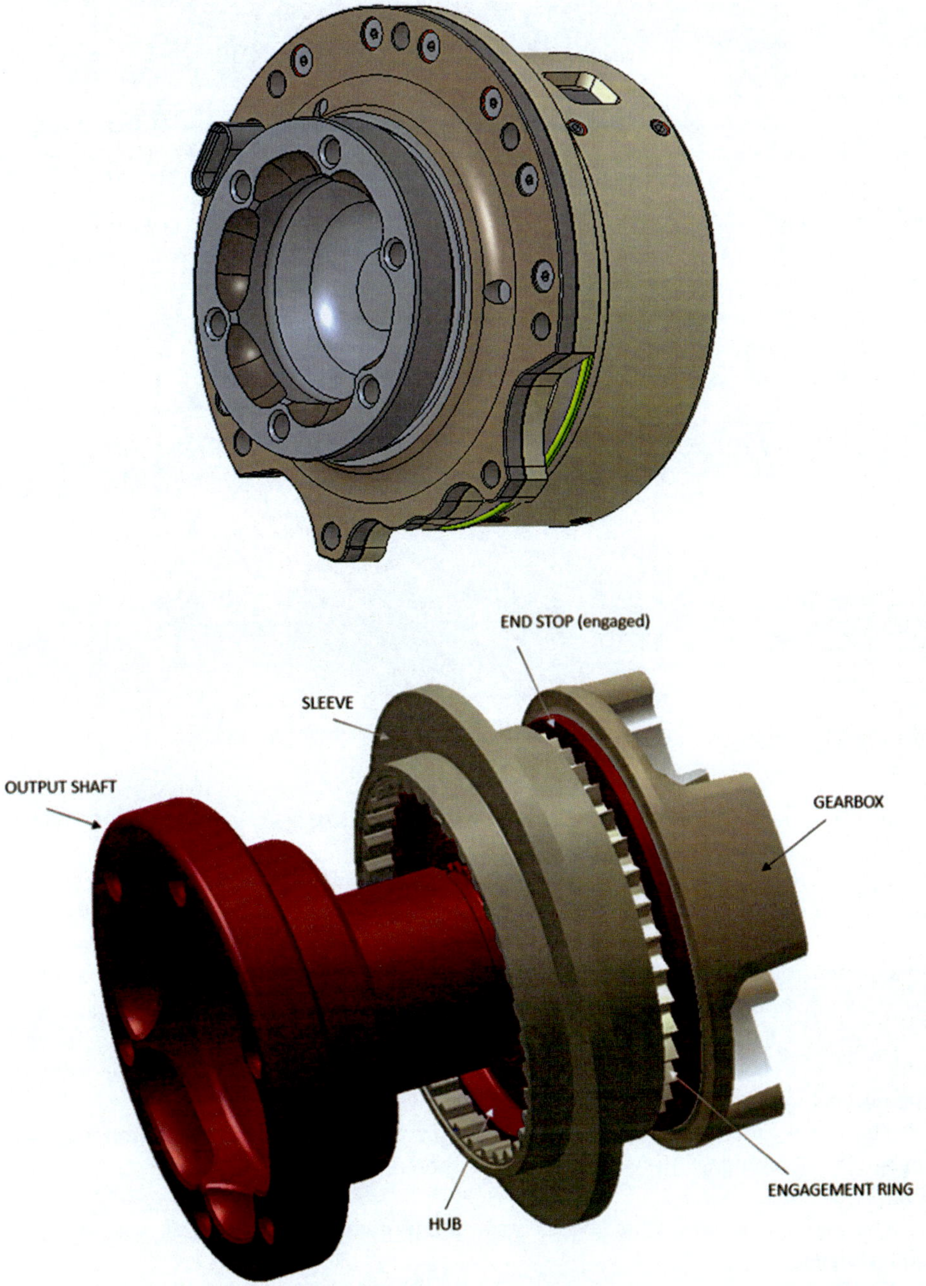

Fig. 7. EMA 3D model geometry.

20 ÷ 30 rpm: this window is guaranteed by the vehicle software control, that comparing output speed sensors and e-machines speeds, drives them to guarantee required target.

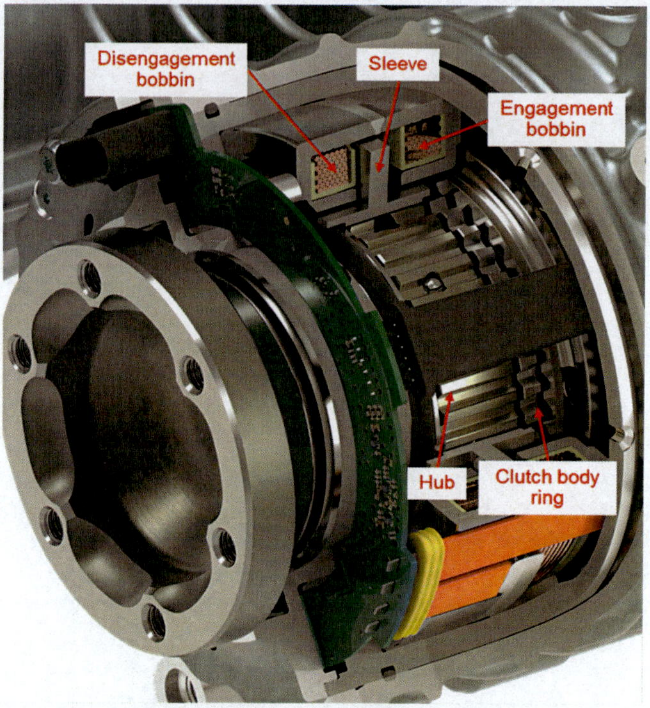

Fig. 8. EMA 3D model section with subcomponents highlight.

In this way, the system is able to reach an actuation time ≤100 ms for temperature above 20 °C (Figs. 15 and 16).

In order to avoid undesired axial sleeve movements, a retention system (see Fig. 17) has been designed for both engaged and disengaged positions (detent): a ball housed into the hub is pressed toward the sleeve that presents two vanes, in order to guarantee a bi-stable position (engaged and disengaged). During actuation, the magnetic force shifts the sleeve that push back the ball into the hub seat till the other vane is reached.

To allow a quick and precise control of the sleeve axial displacement during actuations, EMA has been equipped with an integrated electronic controller and a proportional non-contact inductive position sensor (see Fig. 18):

The signal generated by the position sensor, gives a sleeve feedback position for both engaged and disengaged position, to ensure potential undesired movements tracking (Fig. 19).

The EMA has been validated as stand-alone component according to customer requirements.

EMA patent utility model deposit has been released on 28.11.2017 (Reference number: 202017000136703; Title: "Dispositivo d'innesto ad azionamento elettromagnetico").

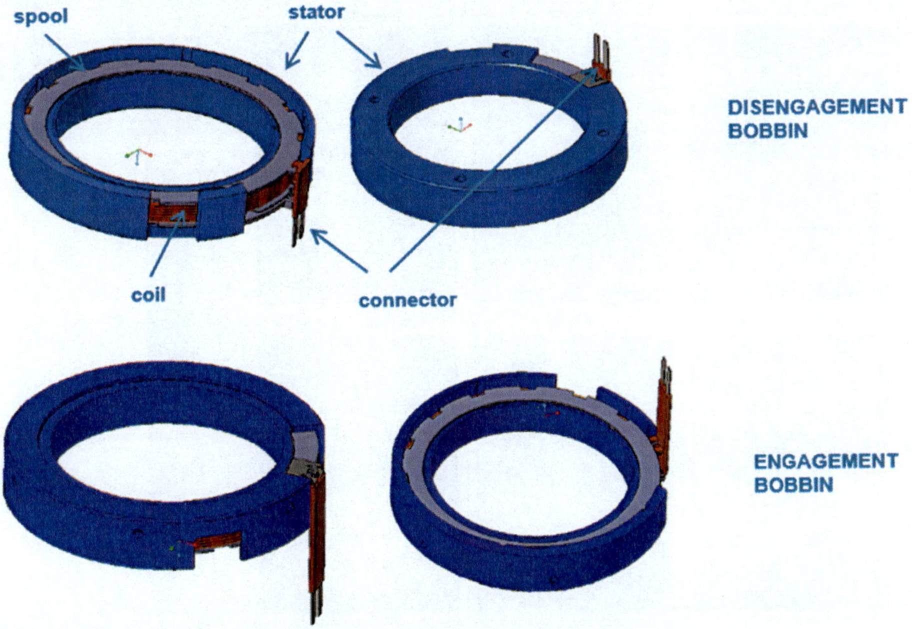

Fig. 9. Bobbin sub-assembly 3d model.

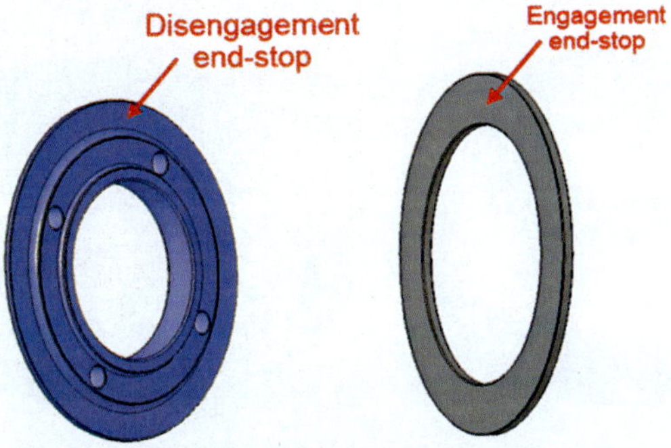

Fig. 10. Engagement and disengagement end-stops.

4 Axle Software Strategy

4.1 Torque Split

Brand new Ferrari SF90 Stradale is equipped with a 1000 hp hybrid powertrain in which a turbo-charged 8-cylinder internal combustion engine is combined with an electric machine on the rear axle (P2) and two independent e-machines on the front axle (P4s; e-FDU):

Several strategies had been developed by Ferrari Controls Department in order to integrate such propellers:

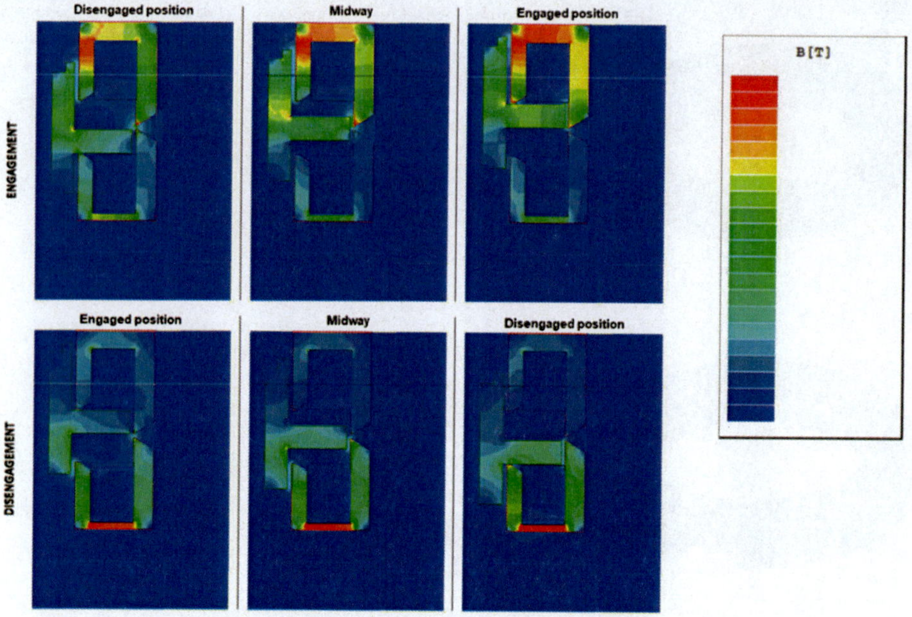

Fig. 11. 2D magnetic flux simulations

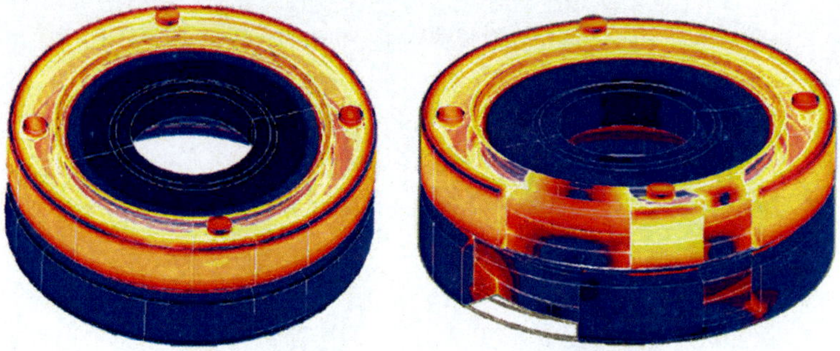

Fig. 12. 3D magnetic flux simulations

- **AWD boost**: engine and e-machines deliver full torque in order to reach maximum performances.
- **FWD eDrive**: engine is shut down and vehicle motion is guaranteed by e-FDU (P4).
- **Torque vectoring**: front axle torque is unbalanced between left and right wheel in order to achieve best vehicle stability in all driving conditions.
- **Reverse gear eDrive**: reverse gear is always done by using e-FDU (P4), permitting to reduce weight by removing conventional mechanical reverse gear from gearbox
- **Regenerative braking**: negative torque requested to e-FDU (P4) when braking is needed, in order to recover electrical energy inside high voltage battery and to prevent brakes aging.

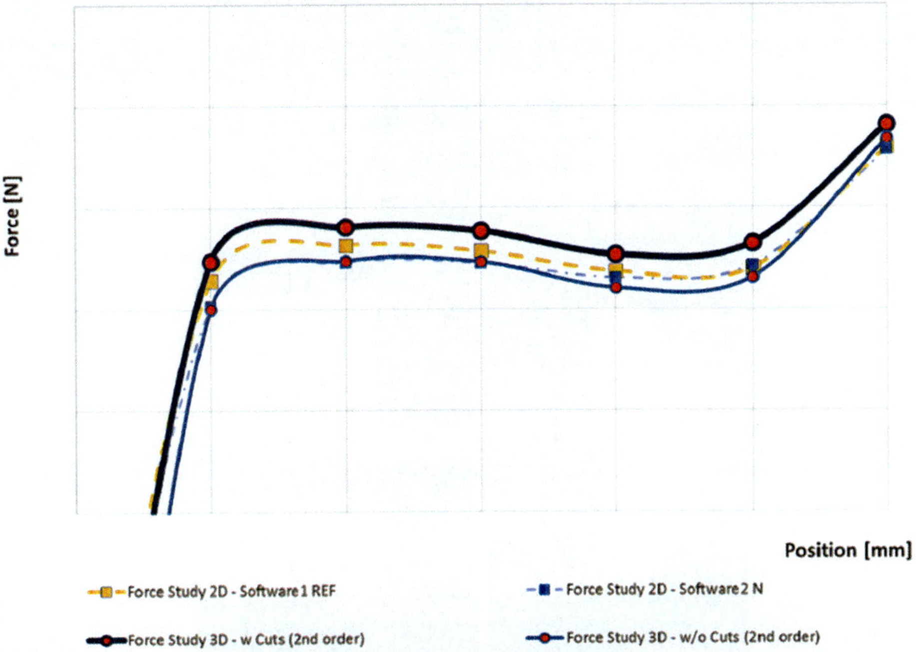

Fig. 13. 2D Force simulation results.

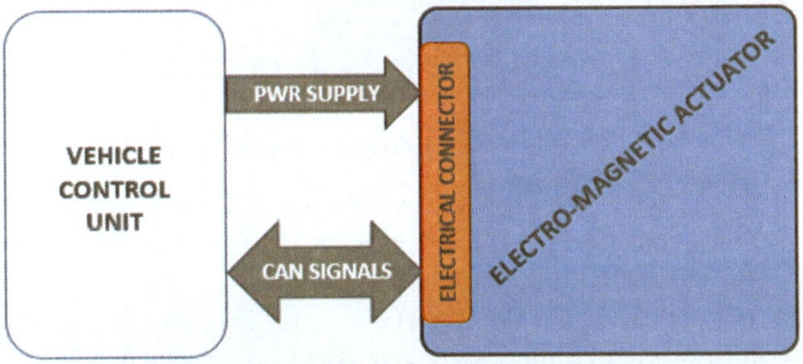

Fig. 14. EMA communication path overview.

Table 1. EMA electrical main references.

	Value	Unit
Supply voltage	9÷16	V
Supply current	30 (max)	A

All these strategies are implemented inside Engine Control Unit (ECU), which communicates with all other nodes composing the powertrain (brakes controller, gearbox

Fig. 15. EMA disengaged and engaged layouts.

unit, the three power inverters, etc.) via several high speed CAN buses and coordinates all the torque requests in order to achieve optimal drivability performances and to guarantee vehicle stability.

In Fig. 20, the powertrain torque split management is presented.

The presence of two independent e-machines in the e-FDU permits to request different torque amounts between left and right side even without the presence of specific clutches.

For example, in case of activation of a specific dynamic control (such like torque vectoring), the torque demand can differ between the two e-machines (see Fig. 21):

4.2 Dog-Ring Disengagement and Engagement

Each e-FDU e-machine can be disconnected from the corresponding wheel via a dog-ring, commanded by an Electro-Magnetic-Actuator (EMA).

The disengagement of the e-FDU is needed when the vehicle is driving at very high vehicle speed in order to avoid e-machines reaching their over-speed value. There is

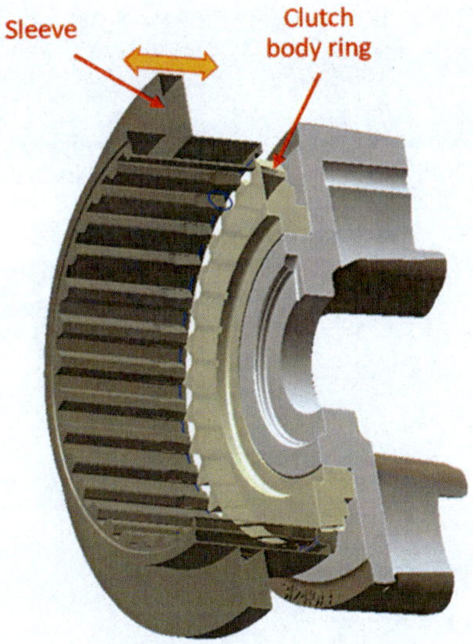

Fig. 16. Sleeve and Clutch body ring indexing detail.

Fig. 17. EMA detent system layout.

also an energy saving strategy that can request front axle disconnection. On the other hand, every time front axle torque delivery is required back, the ECU can command reconnection of the front axle.

Front axle disengagement and engagement are highly sophisticated strategies that involve several control units such as ECU, e-machines, dog-ring actuators (EMAs), and wheel speed sensors.

An overview of engagement/disengagement strategy is presented in Fig. 22.

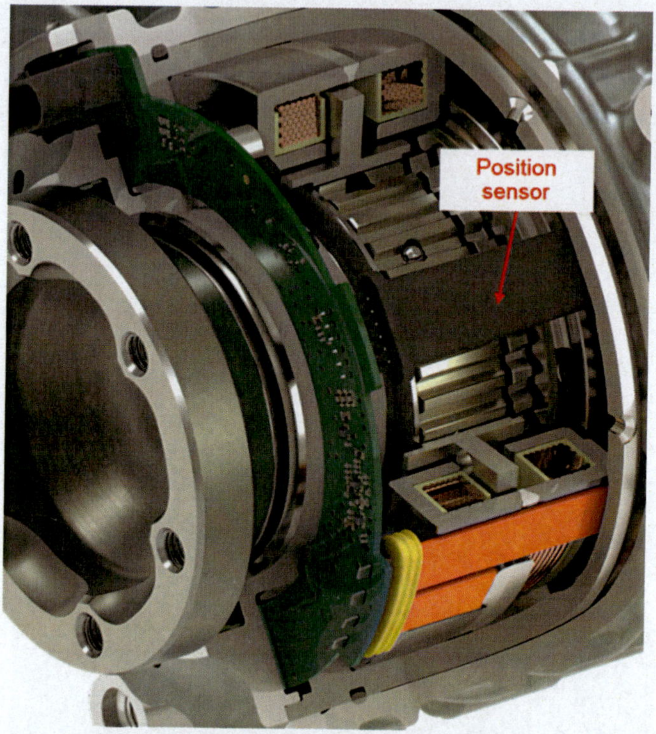

Fig. 18. EMA Position sensor section.

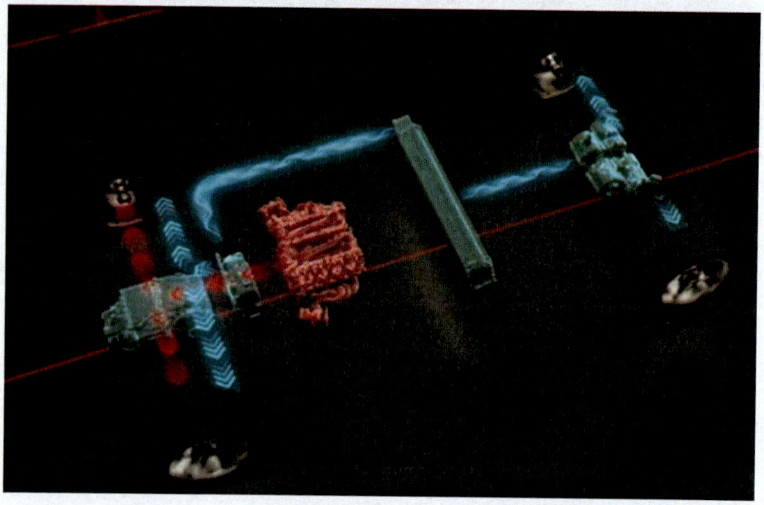

Fig. 19. Powertrain layout of Ferrari SF90 Stradale.

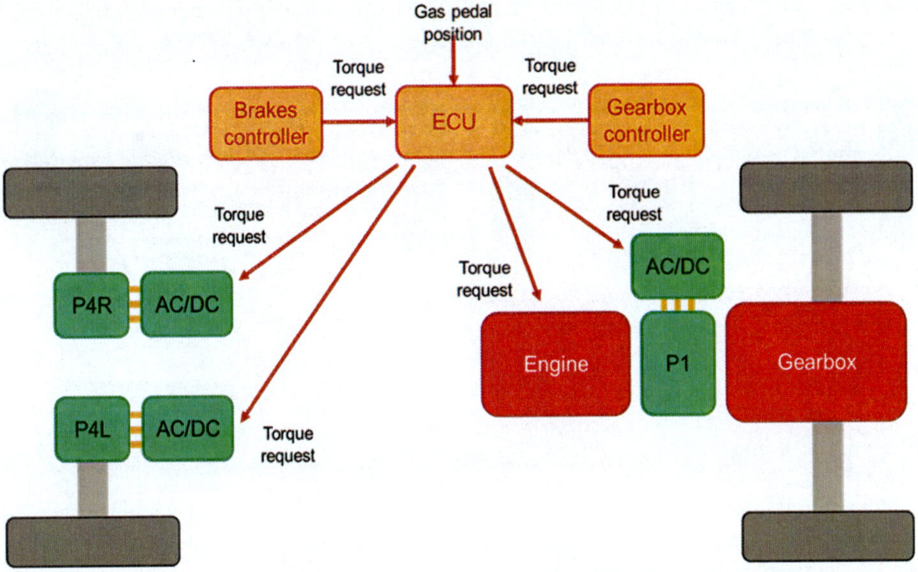

Fig. 20. Powertrain torque split management.

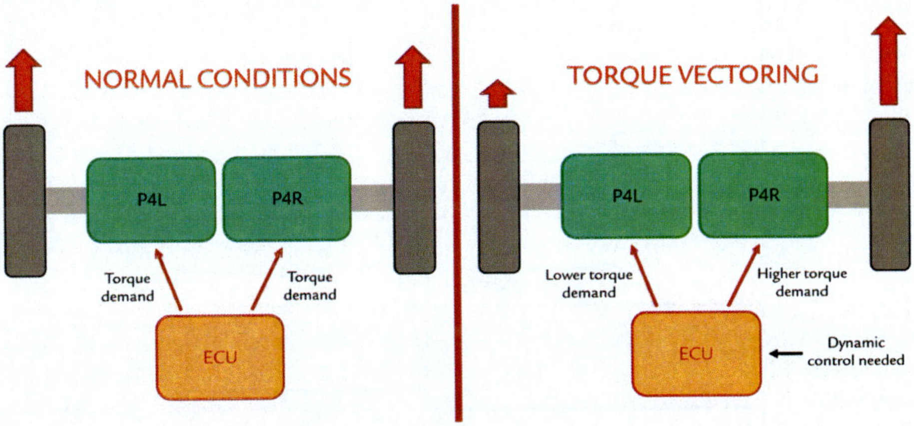

Fig. 21. Torque Vectoring scheme.

The aim of these sophisticated control strategies is to guarantee low engagement/disengagement times (down to 50 ms) and, at the same time, not to affect driver comfort during such maneuver.

4.2.1 Disengagement Maneuver

During disengagement, a specific torque request (depending on P4 speed and acceleration) is demanded during EMA actuation in order to minimize maneuver time up to

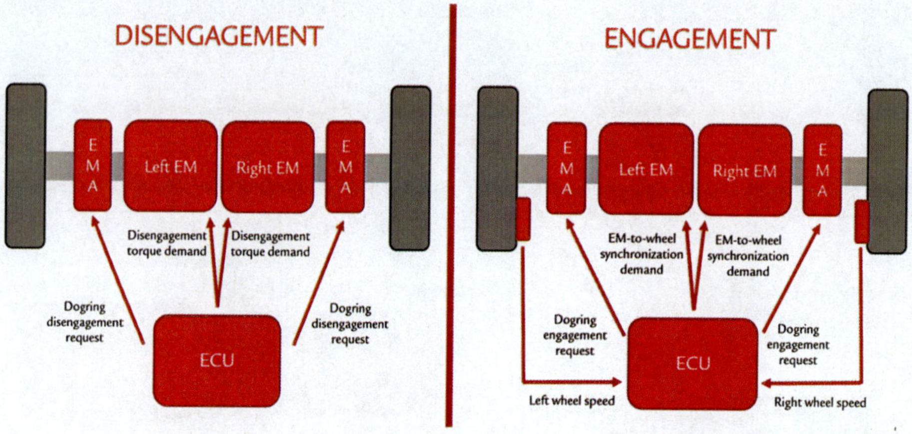

Fig. 22. Overview of engagement/disengagement strategy.

50 ms. In Fig. 23, the expected evolution of P4 velocity, wheel velocity, dog-ring status and P4 torque request are presented.

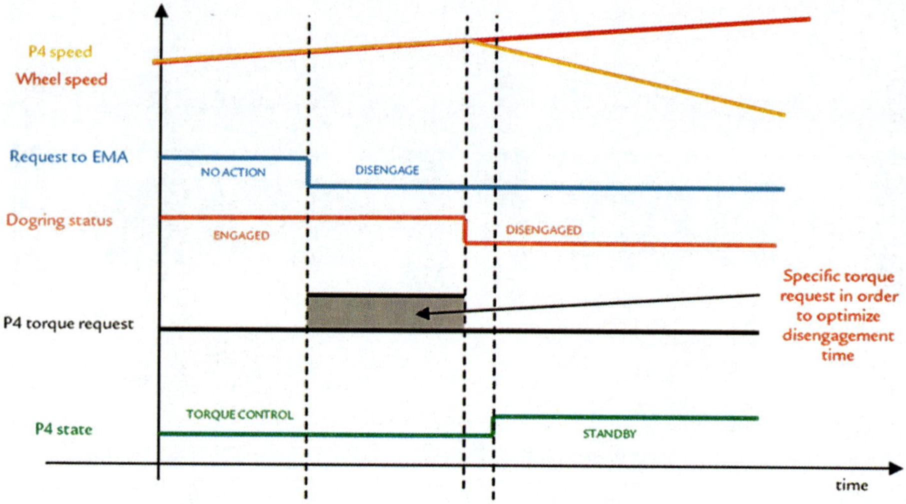

Fig. 23. Disengagement strategy. In orange the P4 speed, in red the wheel speed. In blue the request to EMA, in light red the dog-ring status. In black the P4 torque request and in green the P4 state.

4.2.2 Engagement Maneuver

Due to the mechanical nature of the dog-ring, the disengagement can be achieved only when a speed difference exists between upstream (i.e. P4 side) and downstream (i.e. wheel side) of the dog-ring (Fig. 24).

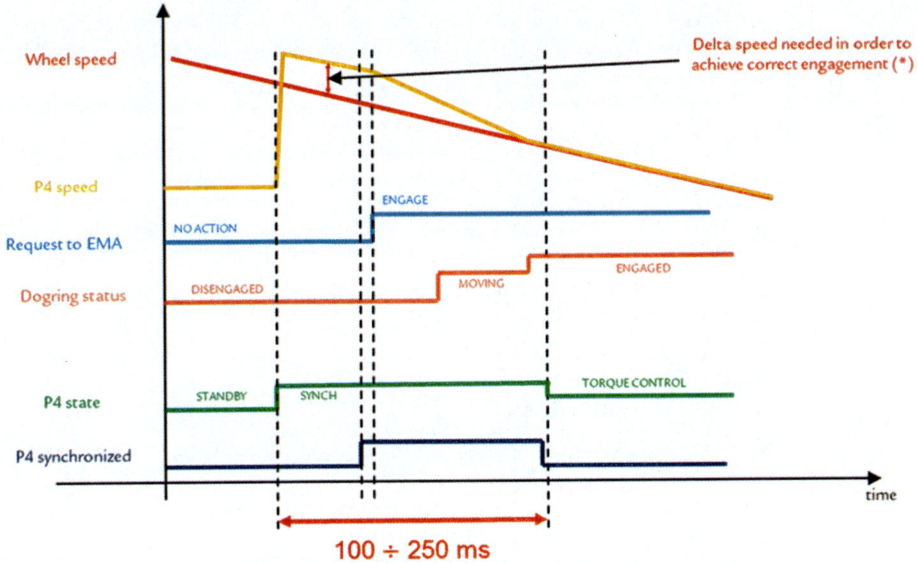

Fig. 24. Disengagement strategy. In orange the P4 speed, in red the wheel speed. In blue the request to EMA, in light red the dog-ring status. In black the P4 torque request and in green the P4 state.

The following has to be considered when tuning the speed difference value for the control strategy:

- The higher is the delta speed, the faster is the engagement, the bigger is the impact on the comfort of the driver.
- The lower is the delta speed, the longer will be the engagement but with no effects on driver comfort.

Therefore, different delta speed values are adopted depending on vehicle conditions.

For this reason, a specific SYNCH control mode has been defined for e-FDU power inverters: in this mode the e-machines are controlled in speed control mode, where the target speed is the wheel speed plus the delta needed for the engagement.

As soon as the e-FDU e-machine has reached this target speed, then a specific "Synchronization flag" is generated, informing ECU that an engagement can be requested to the EMA.

Then, when both EMAs report the "Engaged" state the axle is ready for the torque delivery:

5 1D Model Simulation

5.1 Model Description

In order to evaluate the dynamic behaviour of the e-FDU and to determine the maximum loads that ca be faced during its lifetime, a dynamic simulation model has been created.

The simulation model is a 1D (one dimensional) model and uses the lumped parameter approach to discretize the overall inertia and stiffness distribution of the system.

An overall representation of the model can be found in Fig. 25:

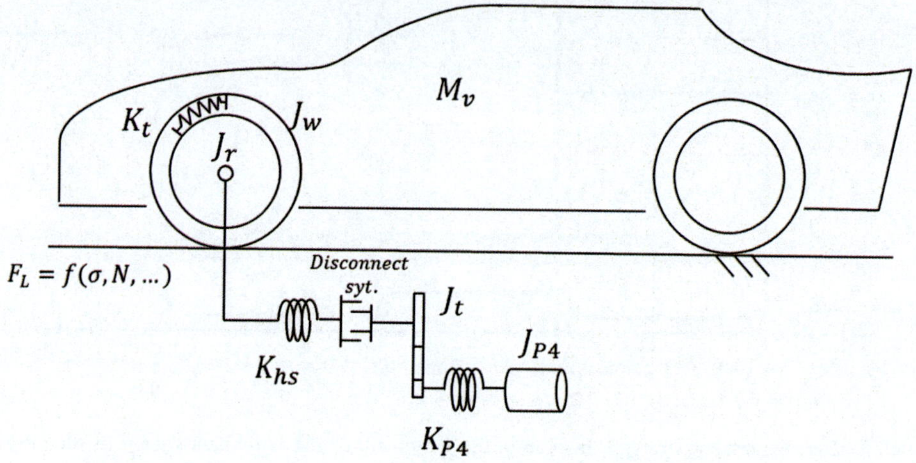

Fig. 25. Schematization of 1D torsional model of vehicle and electric front axle.

In Fig. 25 the main stiffness and inertia that take part in the model are reported. M_v is the mass of the vehicle, $F_L$is the longitudinal force transmitted by the tyre according to Pacejka. J_w is the inertia of the tyre while J_r is the inertia of rim and brake. K_t is the torsional stiffness of the tyre.

K_{hs} is the torsional stiffness of the front half-shaft. The disconnect system can be model as a nonlinear torsional stiffness. When the system is closed the stiffness is modelled like a contact with backlash. This means that stiffness is close to zero inside the "free flight zone", while exponentially increases after the free-flight zone.

J_t is the equivalent inertia of the gear system calculated at wheel axis, J_{P4} is the inertia of the electrical machine calculated at P_4 axis, while K_{P4} is the torsional stiffness of the shaft connecting the electrical machine and the gear system.

Under these conditions the motion equations become:

$$J_{P4}\ddot{\theta}_{P4} + R_{P4}\left(\dot{\theta}_{P4} - \dot{\theta}_t\tau\right) + K_{P4}(\theta_{P4} - \theta_t\tau) = T_{EM} \quad (1)$$

$$J_t\ddot{\theta}_t + \tau R_{P4}\left(\dot{\theta}_t\tau - \dot{\theta}_{P4}\right) + \tau K_{P4}(\theta_t\tau - \theta_{P4}) + R_{hs}\left(\dot{\theta}_t - \dot{\theta}_r\right) + K_{hs}(\theta_t - \theta_r) = 0 \quad (2)$$

$$J_r\ddot{\theta}_r + R_{hs}\left(\dot{\theta}_r - \dot{\theta}_t\right) + K_{hs}(\theta_r - \theta_t) + R_t\left(\dot{\theta}_r - \dot{\theta}_w\right) + K_t(\theta_r - \theta_w) = 0 \quad (3)$$

$$J_w\ddot{\theta}_w + R_t\left(\dot{\theta}_w - \dot{\theta}_r\right) + K_t(\theta_w - \theta_r) = F_L R \quad (4)$$

$$M_v\ddot{x}_v = F_L \quad (5)$$

In Eq. (1) T_{EM} is the applied torque from electrical machine and τ is the multiplication ratio of electrical machine. In Eq. (4) R is radius of the wheel.

Similar simulation models have already been studied in literature, both with 1D torsional approach and with multibody flexible approach [1–3].

5.2 Model Validation

Before proceeding with the analysis of the maximum torque faced by electrical axle, the model has been validated with a maneuver that replicated as much as possible a step excitation.

The maneuver consist in a dog-ring engagement while vehicle is standstill on the road. e-FDU e-machine is rotating a certain speed and a fast synchronization is actuated.

The acquired signal from experimental measurements are:

- Axial position of dog-ring sleeve
- e-FDU speed.

The simulation model receives, as input, the position of dog-ring sleeve and gives the velocity of the e-machines as output. The comparison between numerical simulation and experimental measurement of the dog-ring engagement maneuver is shown below (see Fig. 26):

5.3 Misuse Torque Estimation

One of the aims of the model is to evaluate the abuse torque that the electrical axle may face during its life. The abuse torque must be estimated as precisely as possible in order to produce a design which is, on one hand, as light as possible, and, on the other, strong enough to withstand the maximum load without incurring into failure.

To achieve this goal a series of maneuvers have been simulated in order to estimate the maximum load on the electric axle, in details:

a. ABS Brake
b. Traction loss
c. Cleat test

In Fig. 27 the results about the ABS Brake simulation are presented. For this simulation the control logic of the ABS system has been implemented in the simulation software. The brakes on the wheels are actually slowing down the car, while the resulting dynamic angular velocity excites the P4 axle. The dynamic excitation produces dynamic torque across the half shaft.

On the other hand, to simulate traction loss, the maximum electrical torque for the P4 in forward motion is imposed on the 1D model. Then, a short grip loss for about 0.05 s, is imposed. This maneuver is designed to replicate a driving condition in which the vehicle is crossing a very slippery road under high torque request.

The results of traction loss simulation are presented in
Figure 28:

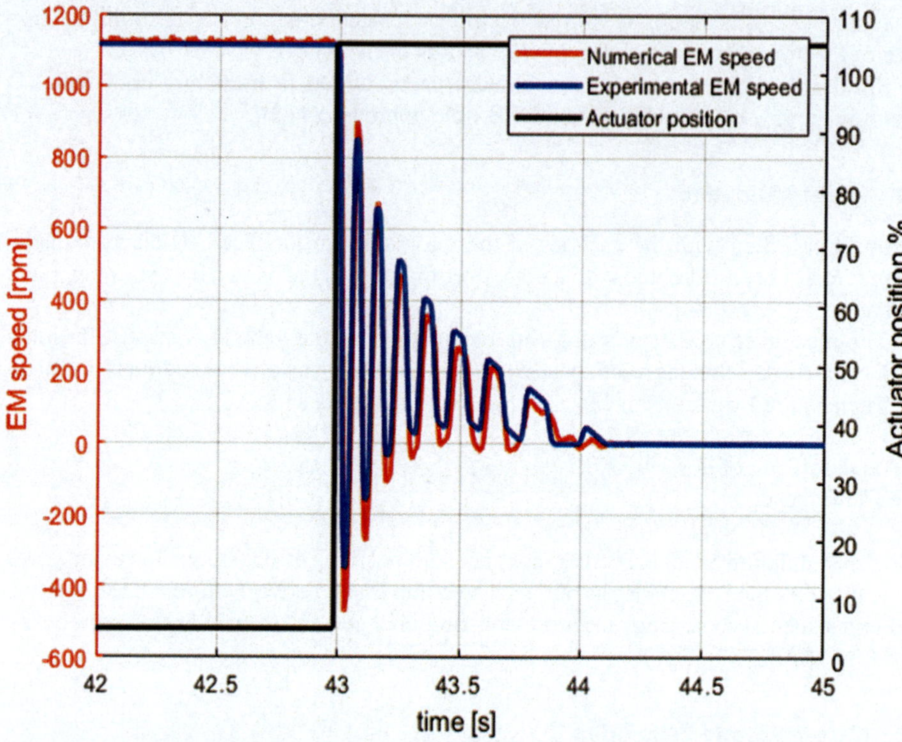

Fig. 26. Fast engagement maneuver. In red the experimental EM angular velocity. In blue the simulated velocity. In black the actuator sensor position.

The last simulation presented is the cleat test. In this test the wheel is approaching a step of 25 mm of height and 100 mm long.

Moreover, to better replicate dynamics of the vehicle for this particular maneuver, the vertical degree of freedom for the suspension and the vertical degree freedom of the tyre in the 1D simulation model are introduced (Fig. 20). To evaluate at best the tyre behaviour approaching an obstacle, the transfer function that must be applied to the wheel approaching the obstacle has been estimated. This result has been obtained using a CD-Tyre formulation of the tyre with a simulation test-rig including a road wheel with the desired obstacle.

The electrical machine in this maneuver is required to provide the maximum possible torque of its characteristic curve.

The results of the simulation are showed in Fig. 29:

To conclude, a summary of misuse simulation is presented in Table 2. As displayed from the table, the traction loss maneuver seems to be the most critical one concerning the abuse torque.

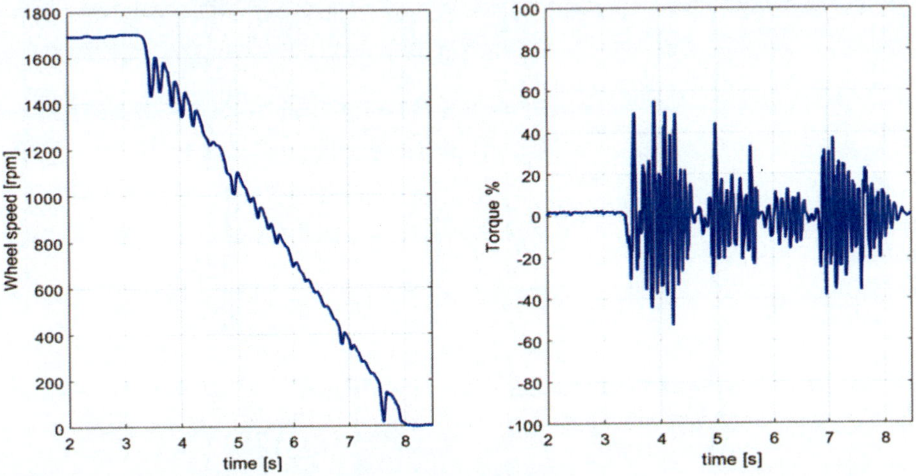

Fig. 27. ABS Brake simulation. The brake system generates the wheel velocity plotted in the left plot. The angular dynamics generates on the half shaft the torque on the right. The torque is normalized with respect the maximum torque experienced in all the simulated maneuvers.

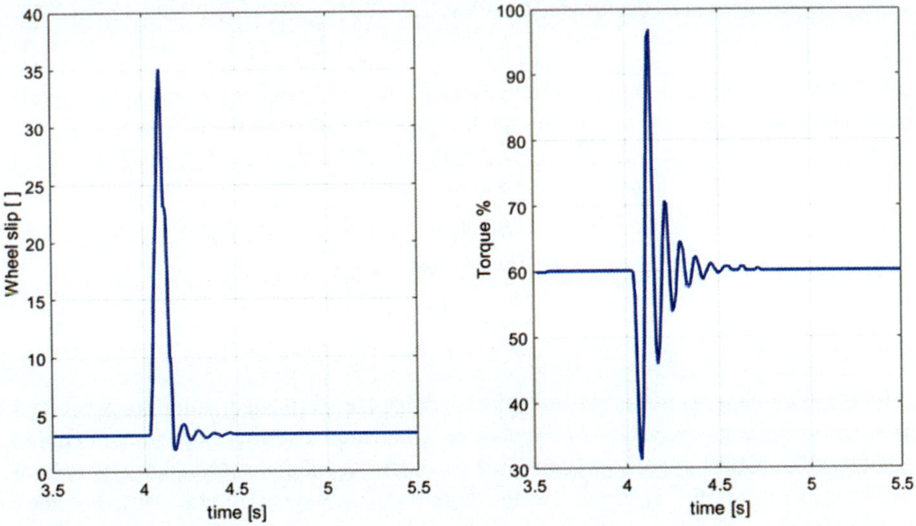

Fig. 28. Traction loss simulation. The wheel slip is plotted on the left, while the dynamic torque generated on the half-shaft is plotted on the right. The torque is normalized with respect the maximum torque experienced in all the simulated maneuver.

5.4 Torque Estimation During Dog-Ring Engagement

Another important aim of the 1D simulation model is to evaluate the dynamic torque exchanged between the wheel and the electrical axle during dog-ring engagement.

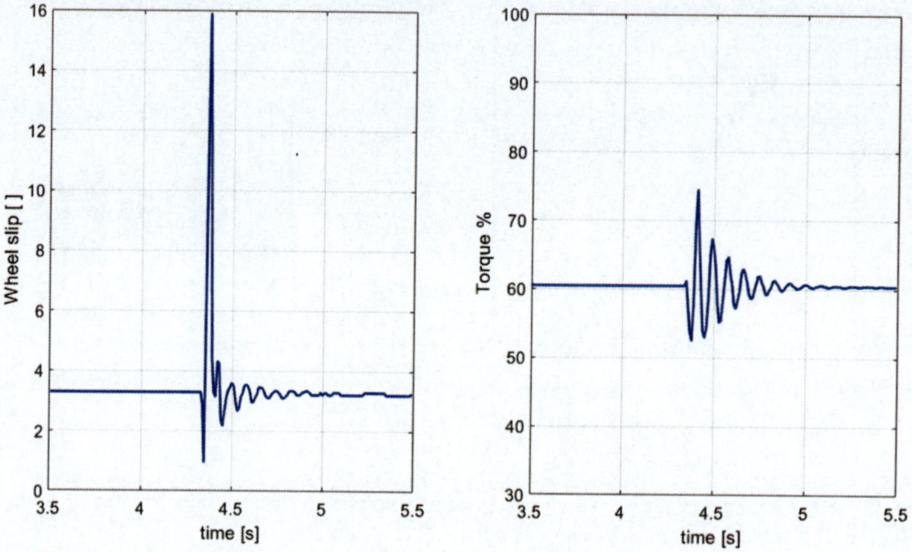

Fig. 29. Cleat test simulation. The wheel slip is plotted on the left, while the dynamic torque generated on the half-shaft is plotted on the right. The torque is normalized with respect the maximum experienced in all the simulated maneuvers.

Table 2 Summary of maximum torque faced by the e-FDU during misuse maneuvers. The torque is normalized with respect the maximum experienced in all the simulated maneuvers.

Maneuver #	Maneuver type	Torque %
1	ABS brake	60
2	Traction lost	100
3	Cleat test	75

Evaluation of dynamic torque exchanged under dog-ring engagement is especially important to evaluate the safety of the vehicle. Undesired dog-ring engagement should avoid any critical behaviour concerning vehicle lateral dynamic, such as large yaw rotation. For this reason, the dynamic torque acting on the wheel should be limited under a certain target.

For this simulation the contact model of disconnect system has been detailed to evaluate precisely contact dynamics.

The 2D geometry of Dog-Ring has been replicated inside the simulation software and a contact detection algorithm has been used to evaluate when the sleeve and hub of the disconnect system start to approach contact phase. Contact forces are modelled with penalty approach.

In Fig. 30 is presented a sensitivity analysis of dynamic torque exchanged at the half-shaft for several relative velocities between wheel and disconnect system:

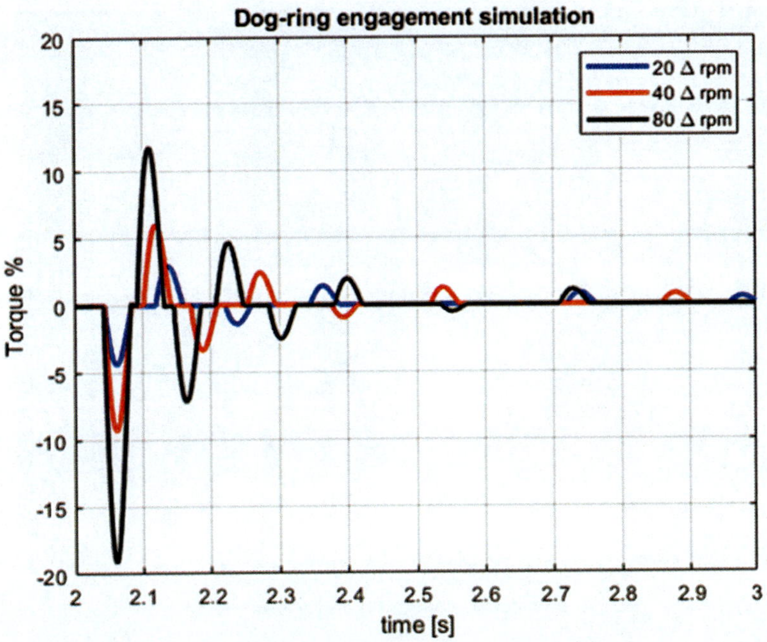

Fig. 30. Dog Clutch engagement simulation. Simulation is repeated with several relative velocities at dog-clutch. The torque is normalized with respect the maximum experienced in all the simulated maneuvers.

As it can be seen in Fig. 30, the maximum torque is increasing with the increase of relative velocity at the dog-ring. Moreover, backlash plays an important role in the simulation, since the peak torque are separated from zones of zero torque. These zones correspond to the cross of "free flight zone" of the backlash. It is important to note that the maximum torque reached during dog-clutch engagement, even with the maximum admissible relative velocity, is well below the misuse torque.

6 Industrialization and Assembly Line

The e-axle assembly line has been designed and developed according the following rules:

1. Safe and Ergonomic
2. Material from outside of Aline
3. Balanced stations
4. Process/Logistic: SEPARATED
5. First Time Quality >99%
6. One Piece Flow
7. «U» shape

The e-FDU assembly line is divided into three different stations (see Fig. 31):

Fig. 31. Assembly line view.

6.1 Station #1 - Subcomponents Assembly

The station #1 is dedicated to subcomponents preparation as: planetary gear set subassembly, bearing pressfit on shafts, snap rings installation, cylindrical gears coupling (see Fig. 32).

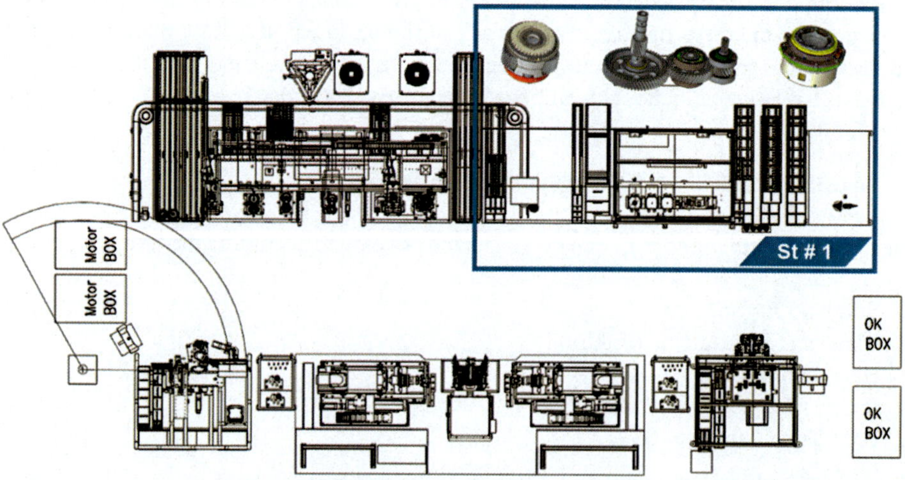

Fig. 32. Station #1 section.

The pressing machine uses automatic pressing tools changeover (revolver, sliding, etc.) and is equipped with all needed sensors or vision systems, in order to ensure presence and correct installation of all managed components (see Fig. 33).

Fig. 33. Pressing machine for bearings installation.

This station includes also EMA pre-assembly completion, in terms of output flange, hub, detents installation and locknut tightening (see Fig. 34).

Fig. 34. EMA assembly area.

A first EOL of the EMA is done from electronic components supplier on a dedicated test bench.

6.2 Station #2 – Housings Assembly and e-Machines Installation

The station # 2 is dedicated to install subcomponents into central housing, heating of bearing seats with induction heater, dowel pins insertion with pressing machine, liquid seal application (manual by guided device with Loctite feeder) and consequential check of pattern with vision system (see Figs. 35 and 36).

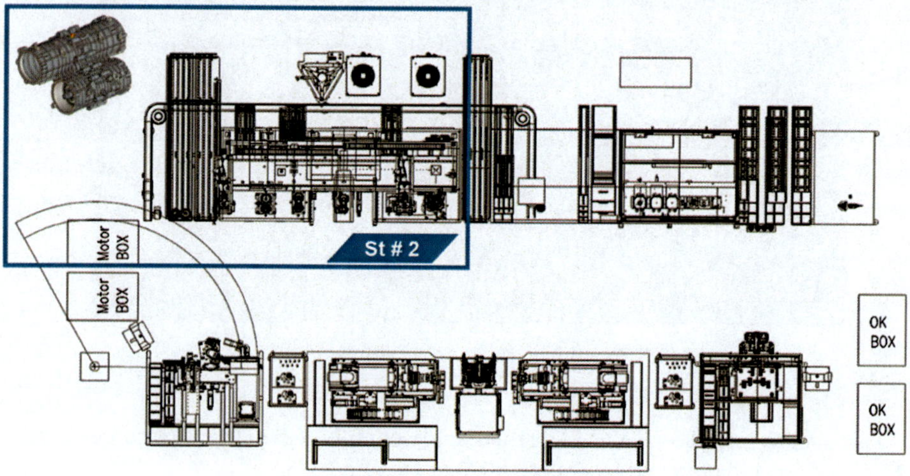

Fig. 35. Station #2 section.

The assembly process continues with intermediate housing and motor housings matching, screws tightening and e-machines installation, managed with specific Dalmec alignment system (see Figs. 37 and 38).

6.3 Station #3 + EOL

Station #3 is dedicated to installation of pre-assembled transmission on revolving station, EMA final installation in the housings with dedicated pick and place, external component assembly (i.e. oil plug, etc.), leakage test of oil and water circuits (this one is done for each transmission side) and EOL test (Figs. 39 and 40).

The EOL test bench is equipped with two dynamos able to provide 1500 Nm braking torque @ 930 rpm installed on a sliding axis in order to be automatically connected and disconnected to the tested unit. A battery emulator of 250 kW, supplies power to the inverter that drives proper e-machines of the e-FDU (see Fig. 41).

The system is able to read, on assembly line server, the e-motor calibration data (resolver) and load it on the inverter. It can also apply different torque\speed ramps

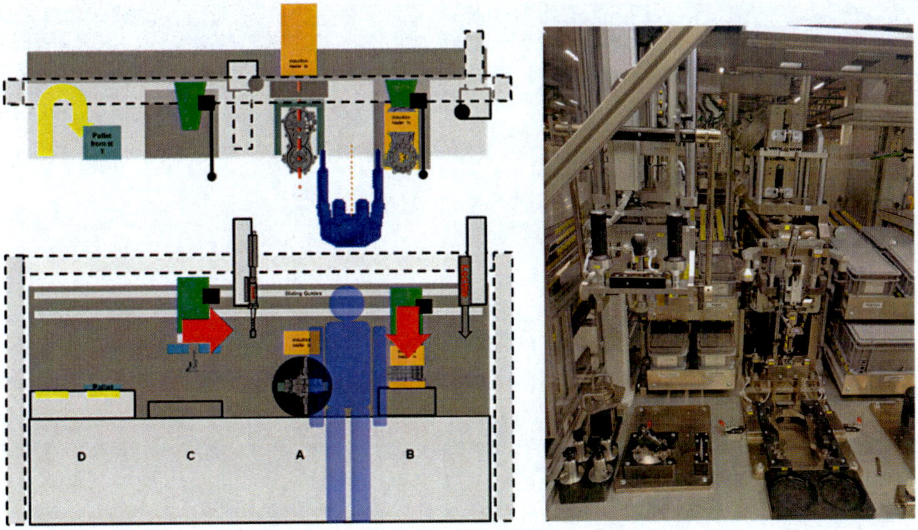

Fig. 36. Station #2 layout and detail.

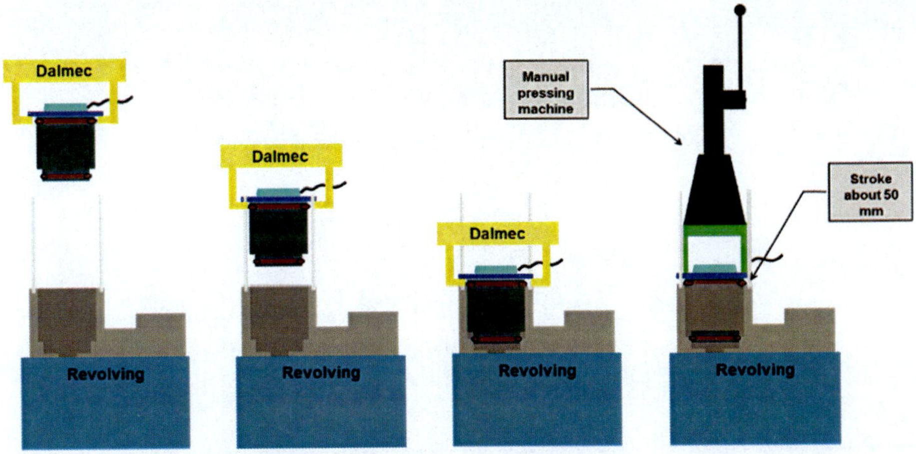

Fig. 37. Dalmec system for e-machine handling layout.

to the e-machines, automatically managing oil and water flow rates, temperatures and pressures.

EMA actuations at different output shaft speeds are included in the EOL test procedure, to verify engagement and disengagement times.

The bench also includes two accelerometers for NVH characterization (see Fig. 42), data post-processing and storage (Fig. 43).

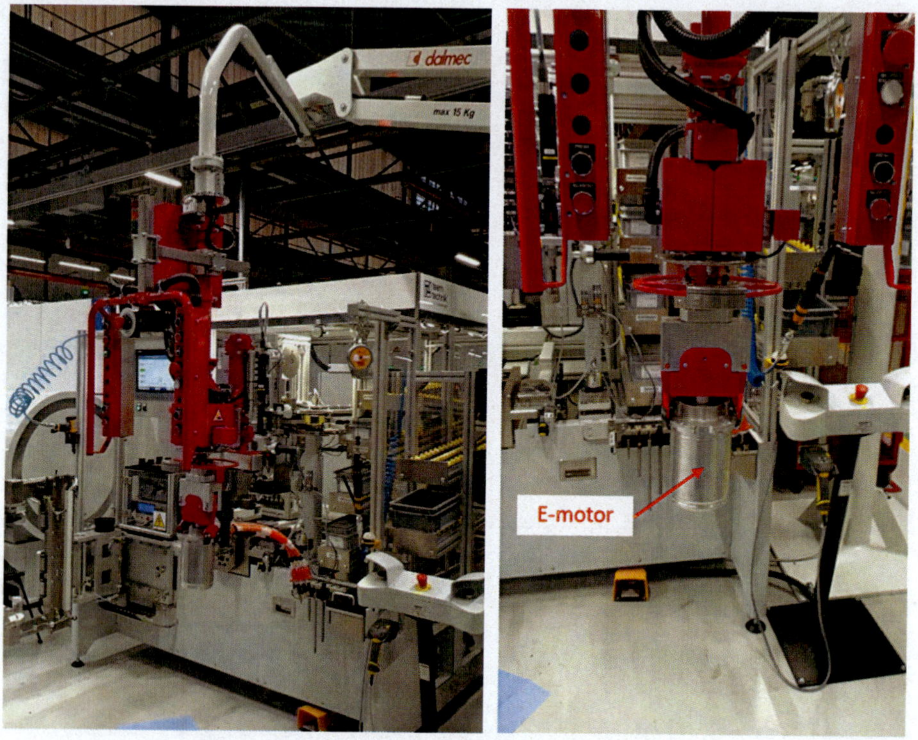

Fig. 38. Dalmec system for e-machine handling

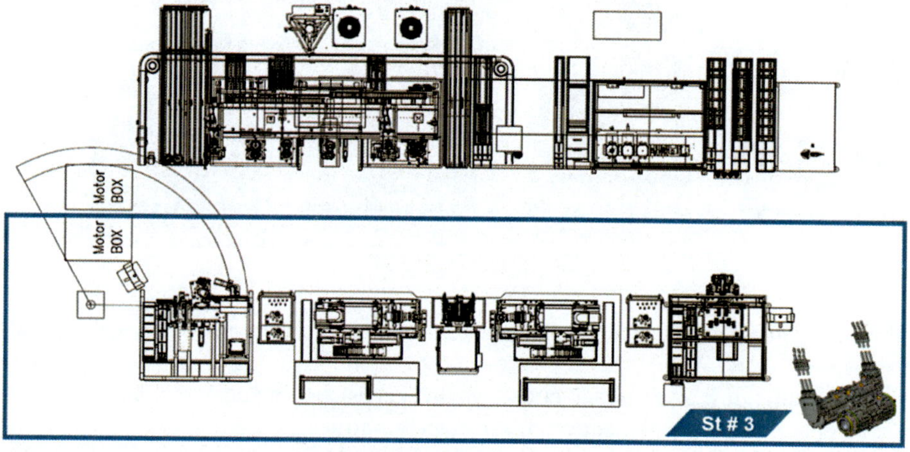

Fig. 39. Station #3 section.

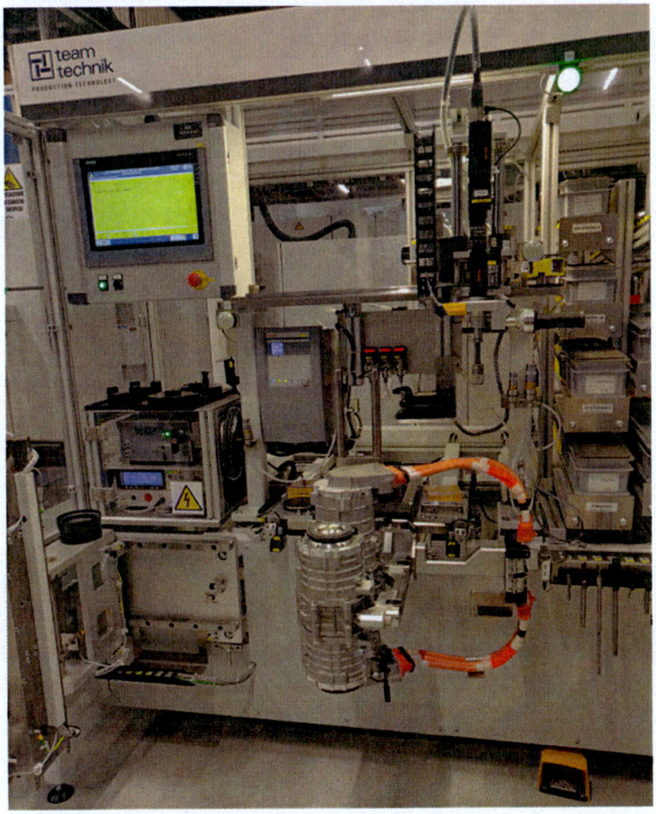

Fig. 40. Leakage test machine (oil and water circuits).

7 NVH and e-FDU Validation

The e-axle cannot be considered as a twin separated gearboxes system. For example, even if both e-FDU sides share the same lubrication oil, but the path is quite different between left and right sides. Dedicated analysis and bench tests have been carried out to optimize lubrication and allow to use same helix gears for each transmission side, with a benefit in terms of costs and parts management. However, this choice, has led to have drive gear flank of the right side gears corresponding to coast gear flank of the left side one: detailed NVH simulation activities with dedicated software tools have been completed to optimize gears microgeometry and reduce noise emissions also at high input speed (25000 rpm). Further dedicated tests have been carried out directly on the vehicle to compare and correlate data with bench results.

Most of experimental design validation tests have been run with original e-machines, in order to verify the complete system behaviour.

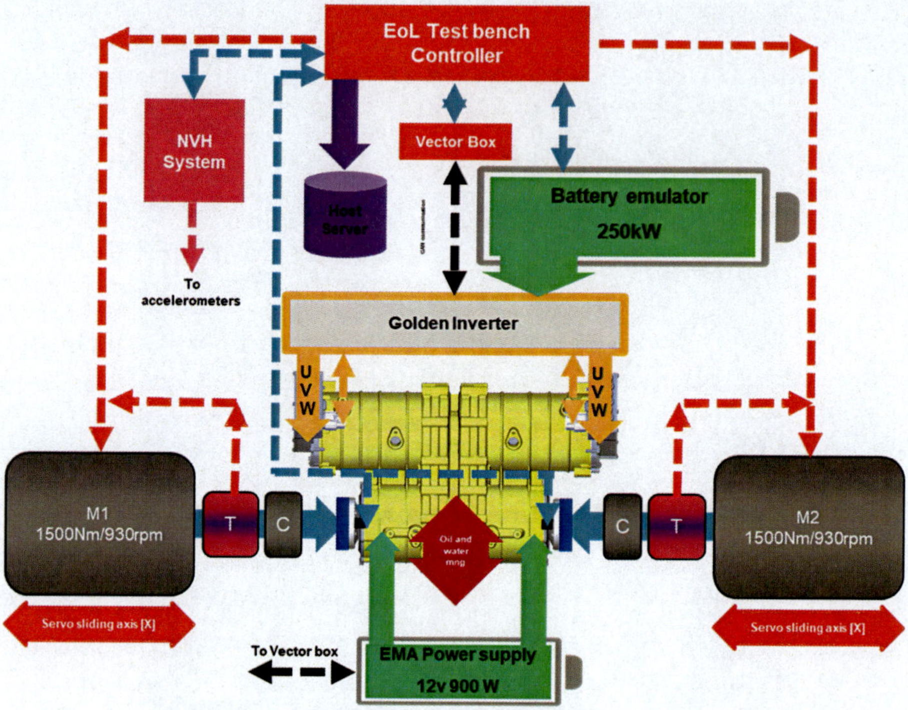

Fig. 41. EOL bench layout.

7.1 Under Load Contact Pattern

The e-machines are powered by a proper battery simulator that drives the inverter and supplies different torques (25%, 50%, 75% and 100% full torque) in symmetric and asymmetric conditions (torque vectoring mode). Gears are coated with special marker that allow to evaluate gear meshing contact pattern (see Fig. 44) and compare test results with numerical simulations done for microgeometry optimization.

7.2 Gears Fatigue Test

The purpose of the test is to verify the e-FDU gears and subcomponents reliability and durability, by performing a fatigue test representative of the mission profile of the vehicle. The transmission is mounted on a proper clamping plate and connected to the 2 dynos of the test rig, via drive shafts. Speeds and torques are measured at input and output shafts by high precision measurement devices, scaled and calibrated to fit exactly the data range of the measurement program (Figs. 45 and 46).

To guarantee e-FDU reliability, the test run 200% of the vehicle mission profile.

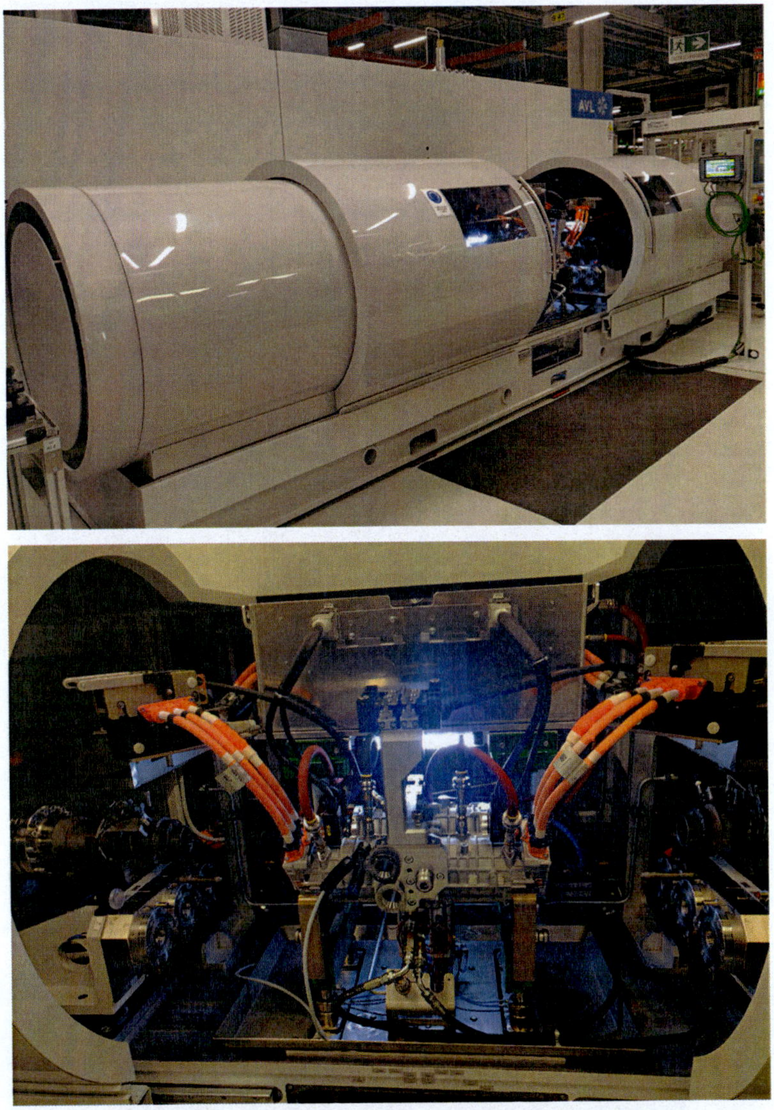

Fig. 42. EOL bench overview.

7.3 EMA Durability Test

The scope of this test is to evaluate EMA performances in terms of actuation time over vehicle lifetime: several actuations have been done to verify EMA functionality at different reference frequencies and oil temperatures.

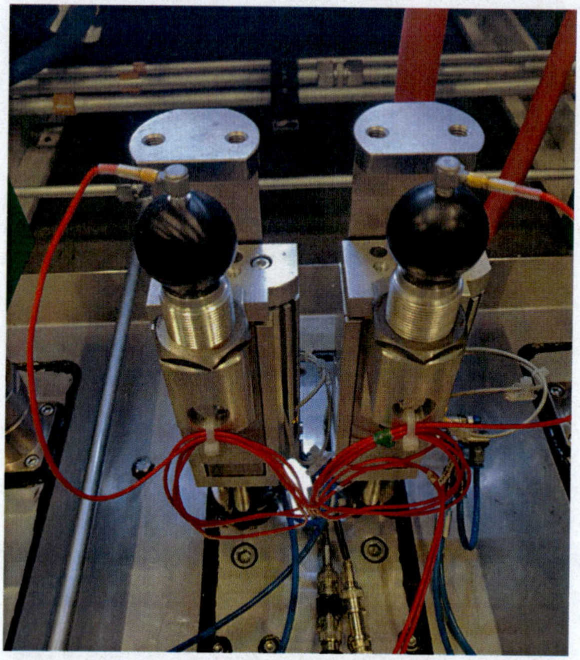

Fig. 43. EOL bench accelerometer for NVH acquisition.

Fig. 44. Example of gear coated with special marker for contact pattern evaluation.

7.4 Efficiency Test

This test is run to verify the efficiency of the e-FDU during symmetrical and torque vectoring working conditions. Power meters have been installed to measure voltage and current on the 3 phases of each motor and DC common line, for both inverters.

Fig. 45. Test bench set up.

Fig. 46. Electrical instruments for voltage and current measurement.

The efficiency values have been recorded in correspondence of the operative points requested as target.

8 Conclusions

Twin independent disconnectable transmissions driven by two high revolution (25000 rpm) and high-power density electric motors (up to 162 kW in total).

Full electric drive in reverse gear.

Vehicle dynamics and torque vectoring optimization.

Specific vehicle control strategies have been developed to meet hybrid functionality and vehicle dynamic performances.

EMA as disconnect device for Ferrari series production high performance PHEV.

Smart actuator with integrated electronic hardware and compact design.

Precise sensor feedback for precise actuation control.

Contactless shifting powered by electromagnetic flux.

Actuation time less than 100 ms @ 20 °C.

Semi-automatic assembly line with complete PLC management (all process controlled and managed by Host server).

Full power (up to 162 kW) EOL of complete powertrain (E-motors + Transmission + Disconnect).

References

1. Furlich J, Blough J, and Robinette D (2017) Torsional vibration analysis of six speed MT transmission and driveline from road to lab. SAE international, 2017-01-1845
2. Galvagno E, Gutierrez P, Velardocchia M, Vigliani A (2017) A theoretical investigation of the influence of powertrain mounts on transmission torsional dynamics. SAE international, 2017-01-1124
3. Parmar V, Di Rocco D, Sopouch M, Albertini P (2014) Multi-physics simulation model for noise and vibration effects in hybrid vehicle powertrain. SAE international, 2014-01-2093

Effective Battery Design and Integration of Cylindrical Cells for High Power Applications

Helmut Kastler(✉) and Kilian Menzl

KREISEL Electric GmbH & Co KG, Kreiselstraße 1, 4261 Rainbach, Austria
helmut.kastler@kreiselelectric.com

Abstract. To fulfill the far-reaching requirements of an effective battery design for high power applications, every single component, including their interactions with the battery module, have to be optimized. Without making compromises on battery safety, designing a compact battery module for Lithium-Ion cells is the main development target. To address customer demands and enable fast adoption of the technology in the markets, tough cost targets have to be met. Even though no design aspect can be ignored, a sufficient battery thermal management system (BTMS) is key to increase safety, fast charging capability and lifetime. Kreisel Electric's battery design based on cylindrical cells and immersion cooling, answers all these questions and proposes new solutions on the design and material side.

Keywords: Battery system · Thermal management · Immersion cooling

1 Introduction

Driven by the vision of integrating the electric drive in all areas of mobility, Kreisel Electric develops the lightest and most efficient battery solution worldwide. The technology has been integrated in e-karts, cars, buses, trucks, boats and airplanes, as well as in stationary storage systems. By working on a wide variety of projects, many skills that are required for holistic system integration, have been established. Furthermore, specific requirements coming from different industries resulted in an extended system integration competence to address exactly the needs of a specific application. The core competence of Kreisel Electric includes battery development and production, integration & testing, prototyping & serial production and software development. As an international solution-provider, Kreisel Electric develops and produces the most-efficient battery storage packs for e-mobility as well as for energy stationary storage systems. As an early mover in battery immersion cooling, extensive research and development efforts have been conducted to optimize thermal performance, safety behavior, durability, environmental impact and system costs. In tight cooperation with established suppliers from the chemical industry, application engineering and research institutions Kreisel Electric steadily redefines the state of the art of immersion cooled battery systems (Fig. 1).

Euroforum Deutschland GmbH (Ed.): *CTI SYMPOSIUM 2019*, pp. 283–293, 2021.
https://doi.org/10.1007/978-3-662-61515-7_24

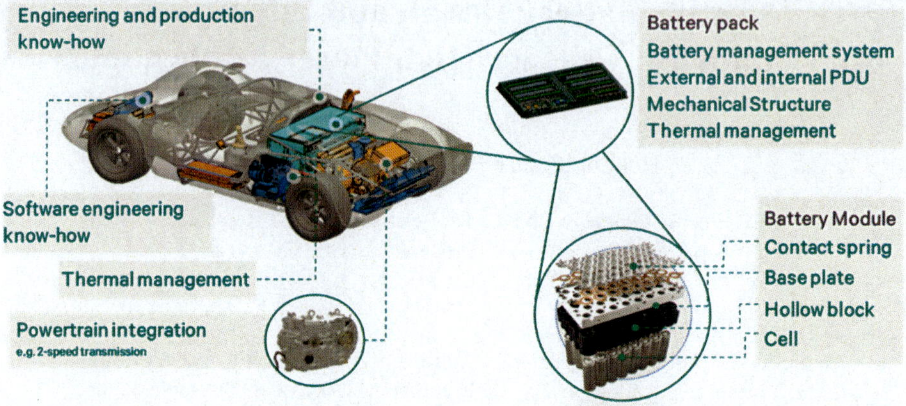

Fig. 1. Holistic solutions for e-mobility with full system integration competence

2 Superior Battery Pack Technology at Competitive Cost

2.1 Simple Patent Protected Architecture Based on Cylindrical Cell

The customer acceptance and competitiveness of battery electric vehicles will be determined by costs, durability and energy density in combination with fast charging capabilities on vehicle level. To meet this targets Kreisel Electric has developed an effective design of arranging cylindrical cells in a secondary traction battery. Based on the cylindrical cells of format 18650 and 21700 battery packs are built in 3 levels (see Fig. 2):

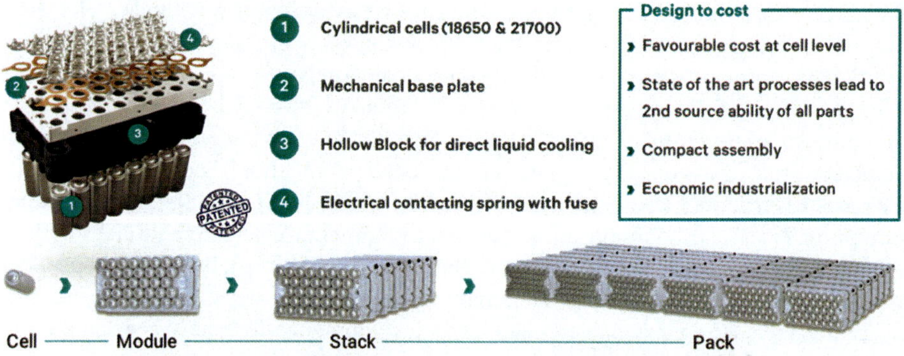

Fig. 2. Simple patent protected architecture based on cylindrical cell

- Level 1: Battery module that consists of several single cells in parallel
- Level 2: Battery stack that consists of several battery modules in series
- Level 3: Battery pack that consists of several battery stacks.

Main elements of a battery modules are:

- Cylindrical cell (18650 of 21700)
- Base plate for mechanical integration to battery structure
- Hollow Block for direct liquid cooling with integrated sealings
- Contact spring with integrated single cell fuse for serial connection and parallel plate for parallel connection within the module

The two key innovations are the driver in enabling the highest energy and power density on the market:

- Active thermal management of battery cells featuring a unique casing-constantly flushed by a nonconductive liquid. This enables very high electric loads with minor impact on cycle life.
- Laser Soldering of battery cells with very low heat input on the cells. This allows fast and cost-efficient manufacturing of the battery packs.

All these design and function related features are the base for a modular arrangement without the necessity of a soldering or bonding process beyond the module level. Thus, building a battery pack is as simple as assembling of building blocks to one complete battery system. This results in competitive battery pack prices due to clear focus on "design to cost" methods throughout the development process.

2.2 Improved Safety Due to Unique Propagation Prevention Features

To be conform with the upcoming testing standards for preventing propagation in case of a thermal runaway of a single cell, the most stringent requirements have been applied to the battery system. Therefore, following test parameters have been defined:

- Cell SOC: Charged to 100% SOC
- Module starting temperature: >40 °C
- Trigger cell temperature >250 C°
- Fluid: Dielectric cooling liquid, static filled – no circulation
- Cell trigger method: Heating wire

To even go beyond that, the starting temperature of the test has been increased up to 50 °C.

The test setup for a thermal runaway test can be seen in Fig. 3. Figure 4 shows the device-under-test condition before and after the thermal runaway event. It can be clearly seen, that no propagation across the cells occurred. This verifies the highest safety behavior.

The Temperature of the surrounding cells stays uncritical as graph in Fig. 5 shows. Several thermistors have been applied to surrounding cells of the trigger cell and none of them comes close to 100 °C.

This positive behavior results from following safety system support features:

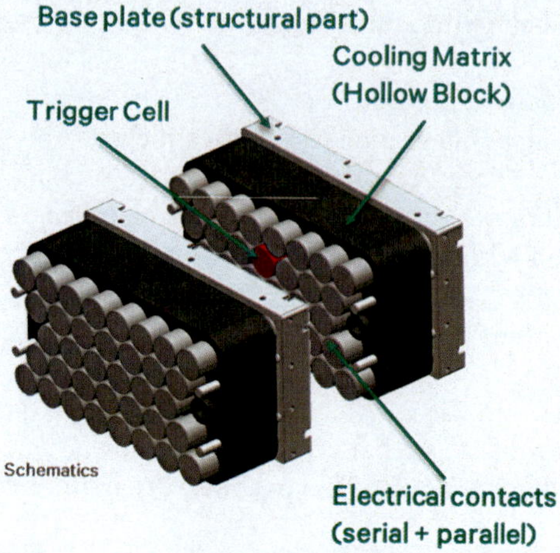

Fig. 3. Test setup

Fig. 4. DUT before/after test

- Mechanical cell propagation protection
- Single cell fuse
- Degassing channel between modules
- Use of non-combustible materials
- Integrated module heat shields
- Nontoxic materials or liquid (except cell)

Still volumetric and gravimetric energy density of 2,7 l/kWh or 4,1 kg/kWh can be achieved without compromises on safety and performance.

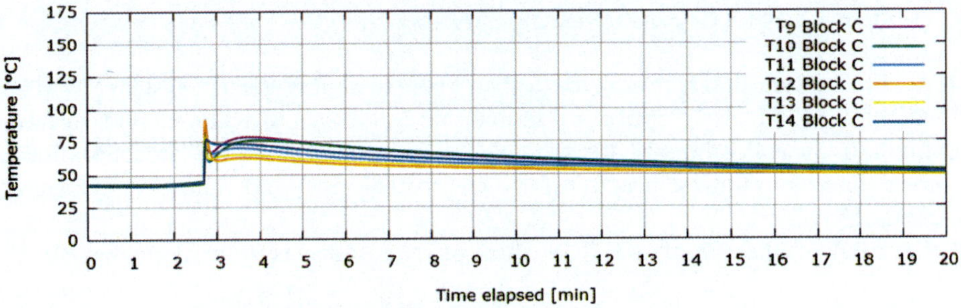

Fig. 5. Temperature evaluation thermal runaway test

3 Key Aspects to Increase Energy and Power Density

To achieve best values in regards to energy and power density without compromises on safety significant design singularities have been considered:

- Compact sizing due to cell arrangement in combination with cooling liquid between cells (immersion cooling) utilizes empty space for thermal management without additional areas outside the battery module.
- Cell to cell connection with laser welding and clamping contacts to reduce ohmic losses to a minimum
- Efficient electric connection of the cells without creating temperature hotspots in the cells. Figure 6 shows that there is no heat transfer of the hot busbars into the colder cells.

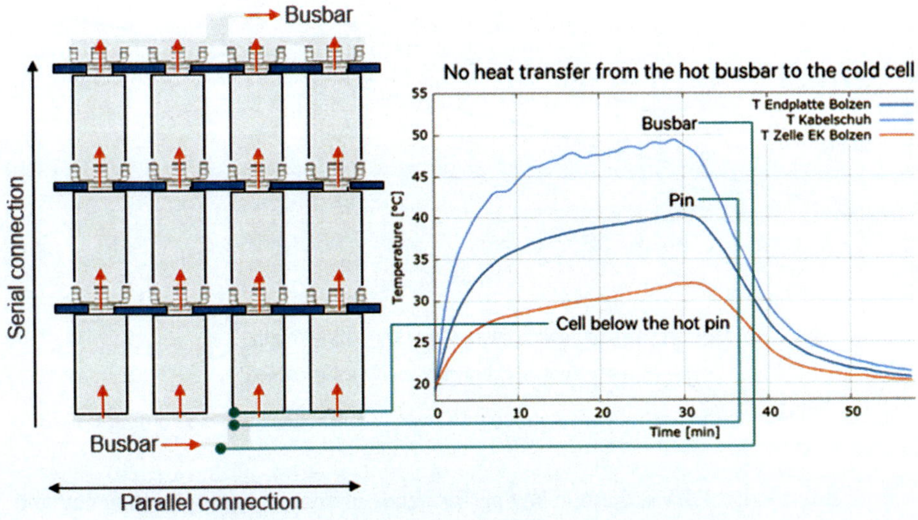

Fig. 6. Cell arrangement and electrical connection

- Dedicated degassing channel per battery module

4 Best in Class Thermal Management

Kreisel Electric's design approach for the battery thermal management system (BTMS) is twofold. First, the use of cylindrical lithium-ion cell format to achieve a high surface to volume ratio and minimize the thermal resistance within the cell, second, direct liquid cooling for optimized heat transfer, increased efficiency and uniform temperature distribution.

The result is a best-in-class BTMS which offers advantages for mass-market BEVs with fast charging capabilities, high-performance HEVs, PHEVs, BEVs and electric motorsport (Figs. 7 and 8).

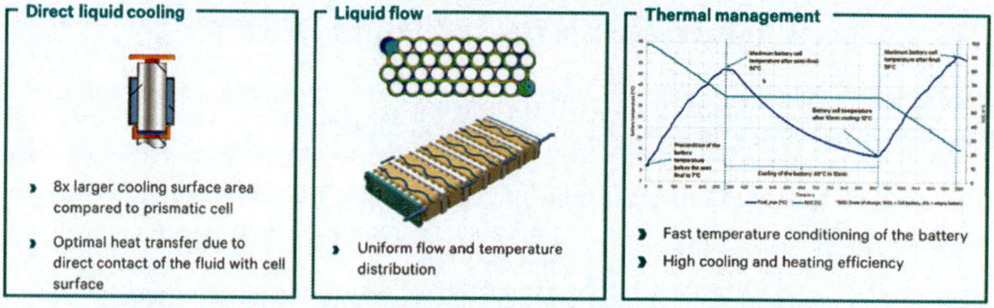

Fig. 7. Advantages of Kreisel Electric's direct cell cooling approach

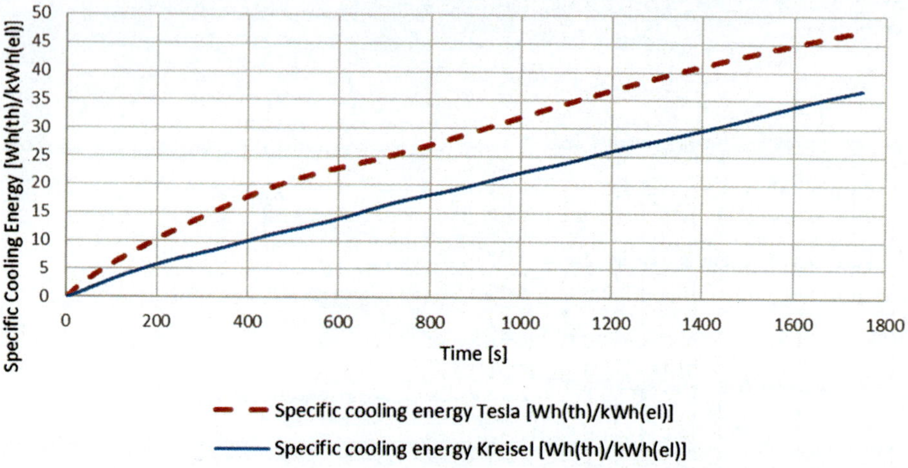

Fig. 8. Comparing the energy consumption of Kreisel Electric and sidewall cooling BTMS

For mass-market BEVs, the two key performance indicators are high efficiency and fast charging. The efficiency of the BTMS can be measured by the thermal energy needed to cool down (or heat up) the battery to a certain temperature level.

Comparing two battery packs of the same size, one with Kreisel Electric's technology and the other with sidewall indirect cooling approach, direct cooling (and heating)

consumes around 29% less thermal energy for cooling and heating. This number was worked out, by cooling both packs from 40 °C to 20 °C with the maximum cooling rate without exceeding a temperature spread of 5 K between the coldest and the hottest cell.

For fast charging, the heat generated by the battery roughly quadruples when the charging rate is doubled (Pv~I^2*R). Therefore, to avoid overheating of the battery, an efficient method of electrically connecting the cells (to keep R low and avoid extra losses) and a high performing BTMS is crucial. By meeting both requirements, Kreisel Electric's battery technology capable handling high charging rates.

The same principles apply to electric motorsport applications and high-performance EVs. The experimental data in Fig. 9 shows the battery cell-, busbar- and fluid inlet temperature during a demanding racing cycle with up to 500 kW drawn from a ~50 kWh battery. It can be seen that the cell and busbar temperatures stay within the limits. Furthermore, the battery can be cooled very efficiently during driving only using a radiator (weight-saving) and cooled down or preconditioned very quickly before and between racing (time-saving).

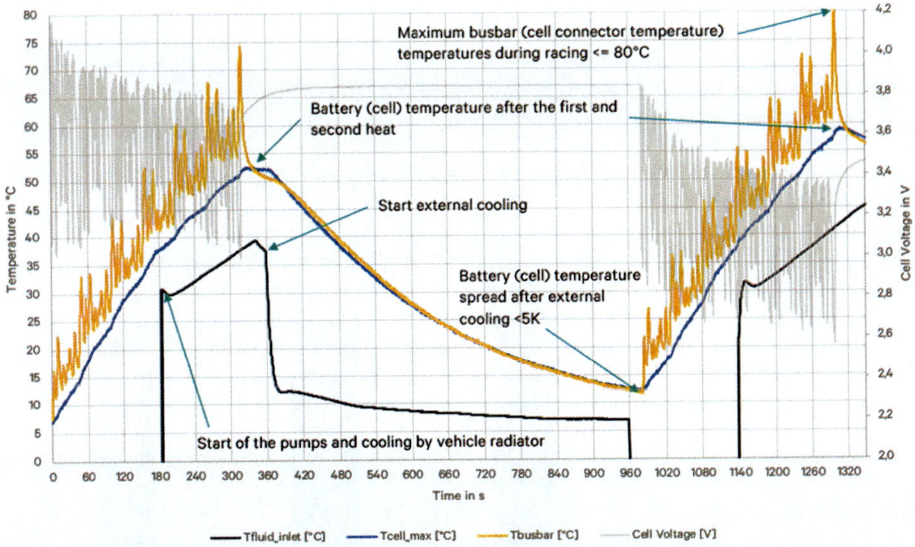

Fig. 9. Cell min. and max., Fluid inlet and busbar temperatures during a racing cycle with up to 500 kW drawn from a battery ~50 kWh Battery

Next, based on a racing profile from the Nürburgring, it was examined how direct cooling stacks up against other technologies in high-performance hybrid applications. With currents up to 36C (~217 kW drawn from a 6 kWh battery), three different BTMS approaches were tested: Passive (no cooling), indirect cooling (sidewall cooling) and Kreisel Electric's direct cooling [1].

It turned out that the battery temperature stayed significantly lower with direct cooling (Fig. 11). The result is an increase in battery life (89% SOH @ 4114 cycles) and safety. The setup with no cooling stopped after 200 s into the cycle due to the dangerous overheating of the cell (Figs. 10 and 13).

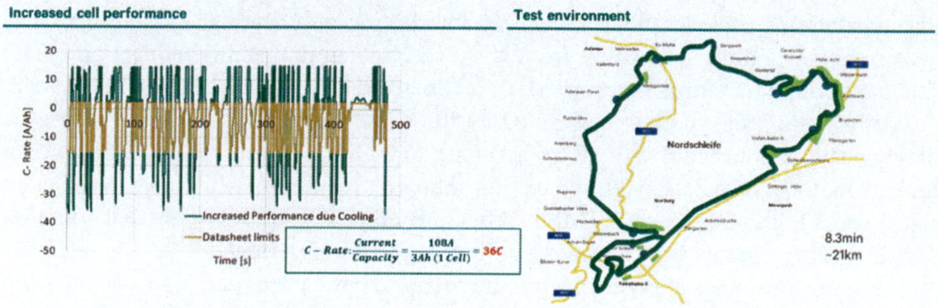

Fig. 10. Nürburgring racing cycle with up to 36C discharge and 15C charging power

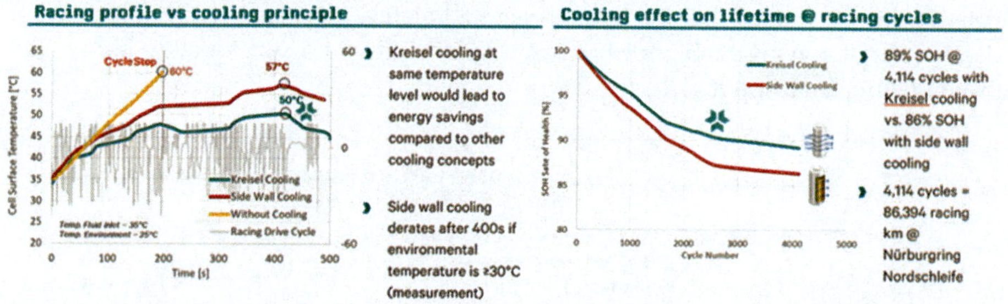

Fig. 11. Temperature and lifetime comparison of Kreisel Electric direct cooling, sidewall cooling and passive BTMS (no cooling)

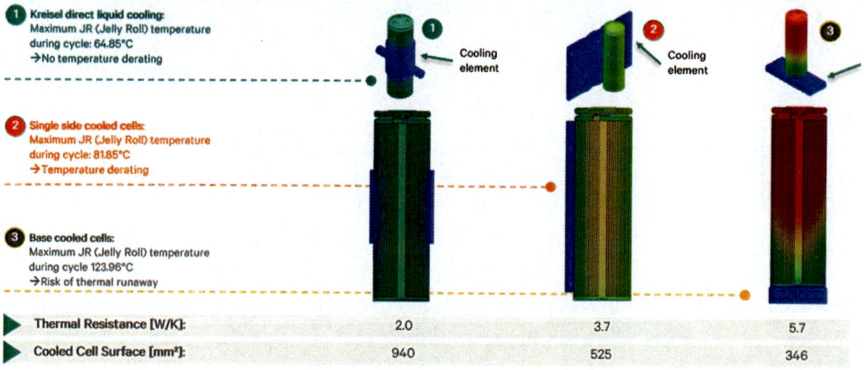

Fig. 12. CFD and conjugated heat transfer simulation of different BTMS approaches [1]

Further, the power loss generated during the racing cycle was applied in a CFD and conjugated heat transfer simulation (see Fig. 12). The jelly roll and current collectors inside the cell were modeled in detail, together with the housing and the different cell cooling elements. For comparison reasons, base cooling added to the investigation.

The results align with the measurements obtained in the test. With direct cooling, the lowest absolute temperature and temperature spread within the cell was observed.

Fig. 13. Testbed to validate the best coolant/material combination

Figure 12 reveals why Kreisel Electric is using the curved surface of the cylinder to transfer the heat. This allows to utilize a larger heat transfer surface compared to base cooling and results in a lower temperature spread within the jelly roll material- despite the lower thermal conductivity across the jelly roll layers.

5 Battery Lifetime Validation

Ensuring a leakage-free system over a lifetime is one of the biggest challenges in the design of immersion-cooled battery modules. The main drivers influencing the system robustness are the coolant, sealing material and sealing geometry.

Regarding the cooling liquid, Kreisel Electric was working on making the switch away from fluorinated fluids. This brings significant economic advantages (up to 90% cost reduction) and required lower system tightness (due to higher viscosity and surface tension of the fluid). The simplification of material selection is an additional plus, together with the incorporation of a redundant sealing design (dual sealing). The result is a leakage-free operation between −40 °C and 85 °C and the increase of system robustness by a factor of 72 in battery module comprising 36 cells. To deal with the drawbacks of the new fluid (e.g. higher viscosity) the design of the flow channels was adapted and extensive testing was done to ensure battery safety.

Key advantages of the new cooling fluid:

- Cost reduction up to 90%*
- Higher heat capacity (~50%*) and thermal conductivity (~50%*)
- Lower Density (~50%*)
- Temperature range −40 °C to 85 °C
- Global Warming Potential ~1
- Non-toxic
- Proven safety in thermal runaway scenario

To find the optimal fluid and material combination and test the new sealing design, the system was put through an extensive application-related material and system validation protocol. Execrated aging, thermal cycling (–40 °C to 85 °C) and thermal shock testing, was part of the program.

The result is an improvement of the helium leakage rate from 1*10E-4 mbar*l/s (before assembling) to <5*10E-6 mbar*l/s after 12000 operating hours (end life). This is caused by the cooling liquid supporting the material characteristics of the sealing over a lifetime.

6 Platform Approach

Due to the highly automated assembling process of the series production, Kreisel Electric's battery technology is particularly fast and thus very economic to manufacture. By using one base battery module in different parallel arrangements without changing the basic design, many development expenses can be saved. By applying this platform approach Kreisel Electric can provide up to 20 different battery module types with 4 different cells for different applications rout of one single production line. A basic module design comprises 36 cells that can be connected in 36/2 × 18/3 × 12/4 × 9 or 6 × 6 cells in parallel (see Fig. 14). With this approach different battery packs can be developed in unrivaled fast time to market even addressing highest demands on quality, safety and cost targets.

Fig. 14. Modular design enables improved customization & industrialization

7 Conclusion

To summarize main points how to find an optimum for an effective battery design for cylindrical cells following key aspects can be highlighted:

- Highest safety due to unique propagation prevention features are given.
- Battery pack costs of <100$/kWh in 2021+ is a realistic target.
- High energy and power density values can be achieved due to specific cell arrangement and electrical connection
- Immersion cooling enables higher C-Rates

- A specific sealing technology that is capable of >15 years of lifetime has already been validated
- New cooling liquids improve system robustness and costs significantly
- A platform approach enables fast time to market and reduced engineering/investment cost

Reference

1. Kastler H (2019) Solutions for an electrified world. EAWD Graz, Austria

Compact E-drive for Trucks—Efficient Utilisation of the E-motor Through Multi-speed Transmission

John W. Kimes(✉)

Sigma Powertrain Inc., Livonia, USA
Jkimes@sigma-pt.com

1 Introduction

The Battery Electric Vehicle (BEV) shows great promise for the future evolution of our transportation system by significantly improving efficiency while reducing cost of operation and harmful emissions. This is still an emerging technology that is gaining broad market acceptance. Key technologies for BEVs have matured rapidly over the last 10 years enabling proven vehicle models which have started to gain a foothold in the public consciousness. This rapid change has been due to advances in battery technology and cost. The now demonstrated benefits of BEV ownership such as lower fuel costs, lower maintenance costs, powertrain responsiveness, and quiet operation have lead to greater penetration in the passenger vehicle market. This is evident in the growth of BEVs in the passenger vehicle market segment worldwide.

Despite the remarkable growth in the passenger vehicle BEV market and demonstrated quantitative benefits, the commercial duty truck BEV markets have yet to experience the same level of acceptance. This is due in large part to difficulties in balancing the interrelated factors of cost, range, weight, and performance for these larger vehicles. Particularly for commercial delivery vehicles, range and cost are primary criteria for acceptance [1]. However, extending range generally means expanding the energy storage system which is a major driver of cost and weight of the vehicle. Furthermore, the added weight of a larger energy storage system not only negatively impacts performance but in turn also decreases the vehicle efficiency (as expressed in MPGe). Ultimately this means that there is a diminishing return on range extension by increased energy storage and more importantly increasing energy storage generates a loss of market interest due to a higher initial investment per vehicle [2].

The medium duty vehicle market is of particular importance however as these are the primary vehicles of the rapidly growing commercial delivery market. The operation of the medium duty vehicles within this market segment also contribute disproportionately to harmful emissions and fuel consumption. Indeed, research shows that even if only 10% of miles covered in a given urban area are attributable to diesel powered commercial delivery vehicles, those miles can account for up to 90% of the total particulate emissions in that area (Bigazzi 2012). Furthermore, medium and heavy-duty vehicles have also been shown to account for 25% of the total petroleum used in the US by the on-highway transportation sector [3]. In light of this, the medium duty commercial vehicle shows

Euroforum Deutschland GmbH (Ed.): *CTI SYMPOSIUM 2019*, pp. 294–301, 2021.
https://doi.org/10.1007/978-3-662-61515-7_25

particular promise as a target for BEV implementation as it holds potential to make a significant impact on overall fleet emission and fuel consumption reductions. The challenge is to improve the performance and efficiency of electric commercial delivery vehicles to bring them more in line with industry expectations while maintaining a neutral impact or net reduction in overall system cost. Therefore, Sigma Powertrain intends to demonstrate the capabilities of the eMax Powertrain to accomplish the range extension and/or reductions in energy storage required for broader market acceptance.

The gearbox will enable increased operational efficiency (MPGe) which means that the vehicle will be able to travel the same distance while utilizing less energy. Furthermore, the incorporation of a multi-ratio gearbox will also allow these vehicles to meet performance targets which are key to broad market acceptance such as gradability, highway speed operation, and stronger acceleration up to highway speed. The multi-ratio gearbox enables these performance increases over baseline without increasing overall motor size and power requirements, which would be necessary with a single ratio gearbox or direct drive powertrain configuration. The additional cost of incorporating the gearbox will be offset by either reduction in energy storage requirement and/or electric motor downsizing.

Historically, most electric vehicles (EV's) operate with a fixed single-ratio gearbox and a single electric motor (e-motor). This simple arrangement in small vehicles is highly beneficial as it takes advantage of the e-motor's wide speed-torque range of efficient operation with minimal powertrain and control complexity. However, as vehicle size and weight increase the e-motor power requirements to meet top-speed, acceleration, and gradeability demands increases non-linearly necessitating a much larger (and disproportionately more expensive) e-motor. Furthermore, when discussing specifically commercial vehicle applications, the majority of the vehicle's typical usage is at low speed and in stop-and-go traffic. When the fixed single-ratio gearbox is sized to meet speed demands, this typically will put the low speed driving operation well outside of the efficient operating range of the motor. The solution to this problem is to incorporate a multi-ratio gearbox in the powertrain of these larger vehicles. A multi-ratio gearbox will allow a step ratio change that enables both high-speed and low-speed efficient operation while improving powertrain performance metrics for a given motor.

In order to meet these challenges Sigma Powertrain Inc. has developed the eMax Powertrain. The eMax is a modular design that is a dual motor 4 mode CVT or a single motor 3-speed transmission that leverages the power density and efficiency of Mean's core product, the Mechanical Diode (MD) and the accurate speed control of the e-motor to deliver a compact, light-weight, highly efficient gearbox for EVs. The eMax is an advanced transmission concept that uses automatic transmission architecture (planetary gear sets) but is not hydraulic. The torque converter, friction clutches, high-pressure oil pump, and valve body have all been eliminated and replaced by controllable two-way clutches from Means Industries. By eliminating all the hydraulic components, the eMax has about the same parasitic losses as a manual transmission. Also, by eliminating the hydraulics and using extremely high power dense controllable two-way clutches, the entire size of the gearbox shrinks significantly. The result is a small, lightweight, highly efficient, high power dense, automatic all electric 4 mode CVT transmission.

2 eMax Overview

The eMax is a completely modular approach for commercial duty truck applications using a common gearbox with interchangeable inputs (e-motors) and outputs (extension housings with various yokes). It is an all-electric powertrain with no hydraulic controls. The eMax utilizes mechanical electric clutches with actuation times less than 20 ms. The clutches are extremely power dense and are core to the eMax concept. Using these types of clutches has produced a design that is about 13 inches long (310 mm) with 4 CVT modes capable of 4000 Nm of input torque. This is enough capacity for a class 8 semi-truck yet physically small enough for a class 1 utility truck. The eMax is designed to maximize efficient in the e-motors.

In the interest of optimizing resources and design proliferation, the eMax was designed as a 1-size-fits-all concept as explained above. Once a class of vehicle is selected, the powertrain system will then match the requirements of that vehicle by sizing the motors, determining the need for 1 or 2 motors, the yoke and extension housing requirements, and finally the need for a final drive or not. These variable inputs will have common bolts patterns for the input motors and extension housings. The output carrier spline will be common for all the various yokes across the classes. This is accomplished without changing the case, gearsets, clutches, and all the barrel parts in the gearbox.

The 2-motor application is a 4-mode CVT (Continuously Variable Torque). Each e-motor operates at 3 and 2 unique ratios respectively, each with independent power flow through a common box thus maximizing system efficiency by keeping more of the drive cycle in the sweet spots for the e-motors. There is no torque interruption during shifting in this configuration.

Alternative, the same gearbox can be configured for a single motor. This is a step ratio 3-speed that will shift similar to an AMT albeit very quickly. Therefore, in this configuration there would be torque interruption during a shift. In a commercial duty application coupled with the short shift times, this is widely acceptable.

2.1 The eMax Powerflow

The eMax utilizes a 4-node lever. The gearset connections in this lever use a ring gear to carrier/ring gear to carrier arrangement. The RC/RC gearset provides good ratio spread with acceptable step sizes. This gearset as shown later is a very mechanically efficient as well. This was also the only 4 node lever that allowed the B-motor to be easily connected to both the output and sun gear 1 via electric clutches. A Ravigneaux or Simpson gearset would not accommodate the eMax lever. This lever yields the following modes, A1B1, A1B2, A2B2, and A3B2 where the nomenclature "A#" means the A motor is in the ratio dictated by the trailing number. The same nomenclature is used for B motor. The following figure shows the eMax block and lever diagrams and the shift table (Figs. 1, 2, and 3).

2.2 Latching Electric Clutches

The eMax shift chart is shown above. Note that Park is achieved by utilizing the two brake clutches B04 and B05. These grounding clutches use magnetically latching electro-mechanical devices. The only time these types of clutches consume electric power is the

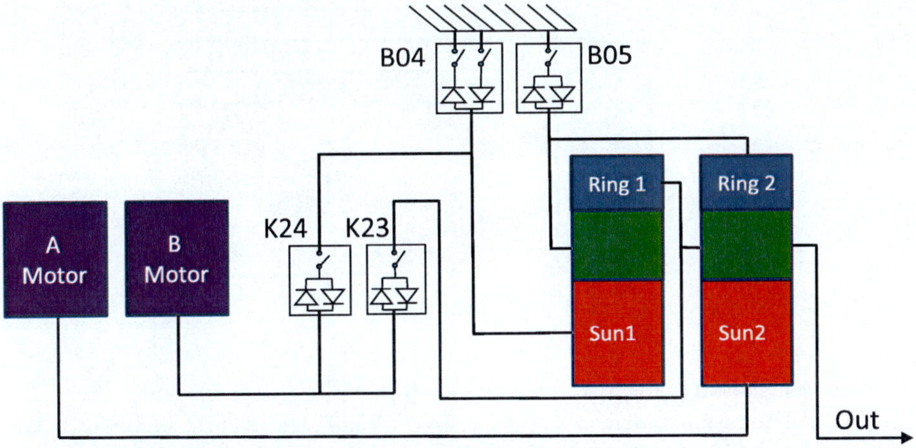

Fig. 1. Block diagram

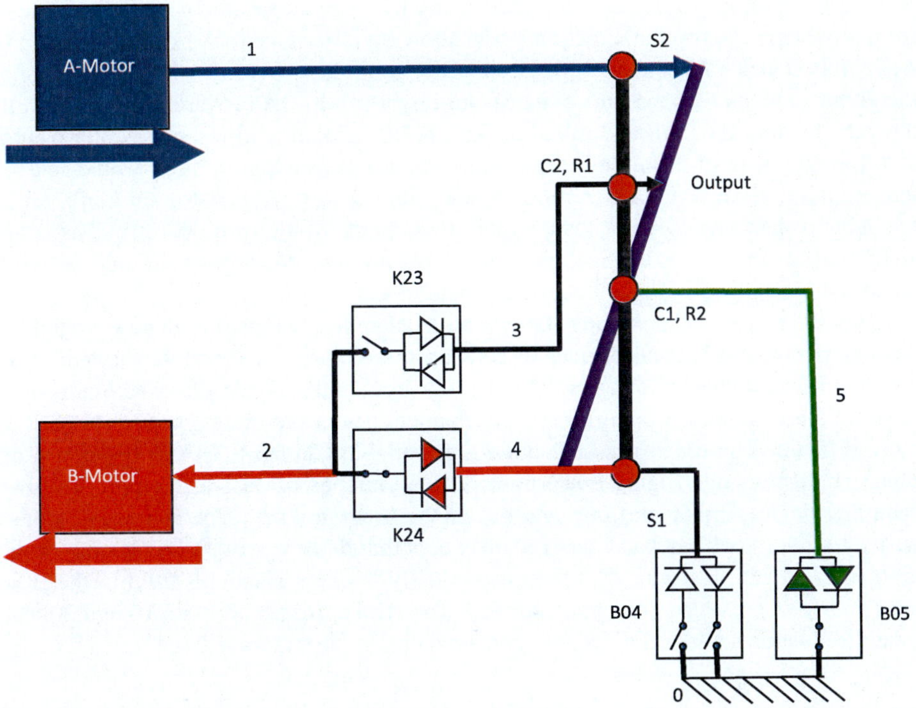

Fig. 2. Lever diagram

transition for ON to OFF and vice versa. Once in a state, the clutch latches maintaining that state without electric power. This function allows for the brakes to be used for Park at extremely high torque capacity.

		Clutches			
		B04	B05	K24	K23
Modes	A1B1		X	X	
	A1B2		X		X
	A2B2	X			X
	A3B2			X	X
	Rev		X	X	
	Park	X	X		
	Hill Hold	1/2	X		

Fig. 3. Shift chart

Moreover, the B04 clutch is a 4-state clutch meaning that it has states of 11, 01, 10, and 00. The case for 11 means that the first "1" indicates that the clutch is locked in the forward direct. The second "1" indicates the clutch is also locked in the reverse direction. So a "10" state is a One-Way clutch and a 11 is a Two-Way clutch. Having independent control of each torque direction in B04 aids the Park algorithm. Also, in A1B1 mode (low gear) the B04 clutch can be in a 01 state. This would prevent the vehicle from moving in the reverse direction while allowing it to move forward, aka Hill-Hold. At vehicle speeds of less than 3 MPH, this B04 could be activated in a 01 state. While the vehicle creeps forward, this clutch would over-run in the forward direction and if the vehicle were on a hill and rolled back, this B04 clutch with respect to B05 would lock the output shaft to ground thus holding on a hill. Forward motion would not be locked, but reverse would be. As the vehicle pulled away from a stop on a hill, once the vehicle speed was greater than 3 mph, this clutch would turn off (00). This is an additional function. This is "free" just requiring software. Note that Park and hill-hold could not be realized if dog clutches were being used.

The B04 and B05 clutches are static or brake clutches. One race is always grounded. Adding electro-mechanical devices to activate these clutches is straight forward. The K23 and K24 clutches are dynamic clutches aka input or disconnect clutches. Both races rotate in these devices. Adding electro-mechanical devices to these types of clutches is more difficult. The eMax uses Dynamic Controllable Clutches (DCCs) developed at Means Industries and Sigma Powertrain. These clutches use a linear motor to move camming devices (plungers) into and out of the locking struts. The action resembles what a tradition park pawl and park rod do except that in the case with DCCs everything is rotating. Again, just like with the magnetically latching solenoids used in the B04 clutch, these DCCs also latch magnetically. The stroke time is between 10 and 20 ms. Like the solenoids above, DCCs only consume energy during transition.

2.3 Electrical and Mechanical Efficiency of the eMax

The Greek symbol sigma $\sum$ in mathematical language means sum. The Sigma Powertrain's eMax is all about summing the outputs of 2 motors into a single output while maximizing efficiency at all the combinations of vehicle speed, throttle position, and transmission mode. There are about 50,000 combinations in our software looking at these 3 parameters. Recall that in every transmission mode the 2 motors are independent power sources. Motor A and Motor B operate independent of each other. So, when a torque demand must be met (determined by percent throttle) that torque command can be met with either Motor A, B, or some combination thereof. The equation for output torque is as follows.

$$T_{out} = T_A(TA_{ratio}) + T_B(TB_{ratio})$$

For a given mode the ratios the motors inputs through are fixed. The table below list the various ratios for the eMax (Table 1).

Table 1. Motor Ratios

Motor and mode	Ratio
A1	3.63
A2	1.73
A3	1
B1	2.64
B2	1

The question then becomes, for a given demanded output torque, what torque values need to be commanded for A and B motors. There are numerous values that would fulfil the equation above. The answer is the torque values of A and B that meet the commanded torque with respect to the optimal efficiency of the system. The system in this case is the inverter efficiencies, motor efficiencies, and the mechanical/mesh efficiencies. Since the motors operate at a fixed ratio in each mode, the motor speed is tied to vehicle speed. So in other words, the motors speed is constant with respect to the vehicle speed. The graph below depicts this constant speed of each motor with a yellow line showing the fixed motor speeds in the mode of A1B1 at a vehicle speed of near 30 MPH.

What is not fixed is torque. The torque level in the graphs below are depicted by a blue dot.

In the graph above, either A or B motor could meet the demanded torque by themselves, however it would not have been the optimal efficiency. By choosing values of torque for A and B that puts the operating condition in the sweet spots of the motors is the correct answer (Fig. 4).

The Sigma software also takes into account the inverter efficiencies and mechanical mesh efficiencies of the system as well. This is all predetermined in a lookup table in the control software. Below is a chart of the mechanical mesh efficiency of the RC/RC 4 node lever (Table 2).

In summary, the Sigma software optimizes the entire powertrain system efficiency. This is an extremely efficient powertrain solution.

3 eMax Summary

Sigma Powertrain is in the process of commercializing our all electric powertrain solutions for commercial duty applications for all classes (1–8). Our approach is a modular

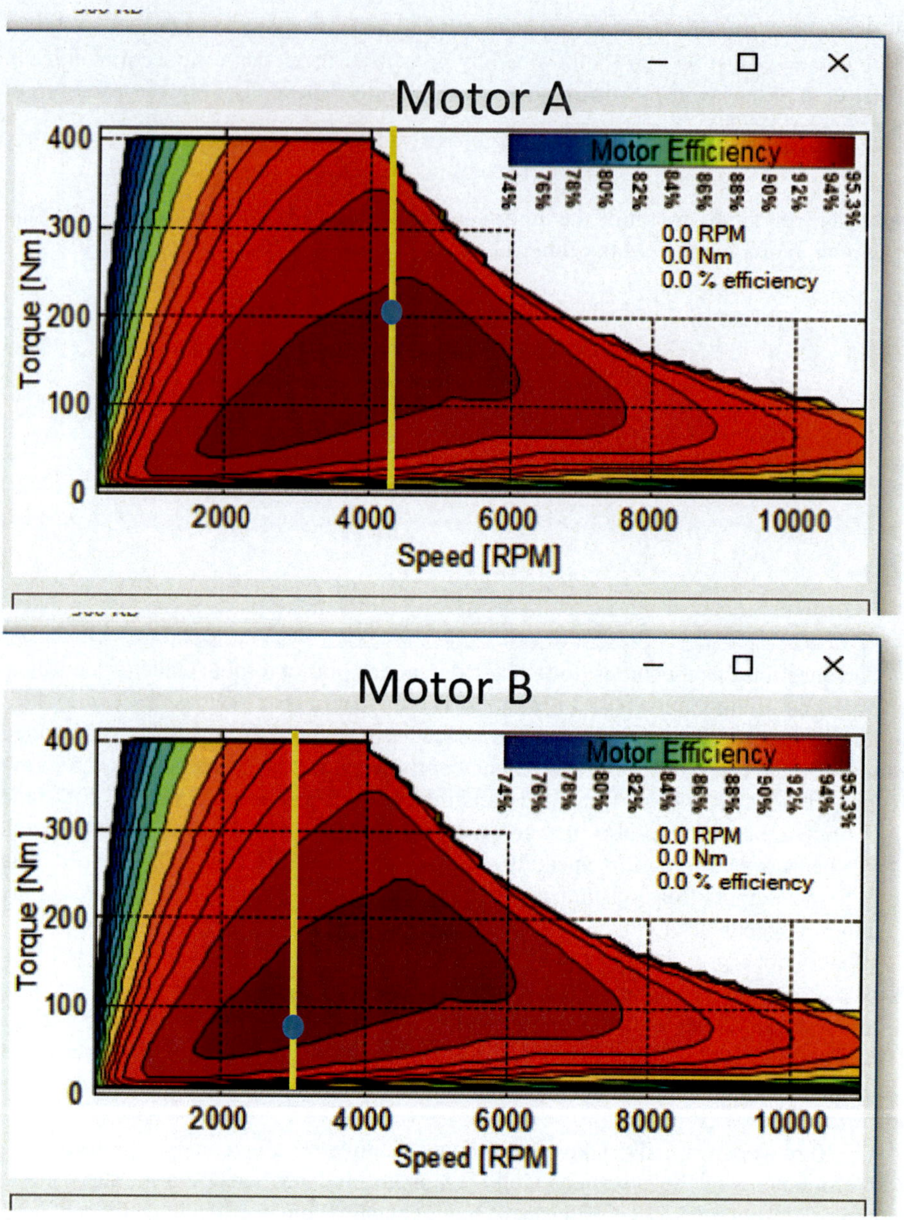

Fig. 4. Efficiency map

scheme with interchangeable input and output modules into a common transmission. The transmission utilizes electric controllable clutches. These clutches are in production at Ford and are being evaluated by advanced groups at major OEMs in Europe, Asia, and North America. We are the innovative force birthing this technology in the auto industry. This technology is the enabling innovation to eliminate hydraulic control and

Table 2. Mesh efficiencies

Trans mode	MOT A ratio	MOT B ratio	MOT A efficiency (%)	MOT B efficiency (%)
A1B1	3.63	−2.64	97.8	97.0
A1B2	3.63	1.00	97.8	100
A2B2	1.73	1.00	97.8	100
A3B2	1.00	1.00	100	100

to create transmissions with high power density. This means a significant improvement in efficiency, torque capacity all in a small package. We have designed the transmission endurance limit to allow 2000 Nm total input torque, 4000 Nm max excursion. This is more than enough to propel class 8 size vehicles yet still be physically small enough to be used in a class 1 pickup truck. The idea is to have one size fits all box with interchangeable add-ons to adapt to the specific needs across all commercial duty applications. This is a multi-mode powertrain. Each e-motor operates at 3 and 2 unique ratios respectively, each with independent power flow through a common box. This creates 7 modes of forward operation. The result of this flexibility is to optimize the efficiency of the powertrain by spending the majority of a given drive cycle in the sweet-spots of the motor(s). There is no perception of ratio change nor is there any "torque-holes" during ratio change. The shift times are less than 250 ms. This is also available in a no-compromise Plug-In Electric Hybrid version. This is our Sigma E-Max Powertrain solution for commercial duty trucks.

WWW.SIGMA-PT.COM.

References

1. EC (2010) Fleet electrification roadmap: revolutionizing transportation and achieving energy security. Electrification Coalition
2. Feng W, Figliozzi M (2013) An economic and technological analysis of the key factors affecting the competitiveness of electric commercial vehicles: a case study from the USA market
3. Davis S, Boundy R (2019) Transportation energy data book: Ed. 37. Table 1.14
4. Beiser C (2019) Compact and effiecient electric propulsion systems enabled by integrated electric controllable clutches. In: CTI Novi symposium

Representative AWD System Evaluation by High-Function Dyno—New Simulation Model of Sand Road

Wataru Kobayashi(✉) and Satoru Okubo

Nissan Motor Co., Ltd., Atsugi, Kanagawa, Japan
w-kobayashi@mail.nissan.co.jp

Abstract. Previously, AWD system evaluations have taken place using actual vehicles on real roads. However, there have always been concerns regarding inaccuracies from test to test due to seasonal differences in road conditions and weather. This problem is especially prevalent on sand road tests, these tests are particularly affected due to the sinking behavior of tires in soft sand. Therefore, when a comparative evaluation is implemented on a vehicle with changes made to the AWD control logics and specification, it is difficult to identify whether any phenomenon observed between tests is a result of the adjustment made to the AWD specification or a change in the road conditions. To solve this problem a new simulation model replicating a sand road has been developed. The model is able to simulate driving a car on sand using Dynamos with low inertia and high response motors. To create this simulation model, 4 characteristics of sand, which are "the running resistance", "the static and dynamic sinkage characteristic of the tires" and "the traction coefficient" were identified and calculated by carrying out full-scale test with an actual vehicle running on sand.

Keywords: AWD system evaluation · Simulation dyno · Driving on sand road · Tire sinkage

1 Introduction

Previously, AWD system evaluations have taken place using actual vehicles on real roads. However, there have always been concerns over poor repeatability quality from test to test due to seasonal differences in road conditions and weather. Therefore, when an evaluation is implemented using a vehicle on real roads, some ingenuity is required in an effort to make the results as accurate as possible, such as carrying out multiple runs.

This problem is especially prevalent on sand roads. It is difficult to evaluate adjustments to the car from test to test due to changes in the road condition and the ensuing impact these changes have on tire sinkage in the sand. Therefore, when making comparative evaluations and changing AWD control logics and specifications on a vehicle, it is difficult to distinguish whether phenomena observed are as a result of the adjustments or a change in the road conditions.

Euroforum Deutschland GmbH (Ed.): *CTI SYMPOSIUM 2019*, pp. 302–312, 2021.
https://doi.org/10.1007/978-3-662-61515-7_26

Recent advances in powertrains have resulted in higher torque responses and high level torque accuracy, this is especially true in Electric powertrains. In order to maintain an efficient, secure and comfortable ride the demand from the cars control systems is increased. An accumulation of small effect in the control system can have a big impact on the cars overall performance, so it is important to address all such effect.

To improve the accuracy of the testing procedure a new simulation model, replicating a sandy road, was created. This model is able to simulate driving on sand using a real car placed on a simulation dyno, with low inertia and high response motors.

2 Constitution of Real Car Simulation Dyno

To replicate a sandy road on a dyno, it is important to accurately simulate all transient effects such as wheel spin (Refer to Fig. 1.). In order to do this dynos with low inertia and high response motors are used. These dynos can replicate a number of transient situations such as slip on Snow and Ice.

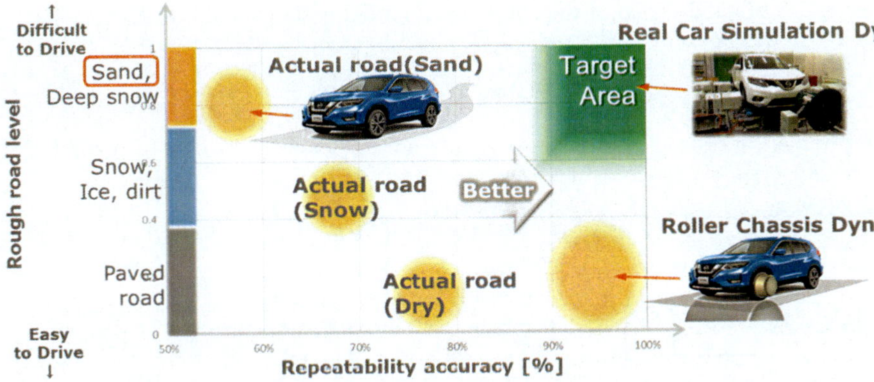

Fig. 1. Positioning of test equipment

The system overview is shown on Fig. 2. An actual vehicle is placed on a test bed, with all 4 wheels attached to 4 individual dynos. Different road surfaces and conditions can be programmed into the system and controlled by the dynos.

It is unnecessary to construct a detailed model of the powertrain because an actual vehicle is able to be used. The road surface and vehicle dynamic models are constructed using Matlab/Simlink.

3 Construction of Sand Road Model

In the case of a paved road, the forces between the vehicle and the road are traction characteristics and running resistance (rolling, air and grade resistance).

In the case of a sandy road, there is the added complication of the tires sinking due to wheel load and wheel spin. As a result, the traction characteristics and running resistance of each of the dynos are changed uniquely.

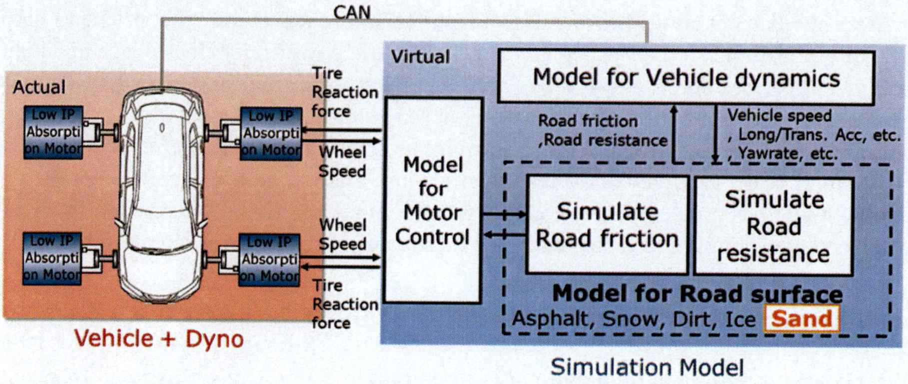

Fig. 2. Overview of real car simulation dynamo system

Four important characteristics, "Running resistance of sand", "Static and Dynamic Sinkage of tire" and "Traction on sand" were programmed as road models to replicate a sand road. A block diagram of the constructed model is showed in Fig. 3.

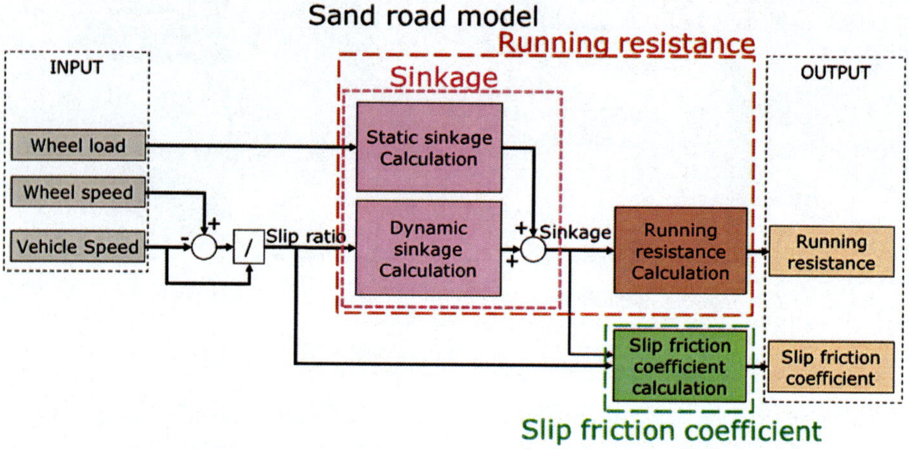

Fig. 3. Sand road sinkage model

With regard to the reproduction accuracy, models were aligned to test results gained from the sand area at the company's proving ground.

3.1 Model Construction of Running Resistance on Sand

The state in which a tire that is run on sand, is showed in Fig. 4.

As a tire sinks into sand to a given depth, the tire rotates forwards creating both friction and slip with the sand.

When a drive torque "T" is applied to the tire, a frictional force is generated between the tire and the sand, at the same time a horizontal tractive force "ma" is created.

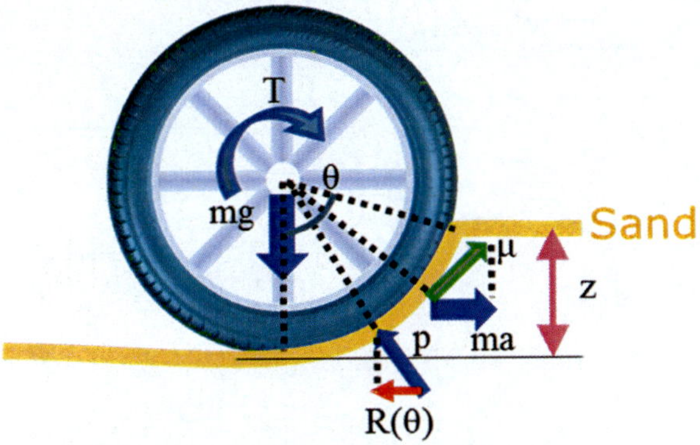

Fig. 4. Tire state at running on sand. z: Tire sinkage[m]; R(θ): Running resistance; T: Driving torque[Nm]; μ: Slip friction coefficient; m: wheel road [kg]; p: pressure between tire and sand; a: Longitudinal acceleration [m/s^2]; g: Gravity [m/s^2]; θ: Contact angle between tire and sand surface [deg]

The horizontal component of surface pressure "p", which is the result of wheel load, is running resistance "R".

The difference between the tractive force and running resistance is acceleration force.

By this relationship, running resistance is fixed by the height of sand wall in front of the tire (Tire sinkage depth "z"), giving:

r: Tire radius [m]
R: Running resistance [-]
b: tire width [m]

The relationship between sinkage and running resistance was verified by vehicle tests on actual roads.

As mentioned above, running resistance is determined by the amount of tire sinkage. In order to verify this a draw bar was attached to the back of a vehicle, creating resistance, which was measured using a load cell. An example of this setup can be seen in Fig. 5, and the results of which can be seen in Fig. 6.

3.2 Model Construction of Tire Sinkage

Tire sinkage in sand has two contributing factors; the static sinkage due to wheel load and the dynamic sinkage caused by wheel spin, as is seen in Fig. 7.

The total sinkage is the sum of these two factors together, as shown below.

$$Z = Z_0 + Z_d$$

Z: Total sinkage [mm]
Z_0: Sinkage due to wheel load [mm]
Z_d: Sinkage due to wheel spin [mm]

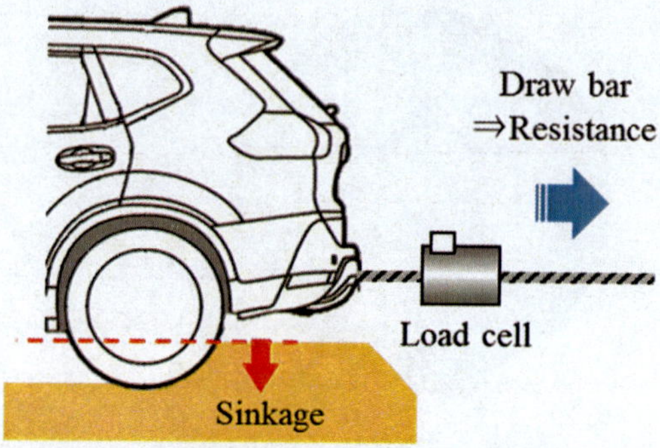

Fig. 5. Test condition of sand resistance measurement

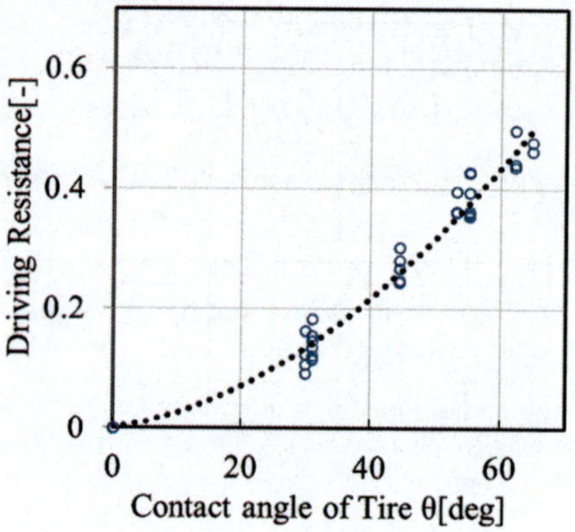

Fig. 6. Characteristic of running resistance on sand

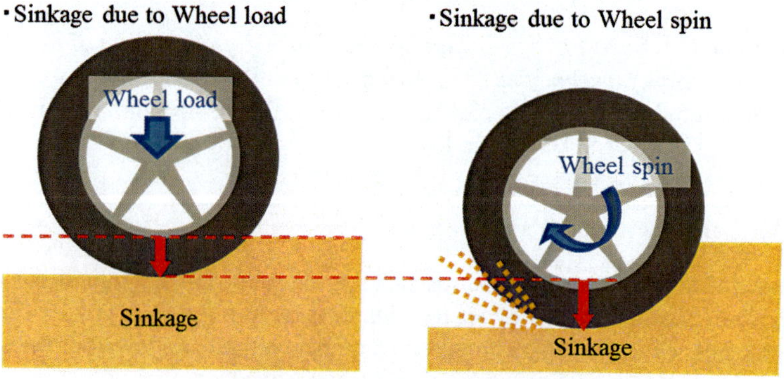

Fig. 7. Type of sinkage behavior

Static sinkage is a phenomenon whereby the tire sinks into the sand due to a load on the wheel, even if there is no tire slipping.

The relational equation is defined below.

a_0, a_1: Coefficient of static sinkage [-]

These characteristics were verified by actual road tests. The measurement method is shown in Fig. 8. To avoid any dynamic sinkage, the tire is moved on to sand slowly and the sinkage is measured using a camera and image analysis.

Fig. 8. Test method for measurement of sinkage due to wheel load

As an example, the result from a C-segment SUV class tire is showed in Fig. 9.

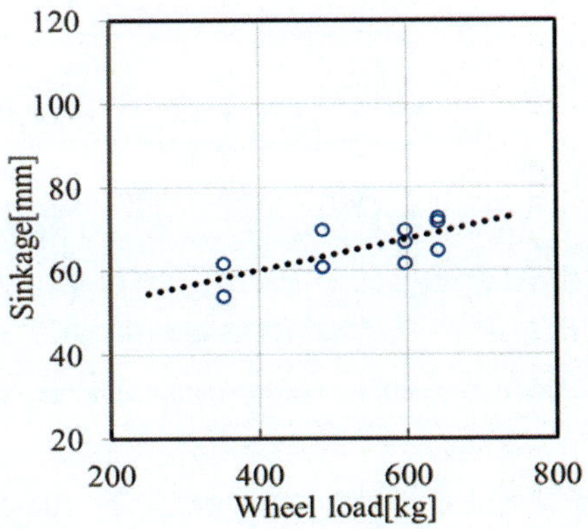

Fig. 9. Characteristic of static tire sinkage with respect to load

Dynamic tire sinkage is a phenomenon resulting in the tire sinking into the sand as a result of wheel spin.

The relational equation is defined as follows.

Z_d: Dynamic tire sinkage

s: Tire slip ratio
C_0, C_1, C_2: Coefficient of dynamic sinkage [-]

In the above equation, tire slip ratio "S" is calculated using the following formula.

$$s = \frac{(V_t - V)}{V}$$

V: Vehicle speed [km/h]
V_t: tire rotation speed [km/h]

These 3 coefficients have been verified by testing vehicles on actual sand.

In the equation above it is noted that dynamic sinkage is linked to the tire slip ratio. This coefficient can be verified by using specialist testing equipment with adjustable parameters. However, in this case, it was acquired by actual vehicle testing on sand.

Because the relationship between tire and sand can be verified directly, corrections are unnecessary. By nature, sand road surfaces are not generally flat, therefore, the measurement method used to collect results has to take this into account. See Fig. 10.

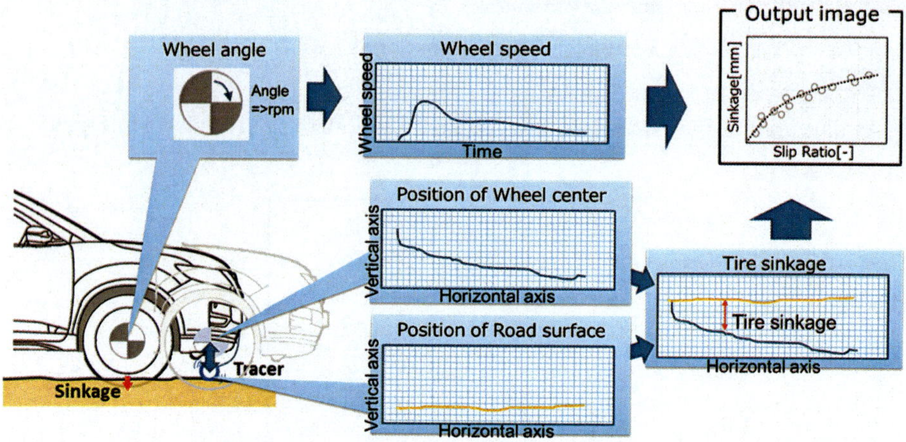

Fig. 10. Test method for measurement of dynamic sinkage due to wheel spin

A high speed camera was also used to measure the profile of the sand surface and the position of the tire as well as the tire rotation speed.

The result of these measurements can be seen in Fig. 11.

3.3 Model Construction of Slip Friction Coefficient on Sand

The traction characteristics observed between the tire and the road surface can be explained. The traction characteristics of sand can be expressed by the tire slip ratio and traction coefficient as is the case with asphalt and snow.

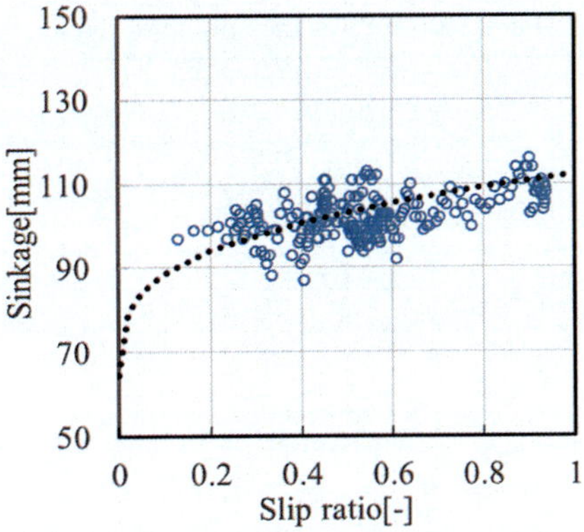

Fig. 11. Slip ratio vs dynamic sinkage

This information can be measured and verified by the data gained from running on sand. The traction coefficient is acquired from a measurement of the torque on the drive shafts using a strain gauge. The final coefficient, the slip ratio, is acquired from data from the high speed camera measuring wheel rotation angle and speed.

The results are shown in Fig. 12.

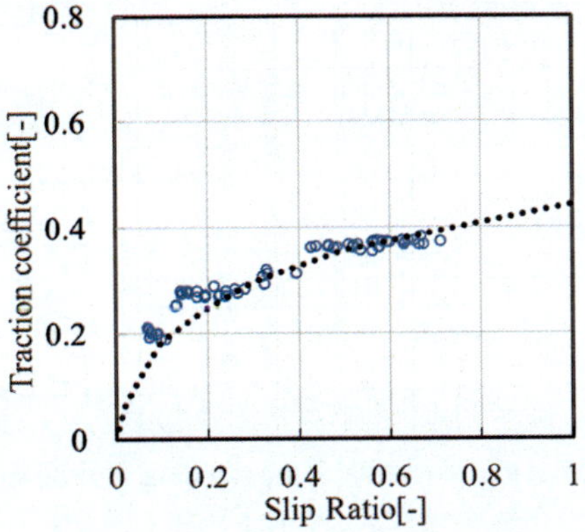

Fig. 12. Slip ratio vs traction coefficient

Asphalt and snow traction coefficients can generally peak at any slip ratio, but it is noted that on sand the traction coefficient gradually increases with slip ratio.

4 Result of Reproduction

The sand model simulator was designed based on the above characteristics which have been verified with vehicle testing. The model was then installed onto vehicle simulation dyno to confirm test to test repeatability.

The conditions set for the real vehicle replication test are shown in Table 1.

Table 1. Parameter for real car simulator dynamo model

Item		Parameter
Vehicle	Vehicle type	C-SUV
	Drive mode	AWD Lock
	Weight	1950 kg
	Tire radius	0.348 m
	Tire width	0.215 m
Sand	Sand type	Sand road in the company proving ground

4.1 Verification of Running on a Sandy Road

As an example, results of a standing start test on a sandy road can be seen in Fig. 13.

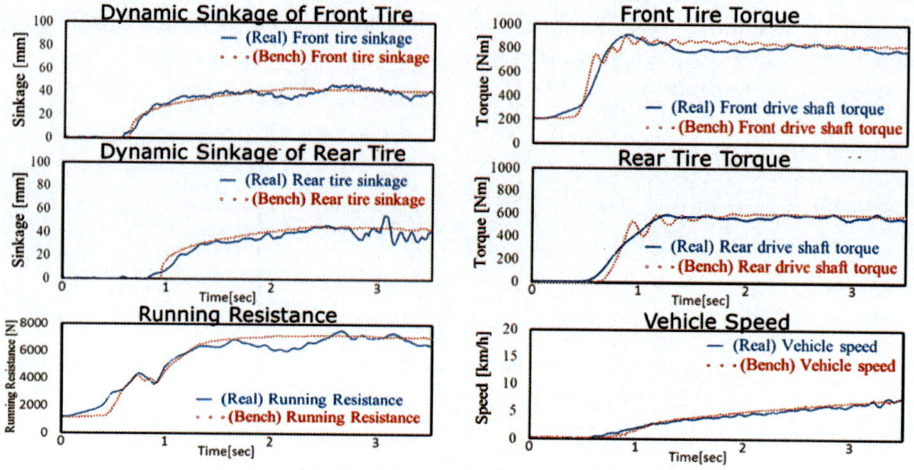

Fig. 13. Reproduction of the standing start on sand

The 3 graphs on the left show the front and rear tire sinkage and running resistance. These can be replicated in the test condition well as is seen in Fig. 13.

Next, the 3 graphs on the right show torque values for the front and rear drive shafts as well as vehicle speed. These results are can also be aligned with replicated actual road test results.

Further study has shown that running resistance varies between different vehicle specifications as shown in Fig. 14. These can also be replicated in the different running resistance condition.

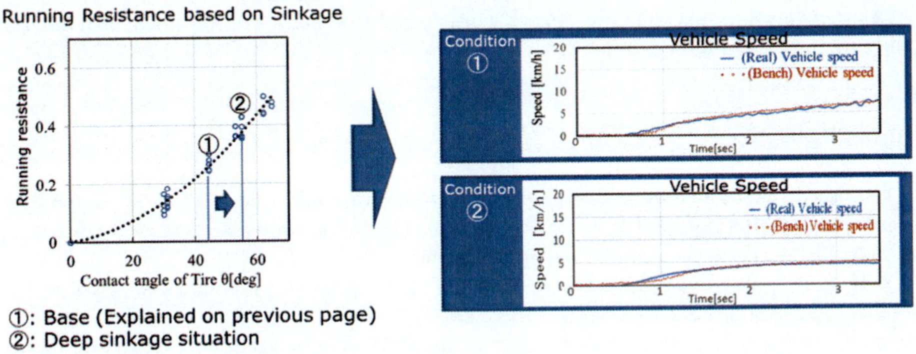

Fig. 14. Reproduction at each resistances

5 Validation Results for This Evaluation Method

The acceleration performance of two AWD logics (A and B) and varying vehicle weight (based on B) are showed in Fig. 15.

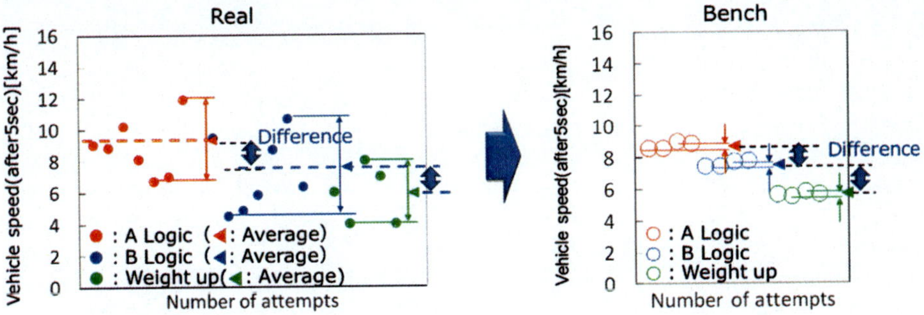

Fig. 15. Effectiveness of new evaluation method by real car simulation dyno

The left hand graph shows actual road testing results, owing to poor repeatability it is very difficult to evaluate. However, as seen on the right hand graph the results from the real car simulation dyno are much more consistent and repeatable giving high evaluation confidence.

6 Conclusion

By identifying the four key characteristics of sand which are; the running resistance, the traction coefficient, the static tire sinkage and the dynamic tire sinkage, a sand road simulator was designed and built.

Using a real car simulation dyno to replicate sand roads enables repeatable tests to be carried under the same road conditions, which is not the case with real sand road testing.

As a further application, any sandy driving condition such as desert or beach can be tested on this simulator.

References

1. Tera Mechanics Research Society (1999) Off-road tire engineering—basic of design and estimation of performance
2. Muro T (1993) Tera mechanics-running mechanics
3. Yamakawa J, Yoshimura S, Watanabe K (2008) Development of tire model on dry sand by model size experiment (first report). J JSAE 20086027
4. Eto R, Yamakawa J (2017) Analysis of traveling wheel on loose sandy soil. In: JSAE, Spring Congress 2017, no. 159
5. Bekker MG (1969) Introduction to terrain-vehicle systems. The University of Michigan Press
6. Ohshio S (2018) Ultra-high accuracy technology for measuring transient transmission efficiency. In: 17th International CTI symposium

Environmental Benefits of Used Batteries from E-Vehicles as Stationary Energy Storage

Juergen Koelch(✉)

EVA Fahrzeugtechnik GmbH, Heidemannstrasse 41a, 80939 Munich, Germany
juergen.koelch@evafahrzeugtechnik.de

Abstract. The batteries aged during the life of an electric vehicle can be reused for a variety of stationary applications. In combination with photovoltaics, higher private consumption rates can be reached at a module level. At the level of complete vehicle high-voltage batteries, DC quick charge stations with low power connections or load management (peak shaving) represent an example for industrial clients. In the megawatt area, used electric vehicle batteries can provided grid support or they can be used to stabilize the energy from wind parks. The environmental benefits of an electric vehicle face repeated criticism. Along with the use of CO_2-intensive, coal-based electricity for vehicle operation, the ecological rucksack of the battery is the primary focus. When the battery is reused at the end of the vehicle's service life, the environmental benefits far exceed those of just regular recycling. The paper will show two already realized battery plants on a megawatt scale and report the experience in planning and construction. Both systems are used for grid control and can support a smart grid in the future.

Keywords: Battery 2nd life · LCA · Life cycle · Reuse · Electric vehicle

1 Introduction

Across the world, more and more renewable energy is coming on line, for example solar and wind-generated electricity (turnaround in energy policy), and the generation of this electricity is subject to significant fluctuations. Expanding energy accumulators to compensate for these fluctuations will become an important topic in the future. Energy accumulators, for example, make it possible to store photovoltaic (PV) electricity during the day, allowing it to be tapped as required during phases with low levels of sunlight.

2 Renewable Electricity, Taking the Example of Photovoltaics

2.1 Falling Photovoltaic Module Prices

Worldwide more and more fluctuating renewable energy, such as solar and wind power, is being strongly increased (so called "Energiewende"). The expansion of energy storage to compensate these energy fluctuations will become an important issue in the future.

Euroforum Deutschland GmbH (Ed.): *CTI SYMPOSIUM 2019*, pp. 313–320, 2021.
https://doi.org/10.1007/978-3-662-61515-7_27

There is the possibility to store e.g. photovoltaic (PV) electricity during the sunny day and give it back to the home grid as needed during night or cloudy days.

Figure 1 shows the global installed photovoltaic (PV) capacity in gigawatts (GW; peak capacity) plotted double-logarithmically against the module price. The module costs account for about half of the total costs of a PV system. The straight line shows the trend for a price reduction of more than 20% approximately for photovoltaic modules if the installed capacity is doubled. At the end of 2018, approx. 500 GW PV capacity was installed worldwide according to Wirth [1]. In the last 15 years (2000–2015) the module price for one kilowatt peak (kWp) fell down from 5,000 to 500 EUR. Eight to ten square meters of module area is required for a PV system with one kilowatt peak capacity (kWp). In Germany, it can be expected that each kWp will generate 800–1,000 kWh per year. If the roof orientation is unfavorable, for example a west or east-facing roof, and assuming a roof slope of 40° then it can be expected that there will be an approx. 25% reduction in yield compared to the ideal south-facing roof. However, it is possible to install twice as much module area compared to a south-facing roof. Even a north-facing roof with a shallow roof slope of 20° can be considered, because the energy sacrifice is approx. 30%.

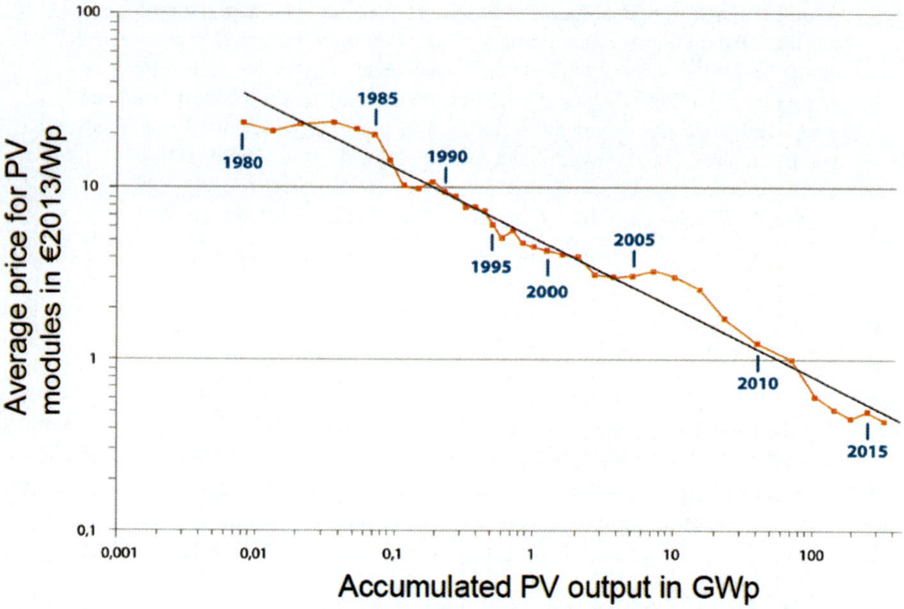

Fig. 1. Trend in historical price development for photovoltaic modules depending on the installed PV capacity. *Source* own presentation according to [1]

2.2 Achieving Grid Parity for Photovoltaic Electricity in Germany

With the introduction of the Renewable Energies Act (REA) in Germany in 2000, all the generated solar electricity was fed into the grid because the payment was significantly higher than the price for domestic electricity. Since 2013, we have now reached the time when it is more economical to cover your own energy consumption using solar-generated electricity first and then feed only the surplus solar electricity into the grid (grid parity). The more the domestic electricity price increases in the future (and the more the price of solar electricity falls), the greater the cost advantage will be for covering your own consumption with solar electricity generated independently (see Fig. 2).

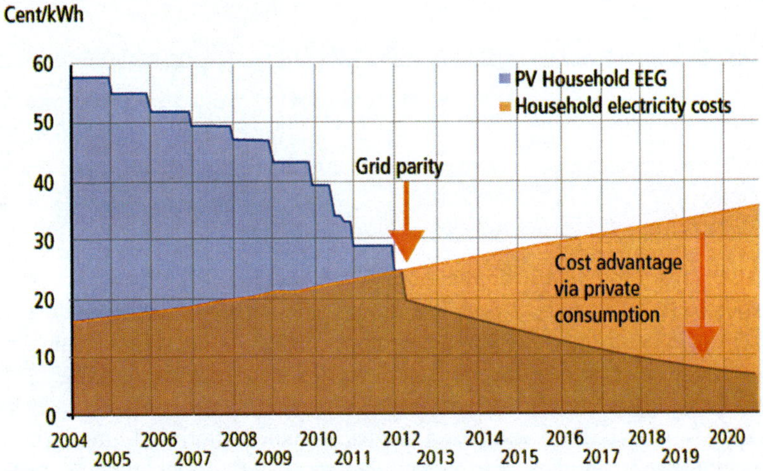

Fig. 2. Comparison between the cost trend in the REA price paid for photovoltaic systems with an installed capacity of less than 10 kWp and domestic electricity prices. *Source* own presentation according to [2]

3 Environmental Balance

Three specific areas can be distinguished for a vehicle when performing an ecological assessment or life cycle assessment (LCA). The first is manufacturing, then usage by the customer and finally, recycling at the end of the vehicle's service life, see Fig. 3. When compared to regular gas or diesel vehicles, greenhouse gases are almost doubled for electric vehicles, above all, for the energy- and raw material-intensive manufacturing of the vehicle battery. [3] The use of old electric vehicle batteries for a stationary application can make the manufacturing effort for a battery that would otherwise be used here obsolete, and thus improve the total environmental balance in a sustainable manner.

4 Applications for Battery 2nd Life (B2L)

For more than one hundred years, lead batteries have been used for storing electricity. The established lead-acid systems are used as standby power supply for hospitals, for example. For mobile applications, the high specific weight represents a huge disadvantage although it is not a decisive factor in stationary applications. The more expensive

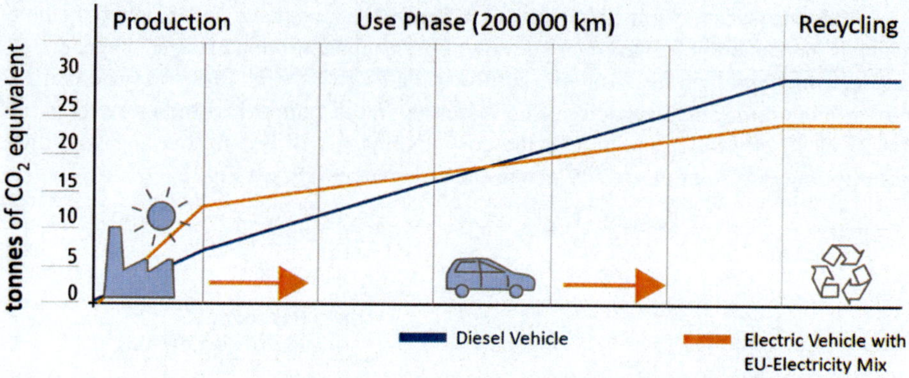

Fig. 3. Life cycle comparison between diesel and electric vehicle (own presentation according to [3])

lithium battery from the mobile application is now increasingly displacing established lead batteries in the stationary sphere too, scoring points significantly with its service life (greater cycle count) and greater depth of discharge. Additional advantages of the lithium battery are its greater efficiency and lower spontaneous discharging.

If there is a significant increase in the use of electric vehicles (exclusively with lithium batteries) the second use of used electric vehicle batteries (Battery 2nd life) in stationary applications could become a significant topic. Various technical variants could be presented for integration into a Smart Home or Smart Grid. In the domestic sphere, it would be feasible to have a direct current connection with the PV system involving a DC/DC converter at the 400 V level. As well as this, an additional inverter can be used for the single-phase integration of the 400 V battery accumulator into the 230 V alternating current system. This variant is highly suitable for subsequent integration of a battery accumulator with an existing PV system, for example when this drops out of the former REA remuneration scheme after a 20-year lifecycle.

The following applications have been identified from a number of possible applications for used electric vehicle batteries. They have been divided by size into three areas.

4.1 Small-Sized Application: Private PV System

When the vehicle high-voltage battery (HVB) is stripped down to the module level, systems can be set up with the modules via parallelization, e.g. for a private photovoltaic (PV) power storage system. They are often designed for a voltage range of 60 V DC as safety extra low voltage (SELV). Protection against contact does not have to be provided within this voltage range. This is currently provided in the single family home with a private user as system operator. In Germany, the generation of PV power has generally been cheaper than household electricity costs (grid parity) since 2013. From this point on, a market has opened up for battery storage in order to increase the share of private consumption and reach a higher economic efficiency of the PV system due to decreasing battery costs. The growing market for electric vehicles will increase the electricity demand on the one hand, but it will also increase the return flow of used vehicle batteries. These batteries are no longer good enough for the mobile application, but are still good enough to be used for a stationary application. If the vehicle high voltage storage (HVS) is split in separate modules they can be used as a small photovoltaic (PV) energy storage system. These are often designed to operate under a voltage range below 60 V DC (safety extra low voltage, SELV). For this voltage range a higher protection against contact needn't be ensured. This means a very low risk for dangerous electrical shocks. To store more PV-energy an array with parallel wiring battery modules can be installed.

4.2 Medium-Size Application: Peak Shaving and Commercial PV System

The total system can be comprised of several HVBs for the power range around 100 kW. High outputs and commercial use with trained personnel allow required voltage levels of 800 V DC. Two HVBs with 400 V DC are connected here in series. Both the top and bottom HVBs are connected together as one unit and can be scaled with regards to output and energy using a suitably designed inverter depending on the application. Applications for large commercial PV systems include peak shaving and load management to smooth out the peak power demands of a company. DC quick charge stations for charging electric vehicles can also be supplied using this battery rack, which do not require a high-power grid connection.

4.3 Large-Size Application: Frequency Stabilization on a Megawatt Scale

Sometimes a minimum power of 1 MW must be able to be provided in the primary control reserve market. [2] For this size, generally the rack construction for the medium-size application is also possible; however this requires a lot of space. A more dense construction is the configuration of individual modules into battery cabinets, which can be set up in containers or suitable halls. The expanded modules of two HVBs are located in one battery cabinet. Along with frequency stabilization, the buffering of wind parks is also a possible application.

5 Design Criteria for Battery 2nd Life Applications in Contrast to the E-Vehicle

A typical vehicle is only in operation approx. 1–2 h per day. The state of charge (SOC) is heavily utilized (completely empty to completely charged) and the temperature range of the battery can vary widely (sometimes extremely). On the other hand, this is different for stationary applications: The battery storage system needs to be operational 24/7, 365 days per year. The SOC range is limited and is set up for lower currents and thus lower outputs when compared to the vehicle. The battery temperature has a comfort zone of 20–25 °C. All of the measures listed should enable the longest possible service life for the second life of the battery.

Due to operation with low currents (C rates) as in vehicles and other design parameters, E/E components such as fuses and protections must be newly designed and implemented for stationary use. This also affects various cable cross sections and new high-voltage cables that cannot be applied from the vehicle to the stationary application.

When the HVB is broken down to the module level, the modules must be properly dimensioned geometrically in battery cabinets. For safety and E/E components, a separate service level is a good idea. It should also be easily accessible for the subsequent replacement of components. The planning of highvoltage and communications lines also plays an important role regarding electromagnetic compatibility (EMC).

HVBs are often liquid-cooled in the e-vehicle. The effort to do this for stationary applications is extremely high so that aircooling is a better alternative. When battery cabinets with individual modules have to be cooled, the air is blown through several, individual fans from the front through each individual module level, see Fig. 4.

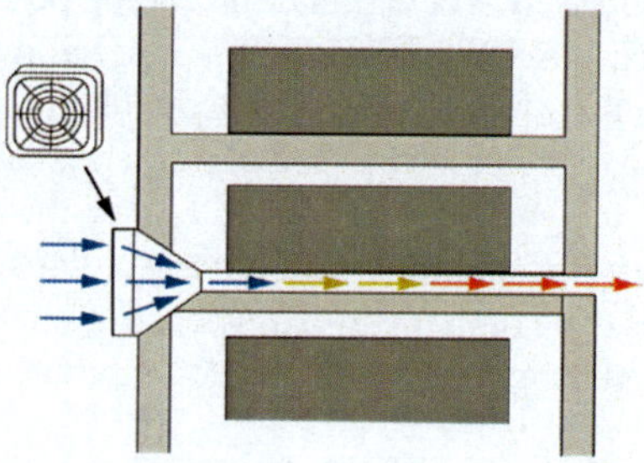

Fig. 4. Air cooling flow of individual modules on the bottom from the front

6 System Architectures with a 800 V and 400 V Voltage Level

6.1 800 V Architecture

Typical vehicle high-voltage batteries (HVB) have a DC voltage of a max. of 400 V DC. This makes it difficult to reach the usual AC voltage level with a standard inverter of 3 × 400 V AC (low-voltage grid). The typical operating range of wide-spread industrial inverters requires double the amount of DC voltage (800 V DC) on the part of the HVB. This results in considerable changes to E/E components, and the maintenance of air and leakage paths. In addition, two similar aged HVBs must be connected to each other in series, see Fig. 5.

6.2 400 V Architecture Without DC/DC Converter

If HVBs are used with a voltage level of 400 V without a series connection of two HVBs, then there is a higher cost on the part of the inverters. Therefore the use of an inverter with lower AC low-voltage (e.g. 3 × 200 V) is required as well as a suitable transformer for the feed-in to the medium-voltage grid at lower AC low-voltage. The difficulty here is that suitable inverters are less readily available on the market for the existing voltage range of the HVB. In addition, there is also a decrease in the efficiency level by a few

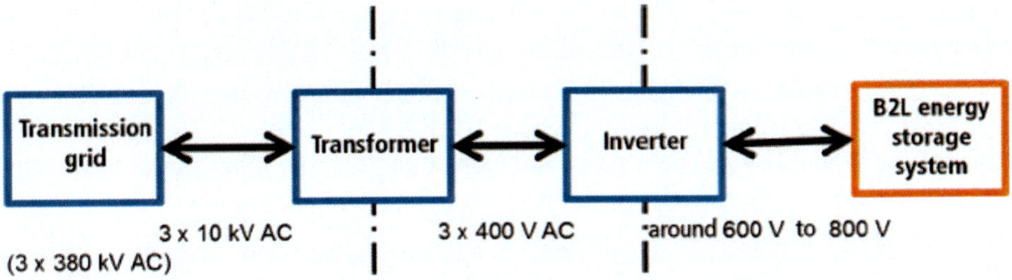

Fig. 5. Wiring from the energy storage system to the transmission grid (800 V system architecture)

percentage points for both the inverter as well as the transformer due to the lower voltage level. Medium-voltage transformers for lower AC low-voltages are available as special types, but they cost more than a standard medium-voltage transformer, see Fig. 6.

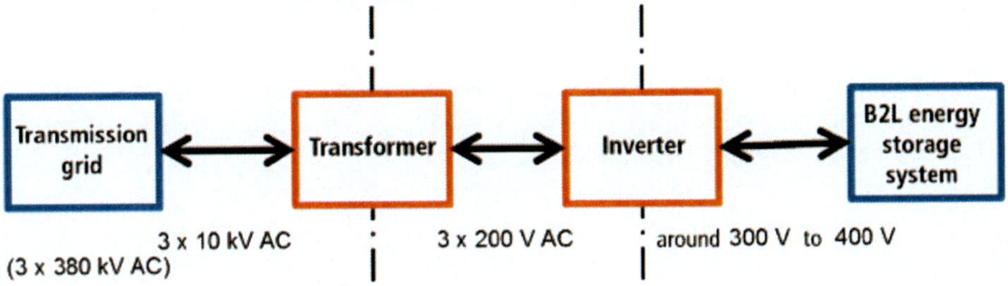

Fig. 6. Wiring from the energy storage system to the transmission grid (400 V system architecture)

6.3 400 V Architecture with a DC/DC Converter

Another variant is the use of an additional DC/DC converter (boost converter) for converting the DC voltage of the HVB to the operating range of a standard inverter, see Fig. 7. This leads to additional costs for the DC/DC converter as well as decreases the efficiency level by several percentage points. In addition, suitable DC/DC converters are also not as readily available on the market.

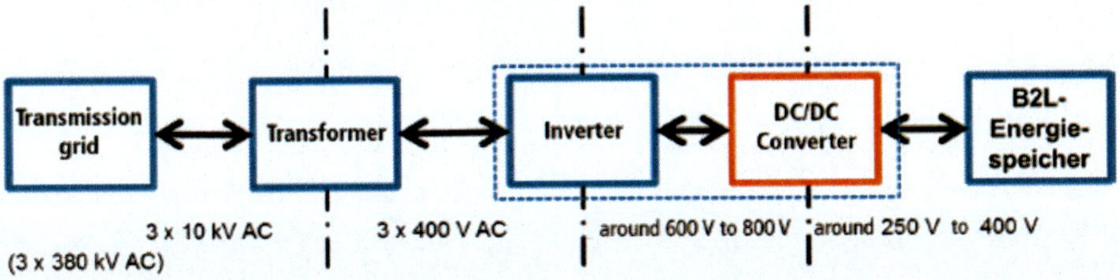

Fig. 7. Wiring from the energy storage system to the transmission grid (400 V architecture with a DC/DC converter)

7 Outlook

The use of used electric vehicle batteries is still a new topic. There are still relatively few electric vehicles on the road and used vehicle batteries for a stationary application will only become available when their range and performance are no longer adequate for the customer. There is little experience in the construction and operation of these systems with corresponding standards and regulations.

References

1. Wirth H, Fraunhofer ISE, Recent facts about photovoltaics in Germany. https://www.ise.fraunhofer.de/content/dam/ise/en/documents/publications/studies/recent-facts-about-photovoltaics-in-germany.pdf. Accessed 10 Oct 2019
2. German transmission system operator's common internet platform: common tendering for primary control reserve. https://www.regelleistung.net/ext/static/prl?lang=en. Accessed 10 Oct 2019
3. Volkswagen, Klimabilanz von E-Fahrzeugen & Life Cycle Engineering, April 2019, p. 11. https://uploads.volkswagen-newsroom.com/system/production/uploaded_files/14448/file/da01b16ac9b580a3c8bc190ea2af27db4e0d4546/Klimabilanz_von_E-Fahrzeugen_Life_Cycle_Engineering.pdf. Accessed 10 Oct 2019

Finite Element Modeling (FEM) and Fatigue Analysis of Hypoid Gears and Laser Welding Joints Installed in a Power Take-off Unit (PTU)

Kibok Lee(✉), Myeongeui, and Jongho Seo

Department of Powertrain Research, Hyundai-Wia Research Institute, 461-18 Sam-Dong Uiwang, Gyeonggido 16082, Korea
kblee@hyundai-wia.com

Abstract. Effective analysis modeling of power take-off unit by using finite element method is developed in order to investigate gear root bending fatigue and welding fatigue. The hypoid gear plays an important role in the PTU system. Without proper information of system stiffness it is impossible to exactly predict fatigue failure life in finite element (FE) analysis. In this study, a simulation model is developed with the focus of affecting gear mesh misalignment, such as gear body flexibility, shafts, bearing stiffness and gearbox stiffness. To validate results this paper presents a correlation study of fatigue failure life between physical test results and fatigue life analysis results. The proposed integrated FE model for predicting the fatigue life of hypoid gear root and welds based on the full-system agreed well with the result of physical durability test.

Keywords: Power take-off unit · Hypoid gear · Gear mesh misalignment · Finite element method · Weld fatigue · Effective notch stress approach

1 Introduction

An all-wheel drive (AWD) system can provide power to all four wheels and distribute under selective controls. The power take-off unit (PTU) in an AWD vehicle is essential machinery. The PTU directly transmits the power thought propeller shaft perpendicularly to rear module, where hypoid gear are used to do it. This paper deals mainly with a simulation modeling for the PTU. They play an important role in the transmittal of a huge high torque. Among the gear trains used in automobile, the PTU has a huge torque more than any other gear train in vehicle. Nevertheless, the vehicle design packages are rarely affordable. Moreover, the gear ratio has been increased due to the customer's requirement to develop the high-performance cars. An increase in gear ratio definitely means a linearly increase rpm speed and reaction moment. Heavily loaded conditions and high speed are the main cause of reducing fatigue life on the PTU system.

Like all mechanical components, internal gears often failed during servicing of generating system for some reasons. In many cases, gear fracture is almost the first encountered issue while operating. Many modes of gear failure have been identified, for example

Euroforum Deutschland GmbH (Ed.): *CTI SYMPOSIUM 2019*, pp. 321–329, 2021.
https://doi.org/10.1007/978-3-662-61515-7_28

tooth root breakage, scuffing, tooth internal fatigue, pitting, wearing, impacting or plastic deformation, Of these, one of the main causes of gear fracture is tooth bending fatigue mode. While two gears are engaged, each tooth repeats bending behavior as long as rotation and then the periodic stress variation happened. These fluctuation sources become a reason for fatigue crack initiation. This mode is called gear root bending fatigue failure mode. Most of research related gear trains has been directed towards normal spur gears or helical gears. Only a few researchers have investigated hypoid gears. Wilcox [1] has firstly attempted for hypoid gears for calculating stresses using an exact solution. Some authors [2–4] conducted stress and strain analysis of a complex structural model including gears. Lim [5] researched the effects of assembly error on hypoid gear mesh characteristics. However, these researches have been ongoing with no inclusion of any parts that are connected to the gear train because of the limitation of analytical model to identify the complex system. They have been employed to investigate only single gear.

The corresponding computer program has been developed. By using commercial software based on analytical method the influence of design parameters and load distribution on deflections and stresses has been analyzed. In analysis results, the influence of design parameters on pinion and gear blank geometry is calculated by the method suggested by the Gleason works [6]. Some difficult issues have been found when using the analytical method for calculating the stress on gears. Since the gear body flexibility and some effects of stiffness components which are shafts, bearings and their gearbox the extended tooth contact can occur outside the theoretical contact line. This cannot find acceptable physical phenomena. The use of FE model was found to be most suitable for checking the extended tooth contact pattern [7]. The finite element method (FEM) has been a reliable and effective tool to study of gear as the CPU performance dramatically increases. It is especially useful to consider various stiffness components and connectors [8, 9]. This allows user to develop Computer-Aided Engineering (CAE) model that takes into account various effects.

Due to either elastic deformations or assembling or gears, bearings, their gearbox it results in gear misalignment upon gear train system. Koide and Oda [10] presented a model for calculating the equivalent misalignment for gears. This modeling consists of calculating the bending and torsional deflections of shafts, bearings and gears excepting a gearbox. They said that one of the causes of tooth bending fatigue failure may be misalignment. Fernandes [11] founded out that the tooth bending fatigue failure may have been due to misalignment of the gear train. Several causes of tooth bending fatigue failure have been identified. These include incorrect assembly or misalignment of the gears, poor design of the gear sets, overloads, inadvertent stress raises or subsurface defects in critical areas [12]. Kramberger et al. [13] performed numerical studies to examine bending fatigue life of spur gears.

Based on the previous literatures, the main objective of this paper is to establish a simulation model for the PTU based on FEM that would enable fatigue failure prediction. By durability testing of the PTU system, the fatigue life was compared with fatigue analysis results from which we developed FE model. This paper demonstrates a correlation study of fatigue failure life between physical durability test results and fatigue analysis results. The effective notch stress approach is a common way to investigate the fatigue of laser beam welds in a structure [14, 15]. Another aim of this paper is to study and try a widely used method for simulating the crack in welds, in order for getting estimation of the welds fatigue life.

In this research fatigue analysis was assessed to find a way to predict the fatigue life of hypoid gears by using the proposed model. As estimated on beforehand, the proposed integration model gave a more accuracy in addition to being competitive with solving time issue. The correlation was carried out for different two designs.

2 Modeling of the Hypoid Gear Root Bending

The simulation model is basically divided into major three components that consist of a hypoid gear pair, bearings and housing [16].

2.1 FE-Modeling of a Hypoid Gear Set

FE-model for gear train is illustrated in Fig. 1. The gear teeth mesh is generated from 3-D geometry of including macro-geometry and micro-geometry. Typical representative of helping us gear design are the Gleason standard and Klingelnberg standard. Either of these, we can have gear dimension based on Klingelnberg standard.

Fig. 1. FE-modeling of a gear train and its detailed elements

The gear train meshes are generated with reliable mesh quality and stable mesh type Hexahedral meshes are economic because the hexahedral meshes are arranged regularly than tetrahedral meshes. It has been well known that the significantly greater accuracy that comes from the hexagonal models when compared to the tetrahedral models. Because of this, the gear train model is generated with almost hexahedral elements and less than 1% pentahedral elements.

2.2 FE-Modeling of Taper Roller Bearings

Taper roller bearings are intended to give the reasonable kinematic behavior for axial and radial direction, which affects misalignment of the total system. 1D-connectors and

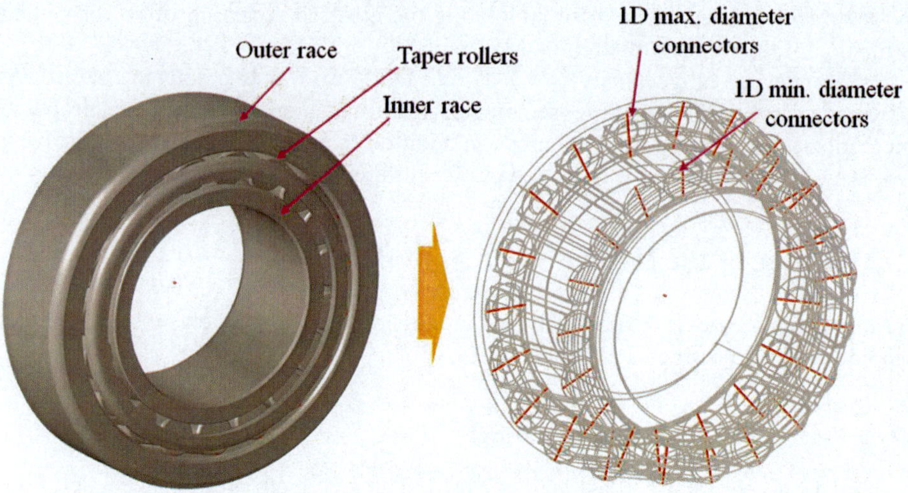

Fig. 2. FE-modeling of a simplified taper roller bearing and 3D geometry.

spring elements are used to generate the stiffness of bearings such as Fig. 2. That is able to facilitate calculation of the simplified bearing models with respect to the relation between loads and deflections for misalignment [17].

For a single roller contact the load and deflection can be expressed as:

$$Q = k_b \delta^n \tag{1}$$

where Q is applied load, k_b is the stiffness factor, δ is deflection and n is an exponent. The expression for the roller raceway deformation can be written in the form:

$$Q = C\left(\frac{L}{L_0}\right)^{n_1}\left(\frac{D}{D_0}\right)^{n_2}\left(\frac{\delta}{\delta_0}\right)^{n_3} \tag{2}$$

The stiffness factor is depend on the effective length of the roller, L and the roller diameter, D, constant value, C and n_1,n_2 and n_3, exponent values. The subscript 0 means unit. We expressed in millimeters on the Table 1. The number of rollers and taper angle are not included in the Eqs. (1) and (2) but are directly oriented in FE model. Houpert includes the effect from the diameter of the roller. The constant factors in Table 1 proposed by the authors are based on experiments.

1D connectors are connected to one center node on roller nodes. Spring of taper roller bearings, unlike cylindrical and ball bearings, are orientated to be sloped to the races, hence the axial direction stiffness naturally works. An extra node on center enables the bearing to rotate axial direction without a constraint.

The characteristic of bearing stiffness is nonlinear. The relation between the load and deflection is shown in Fig. 3. The spring stiffness in the model was calibrated with reference results of full FE-bearing models. There are four different taper roller bearings which are mounting into the gearbox support points.

Table 1. Constant parameters for the load—deflection relationship for a roller-inner/outer race contact [18].

Reference	C[N]	n_1	n_2	n_3
Palmgren	3.73e4	0.8889	0	1.1111
Brandlein et al.	2.65e4	0.9189	0	1.0811
Houpert	1.67e4	0.9910	0.1034	1.1000

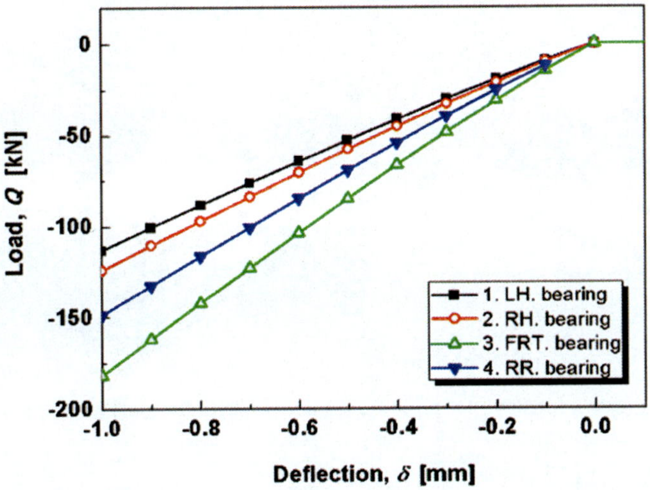

Fig. 3. Comparison of the bearing roller stiffness about a load versus deflection for a line contact between rollers and the inner/outer race.

2.3 FE-Modeling of a Gearbox

Gearboxes have mostly complicated shapes, so we could not generate 3D hexahedral elements. Hence, substructuring is used to resolve problems regarding to the computational cost and correct accuracy [19].

3 Modeling of the Laser Beam Weld (LBW)

Laser beam welding is a welding technique used to join hypoid gear rim into a shaft through the use of a laser. The beam provides a concentrated heating on pieces. The process is frequently used in high volume applications using automation [15], as in the automotive industry. In the PTU processing, the hypoid gear rim is press fitted to the shaft and then laser welded. The welding is proceeding with a concentrated laser beam that heats the metal material so they are melted together.

3.1 Effective Notch Stress Approach for Laser Beam Weld Fatigue Analysis

The effective notch stress approach considers the increase in local stress in the notch region made by the artificial weld toe or root without considering elastic-plastic material deformation. This approach is a common way to investigate the fatigue life of welding joints. The artificial notch diameter and dimensions determined from a standard [14] and then calculated by using a simulation model based on finite element method. The laser beam weld is placed axial behind the gear as shown in Fig. 4.

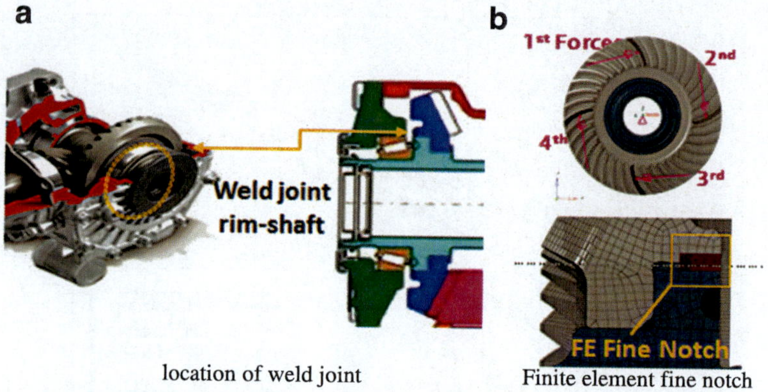

Fig. 4. The artificial notch weld region and mesh on the axial back side face

In order to avoid infinite stress results, a round circle shape with reference radii, instead of the actual sharp toe or root, is usually assumed and followed in Ref. [14]. The present guideline proposed different proposals for reference radii together with associated S-N curves. Recommendations are given for the analysis of the notch stress by the finite element method. Regarding the fatigue strength, appropriate S-N curves are presented for different materials.

3.2 Mesh and Loads on the Effective Notch Stress Approach

According to the welds depth and width a notch with a radius of 0.05 mm is used. Hexagonal elements of first order are used and placed around the notch according to Fig. 4a. The applied gear mesh forces are calculated in Sect. 2 and are based on gear contact points. The load is applied at the nodes in the pitch diameter. To simulate one revolution cycle, four forces are applied in sequence as shown in Fig. 4b.

4 Analysis Environment with the Finite Element Method

We solved the assembled model that we developed with the ABAQUS code [20]. The ABAQUS code's capabilities for the analysis of contact problems together and construction of components in the code have been validated to be useful for the analysis. This code is obviously suitable for solving gear-meshing problems because the ABAQUS solver has special features in contact problems.

The FE modeling of hypoid gears with interactive contact conditions can help engineers simulate real-world phenomena. The boundary conditions include the forces and constraints that are used in the FE-model. The boundary condition for bolting to a test rig, which is similar to the physical test environment, is referred to as a fixed BC. During this testing, the PTU is operated in drive mode, when the engine moves the vehicle forward, which represents reaction torque on the pinion. Therefore, a predetermined constant torque is applied to the pinion axis, which is connected via a spline to represent a flange node with coupling. The forced rotational displacement is set against the direction in which the torque rotates on the hypoid gear axis. An increment to rotate a hypoid gear depends of the pinion gear through kinematic relations, which can be determined with the transmission gear ratio. To avoid contact convergence error and prevent low sampling rate from being too low, for which data would be used for plotting load history data, the increment angle for the hypoid gear axis is limited to 3°. The contact line on tooth faces is moved over the faces in every load step under quasi-static conditions. When varying the applied load level, the contact lines involve nonlinear frictional sliding.

The quasi-static formula is governed by static equilibrium equations. The governing quasi-static FE formulation is defined by means of a non-linear matrix set:

$$\mathrm{K}^t \mathrm{u}^t = \mathrm{f}^t \tag{3}$$

the superscript t means the instant of time.

Equilibrium is checked at each increment of time, t to minimize the residual error, which is computed as follows at iteration n:

$$\mathrm{r}_n^t = \mathrm{K}_{n-1}^t \mathrm{u}_n^t - \mathrm{f}^t \tag{4}$$

The numerical procedure can be solved by the Newton–Raphson method. This method is iterative until it converges linearly.

5 Fatigue Analysis Results

The damage calculation method is implemented into the fatigue software FEMFAT [21] allowing a fast, efficient and high quality assessment of transient-cyclic loaded components. The software is tested on the gearbox's components like hypoid gear root and welded joints. In most cases the results showed good correlation between analysis and measurement in terms of critical location and fatigue life. The method is under evaluation for other applications.

Predicting local component S-N curves was investigated. Figure 5 shows the predicted local component S-N curve plot for hypoid gear root as well as welded joints. The experimental durability tests were performed three times for models A and B under constant torque conditions. A good correlation existed between the measured test results and those of the proposed FE-model, justifying the gearbox stage and modeling method.

The scatter of the fatigue points suggests that the likelihood of accurate estimation is extremely hard. Figure 5a, b show that the established FE model in this study is remarkable, considering that the scatter data of the fatigue range is so vast naturally, so the likelihood of accurate estimation seems to be art.

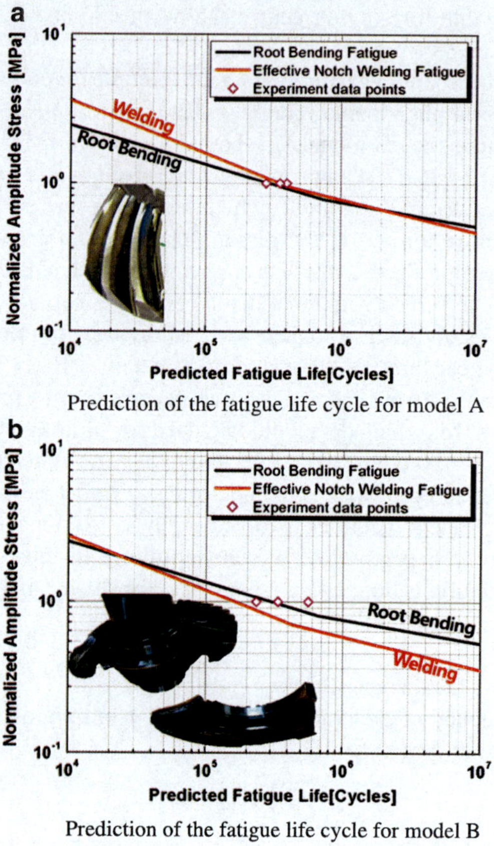

Fig. 5. Prediction of the fatigue life cycle for models A and B.

6 Conclusions

A detailed modeling procedure for predicting the gear-root bending and welded joints fatigue life of the PTU gearboxes was developed in this study. A proper amount of flexible components may significantly affect fatigue life of hypoid gear root and welded joints. This approach includes the FE modeling of the individual hypoid gear, shafts, simplified bearings and a gearbox and welded joints displays how these components can be assembled together through power flow. The influence of their flexibility was researched to determine the effect of gear-mesh misalignment.

When using the FE model, the proposed model for predicting the fatigue life of hypoid gear roots and welded joints based on the full system agreed well with the results of the physical durability-test results. The system torsional stiffness from a conventional model that did not consider misalignment was stiffer than the physical test results and represented low sensitivity to the applied torque. On the other hand, the results of the proposed FE model, which did consider misalignment, were identical with those measured from experiments.

The FE modeling of hypoid gears and welded joints were discussed, and we found that the influence of gear-mesh misalignment for the gearbox stage.

References

1. Wilcox LE (1981) An exact analytical method for calculating stresses in bevel and hypoid gear teeth. In: Proceedings of international symposium on gearing and power transmissions, Tokyo, vol 2, pp 115–121
2. Lu J, Litvin FL, Chen JS (1995) Load share and finite element stress analysis for double circular-arc helical gears. Math Comput 21:13–30
3. Bibel GD, Kumar A, Reddy S, Handschuh R (1995) Contact stress analysis of spiral bevel gears using finite element analysis. Trans ASME J Mech Des 117:235–240
4. Litvin FL, Chen JS, Lu J, Handschuh RF (1996) Application of finite element method for determination of load share, real contact ratio, precision of motion, and stress analysis. Trans ASME J Mech Des 118:561–567
5. Lim TC, Wang J (2005) Effects of assembly errors on hypoid ear mesh and dynamic response. In: ASME international design engineering technical conferences and computers and information in engineering conference, pp 801–806
6. Gleason Works (1971) Method for designing hypoid gear blanks. Rochester, USA
7. Ma H, Pang X, Feng R, Song R, Wen B (2015) Fault features analysis of cracked gear considering the effects of the extended tooth contact. Eng Fatigue Anal 48:105–120
8. Mohammed O, Rantatalo M, Aidanpää J (2013) Improving meth stiffness calculation of cracked gears for the purpose of vibration-based fault analysis. Eng Fatigue Anal 34:235–251
9. Ma H, Song R, Pang X, Wen B (2014) Time-varying meth stiffness calculation of cracked spur gears. Eng Fatigue Anal 44:179–194
10. Koide T, Oda S, Matsuura S, Kubo A (2003) Equivalent misalignment of gears due to deformation of shafts, bearings and gears. JMST Int J Ser C 46(4):1563–1571
11. Fernandes PJL (1996) Tooth bending fatigue failures in gears. Eng Fatigue Anal 2(3):219–225
12. ASM Metals Handbook, Failure of gear, vol 10, Failure analysis and prevention, 8th edn (1975)
13. Kramberger J, Šraml M, Potrč I, Flašker J (2004) Numerical calculation of bending fatigue life of thin-rim spur gears. Eng Fract Mech 71(4–6):647–656
14. Fricke W (2006) IIW recommendations for the fatigue assessment of welded structures by notch stress analysis. In: IIW-2006-09
15. Fredriksson E (2015) Accuracy study in predicting fatigue life for a welding joint. Master thesis, KTH, Stockholm, Sweden
16. Lee K, Song M, Seo J (2019) Finite element modeling and fatigue analysis of hypoid gears installed in a power transfer unit with a correlational study based on an experimental investigation. J Mech Sci Technol 33(6):2797–2807. https://doi.org/10.1007/s12206-019-0526-4
17. Claesson E (2014) Modelling of roller bearings in ABAQUS. Master's thesis, Chalmers University of Technology
18. Harris TA (1991) Rolling bearing analysis, 3rd edn. Wiley, New York, USA, p 1013
19. MSC. Software Corporation (2001) MSC Nastran Superelement User Guide, USA
20. ABAQUS Inc. (2016) ABAQUS Analysis User's Manual, version 2016
21. Magna Powertrain Engineering Center Steyr GmbH&Co KG (2012) FEMFAT manual, 2012 Ed

Highly Efficient Drivetrains for the Mobility of the Future

Andreas Mair(✉)

GKN Automotive, Rienzfeldstrasse 8, 39031 Bruneck, Italy
andreas.mair@gkn.com

Abstract. World population is increasing, urbanization progressing at a constant rate, and by 2030 we will have 10 more megacities in the world. As our megacities continue to expand at a steady rate so would our CO_2 if not counteracted; this is the key driver for the advancement of the electrification agenda. A clear path from zero to 100% electrification has been developed: efficiency increases on traditional drivetrains as well as disconnect architectures have been analyzed and their potential has been assessed. 48 V technology represents a further step on the path of reducing CO_2. PHEV solutions as well as DHT architectures represent technologies for even higher CO_2 reductions, culminating in "zero" CO_2 BEV-powertrains. The potential of all these technologies has been studied in detail. A specific analysis on 2-speed eDrive transmission has been performed and results are presented as well as CO_2 related aspects of the Twinster® technology.

Keywords: CO_2 · xEV · Efficiency

1 Introduction

1.1 Demographics

World population is increasing, but growth rate is reduced. Urbanisation, on the other hand, is still progressing at a constant rate and by 2030, we will have 10 more megacities in the world (see Fig. 1). As our megacities continue to expand at a steady rate so will our CO_2 if transport is not steered in an opposite direction.

1.2 Implications on Transport

Transport CO_2 emissions need to be reduced and this needs to be done at multiple levels and by multiple technical solutions. Efficiency improvement of the drivetrain is a theme which has been followed over years now. So called AWD-disconnect systems are in the market since 2014 and the introduction of hybrid vehicles is progressing on a highly increasing rate. The various CO_2 reduction potentials have been assessed in detail based on various duty cycles. The here shown results are based on the WLTP.

Euroforum Deutschland GmbH (Ed.): *CTI SYMPOSIUM 2019*, pp. 330–336, 2021.
https://doi.org/10.1007/978-3-662-61515-7_29

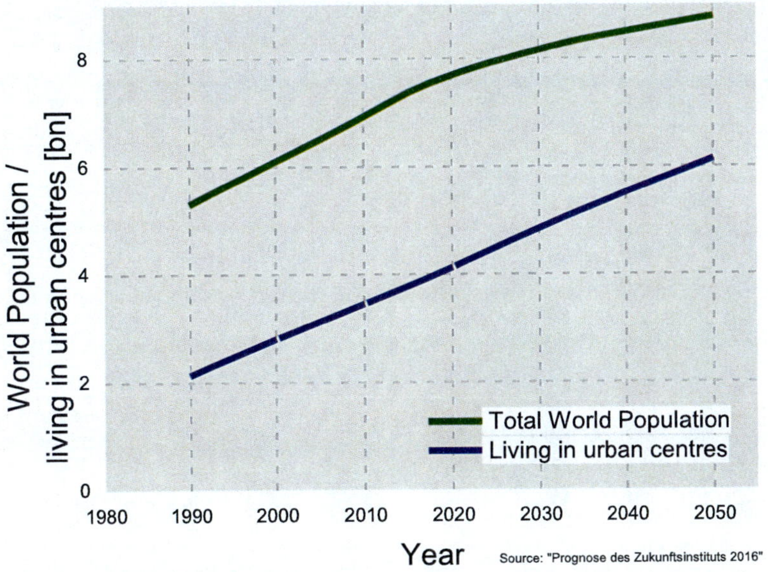

Fig. 1. World population/population of urban centres

2 Solutions

To reach the required CO_2 reduction, the vehicle driveline needs to be analyzed in all detail. This applies for conventional drivetrains as well as for hybridized and electrified drivetrains.

2.1 AWD Solutions

In 2014 GKN introduced the first AWD disconnect system to the market. The principle of this technology is the following: the secondary axle of the vehicle is only switched on when driving situations are requiring it (therefore it is often denominated also as "active connect technology"). Such driving situations comprise increased traction requests, stability improvement or longitudinal or lateral dynamics enhancements. GKN has proven a CO_2 reduction potential of 5% of such drivetrains compared to conventional hang on AWD solutions. The latter have a CO_2 penalty of approx. 6% compared to the respective 2WD version. The compensation of this penalty was the main aim of introducing the disconnect technology. In addition, further improvements of the complete drivetrain have been achieved by specifically reducing churning losses and mechanical friction losses in bearings, seals etc. Here an overall potential of 1.5% has been identified (see Fig. 2) [1].

2.2 xEV Solutions

In 2002 GKN in a collaboration with Nissan brought a first 42 V P4 eAxle to the market. Main purpose of this solution was traction increase on the secondary axle. The complete

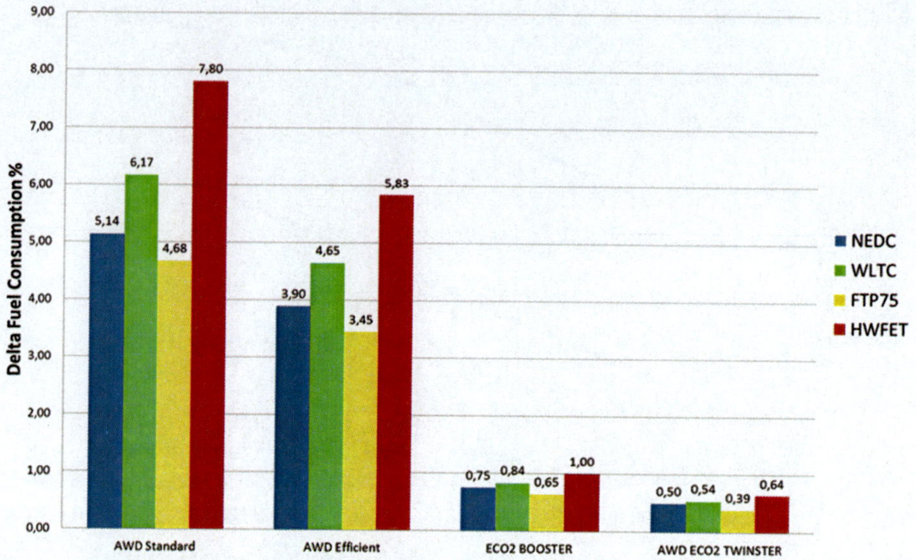

Fig. 2. Difference in fuel consumption compared to standard 2WD analysed for various duty cycles. AWD Efficient: efficiency improved standard hang on AWD drivetrain. ECO2 BOOSTER: disconnect AWD drivetrain with side mounted clutch. AWD ECO2 TWINSTER®: disconnect AWD drivetrain with Twinster® on the secondary axle

potential of 48 V P4 solutions regarding CO_2 reduction has been assessed recently and can be quantified by 14% based on WLTP and compared to the purely ICE driven vehicle (see Fig. 3).

High Voltage DHT is an additional technology which has been assessed in detail. GKN's Multi Mode Transmission featured by the Mitsubishi Outlander PHEV leads, according to the OEM, to CO_2 emissions of the vehicle of 44 g/km. An approximative 10% improvement potential with respect to the actual solution has been identified by optimizing the current transmission layout. Main changes consist in introducing a 2nd speed and a CVT input stage instead of having individual motor and generator [2].

The High Voltage P4-hybrid architecture is by far the most significant for future drivetrains due to:

- its functionalities (electrified AWD, high efficiency, boost, regeneration, creeping, etc.)
- commonalities with e-drive unit for BEV.

These solutions feature transmissions in offset arrangements as well as coaxial arrangements. The latter either realized by a layshaft design or a planetary design. Furthermore, various integration levels have been realized so far. Either discrete architectures, i.e. transmission, e-machine and inverter bolt on, or integrated architectures, i.e. gear set, e-machine active components and inverter PCBs and power modules all

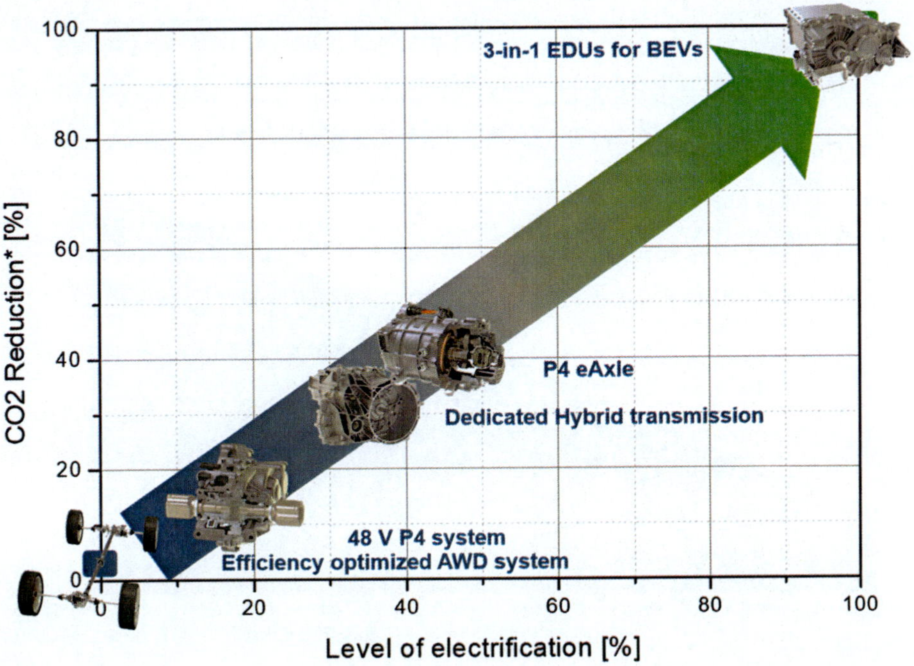

Fig. 3. CO_2 reduction potential of various driveline solutions related to the vehicle driveline electrification level (*) based on WLTP

integrated in one system housing participating on the same cooling circuit and using the same hydraulic circuit.

Measured system efficiencies of such optimized solutions exceed values of 90%. The pure transmission efficiency is even higher than 97% [3]. Typical axle torques are between 1800 and 3000 Nm, typical electrical peak power levels in the range of 50–100 kW.

This technology is culminating in electric drive units for BEVs where we see axle torque levels of 2000 Nm up to 6000 Nm, typical power levels of 50 kW up to 300 kW electrical power.

2.3 Multiple Speeds

Current market forecast figures don't show a significant share of multiple-speed applications by 2026, see Fig. 4. Nonetheless the potential has been analysed and solutions have been developed. Multi-speed-solutions (two and more speeds) enable a better balance of acceleration and top-speed compared to single-speed units, as well as higher system efficiency. Such solutions provide an improved launch performance by using the lower gear and more efficient cruising when in higher gear. Optimised shift strategies keep the eMotor operating close to its efficiency 'sweet spot'. No interruption of torque during shifting is an imperative requirement for BEV applications. The increase in system efficiency provides a higher maximum driving range. On WLTP, a system efficiency improvement of 8% can be expected compared to the single speed versions [4] (see Table 1).

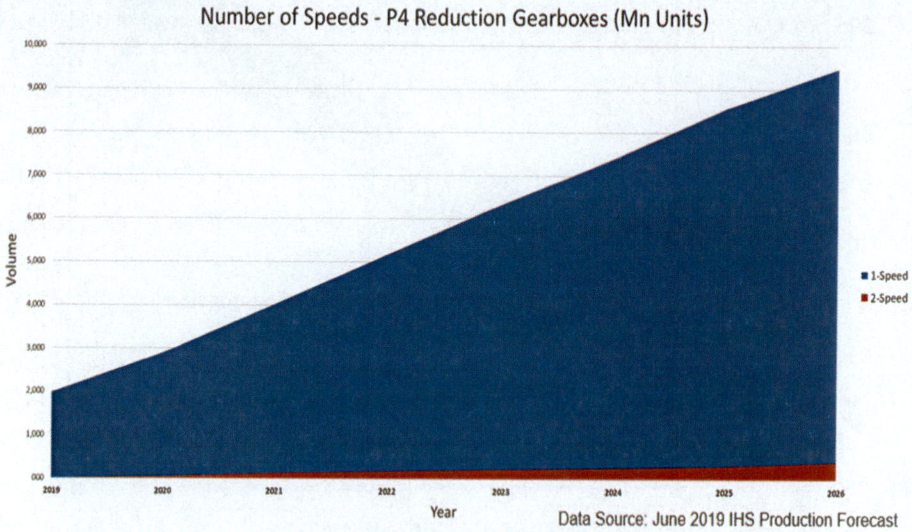

Fig. 4. Forecast: share of 2-speed-P4-transmissions over the years 2019–2026

Table 1. Energy consumption of BEV equipped with GKN eTwinster 2 speed [kWh/100 km]

Cycle	1 speed	2 speed	Improvement %
NEDC	12.4	11.3	**8.9**
WLTP	14.0	12.9	**7.7**

2.4 Twinster®

GKN, with the introduction of the first AWD-disconnect technology has also introduced the Twinster® technology to the market. This technology will be transferred into eDrive applications as well. The proven advantages are:

- Twin clutch design intelligently delivers optimum torque to each wheel depending on driving conditions and driver input (torque vectoring)
- Twinster® enables more stable take-off, acceleration and cornering (eLSD functionality)
- Twinster® allows for torque damping and torque limiting
- Twinster® counters plug-in hybrids' and BEVs' tendency to understeer caused by additional weight of the car.

In addition, it provides a "true"-sailing-functionality to the vehicle, i.e. disconnect of the driveline in proximity to the wheel. By disconnecting the driveline from the road in coast-conditions it improves sailing by 9% (see Fig. 5). This result is based on measurements performed on a vehicle with Twinster® on the primary and secondary axle [5].

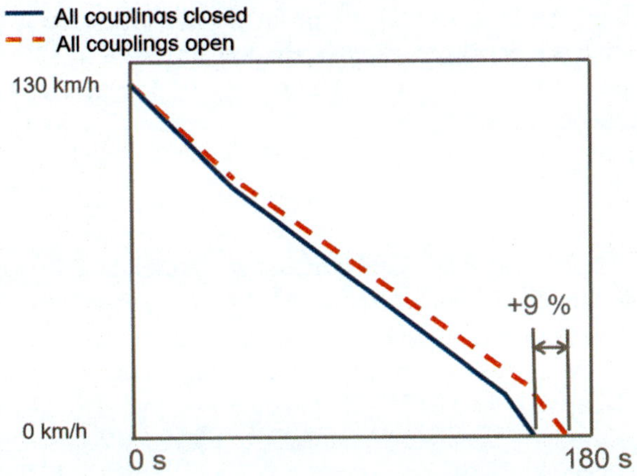

Fig. 5. Coasting test in "Neutral" from 130 to 0 km/h for GKN's GTD 18

3 System Integration

Drivetrain development cannot be done in isolation as complete vehicle system aspects need to be considered.

3.1 Functional Safety

Fail safe open is a basic requirement for AWD-hang-on-architectures and left-to-right-locking-architectures—independently if traditional drivetrain or hybridized or electrical drivetrain. If a driveline-system is affected by a malfunction and fail-safe-open status is not reached, the following will potentially happen [6]:

- driveline wind up during cornering occurs
- damage of driveline components is a potential consequence
- unintentional yaw build-up during acceleration on split-μ is a potential consequence
- during braking or ABS/ESP intervention, vehicle can become instable resulting in a safety critical situation.

During braking or ABS/ESP intervention, all wheels need to be decoupled (clutch open demand), to provide braking performance with optimum vehicle stability control. By locking front to rear, independent brake distribution front/rear is not guaranteed anymore. By locking left to right, brake distribution cannot follow ABS "select low" control principle anymore.

4 Outlook

Multiple drivetrain architectures will compete in the forthcoming decade.

4.1 All-Wheel-Drive

The passenger vehicle AWD-share will still be at a 15–20% level up to 2030. A smaller portion of these vehicles will feature P4 axles, the rest still being equipped with mechanical AWD-systems. Thus, active connect technology will represent a key technology providing CO_2 reduction potentials compared to conventional systems.

4.2 e-Drive

For High Voltage, P2, DHT and P4 will be the most significant hybrid architectures up to 2030. As the P4 solution is from an architectural point of view very close to the main drive unit of the BEV, its volumes will have the highest growth rate. Disconnect-features for P4-drive-systems as well as the Twinster®-technology represent important technical solutions to further improve the xEV's CO_2 balance.

References

1. Hoeck M (2016) One fits all. CTI Berlin
2. Dipl.-Ing. Gassmann T, Dipl.-Ing. Hildebrand W, Dipl.-Ing. Haupt J, Dipl.-Ing. Herber S (2018) Effizientes Multi-Mode-Getriebe für Plug-in-Hybridfahrzeuge. ATZ 12/2018
3. Dipl.-Ing. Schmeink H-J, Dipl.-Ing. (FH) Zaers C, Dr.-Ing. Grunwald A, Dipl.-Ing. (FH) Absenger M, Dipl.-Wirt.-Ing. (FH) Eichholz D, Dipl.-Ing. Sander T (2018) Planetary coaxial drive module—induction motor driven semi-integrated-electric-drive-module. Aachen-Colloquium
4. Dipl.-Ing. Gassmann T (2017) GKN 2-speed transmission technology for hybrid and electric vehicle. Tag des Hybrids Aachen
5. Dr. Kaiser G (2019) GKN all wheel twinster hybrid vehicle. In: Electrification & all-wheel drive congress
6. Hoeck M (2019) Twin AWD. In: EAWD congress

The 8G-DCT Plug-in Hybrid Transmission for the Mercedes-Benz Compact Car Family

Matthias Maisch(✉), Martin Hörz, and Daniel Jesser

Mercedes-Benz AG, Stuttgart, Germany
matthias.maisch@daimler.com

Abstract. With the new 8G-DCT plug-in hybrid transmission, Mercedes-Benz has electrified the modular 8G-DCT transmission family with a powerful and efficient plug-in hybrid powerhead. The plug-in hybrid transmission is launched as part of the new generation of the Mercedes-Benz compact car platform. Therefore, additional requirements concerning fuel consumption, pure electric driving possibility, torque capacity, minimal space, modularity and shifting performance are considered. This allows to fulfill increasing customer requirements in hybrid powertrains. The main elements of the highly compact and modular plug-in hybrid powerhead are an integrated dual clutch, an additional low-loss disconnect clutch and a new powerful 75 kW electric motor. The powerhead is characterized by its high functionality and extremely high efficiency in combination with the high torque of the electric motor. It provides all hybrid functionalities, such as boost, energy recuperation and pure electric driving. Special emphasis was placed on optimized efficiency of the powerhead by reducing the drag torque of the disconnect clutch as well as by means of on-demand cooling and lubrication. The actuation and cooling of the hybrid powerhead required only minimal adaptations to the basic transmission. The hybrid powerhead also allows starting the combustion engine in a very smooth manner, almost imperceptible for the driver. Within the transmission family the 8G-DCT plug-in hybrid transmission is therefore available for future use in electrified front-wheel drive compact car series.

1 Hybrid Transmission Development by Mercedes-Benz

Hybrid technology plays an important role within the Mercedes-Benz passenger car strategy. Since 2009, hybrid powertrains with hybrid transmissions have formed part of the Mercedes-Benz premium segment. Hereby Mercedes-Benz looks back on over ten years of successful hybrid development [1, 2]. The experience and valuable knowledge that were obtained in this time resulted in the development of the current generation of hybrid powertrains and transmissions [3]. Until now, all these hybrid transmissions are based on planetary gear set transmissions (e.g. 7G-TRONIC, 9G-TRONIC), which are used with a hybrid powerhead in the medium and luxury class vehicles for rear-wheel drive powertrains in 4×2 and 4×4 applications.

Because requirements concerning consumption and pollutant emissions are increasing constantly and furthermore pure electrical driving performance is required, the new Mercedes-Benz compact car family will also be equipped with a hybrid powertrain

Euroforum Deutschland GmbH (Ed.): *CTI SYMPOSIUM 2019*, pp. 337–345, 2021.
https://doi.org/10.1007/978-3-662-61515-7_30

including a hybridized transmission for the front-wheel drive application. For this purpose the development of the new family of dual clutch transmissions is the optimal technical basis, Fig. 1.

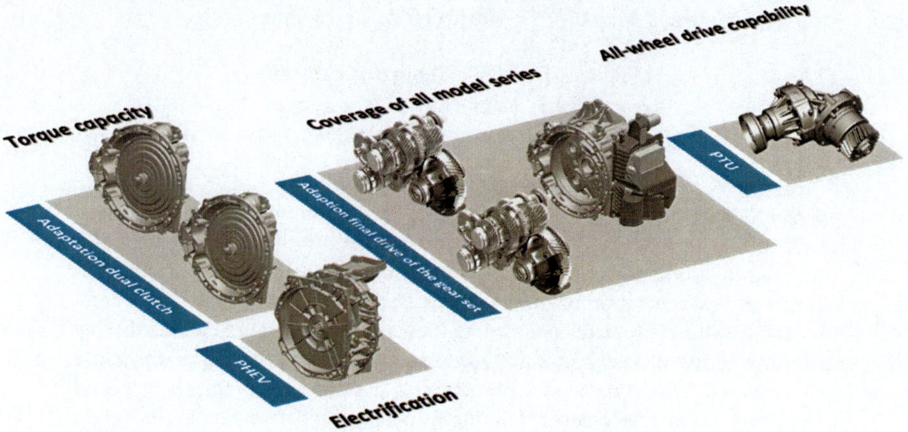

Fig. 1. Modularity of the new 8G-DCT front wheel-drive transmission family

2 New 8G-DCT Dual Clutch Transmission Family

The new 8-speed dual clutch transmission family for front-wheel drive application was developed and launched as part of the next generation of Mercedes-Benz compact vehicles [4]. The main development targets were set to meet the increasing requirements regarding fuel consumption, torque capacity, weight, package and modularity. All transmissions achieve very high efficiency and make a valuable contribution to CO_2 reduction within the new compact car platform.

To meet the Mercedes-Benz specific driving comfort at low speed all drivetrains are equipped with a dual mass flywheel with pendulum technology [5]. This transmission platform sets high standards with regard to performance, power density and efficiency.

2.1 Modularity as One Main Development Focus

In the development of the 8G-DCT one focus was set on the modular design of the transmission in order to cover all variants based on the new compact vehicle platform. In addition, the future electrification, the all-wheel drive capability and the adaptation to different model series of the platform were taken into account [6].

Due to the systematic modular design of the 8G-DCT, the transmission can meet the requirements of all model series of the compact vehicle platform. While the future-oriented design of the 8G-DCT makes it possible to electrify the powertrain, the torque capacity can be specifically increased by adapting the dual clutch. This is achieved by varying the number of friction plates and the use of high-performance materials. In order to meet the torque requirements of the various model series of the Mercedes-Benz compact cars, the axle ratio in the gear set can be adapted to the vehicle [6].

The high modularity also enables the all-wheel drive capability of the transmission. The power take off unit (PTU) acts as transfer case and, in conjunction with the rear axle drive, enables the familiar 4MATIC drive to be implemented in the latest generation of compact vehicles [6]. The integration of a hybrid powerhead between the combustion engine and the transmission enables an electrified version of the 8G-DCT.

2.2 Technical Highlights of the 8G-DCT Basic Transmission

The transmission consists of a compact, classic 3-shaft design with eight forward and one reverse gear. One standard gear set with two various differential ratios is intended for all variants of the vehicle platform. The spread of 8.81 ensures optimum gear steps with outstanding driving performance. At the same time due to the high ratio, it is possible to reduce fuel consumption and emissions, while increasing driving comfort. The engineers paid attention to the reduction of mechanical and electrical losses in each component of the transmission. These analyses led to a design with a fixed-lot bearing for the gear set concept, the implementation of dry sump lubrication system, the directly controlled electro-hydraulic control and the use of a demand-driven two-pump concept. The weight targets are achieved by a consistently implemented lightweight design of the cases through intensive digital analyses. It is the first time we enabled the use of high-performance plastics for the inner shift module, the oil sump spider and the oil pan [5].

Another innovation is the modular and independent design of the actuators. For example, the internal gearshift, the parking lock actuator and the electrohydraulic control are each independently testable modules. The transmission oil pan is made out of high performance plastic components. Primary and secondary filters, another pressure oil filter and the transmission cooler are combined to form an oil pan module system [5].

3 The 8G-DCT Plug-in Hybrid Transmission

Developing a powerful and compact plug-in hybrid transmission with very high efficiency and outstanding shifting and driving comfort was an elementary goal in developing the 8G-DCT plug-in hybrid transmission. Guaranteeing a very high powertrain performance with all the necessary hybrid functions for the future Mercedes-Benz plug-in hybrid powertrain was the key focus here. These goals were achieved through the complete new development of the hybrid powerhead as well as the minimal modification of the main transmission, specifically for the plug-in hybrid. The 8G-DCT transmission family was developed from the beginning up with hybridization in mind, which led to the fact that only minimal changes had to be made to the main transmission.

Due to the modular design of the 8G-DCT, several parts can be used both in the basic and in the hybrid transmission without any modification. That includes the oil cooler, the parking lock, the inner shifter, the transmission control unit as well as the gear set. That ensures low part costs and a lower variance at the production line.

For other parts, minimal modifications are necessary:

- Electro-hydraulic control unit: In the hybrid version the electro-hydraulic control unit has one additional pressure control valve to actuate the disconnect clutch.
- Housings: The transmission and the clutch housing have additional cooling channels to provide the stator with cooling oil. Further channels were added to provide the actuation and the cooling of the separation clutch.
- Oil pan: Because of the different cooling circuit of the total vehicle, the oil pan in the hybrid version has an additional thermostatic valve.

3.1 Design of the Hybrid Powerhead

To enable an electrified version of the 8G-DCT, the new developed hybrid powerhead expands the main transmission, Fig. 2. During the development of the new hybrid powerhead, considerable attention was paid to integrate all the subcomponents with high power density to achieve minimal axial space and meet the package restrictions in the car package. Through the optimum constructive and functional integration of the disconnect clutch, dual clutch and electric motor an extremely compact design is realized. Thanks to the modular and highly compact hybrid powerhead, a minimum lengthening of the transmission of only 56 mm was achieved compared to the 8G-DCT transmission.

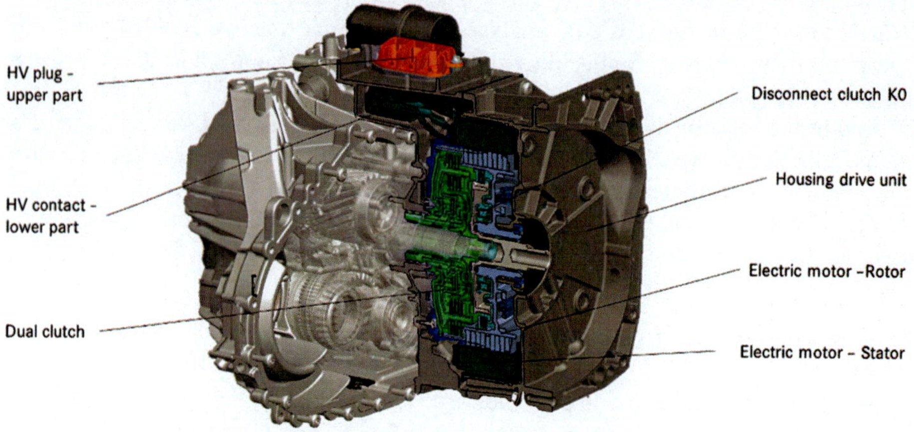

Fig. 2. Design and main components of the plug-in hybrid powerhead

This is realized by the usage of a third housing. Within this drive unit housing the electric motor, the dual clutch, as well as the additional disconnect clutch are integrated. The hybrid powerhead is designed as a wet system and is operated with transmission oil. To this end, the hybrid powerhead is connected to the main transmission's hydraulic circuit and is actuated via the mechatronic control unit. This means that the powerhead with the exception of the electric motor does not need any actuators or control devices itself.

3.2 Dual Clutch

Due to package optimizations in the hybrid powerhead it was necessary to adapt the dual clutch of the basic transmission. Therefore, the dual clutch is integrated in the rotor of the electrical motor and a minimal package in axial length was derived. The dual clutch consist mainly of identical parts from the basic transmission. Furthermore, the friction plates and friction system and in addition the hydraulic actuation and cooling system are derived from the basic transmission. The dual clutch is mounted as one closed module in the hybrid powerhead during the assembly of the powerhead. With this design of the dual clutch a maximum torque capacity of 450 Nm with a high thermal robustness is reached. In addition, this leads to a high comfortable drive-off performance and an impressive trailer load at 1600 kg (braked).

3.3 Disconnect Clutch

The disconnect clutch is realized in the form of an open hydraulic multi-plate clutch. It consists of two pairs of plates with very low drag torque and is actuated hydraulically. On the one hand the disconnect clutch allows the combustion engine to be disconnected from the drivetrain for independent operation of the electric motor and the transmission in electric mode. On the other hand, it allows the combustion engine to be started very smoothly. For this purpose the disconnect clutch was designed and optimized to maximum thermal robustness and extremely fast actuation. It allows a demand based lubrication in electric driving mode up to maximum lubrication for possible frequent starting of the combustion engine in stop-and-go traffic.

3.4 Electric Motor

As a core part of the hybrid powerhead, the chosen electric motor type is a permanent synchronous motor with buried magnets and single tooth winding technology. The electric motor is highly integrated, as i.e. the rotor is both: clutch and electromagnetic device at the same time, Fig. 3. The electric motor is extremely compact. Combining this high degree of integration, modern active materials and winding technology with minimized winding end length it leads to a design with excellent performance without compromises in efficiency. The electric motor, which is integrated in the cooling circuit of the clutch, runs in an oil spray environment. This concept allows the use of permanent magnets with highest energy density.

For use in the 8G-DCT plug-in hybrid transmission the electric motor can deliver a maximum power of 75 kW and a maximum torque of 300 Nm in boost and energy recuperation mode. This high torque is available immediately from zero rpm and allows high starting-off agility as well as high dynamic and sporty performance. Due to the high performance of the electric motor, it was possible to omit the combustion engine starter. So all combustion engine starts derive from the high voltage electrical motor.

3.5 Cooling System

The thermal management of the 8G-DCT hybrid transmission guarantees that the transmission is always operated in the optimal temperature range. During the start-up a thermal

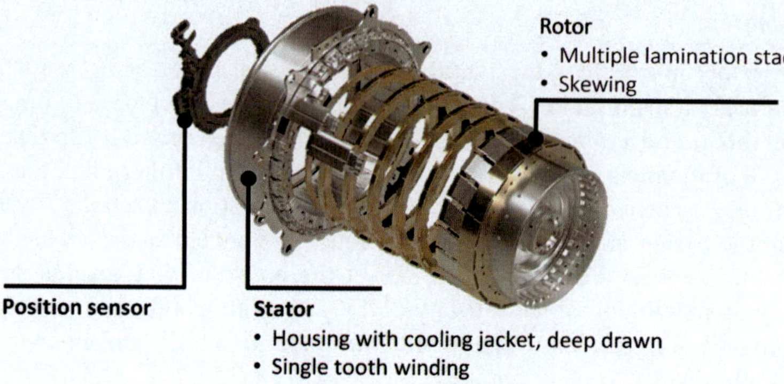

Fig. 3. Electric motor

valve inside the transmission provides optimal and fast heating-up to operating temperature and ensures that no thermic losses slow down the heating-up of the transmission. This rapidly increases the efficiency of the transmission during the warm-up phase. The thermal valve is integrated into the oil labyrinth in a way that it is permanently subject to the oil flow and does not open the oil flow to the transmission oil/water heat exchanger until the oil has reached the operating temperature.

The heat exchanger is supplied with coolant from a low-temperature circuit that is also responsible for cooling the power electronics. In combination with the on-demand actuation of the electric coolant pump, shutter and fan, an optimal cooling capacity in the circuit with maximized efficiency for the common vehicle is guaranteed. Internally the 8G-DCT hybrid transmission offers the possibility to regulate the oil demand of the basic transmission and the components of the hybrid powerhead in line with the demand of lubrication and cooling.

The electric components rotor and stator are supplied with oil separately. The electric engine is arranged in the oil circuit in a way that it profits from the cooling and lubrication demands of the other components and is cooled preventively for challenging driving situations. Therefore, it is able to reach a high electric peak and continuous output and to operate the transmission in an efficient temperature range at all times.

3.6 Combustion Engine Start

For the first time on a Mercedes-Benz plug-in vehicle, the combustion engine is started only by the high voltage electric motor—the compact hybrids do not have a separate 12-volt starter [8]. Therefore a major focus in the 8G-DCT plug-in hybrid transmission was placed on a quick and imperceptible combustion engine start by means of the electric motor in the hybrid powerhead. The electric motor is located between the disconnect clutch and dual-clutch, each of which can be actuated independently. Thanks to this layout, the combustion engine can be started in many different ways, depending on the driving situation. The entire start process is a complex interaction between engine control, transmission control, power electronics and the central powertrain controller (CPC).

A high level of start comfort is characterized by a very smooth and quick engine start in response to the driver's request, which makes it inaudible and takes place with no discernible acceleration amplitudes. In the 8G-DCT plug-in hybrid transmission, an extremely quick hydraulic activation of the disconnect clutch was achieved by means of a systematic hydraulic design. As a result, a minimum filling and application time was achieved for the clutch.

With the mentioned P2 configuration of the plug-in transmission it is possible to use the high torque of the electrical motor and the additional inertia to start the combustion engine. A high slip between the active clutch and electric motor is set with a speed control of the electric motor. During the subsequent engine start, slip is intentionally reduced and controlled to a lower slip. This requires less torque reserve for the engine start, because a part of the torque is generated by the slip reduction and therefore from the inertia.

4 Plug-in Hybrid Vehicle

4.1 Plug-in Hybrid Powertrain [7]

Mercedes-Benz Cars is pushing ahead with the development of its plug-in hybrids under the EQ Power label. The new models from the compact-car family with the third-generation hybrid drive are now celebrating their premiere with the new A 250 e (combined fuel consumption 1.6-1.4 l/100 km, combined CO_2 emissions 36–32 g/km, combined electrical consumption 15.3–14.6 kWh/100 km)*, A 250 e Saloon (combined fuel consumption 1.5–1.3 l/100 km, combined CO_2 emissions 35–31 g/km, combined electrical consumption 15.2–14.5 kWh/100 km)* and B 250 e (combined fuel consumption 1.6–1.4 l/100 km, combined CO_2 emissions 36–32 g/km, combined electrical consumption 15.4–14.7 kWh/100 km)*. The EQ Power car series for the front-wheel drive powertrains underscores the plug-in initiative of Mercedes-Benz: the company aims to extend its offerings to more than 20 model variants by 2020.

For the new compact vehicles with EQ Power (electric) driving pleasure and suitability for everyday use are to the fore. This is illustrated by the model highlights:

- Electric operating ranges of 70–75 km (NEDC)*
- Electric output 75 kW
- System output 160 kW, System torque 450 Nm
- Top speed 140 km/h (electric)
- Acceleration 0–100 km/h in up to 6.6 s
- Hardly any restrictions on the load compartment.

The electric motor achieves 75 kW. Together with the 1.33-litre four-cylinder engine this adds up to a system output of 160 kW (218 hp) and a system torque of 450 Nm. Thanks to the EQ Power of the electric motor the vehicles react immediately to a step on the accelerator, and the performance is impressive.

The lithium-ion high-voltage battery has a total capacity of approx. 15.6 kWh. It can be charged at an external electric energy source. It operates with either alternating or direct current. A corresponding vehicle socket is located in the right-hand side wall of the vehicles. This means that the compact plug-in hybrids can be charged at a 7.4 kW wallbox with alternating current (AC) within 1 h 45 min from 10–100% SoC (Status of Charge). For direct-current charging (DC) the battery can be charged from 10–80% SoC in around 25 min. The batteries are supplied by the wholly owned Daimler subsidiary Deutsche ACCUMOTIVE. The high-voltage battery is water-cooled and weighs approx. 150 kg.

4.2 Packaging

One big challenge was to reach the ambitious packaging requirements of the front vehicle, which was solved with a minimal axial space of the combustion engine and the hybrid transmission, Fig. 4.

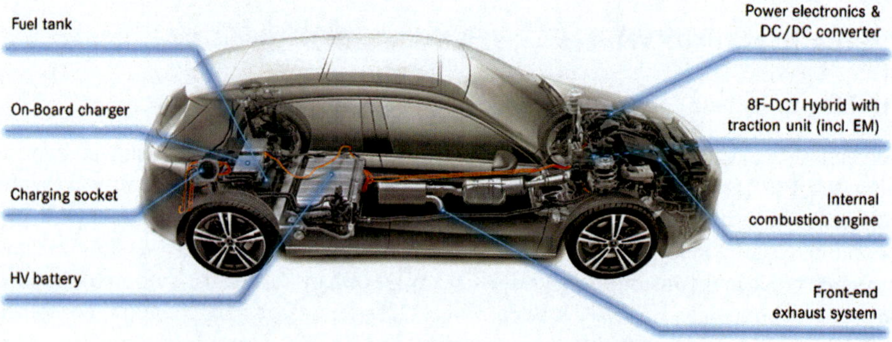

Fig. 4. Plug-in hybrid powertrain with 8G-DCT

An innovative exhaust system enables ingenious packaging: rather than extending to the end of the vehicle, the exhaust ends in a centrally positioned outlet under the vehicle floor, with the rear silencer housed in the transmission tunnel. Integrating the fuel tank into the axle installation space creates room beneath the rear seats for the high-voltage battery. This results in only minimal reduction in boot capacity [8].

4.3 Operating Modes

As the compact vehicles use third-generation plug-in technology, all of its functions are also available. These include in particular the intelligent, route-based operating strategy, taking factors such as navigation data, speed regulations and route into account. The operating strategy considers the entire planned route and prioritizes the electric driving mode for the most sensible route sections in each case.

With the launch of MBUX (Mercedes-Benz User Experience) the previous plug-in operating modes of all EQ Power models have been converted to drive programs. That means that in every Mercedes-Benz plug-in hybrid the new drive programs "Electric" and "Battery Level" are available. This is the case from the outset for the compact models. Maximum e-performance can be experienced in "Electric". The combustion engine is only engaged if the driver uses kickdown on the accelerator pedal. In the "Electric" program the recuperation strength can also be selected via paddles behind the steering wheel. These enable the selection of five different recuperation levels (DAUTO, D+, D, D- and D–).

Comfort, ECO and Sport modes are also available. According to the given requirements, the driver is thus able to give priority to electric driving, place the emphasis on driving dynamics in combined drive mode or give preference to combustion mode in order to save electric range, for example [7].

End Notes

1. *The stated figures are the measured "NEDC CO2 figures" in accordance with Article 2 No. 1 Implementing Regulation (EU) 2017/1153. The fuel consumption figures were calculated based on these figures. The range and the electrical consumption have been determined on the basis of Regulation (EC) No. 692/2008. A different value is applied in accordance with EmoG. A higher value may apply as the basis for calculating the motor vehicle tax.
2. Further information on the official fuel consumption and the official specific CO_2 emissions of new passenger cars can be found in the "Guide to Fuel Consumption, CO_2 Emissions and Electricity Consumption" for new passenger cars, which is available free of charge at all sales outlets and from Deutsche Automobil Treuhand GmbH at www.dat.de.

References

1. Mürwald M, Keller U, Strenkert J, Maisch M, Nietfeld F, Schmiedler S (2016) Die neue Generation Hybrid-Antriebe von Mercedes-Benz. In: 37rd international Vienna motor symposium
2. Maisch M, Homm M, Keller U (2016) The higly efficient 9GH-TRONIC plug-in hybrid transmission by mercedes-benz. VDI Tagung Getriebe in Fahrzeugen
3. Gödecke T, Ascher F, Dickgießer D, Engbroks L, Schürmann T, Voigtländer R (2019) EQ power plug-in hybrid – EQ power plug-in hybrid – electrified from A to S. Wiener Motorensymposium, 2019
4. Harsch A, Kemmner B, Ertel A, Rink A (2018) Die nächste Generation DCT-Getriebe für die Mercedes-Benz Kompaktfahrzeugfamilie. In: 17th international CTI symposium, Berlin
5. Dörr C, Lessi S, Adolph M, Marx V, Veik F (2018) The drivetrain for the new Mercedes-Benz compact cars. In: 17th international CTI symposium, Berlin
6. Harsch A, Clausen G, Rink A (2019) Modular system 8G-DCT with all-wheel drive. VDI-Kongress "Dritev - Getriebe in Fahrzeugen"
7. Mercedes-Benz Press Information (2019) EQ power for the first time also for A- and B-class, 19 Sept 2019. Daimler media. https://media.daimler.com
8. Mercedes-Benz press Information (2019) EQ power: The plug-in hybrid family, Sept 2019. Daimler media. https://media.daimler.com

Solutions for Increased Power Density at Shifting Clutches

Harald Merkel(✉)

Technical Specialist, BorgWarner Transmission Systems, Heidelberg, Germany
hmerkel@borgwarner.com

Abstract. Modern vehicle and drivetrain concepts face many integration problems due to the various functions within the system. Hybrid and Electric vehicles present new challenges, especially in terms of weight reduction and installation space. BorgWarner has developed special solutions to increase the power density of shifting clutches. New friction materials with increased static and dynamic coefficient of friction have been developed to allow a compact clutch design. BorgWarner has also developed new high performance friction materials which can handle higher interface temperatures without sacrificing robustness against hot spotting. Thanks to these advantages, the power and energy limits of clutches can be increased. A new friction plate design has been developed using plastic core plates and a new manufacturing concept. The design advantage is getting the same performance in a smaller packaging space with lower weight. BorgWarner has also developed a new clutch concept that allows high torque transmission in an extremely compact installation space. The new clutch design combines a friction plate clutch with a mechanical lock device.

1 Introduction

Friction clutches are the most complex elements in a drivetrain. The mechanism of functionality is until today not really understood. At the same time there is a tendency to go to the limits of the system by layout. This means an increased power density by installation space and mass. The consequence is an increased demand for friction A Subsection Sample performance and wear reduction by increased requirements for thermal and mechanical robustness.

As an enabler for increased power density of friction clutches there is a need for improvements in the friction materials and in the component designs.

2 Transmission Design Trends and Friction Material Requirements

Recent development of 9–11 speed transmission for combustion engines and transmissions for Hybrid and EV vehicles have increased a need for friction materials to handle the increased power density and increased energy levels. Due to reduced friction in the powertrain there are increased requirements for NVH robustness for the clutch system. To support the demand for increased power density the package size and the weight has to be reduced.

Transmission Trends and Friction Material Requirements

Euroforum Deutschland GmbH (Ed.): *CTI SYMPOSIUM 2019*, pp. 346–355, 2021.
https://doi.org/10.1007/978-3-662-61515-7_31

- Trend to 8-10 speed AT
- DCT & CVT
- Hybrid
- BEV

The demands for the friction clutches are:

- Higher Torque Capacity
- NVHAnti-Shudder Robustness
- Reduced/Minimum Drag Loss
- High Temperature Durability
- Great Shift Quality
- Reduced package size
- Reduced weight

3 Increased Power Density

Increased power density means the reduction of the installation space and reduction of the mass. To achieve these objectives there are some needs to improve the friction system. The improvements has to be in thermal robustness and hot spot resistance. An enabler for higher power density is the increase in the static and also dynamic coefficient of friction. This allows the reduction in diameter or in the number of friction surfaces. Additional weight reduction can be achieved by replacement of traditional materials. The alternative materials are Aluminium and plastic.

4 Friction Materials Solutions

For different applications BW has developed special friction materials. For shifting clutches in Step AT or shift elements for EV vehicles BW developed a friction materials with high static COF and with high dynamic COF. For shifting clutches with extremely high shifting power the new Hot Spot resistant friction material BW5002 is available.

For launch clutch applications like DCT clutches the new friction material BW6970 is now available. The BW 6970 is able to handle interface temperature up to 440 °C (Fig. 1).

5 BW4390 High μ Static Friction Material

The BW4390 is a new friction material development. Based on a serial friction material a special refinement is increasing the static coefficient of friction.

Figure 2 shows the static COF of BW 4390 under different testing conditions.

The chart shows the influence of pressure, temperature and cooling oil flow on the COF

The static COF is strongly dependent by pressure and temperature.

In direct comparison BW4390 shows a 25% increase in the static COF compared to a conventional friction material.

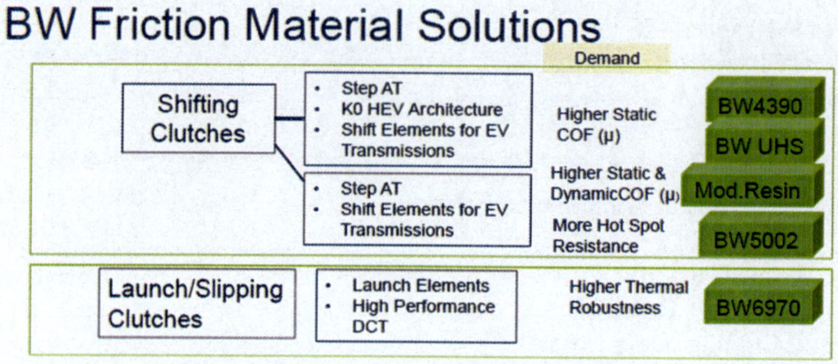

Fig. 1. BW friction material solutions

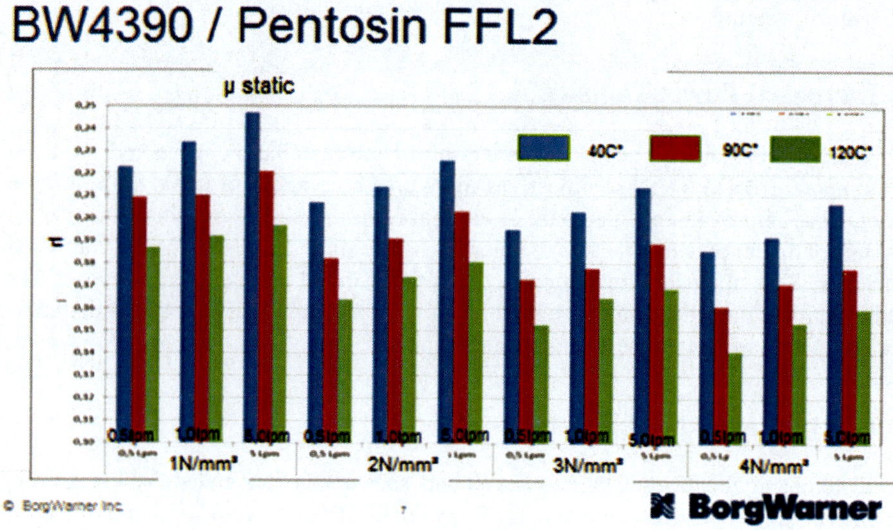

Fig. 2. BW4329

6 Ultra High Static Friction Material

Ultra high static friction material is a new material development. Developed for static only clutches which have a very limited slip speed. The so called static only clutches has a very low differential speed. Applications can be in multi speed BEV's or disconnecting clutches at PHEV transmissions. This Ultra High Static clutches can be used as a replacement for a dog clutch. A multi-plate friction clutch would allow for near. speed matching that is not as precise as a dog clutch would require. The static COF which is achievable is about 0,4. The benefits of the UHS materials are:

- The perfect speed matching is not required
- The clutch can release while carrying torque
- Significant plate count reduction compared to conventional friction materials.

This gives additional advantages:

- Reduced system cost
- Reduced axial length of clutch and transmission
- Reduced mass
- Reduced drag loss
- Improved response time due to shorter piston stroke.

7 Modified Resin

Conventional friction material is saturated with a phenolic resin to improve the strength of the material to make it resistant against surface pressure and wear.

The modification of the phenolic resin with additional organic components shows an increase in the static and dynamic COF. The advantage is that such a material with modified resin is also usable as a shifting clutch because of the neural friction curve (Fig. 3).

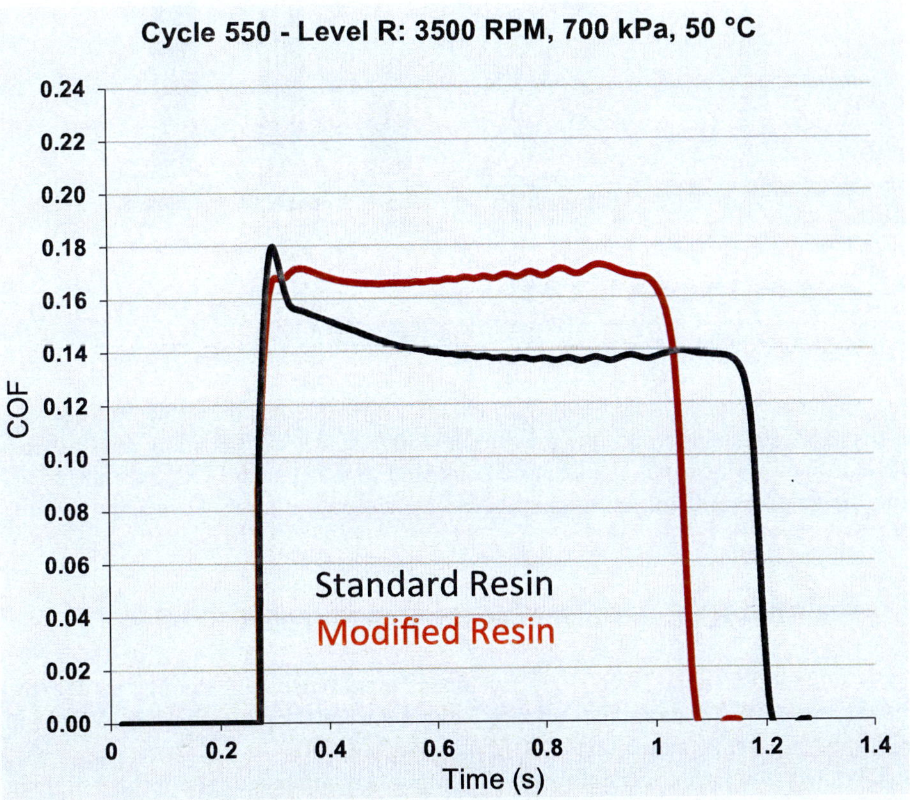

Fig. 3. Comparison of different resin

8 BW5002—Hot Spot Resistance

BW 5002 is a new friction material development.

BW 5002 has an increased pore structure compared to a standard Hot Spot material. The pore size and the number of pores are larger than at the conventional Hot Spot material. Elasticity is an important physical property for a hot spot resistant friction material.

Figure 4 shows the elasticity comparison between a standard shifting clutch material, a standard Hot Spot resistant material and BW 5002.

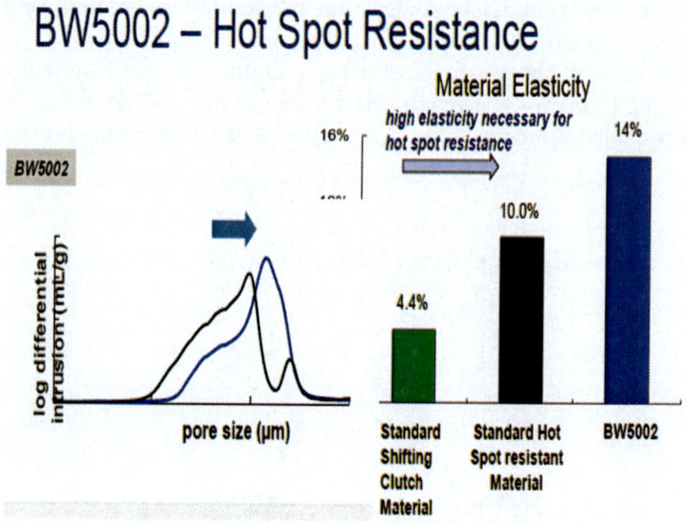

Fig. 4. Elasticity and porosity of BW5002

Compared to the standard Hot Spot material BW5002 has desirable curve shape. Also at low temperature and low pressure BW5002 shows a descending curve shape. The Hot Spot performance is increased compared to the standard Hot Spot materials. With the new BW5002 differential speeds at the shift event up to 97 m/s are possible (Fig. 5).

9 BW6970—High Performance Friction Material

The new friction material BW6970 is developed for extremely high thermal input in the friction system. The material is developed for slipping clutch applications like WSC and DCT. Interface temperatures up to 440 °C are possible.

BW6970 shows a high elasticity and a larger porosity compared to traditional slipping clutch materials. The benefit is to handle high differential speed and high interface temperature. Due to the pore structure and high temperature ingredients the material can maintain a positive µ/v characteristic over life time (Fig. 6).

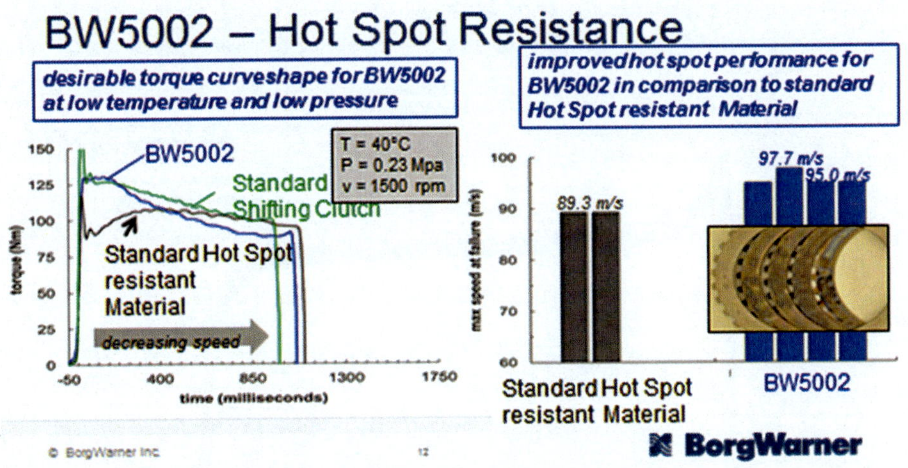

Fig. 5. Hot Spot resistance test results

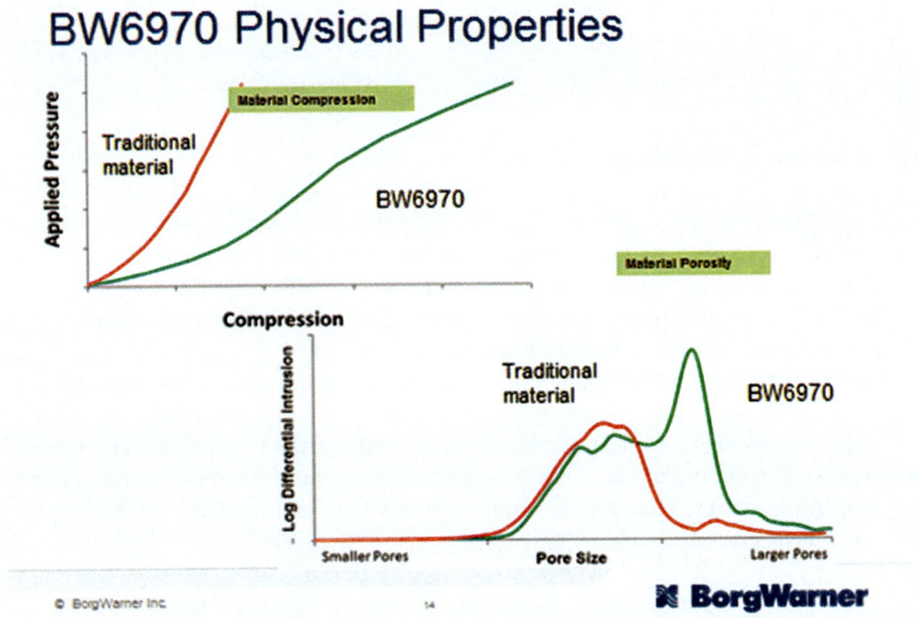

Fig. 6. Physical properties of BW6970

The High Speed Launch Test shows the friction performance of the new BW6970 in a 15000 cycle durability test. Compared to a traditional slipping clutch material the BW6970 shows almost no drop in the COF after 15000 high energy shifts in the durability test (Fig. 7).

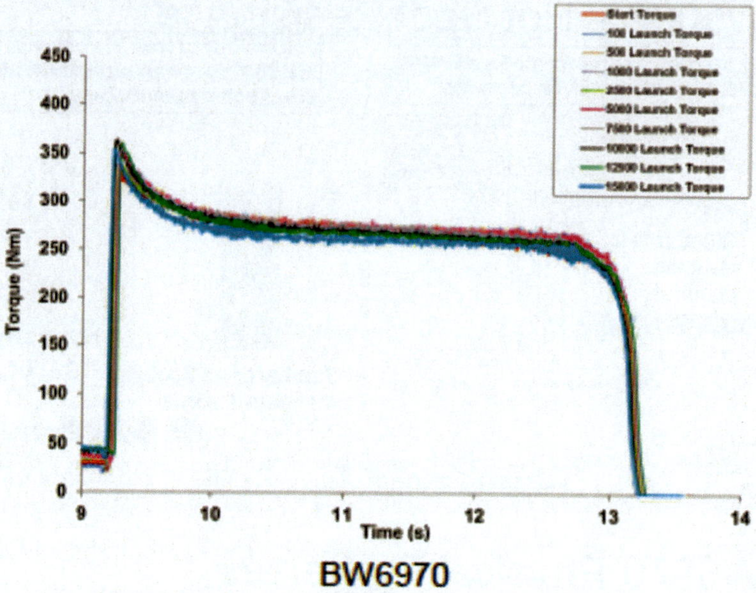

Fig. 7. BW 6970 test results

10 Molded Core Plate

The traditional core plate consists of a stamped metal core covered with glue and two friction linings. To produce this friction plate a lot of process steps are necessary.

Stamping, etching, coating with glue, bonding and sizing.

The Molded friction plate consists of a plastic core with enclosed friction pads.

This is a process with significant lower process steps.

Further advantages include the significant lower weight, inertia and shorter axial (Fig. 8).

Figure 9 compares the manufacturing process steps between a conventional friction plate and a molded core plate. There is significant lower amount of process steps with the molded core plate. The surface preparation and the coating of the steel core with glue can be eliminated.

11 Aluminum Core Plate

Aluminum core plates are an enabler for significant weight reduction. The advantage is not only the 65% weight reduction of the core plate itself but the possibility to manufacture the mating parts like hubs or housings with aluminum without having the risk of hammering problems. Another advantage is the non magnetic characteristic of the Aluminum core plate which could be an advantage in proximity to electronics.

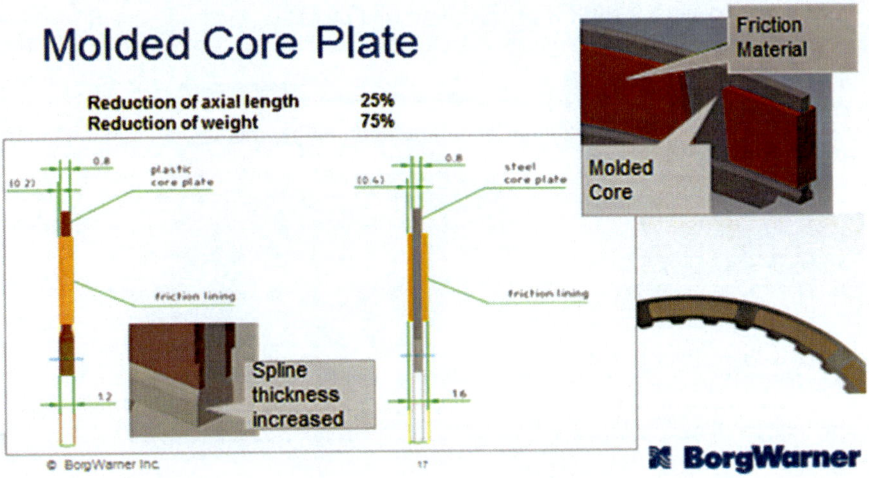

Fig. 8. Molded core plate design

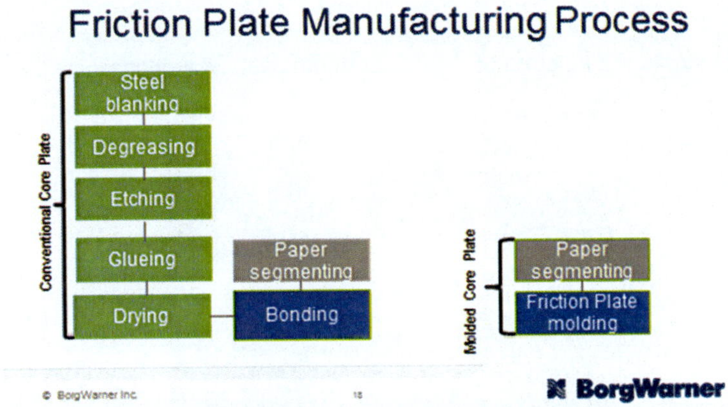

Fig. 9. Process flow comparison

12 High Torque Low Drag System (HTLD)

The HTLD is a system which is combining a friction clutch for synchronization with a dog clutch for transmitting high torque. The advantages are the extremely high torque density of the clutch system and the ability of the system to switch between synchronization and form closure without torque interruption. Due to the low number of friction plates and the low oil flow the system has a significant low drag torque. Due to the form closure the toque capacity is significant higher than with a friction clutch. In this example the clutch can transmit 4000 Nm with a significant compact package (Fig. 10).

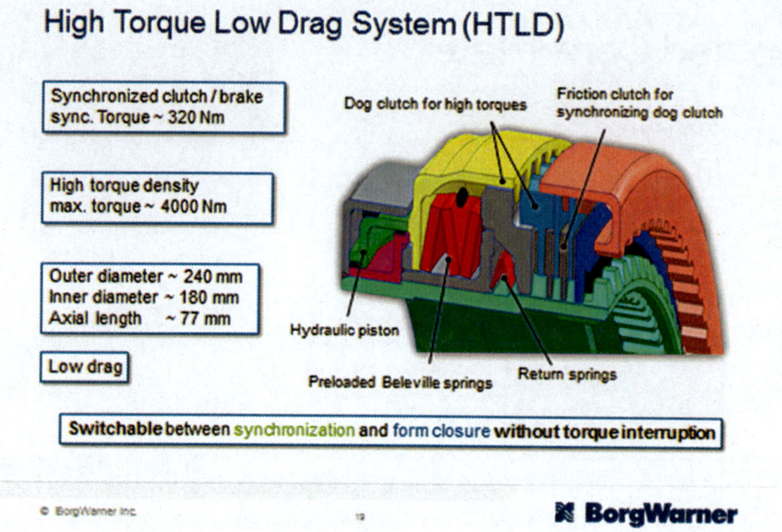

Fig. 10. HTLD clutch system

Figure 11 shows the function of the clutch in three different modes.

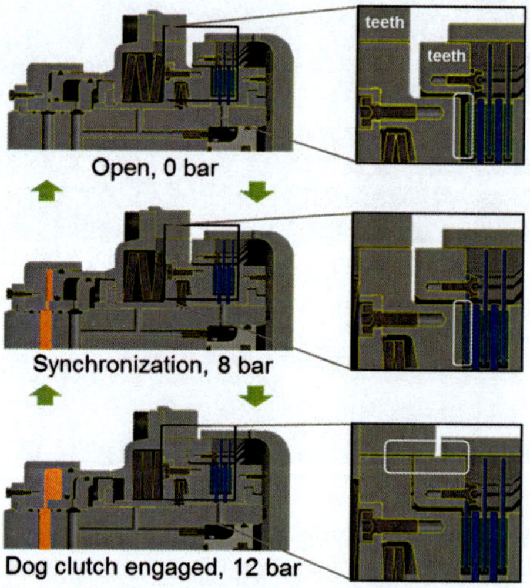

Fig. 11. operation mode

1. Open clutch mode
   Friction clutch and dog clutch are not engaged.
2. Synchronization
   Piston is pressurized with 8 bar, friction clutch is slipping with synchron speed.

3. Dog clutch engaged
   Dog clutch is engaged at 12 bar apply pressure. Torque is transmitted.

The system shows very low drag torque at different oil flow rates. The max drag torque was about 2.5 Nm at max. oil flow.

Figure 12 shows a disengagement of the dog clutch. What is plotted here is the switch from dog clutch engagement to controlled slip. There is no torque interruption after the dog clutch has disengaged.

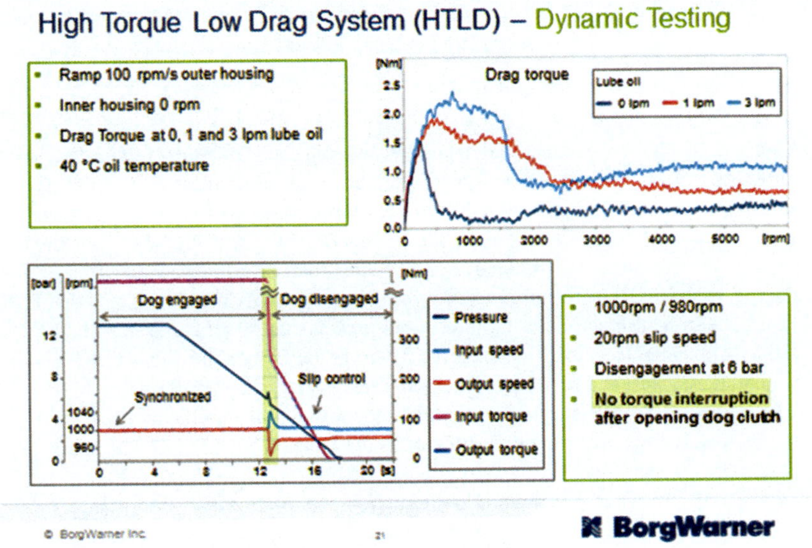

Fig. 12. Dynamic test results

Figure 12 shows a disengagement of the dog clutch. What is plotted here is the switch from dog clutch engagement to controlled slip. There is no torque interruption after the dog clutch has disengaged.

13 Summary

New friction materials with increased thermal robustness and Hot Spot resistance and materials with higher static and dynamic COF are enablers for increased power density

Alternative materials and design concepts for friction plates support weight and package size reduction

The High Torque Low Drag System (HTLD) combines a friction clutch with a dog clutch with the advantage of high torque density and low drag losses

A Brief Evaluation of Freewheeling Motor at P4 Position: Retrofit Approach to Electrification

Jérôme Mortal[1](✉) and Ashwin Charles[2]

[1] Director, Valeo Special Vehicles Application, Powertrain Systems Business Group, Isle D'abeau, France
jerome.mortal@valeo.com

[2] Simulation Engineer, Valeo Special Vehicles Application, Powertrain Systems Business Group, Isle D'abeau, France

Abstract. Several approaches are taken towards the electrification of the future's automotive fleet. This Paper explores a possibility for electrification of existing on-road vehicles with high residual value. Notion of this paper is that electrification of certain sectors of the vehicle fleet is economically feasible when existing vehicles are retrofitted with electric parts rather than being replaced with pure electric vehicles. Retrofit refers to addition of components which was not there during the time of manufacturing. The paper presents the benefits of retrofitting a "P0 hybrid" vehicle with a freewheeling motor at the P4 position. This Config is evaluated by a detailed simulation on CO_2 savings and performance using MATLAB and AMESim. We try to contemplate the amount of modifications needed on the vehicle for retrofitting and summaries the benefits.

Keywords: Freewheel · Hybrid · Co-simulation · Retrofit · HEV · Modelling · Power-train

1 Introduction

1.1 Market Demand for Hybrid and Electric Power-Train

For the past 5 years, the automotive market has turned its focus towards the electrification of powertrain systems. Driven by legislative measures taken by governments, especially by the economic giants like the USA, China, EU, Japan, and In-dia. Meeting the emission targets is becoming difficult and cost-sensitive with only ICE as a vehicle's powerhouse. If the manufacturers do not upgrade their technology to match the trend, then they tend to be left behind in the market. As per 2018 sales statistics 1.3 million BEV's units, 0.5 million PHEV's, 1.4 million HEV's have been sold with a 76% increase in sales of New Mobility since the year before. Manufacturers are on the verge of shutting down combustion only power-train systems. To adapt automakers changing market, automotive suppliers are shifting towards e-motors and electrified drive system supplies.

Euroforum Deutschland GmbH (Ed.): *CTI SYMPOSIUM 2019*, pp. 356–368, 2021.
https://doi.org/10.1007/978-3-662-61515-7_32

1.2 New Mobility in Passenger and Utility Vehicles

We will explore a few of the major fields where electrification is applied. Several multinational automakers and emerging local companies are investing in sector of electric quadricycles and scooters with limited seater capacity and low top speed, made for urban traffic. Sector of utility vehicles have taken a major dive towards electric mobility. Airports, manufacturing facilities, storage units, and other logistic companies extensively use very low speed and high torque utility vehicles for moving loads. Electric utility vehicles ensure workplace air quality and high work efficiency. Micro Mobility is picking up its pace in major cities as an innovative adoption of technology. Several establishments are investing in this division. Light E-scooters and torque assisted bicycles have become the basis for many startups, drones are remotely controlled to deliver goods, safety inspection, monitor large fields, aerial photography.

The sector of passenger vehicles took a turn since major automakers started commercial production of electric and hybrid vehicles since 2010. The decreasing cost of battery production and increased CO_2 savings became a driving force. The technology for new mobility poses a challenge. Electric Vehicle's range is limited by the battery capacity, increasing battery size would mean higher cost and increased charging time. Lack of high-power source charging facility makes it quite challenging. In countries where electricity production is majorly based on thermal sources, EVs' pollute as much or even more than a gasoline vehicle in its life cycle. At least for the next few coming years, we can say hybrid vehicles outweigh EV's in terms of life-cycle CO_2 benefits and cost. Mild Hybrids are more affordable compared to electric vehi-cles. They comply with the current emission standards. They have Higher energy density compared to EV's. There are multiple hybrid Configs with varied abilities (Fig. 1).

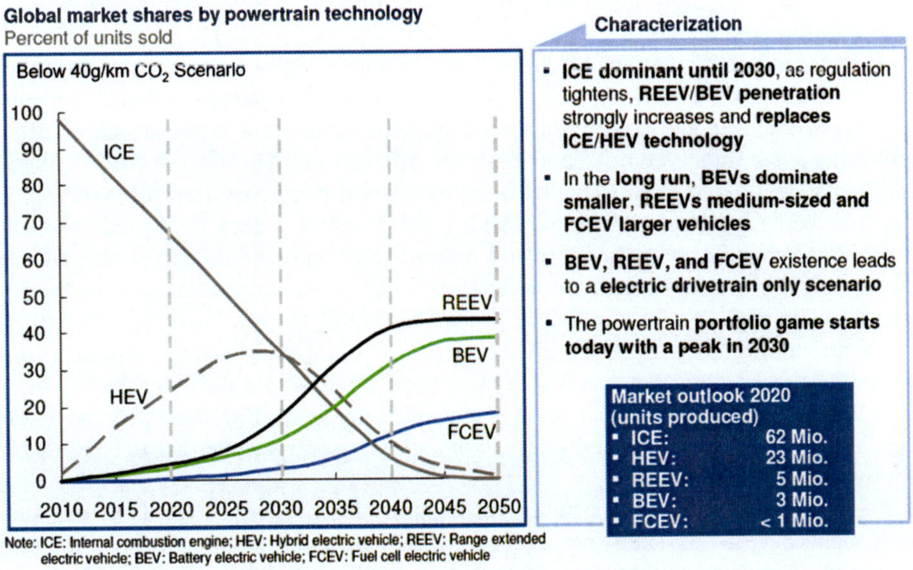

Fig. 1. Market prediction. Source: McKinsey.

Series hybrid, parallel hybrid, series/parallel hybrid are few major classifications. Mild Hybrid Configs are capable of giving 10-20% of CO_2 reduction compared to a conventional ICE system. PHEV give a high electric range in addition to HEV functionality, but the cost is halfway to buying an EV. The below figure gives us the predicted market trend of powertrain systems.

1.3 Electrification of the Existing Vehicle Fleet

Some companies pioneer in taking up an old model vehicle and replacing its combustion powertrain system with an electric powertrain and a battery. But this kind of replacement is expensive because of the battery cost and also because they are done in small scale. This paper takes a glimpse into the idea of converting a conventional car into a hybrid and to find a suitable market for it. We focus on vehicle with high residual values like ambulances, patrol cars, military vehicles where the purpose oriented modifications made lead to increased residual value. Take ambulances for example, a new ambulance would cost 70 k€ to 250 k€ based on the facilities. There sale value on these ambulances are very high even after depreciation. These kinds of vehicles whose lifetime is long, will be the target market for retrofits. To prove the concept's benefits we will try to simulate a P0P4 hybrid, where the P4 motor is freewheeling and assumed to be attached to the rear wheel shaft with a differential, in parallel to engine torque. We will then speculate on the complexities of compatibility and compactness.

2 Retrofitting by Adding: Why Is It Better?

With introduction, we focused on the market and causality of electrification. In this section, we will highlight the purpose on why retrofitting is better and how will it be possible within the market.

Why? Today only way for electrification is to either replace the car or replace the entire powertrain system within the car. This is where adding a retro fit is more practical. No kind of replacement is required. Fewer engineering modifications are needed for retrofitting a rear motor, compared to replacing a powertrain. This retrofit will act like a simple plug and play system. This leads to lower installation cost. Retrofit system can vary from being extremely simple to complex. The simplest system will be independent of existing Vehicle's control unit, it will involve addition of motor in the rear which only interacts with pedal and does not interfere with any other component. This simple system can be applied to any vehicle.

Certification Process: The certification process followed by some governments do not facilitate technology like retrofits or electric cars. For example, the UK government is more lenient compared to French government with regulations. In vehicles, the safety of each component individually and safety of the whole system together has to be ensured before getting to certification.

Safety Aspect of simple system: As mentioned above, in a hybrid system, the components have to pass safety procedure separately and then component along with the system has to go through safety procedures once more for validation. The idea of simple retrofit system is that this retrofit equipment's safety is independent of type of vehicle using it and will not require 2 step validation. The simple system will be validated against a vast range of cars to make the retrofit robust. By avoiding unnecessary procedures and making a robust system, we can cut major costs and make the product more affordable.

3 Power-Train Simulation

3.1 System Modeling

We simulate our hybrid system Config using a Co-simulation tool, an interface between AMESim and Matlab. In this platform we create a model which can simulate multiple vehicle Configs including conventional ICE, HEV and EV. We modified the P0P4 architecture by adding a freewheel in between the P4 positioned motor and wheel shaft (Fig. 2).

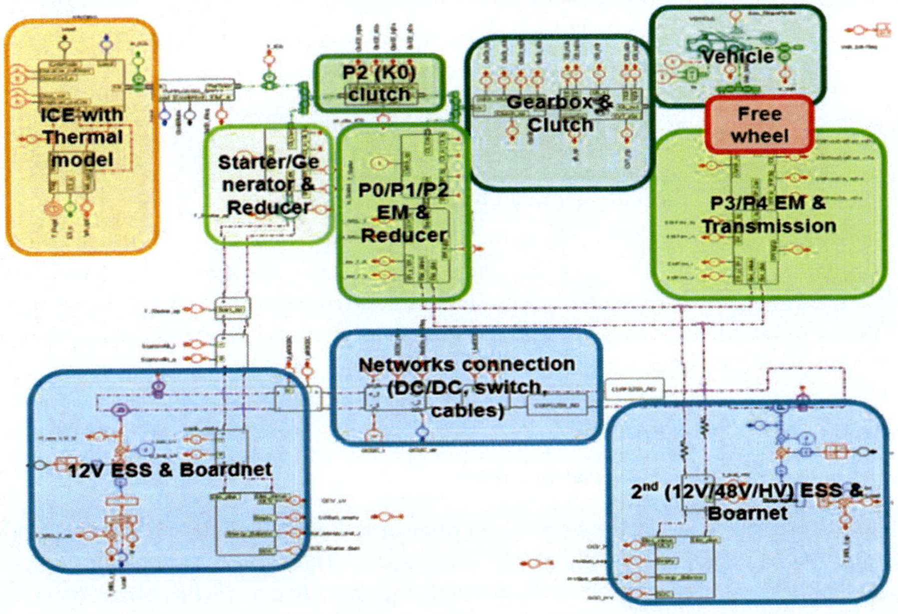

Fig. 2. Model architecture.

To improve the accuracy of the simulation results the model considers every parameter which contributes to energy consumption. The battery dynamics, DCDC dynamics, motor losses, vehicle losses, gear-box losses, auxiliary consumption, etc. characteristics are included. The hybrid vehicle is modeled by Valeo in AMESim using the inbuilt components, and the control strategy is carried out by a Matlab Simulink model. The tool simulates multiple hybrid Configs based on varied motor positions and types of energy sources present. The model calculates vehicle performance and consumption on a driving cycle. We use a 48 V system for mild hybrid Config for our study. There are 2 batteries that we use, a 12-volt battery for auxiliary system and a 48 V battery for high voltage grid (for motor and generator). We focus mainly on NEDC cycle for this study. This model takes into account the driving cycle procedures for results precision. The driver model is based on a PID feedback control with a feed-forward on torque. The model evaluates the driver torque demand and splits it between motor and engine based on optimal control strategy. The optimal control strategy is based on Pontryagin's Minimum principle where we evaluate which torque split costs less in terms of fuel consumption. Hamiltonian (H) is calculated for each torque split, and the torque split between engine and motor with the lowest Hamiltonian is chosen and implemented in the hybrid system. A precalculated map of best *lambda* as a function of DE and SOC, produced from the model before starting the simulation. The simulation follows an online approach where the *lambda* is changed only every 15 s (Fig. 3). The system assumes the

Pontryagin's Principle: $H = \lambda \Delta E + Qfuel$

$\Delta E = change\ in\ batteryEnergy$

$\lambda \Delta E = Change\ of\ batteryEnergy\ in\ terms\ of\ fuel\ flow\ rate$

$Qfuel = Fuel\ flow\ rate$

Fig. 3. Pontryagin's minimum principle.

Energy variation for the last 15 s will remain the same as the next 15 s. There are penalties kept in place for gear shifts, clutch disconnect and connect, switch on and switch off the e-motor and ICE. These penalties ensure that strategy takes transient consumption into the account. The optimal strategy evaluates torque split every 2 s which might cause problems in case of drivability, the optimal strategy is not tuned for drivability as it's not our focus of observation. We will not describe the component model or the strategy in detail.

3.2 Features of Different Architectures

P0 architecture: A motor is placed at the P0 position of powertrain which is capable of regeneration, generation, fully electric drive, and torque assist. This motor faces a considerable amount of belt friction losses which makes full electric drive less efficient. It facilitates the start and stop system of the engine.

P0P4 (P4 with clutch) architecture: To P0 architecture another e-motor is added at the P4 position of the powertrain. A clutch is placed in between the P4 motor and the transmission with a single reducing ratio. This motor is capable of full-electric drive, regeneration and torque assist.

P0P4 (P4 with freewheel) architecture: To P0 architecture another e-motor is added at the P4 position of the powertrain. A freewheel is placed in between the P4 motor and the transmission with a single reducing ratio. This motor is capable of full-electric drive and torque assist. Re-generation is only possible at P0 and not at P4 as freewheel will start slipping when there is deceleration and deceleration value is less than the motors natural deceleration rate. The slipping happens when there is a negative torque (Fig. 4).

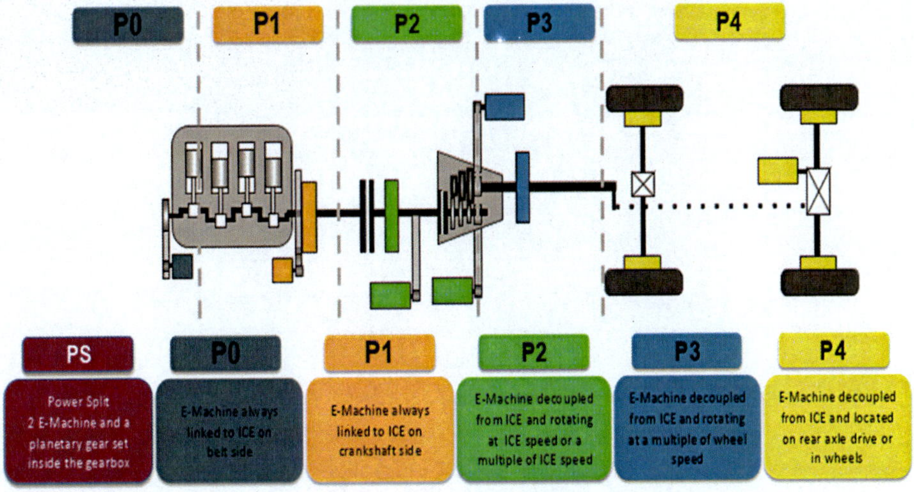

Fig. 4. Different motor positions. *check appendix for bigger images and glossary.

4 Simulation Results

4.1 Results

We present P0P4 freewheel results exclusively and the rest of the architecture simulation will only be used for comparison. We will evaluate the benefits retrofittable P4 motor + 48v battery system as go through the simulation results.

Input parameter overview: We give an overview of parameters used for simulation and we will avoid going into the details of parameters.

For example, motor input will include max torque operating points and efficiency map for varied voltages, details as such is out of scope for this thesis. Motor power is defined as the average peak power at an operating voltage of 48 Volts. Motors use single reducer ratios.

Results: overall CO_2 emission comparison: The goal of the model simulation is to evaluate the CO_2 benefits and compare it with known Configs. We will establish the CO_2 impacts of needed Configs then we will investigate the reasons for the positive and negative impacts. Figure 5 gives CO_2 emission over a NEDC cycle. If we notice, we see the P0P4 freewheel Config is better than P0 and less effective than P0P4 clutch system. We cannot conclude that freewheel is not effective enough, the objective is to justify the balance between benefits, cost of installation and complexity for it to be used as a retrofit. We will investigate what causes the difference in consumption (Fig. 6).

Results:

Energy mode comparison (P0 & P0P4 freewheel): We compare the mode changes of P0 and P0P4 free wheel hybrid system to explore the benefits.In both the Configs the engine is forced to on, for the sake of activating catalyst and steady coolant temperature. The blue region marked in the mode changes signify ZEV mode or full electric mode. We notice that for freewheel Config, the ZEV mode operations are more compared to P0

Parameter		Value	unit
Coastdown Coefficients	F0	0.41	N/(m/s2)
	F1	1.9	N/(m/s)
	F2	121.96	N
Mass		1570	kg
Engine Capacity		1.198	L
P0 motor power		4	kW
P4 motor power		15	kW
Turbocharged		Yes	
GearBox		6 Ratio AT	

Fig. 5. Input Data.

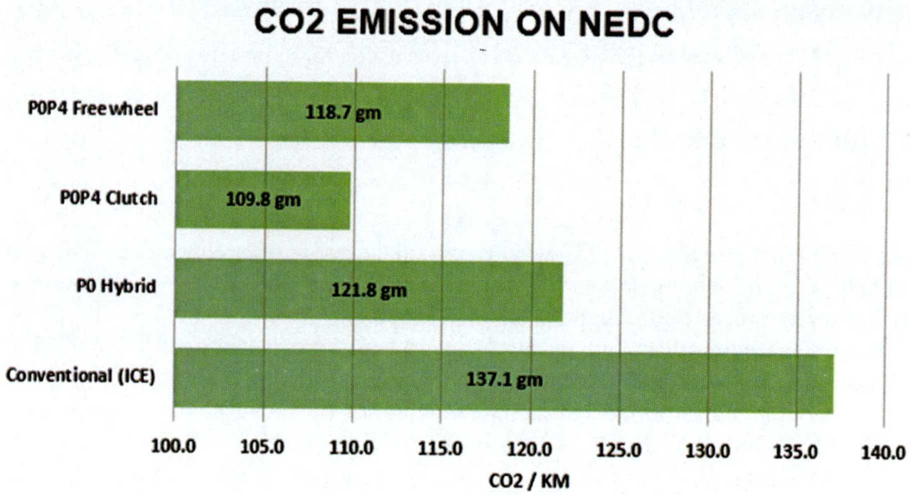

Fig. 6. CO_2 emission data.

Config. The P0 motor has to overcome belt friction to use ZEV mode. So ZEV mode is only activated during high load and low RPM of engine for P0. But in freewheel Config there is no such losses, so we see more frequent ZEV modes and P0 in free-wheel Config is only used for torque assists for engine. If the freewheel realises negative torque with deceleration greater than the motor's natural deceleration the freewheel is still connected. In the mode it's depicted as regenEMr, but the motor doesn't regenerate.

Results:

Battery Energy comparison (P0P4 free-wheel & clutch & P0): With Fig. 7 we can establish how both configs uses the available battery. To maintain the same SOC at the end of NEDC cycle, the hybrid system only uses a range of the battery's available energy which is how we size our battery capacity. The energy difference of operation for P0P4 clutch Config is around 400 kJ whereas for freewheel Config its around 300 kJ. For P0 Config its around 200 kJ. This doesn't mean that the energy's presence is a must, the optimal strategy can adapt if the battery is small. But if the battery has the extra energy (100 kJ difference between P0 & freewheel) the freewheel Config can be completely utilized. So, if decide we to retrofit a P0 Config with motor at P4 we can increase the battery capacity or keep the same 48 V battery.

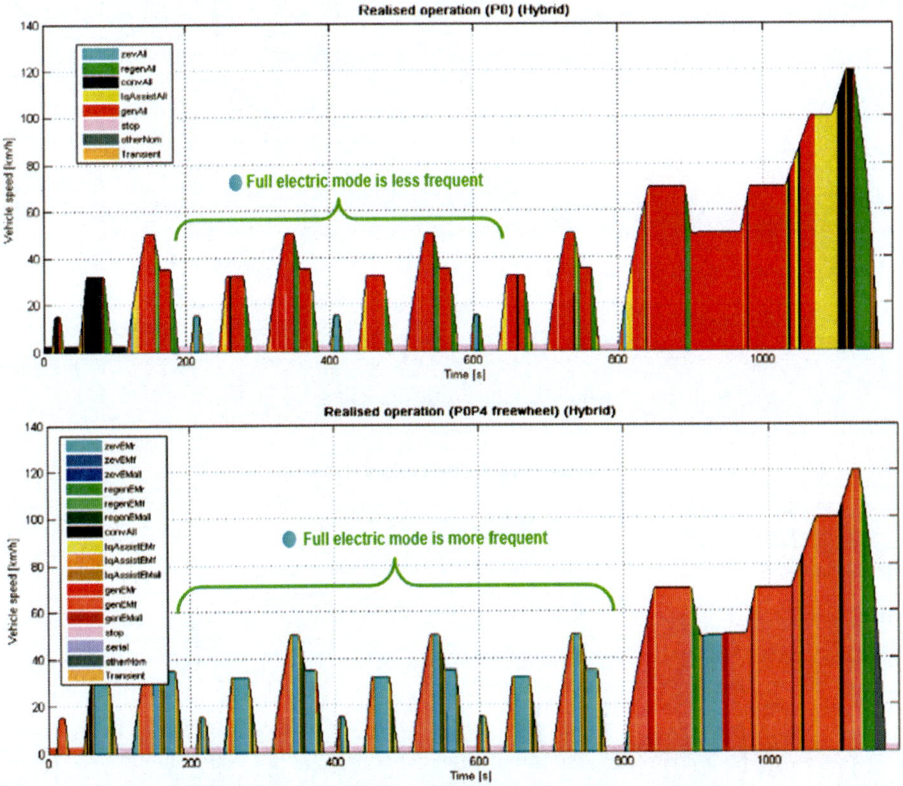

Fig. 7. Hybrid Mode operations [P0 & P0P4 freewheel].

Results:

RPM comparison (P0P4 freewheel & Clutch): With Fig. 8 we will observe the behavior of Free-wheel and how it contributes to the consumption by looking at P4 motor's RPM. At this point we have to mention that the optimal strategy is tuned for P0P4 clutch system and not the freewheel system. The optimal strategy almost chooses the same point of time to use the P4 motor, but with few differences. The optimal strategy switches off the P4 motor if the motor is disconnected, in both clutch and freewheel Configs (Fig. 9).

If we look at the region marked 1 in both freewheel and clutch system, the P4 motor is connected to the wheels. The consumption is almost the same. But if we take the region marked 2 the freewheel disconnects and reconnects again, the angular acceleration of

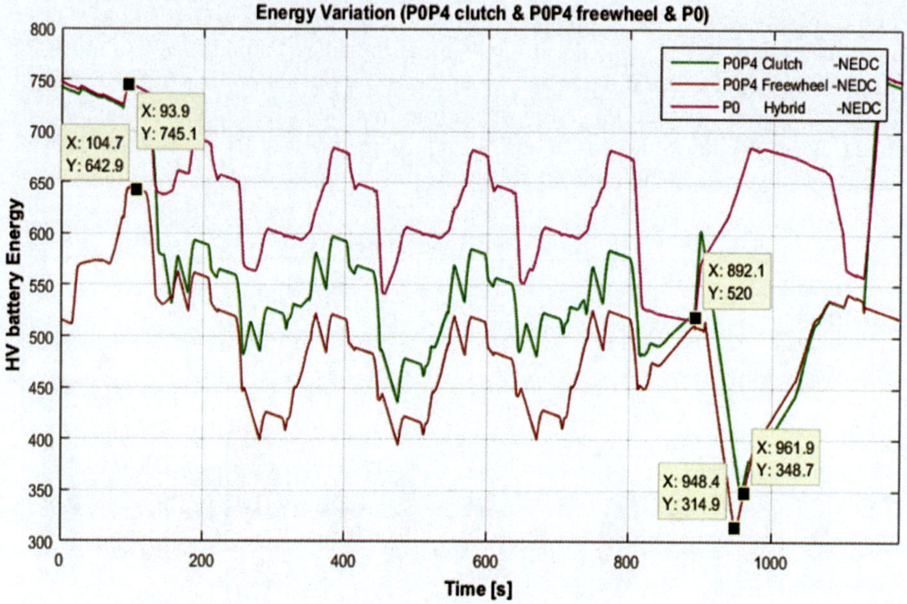

Fig. 8. Battery Energy Variation [P0P4 freewheel & clutch & P0].

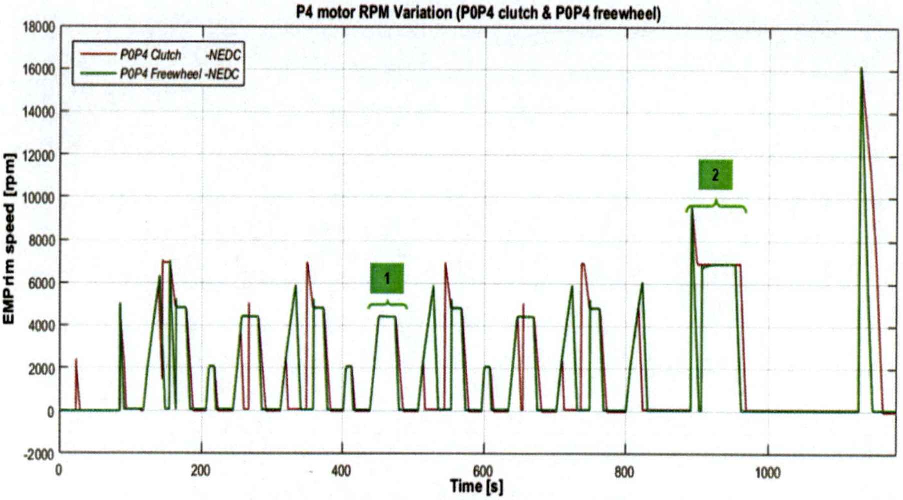

Fig. 9. RPM Variation [P0P4 freewheel & clutch].

motor to equalize RPM between wheel and motor increases the energy consumption. So, we can say that we can attain better CO_2 benefits if we optimize the strategy. For example, we can continue running the motor for some more time before switching off the motor in free-wheel. But we need varied simulation as well as measurement data to optimize the strategy.

Results:

Torque comparison (P0P4 freewheel & Clutch). The P0P4 clutch Config has higher CO_2 benefits because of its recuperation or regeneration. P4 motor regenerates and directs the energy to battery which freewheel system is not capable. When there are sudden transition the freewheel realises low magnitude negative torque but this torque is not regenerated. But the high magnitude negative torque realised by clutch is regenerated (Fig. 10).

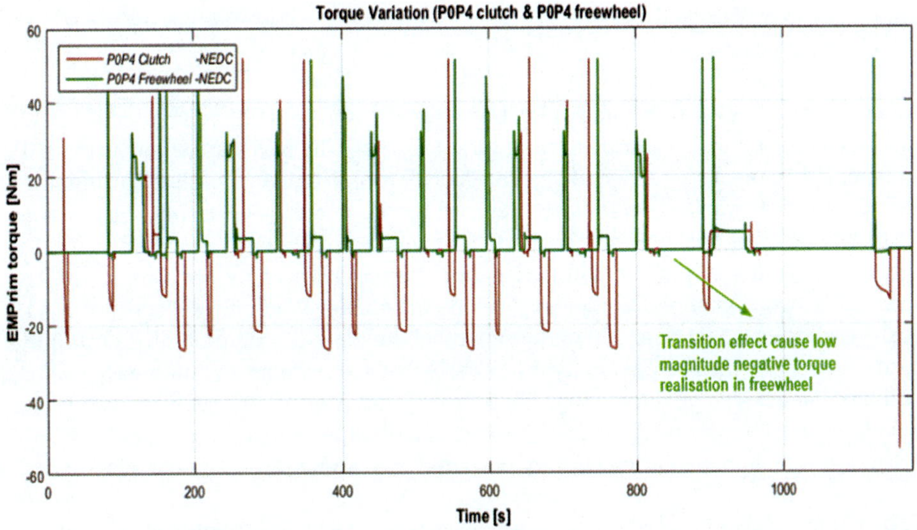

Fig. 10. Torque Variation [P0P4 freewheel & clutch].

In free-wheel system the regeneration is done by P0 motor, only if it meets respective conditions. For example: If engine is forced to be on the gearbox will be forced to be in neutral when braking, where P0 can't regenerate.

4.2 Results Inference

Obviously a freewheel Config is not the best among the shown results. The objective is not to evaluate which Config is most efficient, but to find which one can be adapted to be a better retrofit. If we take P0P4 clutch system, to retrofit a clutch system into an existing vehicle design without any change is impractical. The design changes in clutch system will include both mechanical, electrical and software changes extensively. The clutch opening and closing would itself take considerable amount of torque. In case of freewheel, the motor can be permanently coupled with a freewheel and a single reducer to act as a single component for simplicity. The energy management strategy should be designed in a way it's almost independent of existing system or demands only few changes. Example for simple strategy: Retrofit can be tuned to only enable full electric mode. In this case strategy would be to charge the battery at high efficiency point of engine using generator and use full-electric mode when battery is capable. From a software perspective, there is no torque split between motor and engine, only if engine is made neutral and generator is engaged. Example for complex strategy: This would involve torque split between engine and motor [Torque assist] which would mean making changes in ICE control unit in terms of operation. We might need a dedicated hybrid control unit installation. Complex strategy would mean higher CO_2 benefits but also would mean higher development cost. So, depending on available market the control strategy will be defined.

5 Conclusion

We show the benefits of adding a P4 freewheel system to a P0 vehicle or a vehicle which is equipped with a generator through simulation. The next step would be to verify the simulation results with experimental data. We plan to force a clutch to mimic a freewheel which would give the first step results. This freewheel retrofit can be tuned for performance or consumption depending on customer need and cost. This retrofit is completely capable of 4WD but implementing this would mean higher complexity. The trade-off between cost and complexity seems inevitable. We believe electrification of existing vehicles would be an alternative approach for emission control. If the retrofitting culture grows it could trigger a modular design of cars, where the automaker would facilitate the design for retrofitting. This thesis lacks a detailed feasibility study. But from the current resale price of these high residual value vehicles and cost of a hybrid or electric vehicle retrofitting is highly plausible and profitable.

Acknowledgements. First and foremost, we would like to thank the rest of Valeo's simulation team. Their astonishing work in developing the hybrid simulation platform which has lead to multiple consumption and drivability studies throughout the years. These kinds of studies have been performed for multiple automakers. We would like to thank Valeo's SVA team for the review of our work and the support.

Supplementary Note 1: Glossary

PHEV	Plug-in hybrid electric vehicle
HEV	Hybrid vehicle
EV	Electric vehicle
4WD	All wheel drive
ICE	Internal combustion engine
NEDC	New european driving cycle

Supplementary Note 2: Appendix

See Figs. 4, 7, 8, 9, and 10.

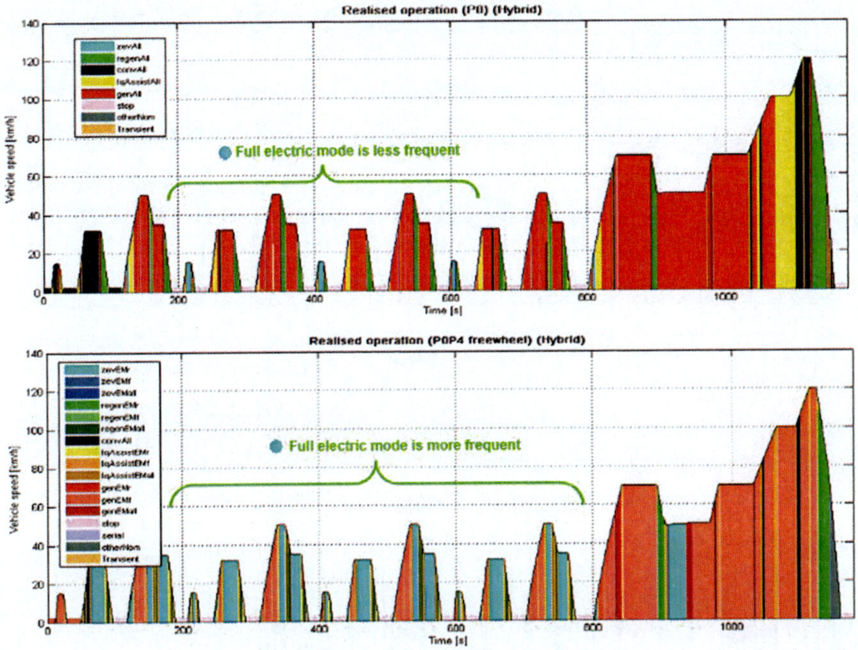

Fig. 4. Different motor positions

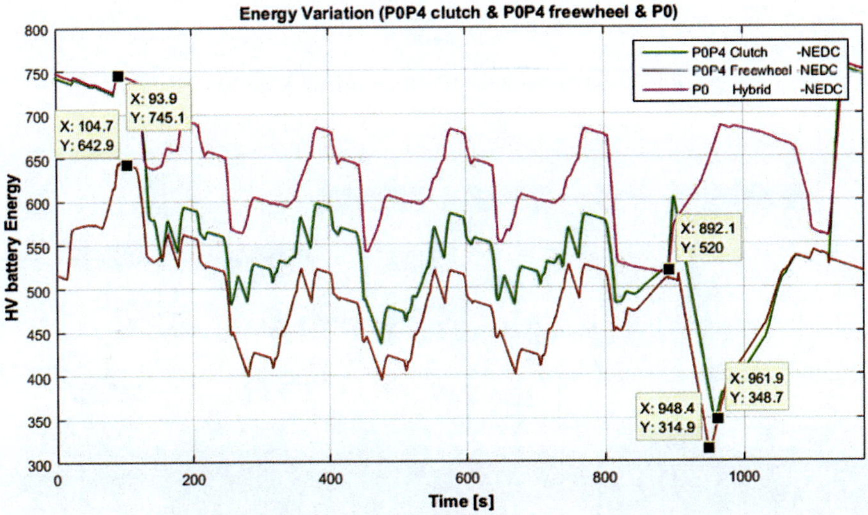

Fig. 7. Hybrid Mode operations [P0 & P0P4 freewheel].

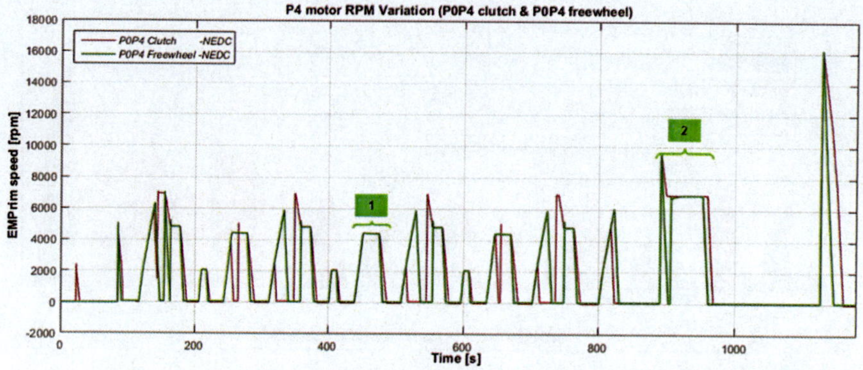

Fig. 8. Battery Energy Variation [P0P4 freewheel & clutch & P0]

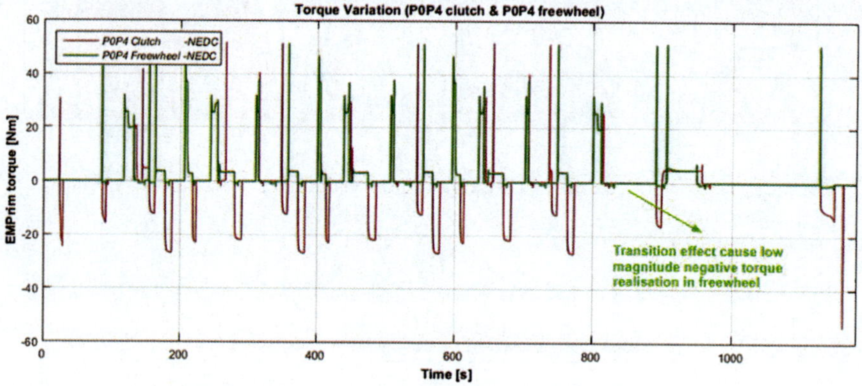

Fig. 9. RPM Variation [P0P4 freewheel & clutch]

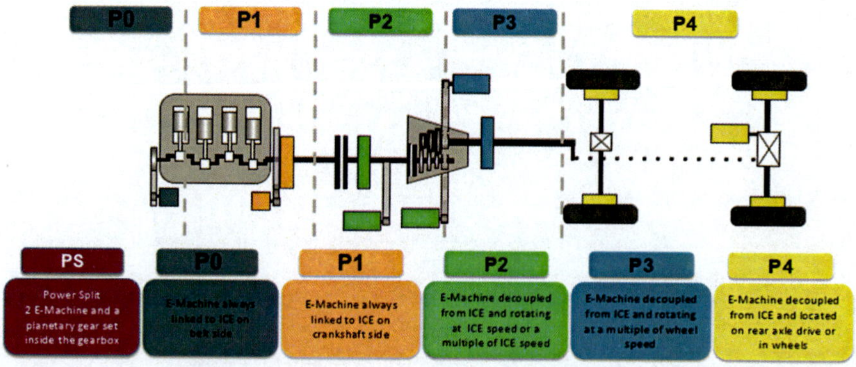

Fig. 10. Torque Variation [P0P4 freewheel & clutch]

Laser Joining of Copper-Copper and Copper-Aluminum Application in the E-Mobility

Stefan Mücke(✉)

Scansonic MI GmbH, Schwarze-Pumpe-Weg 16, 12681 Berlin, Germany
stefan.muecke@berlin.industrial.group

Abstract. The rise of e-mobility has particularly increased the use of copper and aluminum in the series production of vehicles. The production of electric motors in particular focuses on copper-copper connections in hairpins. Due to the late position of hairpin welding in the stator's value chain, defective welds lead to irreparable damage to the stator and thus to increased scrap costs. In addition, the focus is also on copper-aluminum connections, particularly in the area of battery contacting with high demands on the connection cross-section. The presentation will highlight the possibilities of using precise beam guidance, high deflection speeds and beam shaping to safely and robustly design automated welding processes in series production. Different welding strategies will be discussed using the example of the copper hairpin welding process and the welding of aluminum and copper contacts in battery production. Furthermore, the possibility of using beam forming technology in the beam source and the welding optics to make the processes safe and robust and to additionally improve the appearance of the welds on copper and aluminum will be discussed. Furthermore an outlook is been given which technical trends for welding especially copper in the laser branch, as well as the possibilities to use intelligent optics to ensure robust joining processes in series production.

Keywords: Copper welding · Cu-Cu-welds · Aluminum welding · Al-Al-welds · Al-Cu-welds · Beam shaping · Beam forming · Laser · Fast laser beam movement · RLW-S · Fast scanners · Stable process · Reduced spatter · No spatter · Welding process · Scansonic · High speed beam movement · Laser process · Hairpin · Hair-pin · Stator · Electric motor · Electric engine · Process knowledge

1 Cu-Cu Applications

1.1 Basics of Welding Copper

Welding copper with laser can become quite a challenge due to many reasons. First of all the material itself challenges the whole laser welding process just with its material properties. Most of the lasers used in welding metals especially in the automotive industry use wavelengths in the near IR spectrum between 1000 and 1100 nm. As the absorption

Euroforum Deutschland GmbH (Ed.): *CTI SYMPOSIUM 2019*, pp. 369–372, 2021.
https://doi.org/10.1007/978-3-662-61515-7_33

rate of copper is especially for wavelength above 800 nm pretty low only laser sources with high power can be used for this kind of application as copper in solid state reflects about 95% of the complete power. [1]

Second the heat conductivity provides the next challenge as the power provided by the laser is used to heat up the material until it's melted. So it is essential to check where and how far away the heat send into the material is conducted, to avoid damages on the whole unit. [2]

These points together crate a very narrow process window in which deep welding is possible. Too much laser power causes so called blow outs, pores and holes, while not enough laser power causes just heat conduction welding which often doesn't fit the requirements in strengths of the welding seam. [3]

One way to keep the molten pool stable during laser welding and thus avoid pores is the speed at which the laser beam moves over the copper. The IFSW of Stuttgart, Germany showed with their tests in 2019 that the melting pool is much more stable at a laser movement speed of 30 m/s than at e.g. 12 m/s. [4]

1.2 Realization of Requirements

As a developer and producer of only laser optics Scansonic MI is related to the laser sources the customers are using. Therefore the optics need to cope with wavelengths between 1000 and 1100 nm. As also mentioned above to weld copper high power is needed. So the optic and its elements are designed to cope with laser power up to 8 kW. But the real advantage of the Scansonic optics RLW-S are the scanner drives. With scan frequencies up to 1000 Hz and a process speed of max. 1 m/s they are the fastest ones on the market. This speed makes it possible to create a stable key-hole during the welding process and thus to reduce pores and spatter significantly.

Additionally the optic is equipped with an adaptable collimation to fine-tune the energy input. The image processing paired with a triangulation system allows the optic to adapt to the challenges that are caused by the mechanical steps and tool wear in series production.

These features in combination allow to adapt the needed energy input for a robust welding process to real life conditions.

1.3 Applications

The major application for Cu-Cu application is for sure the welding of the copper in stators. No matter if copper rods have to be contacted like in stators with hairpin, I-pin or even with wave-winding concept, or simple copper wires like in traditional winded stators. Additionally the late position of welding in the process chain of producing stator makes a robust process even more cost effective on the whole production line.

But also contacting the power electrics or batteries the same know-how and methods can be used.

2 Al-Cu and Al-Al Applications

2.1 Basics of Welding Al-Cu

Speaking about E-Mobility means also that especially in the area of batteries it is almost impossible to avoid that the materials copper and aluminum have to be combined. Due to the electrical properties of copper and the low density of aluminum there is no way around both materials.

The difference between both materials in just the melting temperature only allows to weld aluminum to copper. The other way around the lower aluminum won't be able to cope with the energy, which is needed to melt copper, therefore a robust welding process is not possible unless the aluminum has been coated. [5]

Intermetallic phases between copper and aluminum are also nothing that should be achieved as high hardness is paired with high porosity. Also the specific electrical resistance of these phases is so much higher that the caused heat can cause major damage to the whole part. [6]

2.2 Basics of Al-Al Welding

When welding aluminum and aluminum together some characteristics of the material have to be considered.

First of all an exact positioning of the parts that should be welded together is necessary. Test and researches have proven, that especially the length of the flange at filet welds, a continuous welding depth and a defined heat input are the key factors for a robust welding result when it comes to weld aluminum parts together. [7]

2.3 New Welding Strategies

Using an optic that can deliver highly precise and fast laser beam movements offers completely new possibilities for developing welding strategies.

One possibility is the so called screw- or spiral-weld. In this strategy the laser beam is guided in a precise spiral instead of creating one big welding spot. This method has the big advantage that an almost identical traverse section in the weld can be achieved by much less heat input. An additional advantage of this method is also that the risk that the big weld pool just drops down on the lower sheet is almost completely eliminated. Using a fast scanner with this strategy also allows to put an additional oscillation on the spiral to make the heat input into the material much more stable and homogenous. This strategy is especially usable for welding foils or thin material together or materials where only heat conduction welding should be done to avoid intermetallic phases

For joining thicker metal sheets (e.g. contacting bus bars) replacing the traditional I-seam with a fillet weld brings many advantages. First the positioning is easier as it is possible to see what is below the upper part. Next there is the possibility to detect the exact welding spot e.g. via pattern recognition. The welding process itself also is more homogenous through oscillation and most important the quality control afterwards is much easier as it can be seen if the welding seam really connects both metal parts without destructive testing.

3 Future Outlook

3.1 Beam Forming

Forming the laser beam causes some advantages for welding copper or aluminum. Through different foci and power intensities material can be warmed up before starting desired welding process or seams can be smoothened right after welding. This happens all at once and this reduces cycle times which causes more cost efficiency. There are two ways to do laser beam forming. First in the laser beam source itself and send the formed beam through the fiber to the welding optics. Or form a conventional beam through a special module in the optics. Both methods have their pros and cons. As beam forming in the laser source is highly precise and flexible it comes for a price because the producing the beam source itself is more complex as well as special fibers are needed. Forming the beam in the optics is overall cheaper but it has to be considered if the desired geometry is still available after the Laser beam is adjusted in x-or y- direction.

3.2 Further Features in Series Production

All the currently used welding optics are or can be equipped with a number of sensors and recognition hardware. But for what can this equipment also be used to make the whole process of welding more stable and cost effective for series production?

It is very desirable that the welding tools not only control the process itself. To reduce costs and to avoid loads of damaged parts it is important that the situation is checked/measured before the welding itself. Even detect if it is possible to weld the part or not.

To control the welding is neither important than the pre-check. Controlling the parameters and deliver a constant quality on a high level should be the target which can be achieved by reducing shifts of focus and scanners for example.

Third are the post-weld-checks. These are the last possibilities to keep a damaged part away from the final product. Of course only x-ray or destructible test can guarantee 100% if the welding was done in the right way but the welding seam or even the form and the position of the welding beads can deliver indicators for a good or poor quality welding seam.

References

1. NASA (1969)
2. Blom et al (2003) SPIE 2003
3. Hermann D, Herzog P (2013) Laser welding of copper, 01 Jan 2013
4. Universität Stuttgart, Institut für Strahlwerkzeuge (IFSW), KickOff-Treffen DVS HighSpeed, 17 July 2019
5. Ay, Celik (1999) Comparison of properties of friction and diffusion welded joints made between the pure aluminium and copper bars
6. Petzold, Bergmann, Schürer, Schneider (2013) Einfluss intermetallischer Phasen auf die Langzeitstabilität von ultraschallgeschweißten Kupfer-Aluminium-Kontakten; Metall, Nov 2013. LNCS Homepage. http://www.springer.com/lncs. Accessed 21 Nov 2016
7. Bachhofer A (2000) Schneiden und Schweißen von Aluminiumwerkstoffen mit Festkörperlasern für den Karosseriebau, München

ELIKA Automotive

Danilo Persici and Michelangelo Musiani(✉)

Marzocchi Pompe S.P.a, Casalecchio Di Reno (BO) 40033, Italy
marketing@marzocchipompe.com

Abstract. Marzocchi Pompe was established in 1961 by Guglielmo and Stefano Marzocchi, in Casalecchio di Reno, in the outskirts of Bologna. Today the Company is the head of an industrial group that employs more than 400 people. This group, owned and led by Paolo Marzocchi, is specialized in designing and producting of hydraulic external gear pumps and motors. As of 16th July 2019 Marzocchi, is listed on the Milan Stock Exchange (: MARP). Over the years, Marzocchi Pompe has expanded the Company and increased its product range to reach its present position as one of the most important Italian players in the hydraulic pump market. Marzocchi Pompe is the manufacturer with the broadest range of displacement reaching as low as 0,12 cc/rev and up to 200 cc/rev. The product includes micropumps (families 0,25, 0,5, 1P; displacements range 0,12–8 cc/rev), pumps and motors (families 1, 2, 3 4; displacement range up to 200 cc/rev) that could be also supplied in multiple pumps, with integrated valve, with special codesigned interfaces tailored on Customer needs, in low noise solutions. The families 0,25–0,5–1P–1–2–3–4 identifies different, increasing, specific displacement (in simple terms, "the size of the gear"). Among this wide variety of products, there are 2 main areas where Marzocchi is targeting, pulled by the market: special micro-pumps conceived for automotive applications and low-noise pumps for all the markets where noise reduction is a requirement (automotive market included).

Keywords: #silent · #HighlyEfficient · #GearPump

1 Introduction

1.1 Marzocchi Pompe Background in the Automotive Sector

Marzocchi Pompe has been, for the past 50 years, a leading supplier of External Gear Pumps in the Industrial and Off-Highway Mobile applications. Perhaps not everybody is aware that it is also a key player in the on-highway Automotive Sector.

Gear pumps are volumetric machines widely used in hydraulic system design since a long time mainly because of their unbeatable cost/efficiencies ratio and for their simple construction and compactness (Figs. 1, 2, and 3).

In the mid-nineties Marzocchi micropumps were chosen by TIER1 manufacturers for several new high-volume applications used then in cars belonging to different range and produced by different car makers. In the 1995, in fact, Marzocchi started developing and refine micropumps to be used in these projects for automotive market.

Euroforum Deutschland GmbH (Ed.): *CTI SYMPOSIUM 2019*, pp. 373–397, 2021.
https://doi.org/10.1007/978-3-662-61515-7_34

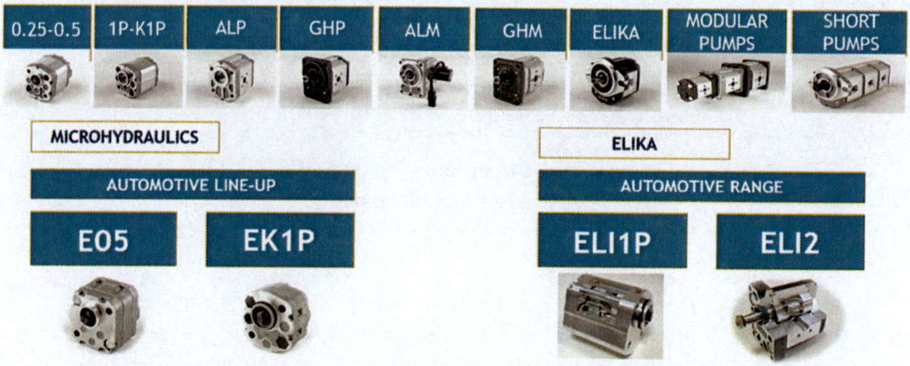

Fig. 1. Marzocchi Pompe's wide range od Gear Pumps and Motors

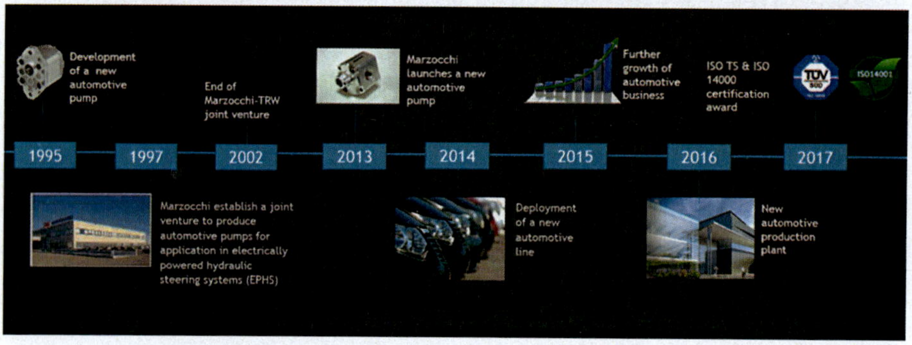

Fig. 2. Marzocchi Pompe's evolution in the automotive sector

Motor:	160 W 12 V DC 20 A permanent magnets
Pump:	Marzocchi U0.25D24KA
Displacement:	0.260 cm³/rev
Minimum speed:	2500 rpm
Maximum speed:	4000 rpm
Max pressure:	P = 70 bar
Weight:	1,20 Kg

Fig. 3. Small motor-pump with high-speed DC motor

Because of the high volumes, a new Company, called M.A.P. (Marzocchi Automotive Pumps), was established in 1997 and a new manufacturing facility located in Ostellato—Ferrara, Italy—was built. After a few years, one of the Customers, the US based TRW, entered as a partner into a dedicated joint-venture called T.M.A.P. (TRW Marzocchi Automotive Pumps). At the end of 2002, TRW took over the complete ownership of this Company and in order to go on with the production of the same size of pumps in quantity over four millions units. Since then, Marzocchi Pompe started re-incorporating its individual capacity to supply the automotive market within its own original Company and Location in Bologna, Italy and several automotive Customers are now using our products manufactured there (Figs. 4 and 5).

Motor (not shown):	**1000 W 12 V DC 90 A**
Pump:	**Marzocchi 05–D-2.00**
Displacement:	**1.50 cm^3/rev**
Minimum speed:	**1500 rpm**
Maximum speed:	**3000 rpm**
Max pressure:	**P = 100 bar**
Weight:	**2,00 Kg**

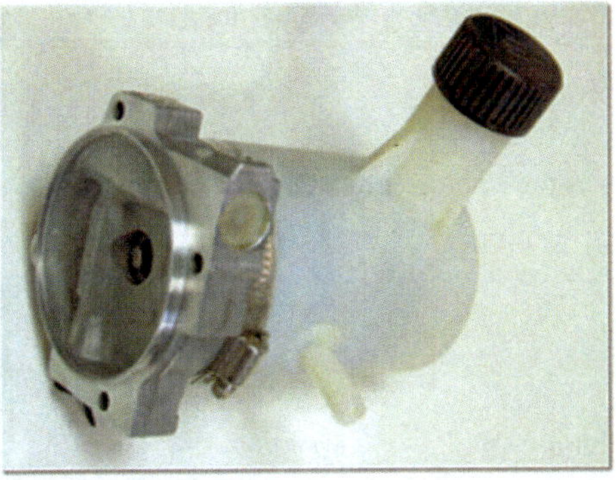

Fig. 4. Hydraulic power-unit

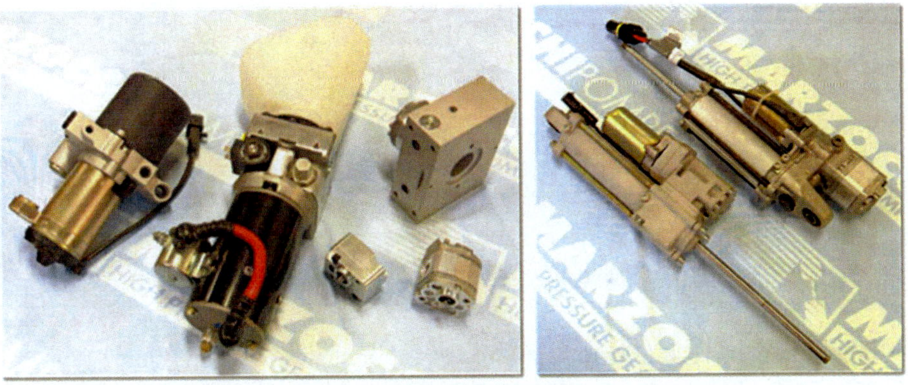

Fig. 5. Complete power-units and electro-lift systems

Marzocchi Pompe is available to supply the automotive market not only with pumps but also with integrated solutions, completed with motor and manifold with valves, actuator systems, complete power units with reservoir. On one side, a range of products with very high performances while on the other side, an enviable product know-how allow our engineers to develop a smaller size, reliable and affordable in cost new products as well as micro systems integrated into the gear pump. The culture of micro hydraulic is within Marzocchi DNA (Figs. 6 and 7).

Pump:	**Marzocchi UK 0,5**
Displacement:	**0.50 – 1.50 cm³/rev**
Minimum speed:	**1500 rpm**
Maximum speed:	**7000 rpm**
Max pressure:	**P1 = 30 bar**
Weight:	**0,26 Kg**

Fig. 6. Fuel Supply—racing injection systems

Pump:	**Marzocchi 0,5–D–0.75**
Displacement:	**0.650 cm³/rev**
Minimum speed:	**1500 rpm**
Maximum speed:	**7000 rpm**
Max pressure:	**P1 = 30 bar**
Weight:	**0,49 Kg**

Fig. 7. Racing gearshift—pump with titanium flange

Applications more and more sophisticated and brought to the limit allowed us to further enhance our experience in special micropumps with high performances. We can find them in different applications such as:

- F1
- Rally (World Champion)
- Grand Raid
- DTM
- IMSA.

1.2 The New E05 Automotive Gear Pump

As mentioned before, Marzocchi Pompe is the manufacturer with the broadest range of displacements, with significant strength in small pumps with displacements as low as 0,12 cc/rev which are widely appreciated in all those applications where a mini power pack is required. The top characteristics of quality, reliability of the Marzocchi products allowed the Company to gain an interesting share in the automotive market.

Following the experience of previously developed EK1P ("E" version of 1P family), Marzocchi decided to develop the automotive version of 0,5 pump, called "E05". The project was launched after the need of the market and after an explicit request of a specific automotive Customer.

The automotive applications generally require medium-high volumes and are strongly interested in high quality and high efficiency, low cost, low weight, reduced dimensions and low noise.

Starting from the solid base of a 05 pump, Marzocchi absorbed all the automotive targets above transferring them into the new E05 pump.

This makes E05 pump family extremely suitable to be integrated into assemblies of automatic transmissions, semi-automatic clutches, electro-hydraulics power steering, AWD systems, assistance in hybrid-type of propulsion, etc. (Figure 8).

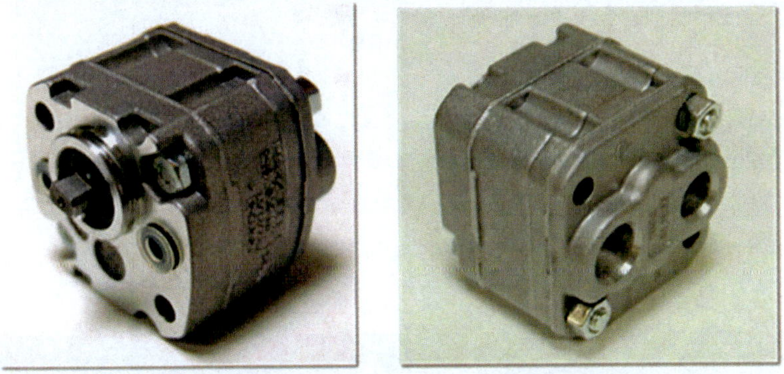

Fig. 8. New E05 automotive micro gear pump

Standard applications of Marzocchi products are easily requiring up to 300 bars while the limited operating pressure of the automotive application, usually not exceeding 140 bar. For this reason design and process engineers could introduce several design and process optimizations with the goals, on one side, to maintain and even enhance very high efficiencies, low noise characteristics, reducing overall sizes as well, and, on the other side, to decrease the manufacturing costs thanks to an improved of automatization of the production and assembly. Despite the small size, E05 pumps are internally bi-compensated; the compensation system always maintains the compensation plates in contact with the gears ensuring, in all operating conditions, a drastic reduction of internal leakage, adequate lubrication of the moving parts and excellent volumetric and mechanical efficiency. The compensation system was designed to reduce the friction between components, increasing the mechanical efficiency of the system. In applications with suitable working conditions, it is possible to use a mono compensated version of the E05 pump, removing 2 internal components, thus further reducing the dimensions and cost.

Now Marzocchi Pompe consolidated the production of pumps for the automotive industry in NEPLA, a new plant built in 2016 in Zola Predosa, Bologna. NEPLA is a company established from the ground up in compliance with the standards IATF 16949 and ISO 14001. NEPLA is divided into two divisions: one manufactures all the gears for the entire Marzocchi pump and motor range; the other one assembles and tests the pumps for automotive applications, on specifically designed lines. In NEPLA, the high level of automation, precision and repeatability of manufacturing and assembly allows automotive product to reach very high and constant level of performances.

The NEPLA production, following the IATF guidelines, grants the full traceability of the components.

E05 pump technical features were shown in detail by Marzocchi in recent CTI Symposiums held in Novi and Shanghai in 2019 (Figs. 9 and 10, Table 1).

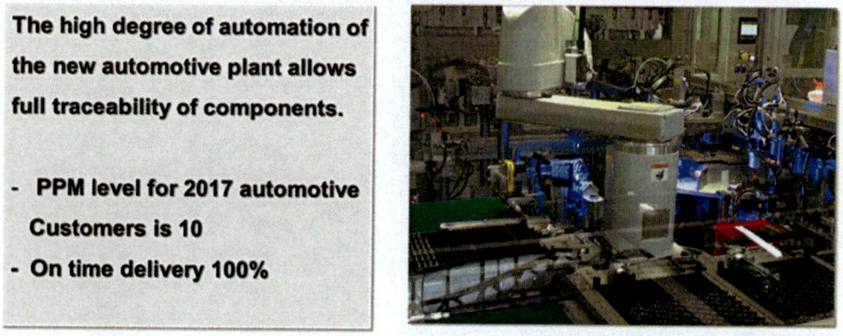

Fig. 9. Marzocchi automotive plant

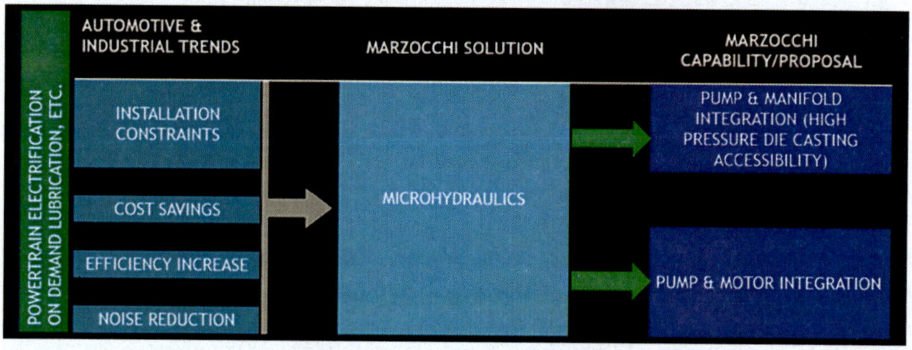

Fig. 10. Marzocchi system

Table 1. Automotive gear pump production

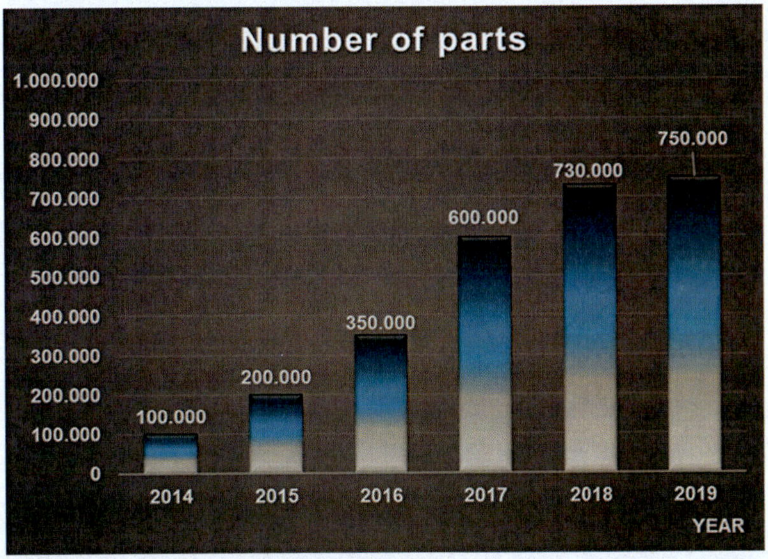

2 The New ELIKA® Gear Pump

2.1 Introduction

In the 2012 Marzocchi Pompe commercialized the first samples of ELIKA pump (ELI2 family), his answer to the problem of gear pumps noise, anticipating the increasing demand of the market for low noise solutions.

After few years and thousands of pumps delivered, Marzocchi recognized that customers definitely appreciated the characteristics of the new product, in terms of reliability, efficiencies and low noisiness. Following to this success and the continuous demands of the market, Marzocchi Pompe continued to expand the ELIKA family developing new ELIKA families widening the range of displacements and versions.

Starting from the 2012 when only ELI2 (displacement range from 7 to 35 cm3/rev) was available, Marzocchi expanded the low noise solutions adding ELI3 first (range from 20 to 87 cm3/rev), and finally ELI4 (range from 87 to 200 cm3/rev).

The edge of ELIKA pump performances and noise versus the standard straight gear pumps increases with the displacement. In particular, the noise emission tends to increase with the displacement; for this reason, the higher is the displacement, the bigger is the noise reduction allowed by ELIKA. This led Marzocchi to expand ELIKA family towards bigger families first.

Marzocchi extended the ELIKA family developing a solution granting the possibility of having multiple EILKA pump. The connection system between pump stages is extremely compact and it makes the pumps interchangeable with the normal multiple pumps, both in terms of dimensions than as torque applicable to the stages. The connection system used in ELIKA multiple pumps not only transmits the torque to later stages, but also ensures the correct axial balancing at the helical gears.

This solution is patented and it was awarded the Technical Mention Prize at EIMA 2018.

Following the success of ELIKA and the increasing demand of automotive market for low noise solutions (due to "silent" electric and hybrid powertrains), Marzocchi decided to extend the ELIKA pump to smaller family, approaching the displacement range usually more common for automotive application.

Therefore, Marzocchi developed the ELI1P family (range from 2.1 to 8,1 cm3/rev). The A-Sample (full material protos) phase has already been successfully completed. By the end of 2019, the first diecasted prototypes will be delivered to Customers. The passage to the serial production is planned for the first half of 2020 (Table 2).

Table 2. Elika's available displacements

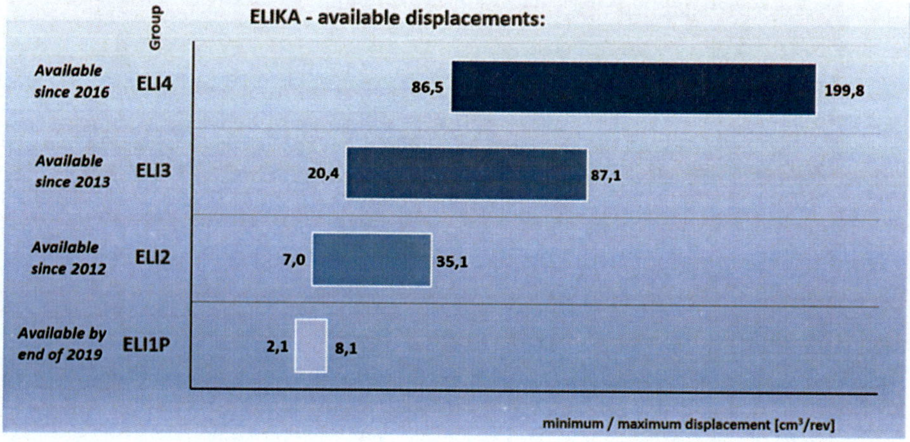

2.2 Elika Pump Development

In the past years, Marzocchi Pompe analysed the noise emission matter in order to optimize the behaviour of its product and, in parallel, to develop a new product line with different technical solution greatly improving the noise emission. For this reason, Marzocchi studied and prototyped different solutions in order to reduce noise emissions submitting them to comparative measurements.

The starting point of Marzocchi activity regarded the analysis of existing solution: the traditional straight external gear pump, in particular the main source of noise and pressure pulsation.

The origin of noise and pulsation stays in the meshing of the gears in the central axis of the pump.

During the meshing, some oil is trapped between the teeth; proceeding with the rotation, the volume where the oil is trapped decreases causing localized peaks of pressure with negative drawback on the noise emission, pressure pulsation and efficiencies as well (Fig. 11).

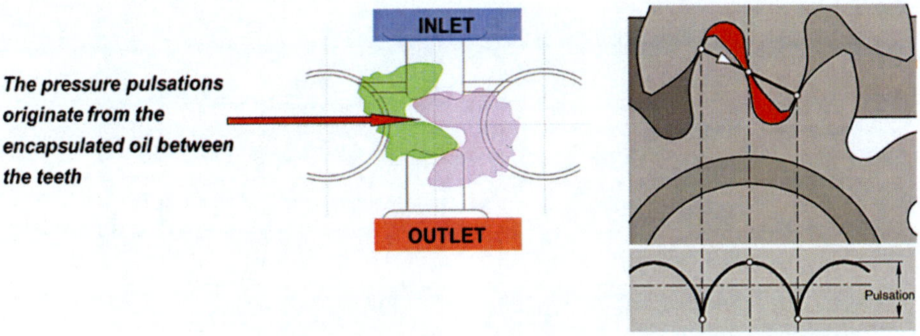

Fig. 11. Elika's teeth profile

In the traditional straight gear pump, this effect is contained by 2 machining on the bushing, limiting the volume of oil encapsulated and therefore the consequent negative effects of the compression (Fig. 12).

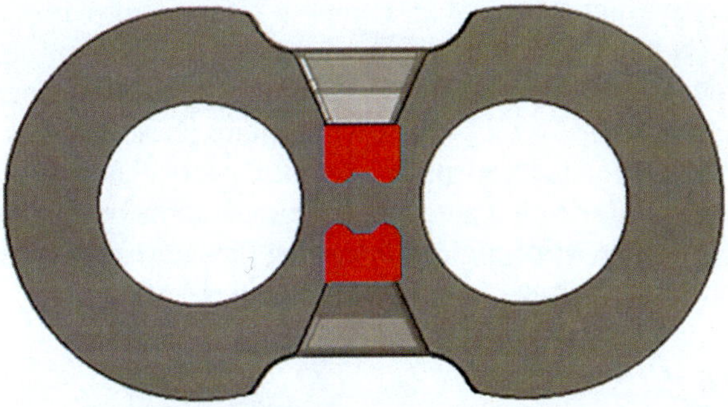

Fig. 12. Bushing profile

Despite the optimized discharge profile, due to intrinsic characteristics of straight gear pump, the resultant behaviour in terms of noise and pulsation could not fulfil the requirements of some high demanding application.

Therefore, Marzocchi kept studying, developing and testing alternative technologies to reduce the noise level of gear pumps, for example double-contact gears, helical gears with not conjugated profile, and much more, without finding an effective and satisfactory solution, until the ELIKA design.

Heart of the product is the particular shape of the ELIKA conjugated profile that eliminates the phenomenon of encapsulation, typical of normal gear pumps, deleting the main source of noise and pulsation.

The shape of conjugated teeth profile requires an helicoidal gear to properly transmit the torque to the driven gear, guaranteeing the contact in the suitable portion of the profile (Fig. 13).

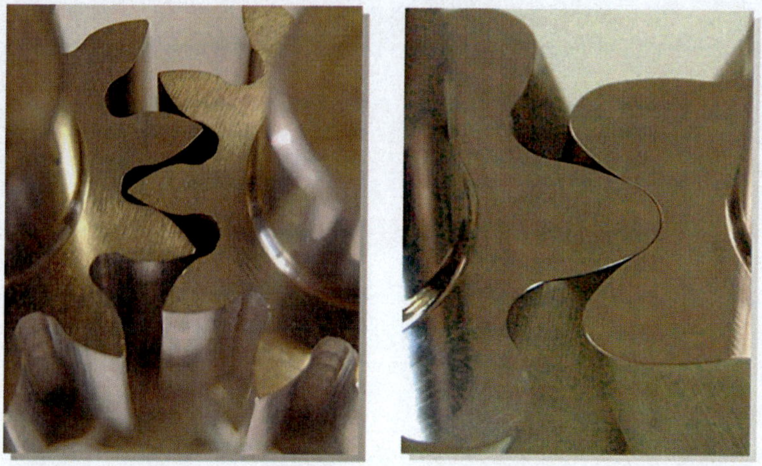

Fig. 13. Traditional tooth spline (right) vs ELIKA Tooth profile (left)

The total elimination of the encapsulated volume produces a significant reduction in internal pressure oscillation allowing a smooth functioning with positive effect on the noise produced by the pump; the lower pressure pulsation has a subsequent positive effect on noise reducing the vibration transmitted to the other components connected to the pump (hoses, tank and valves). During the meshing phase, there is continuous, not interrupted, contact between the teeth profiles of the 2 gears; this not only guarantees the smooth functioning, but above all it guarantees the separation line between high and low pressure area, whatever the helix angle, minimizing the internal leakage and improving volumetric efficiencies, in particular at low speed (Fig. 14).

Fig. 14. Different meshing conditions

The definition of the special Marzocchi tooth profile was done in cooperation with the Faculty of Engineering of Bologna University, thanks to the creation of a dedicated software for assisted design (Fig. 15).

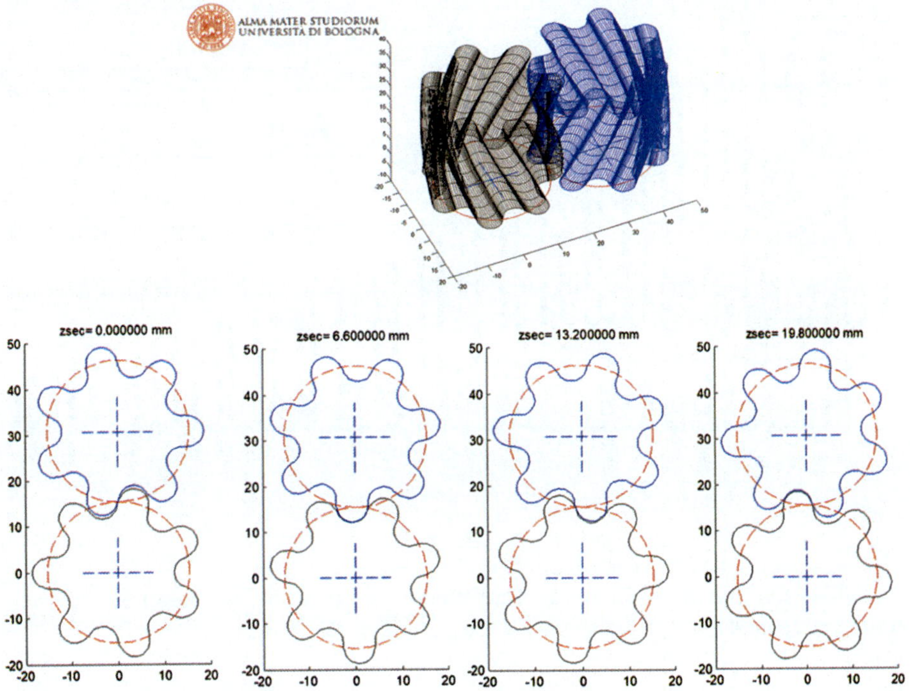

Fig. 15. The output of the dedicated software created together with Alma Mater Studiorum University of Bologna, Engineering Faculty

The developed software for the virtual modelling of the profile has 3 main functions:

- to model and visualize the splines of two gears during meshing;
- to define the optimal sequence of points belonging to the contour of the profile for the machining (grinding) of the profile;
- to show a continuous simulation of gear set by modelling and visualizing the surfaces of the gears in meshing conditions.

The simulation and following experiments led to the definition of a specific tooth profile able to produce the best performances and acoustic characteristics.

The helical gears ensure the continuity of motion despite the low number of teeth. The low number of teeth greatly reduces the fundamental frequencies of the pump noise making the sound particularly pleasant (Table 3).

The picture above shows the comparison of the fundamental frequencies between external gear pump, dual flank gear pump, ELIKA pump.

Table 3. Noise frequency distribution

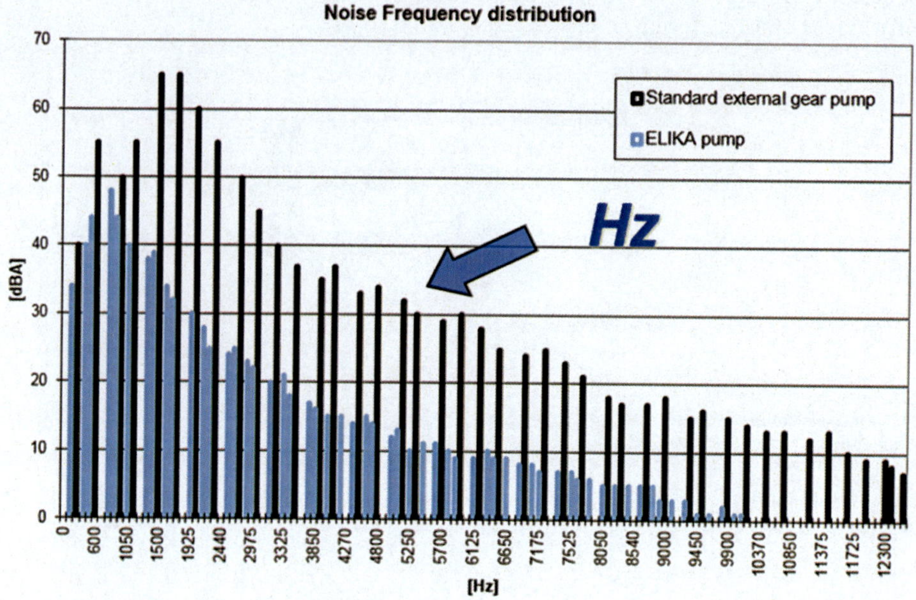

ELIKA solution allowed to minimize both the pressure oscillations and their frequency (Table 4).

Table 4. Pressure ripple [bar]

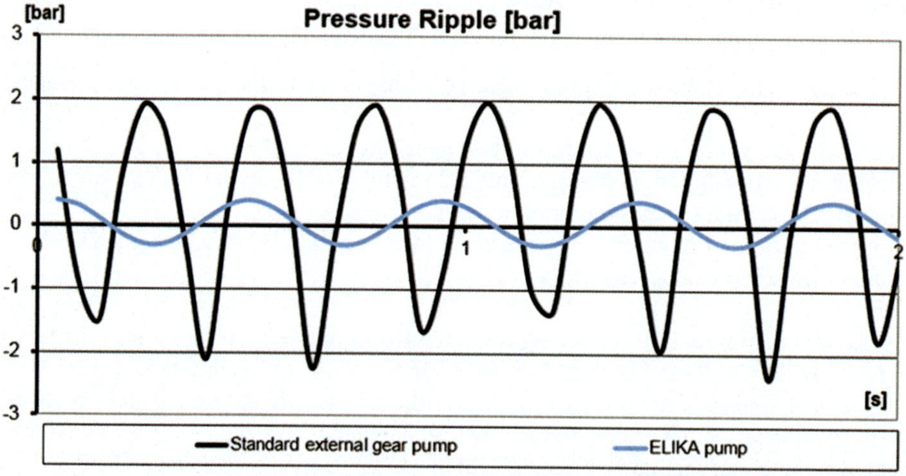

2.3 ELIKA Axial Compensation

The helicoidal gears have the drawback to create an axial force, proportional to the pressure, pushing the gears towards the back of the pump. If not properly balanced, this force would prematurely wear the pump.

The axial force depends on the pressure pushing on the side of the gears and on the torque, partially transformed by helix into an axial force (Fig. 16).

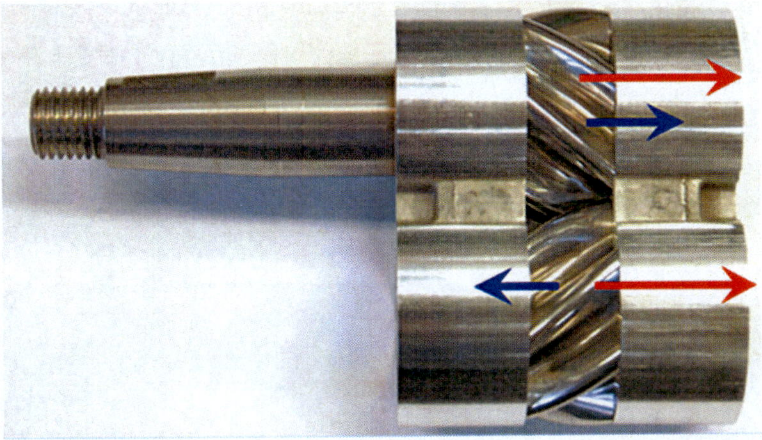

Fig. 16. Elika helical rotors

Component of transmission (blue colour) that are counteracting while component of pressure (red colour) that are concordant.

The result is 2 axial forces, proportional to the pressure, both pushing the gear towards the back of the pump; the driving gear force is higher than the driven gear one.

The solution to balance these forces is 2 pistons pushing on the stems of the gears.

These 2 pistons are fed by the working pressure and have areas proportional to the different forces on the gears. This allow a perfect hydraulic balance of the axial forces (Fig. 17).

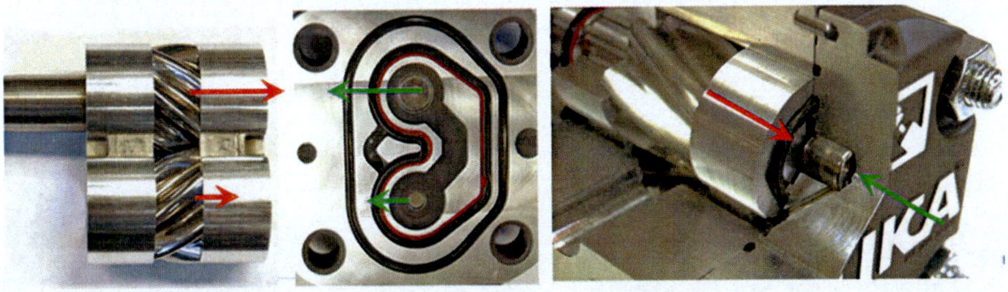

Fig. 17. The perfect hydraulic balance of the axial forces

If the application requires the pump to run only at very low pressure, the need of this axial compensation disappears. Marzocchi developed an ELIKA version of the pump devoted to this kind of applications with helical gears, granting noise benefits, but with no axial piston compensation system. This pump line is called FTP (Fluid Transportation Pump)

2.4 ELIKA Pumps Achievements

The ELIKA solution grants a significant improvement if compared with traditional straight gear external pumps in terms of volumetric efficiency (in particular at low speed), noise and pressure pulsation (Tables 5, 6, 7, 8, 9, 10, 11, 12, and 13).

Table 5. ELI2/GHP2 volumetric efficiency comparison

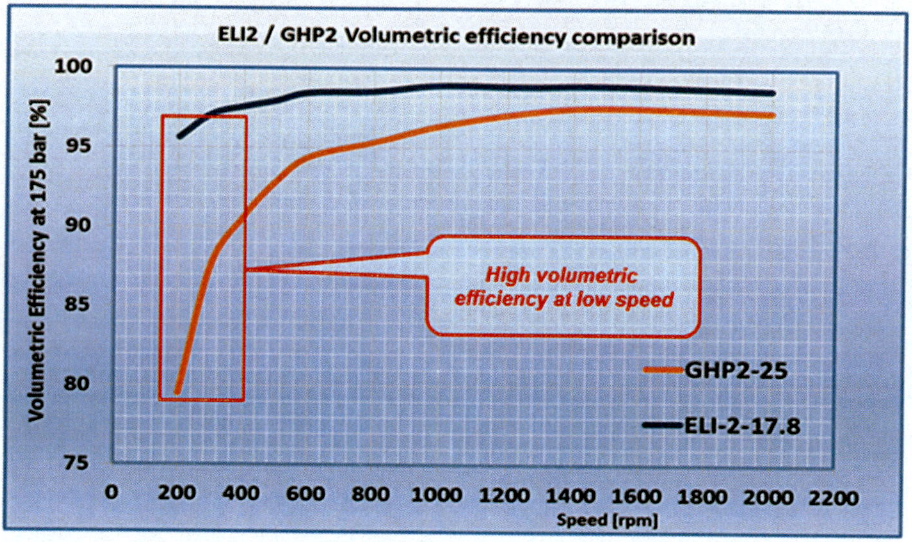

Table 6. ELI2/GHP2 average noise level comparison

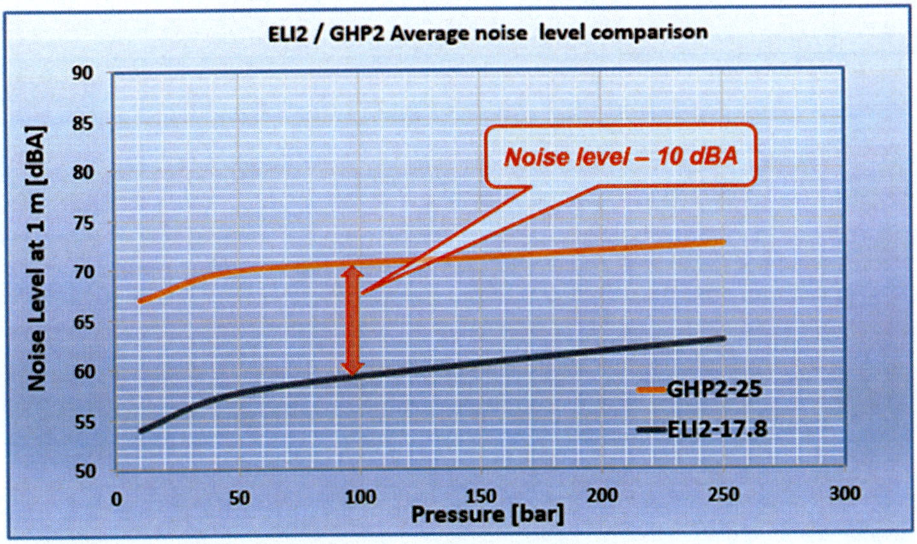

Table 7. ELI2/GHP2 pressure ripple comparison at 1500 rpm

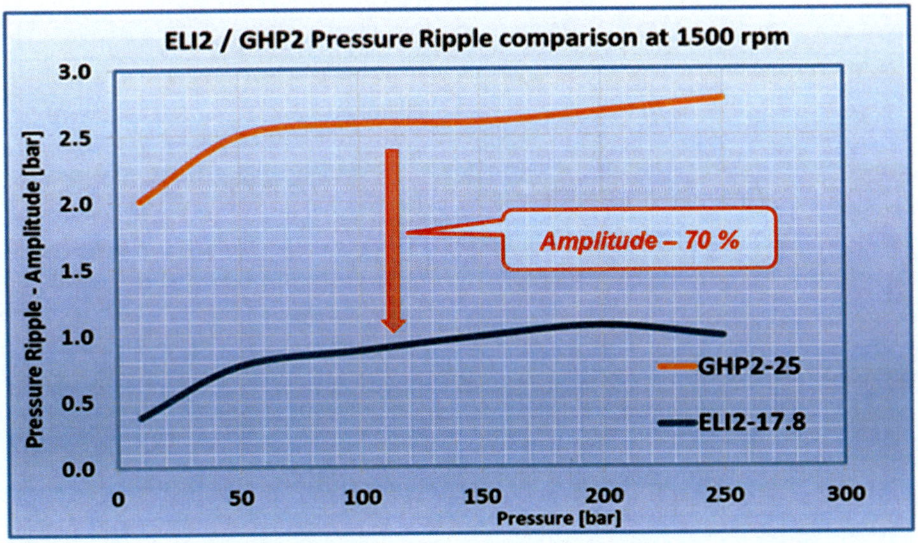

Table 8. ELI3/GHP3 volumetric efficiency comparison

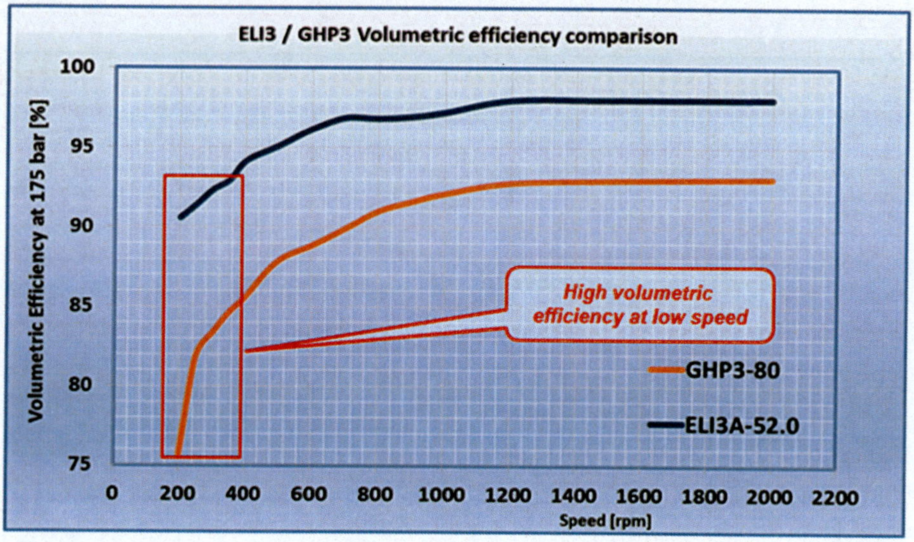

Table 9. ELI3/GHP3 average noise level comparison

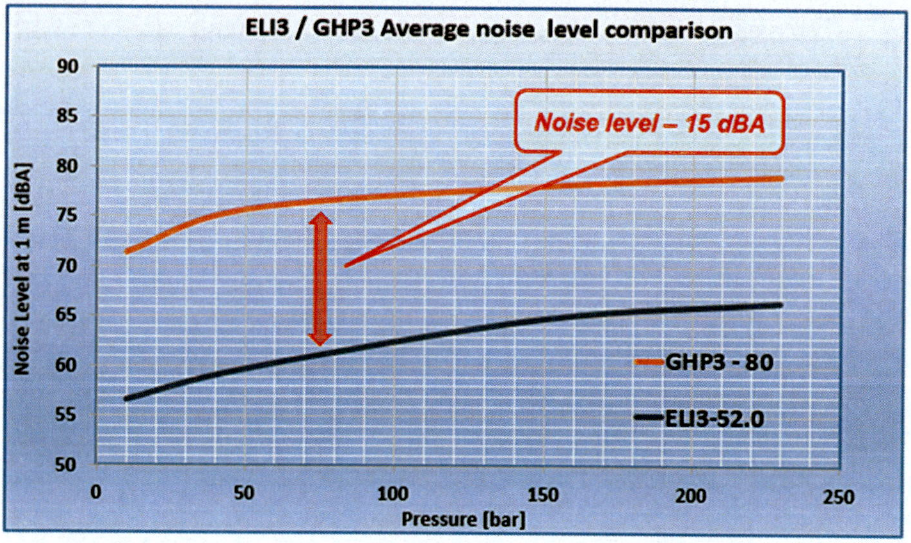

Table 10. ELI3/GHP3 pressure ripple comparison at 1500 rpm

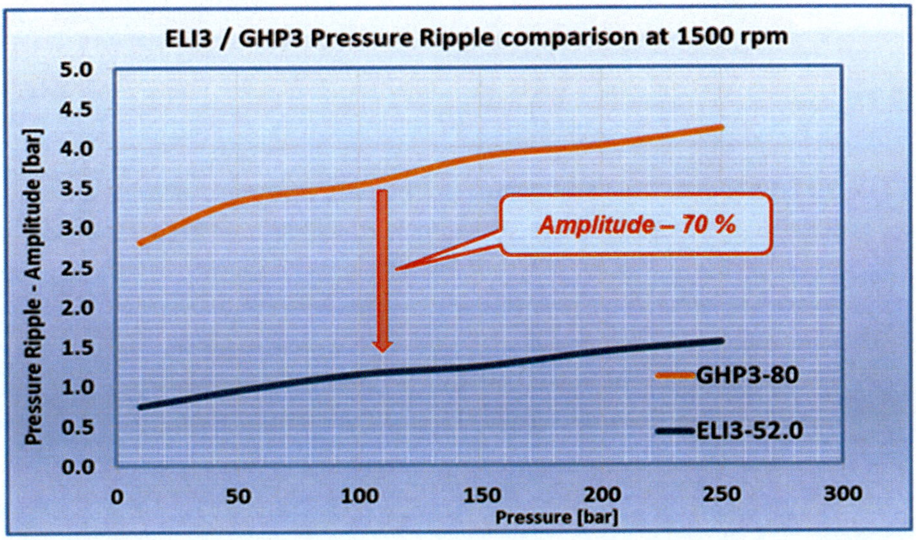

Table 11. ELI4/ALP4 volumetric efficiency comparison

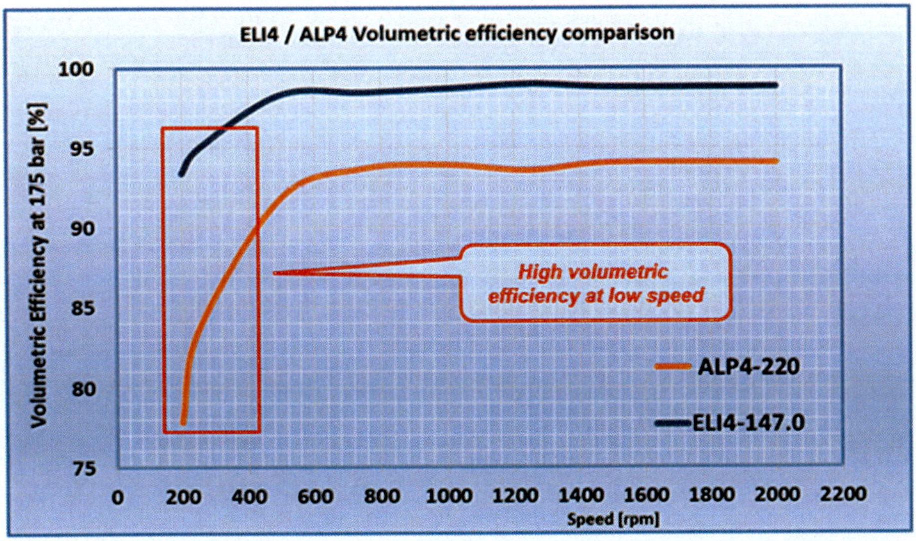

Table 12. ELI4/ALP4 Average noise level comparison

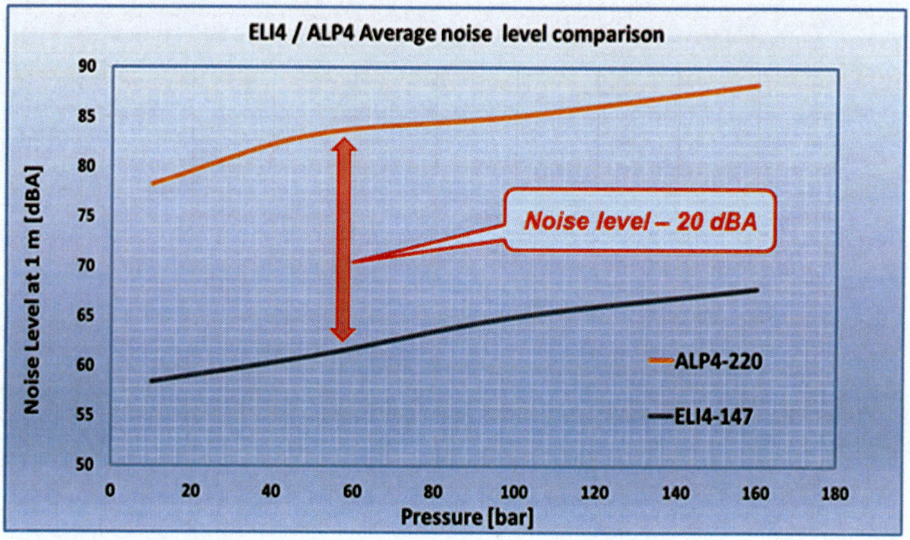

Table 13. ELI4/ALP4 pressure ripple comparison at 1500 rpm

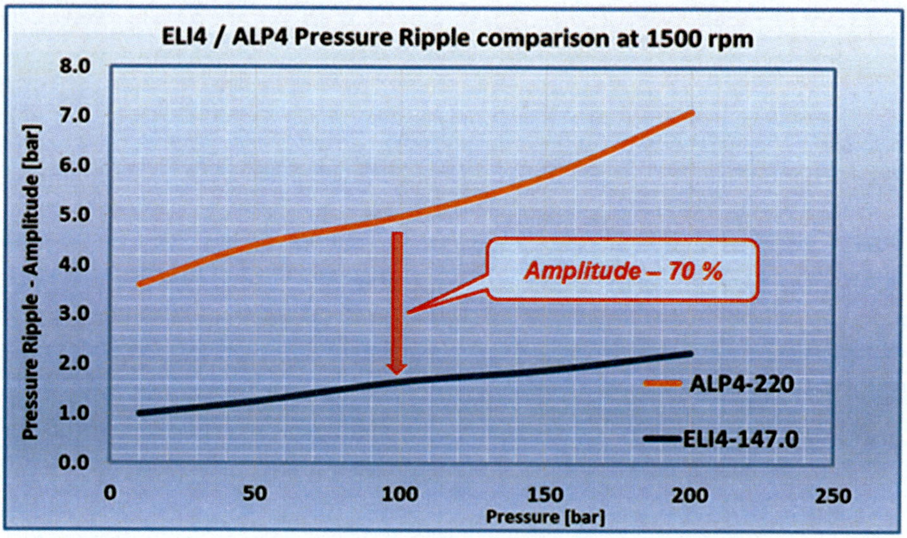

2.5 ELI1P Development

The pictures above confirms how the noise and the resultant noise save thanks to ELIKA solution increase with the size of the pump.

In the recent years the development of fully electric and hybrid vehicles is constantly increasing. In these vehicles, the use of endothermic engines disappears or it is reduced to a minimum, erasing the main source of noise; therefore the noise of the hydraulic circuits becomes predominant leading to the need of silent pumps.

The very low noise level generated by ELIKA pumps makes this product particularly suitable for those applications where, screw pumps, vane pumps or internal gear pumps are generally used. As an additional advantage compared to other solutions, ELIKA allows to obtain superior volumetric and mechanical performances, typical of external gear pumps.

Marzocchi then developed the ELIKA solution for the family 1P, going to embrace a displacement range usually more often used in the automotive applications (Figs. 18 and 19, Table 14).

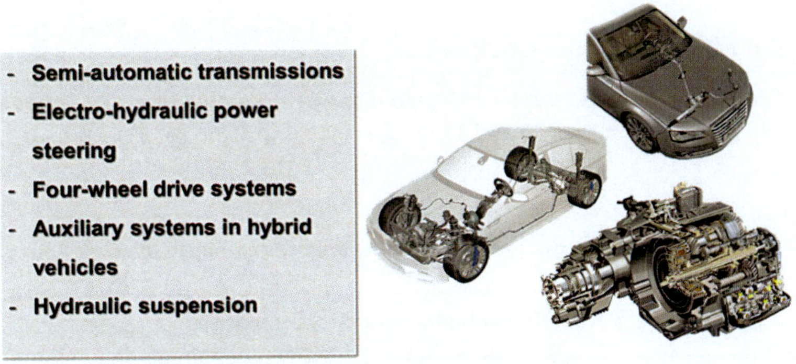

Fig. 18. Automotive applications

Fig. 19. ELI1 and ELI1P pumps

Table 14. Here below the ELI1P family with the envisaged displacement.

Pump Type	Displ. [cm^3/rev]	Flow at 1500 [l/min]	Operating pressures		
			P1 max cont. [bar]	P2 max interm. [bar]	P3 max peak [bar]
ELI1P-2,1	2,1	3,2	230	245	260
ELI1P-2,7	2,7	4,0	230	245	260
ELI1P-3,2	3,2	4,8	210	225	240
ELI1P-3,7	3,7	5,6	210	225	240
ELI1P-4,0	4,0	6,0	210	225	240
ELI1P-4,3	4,3	6,4	210	225	240
ELI1P-4,8	4,8	7,2	190	205	220
ELI1P-5,9	5,9	8,8	190	205	220
ELI1P-8,1	8,1	12,2	160	175	190

2.6 ELI1P Efficiency, Noise Comparison

The table and the following graphs compare the measurements of a straight gear pump 1P-6,7 P185 (4.2 cm3/rev), and an ELIKA ELI1P-D-4 (4.0 cm3/rev) (Tables 15, 16 and 17).

Table 15. ELI1P/1P volumetric efficiency comparison

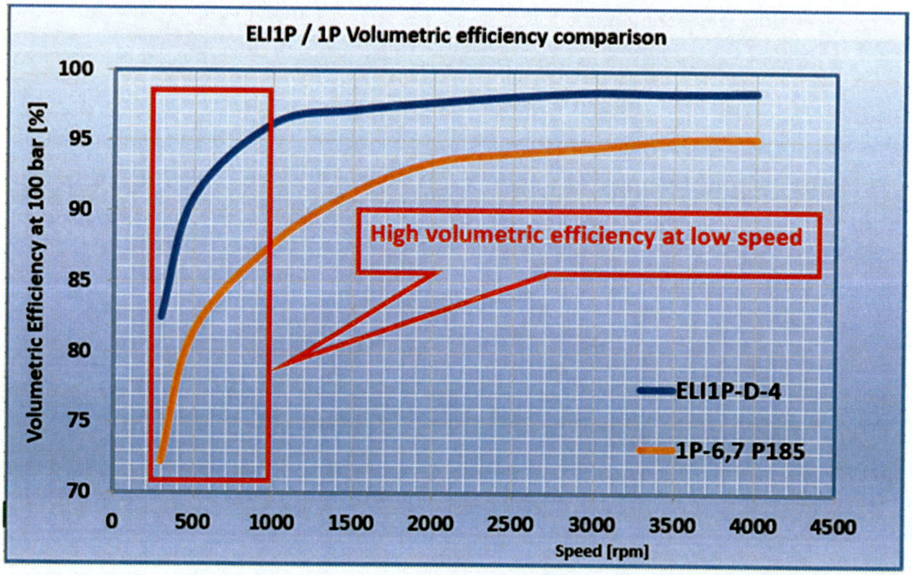

Table 16. ELI1P/1P average noise level comparison

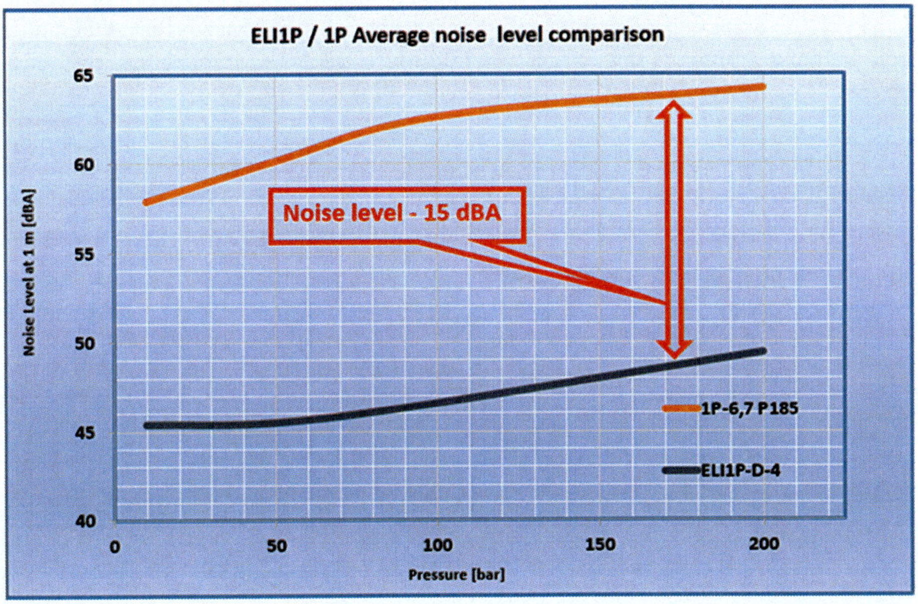

Table 17. ELI1P/1P pressure ripple comparison at 1500 rpm

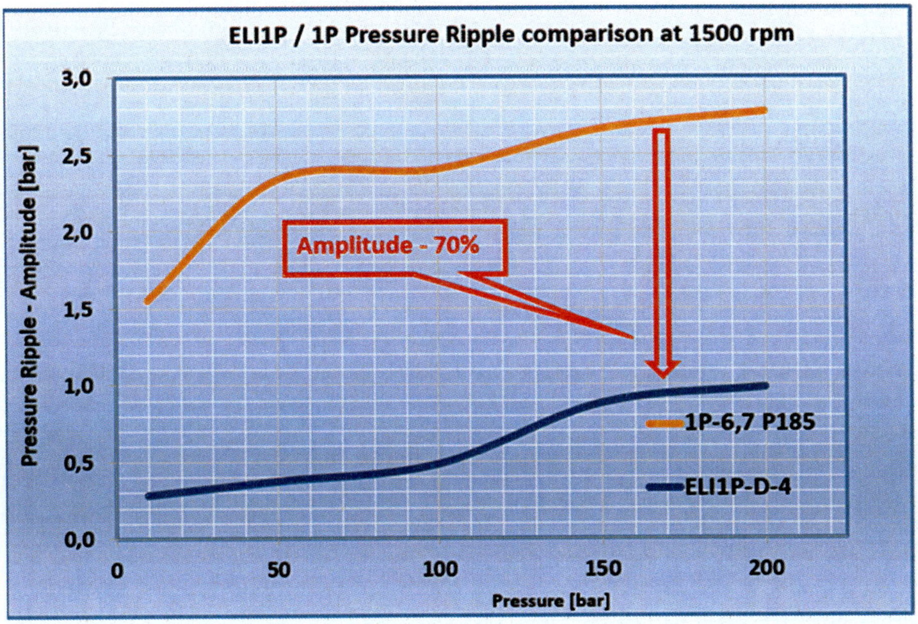

In the frequency analysis it is necessary to bear in mind that the different number of teeth: (Z = 7) for ELIKA, (Z = 9) for 1P. The first, with the same rotation speed, will have lower noise and pulsation frequency. This phenomenon means that at the same speed the noise emitted is not only lower, but it also has lower frequencies, resulting therefore less annoying (Tables 18, 19, 20, 21 and 22).

Table 18. ELIKA ELI1P vs 1P data

Pump ELIKA ELI1P-D-4				Pump MARZOCCHI 1P-6,7 P185			
Frequency	2000 rpm 100 bar	3000 rpm 100 bar	4000 rpm 100 bar	Frequency	1P 2000 rpm 100 bar	1P 3000 rpm 100 bar	1P 4000 rpm 100 bar
Hz	dB(A)			Hz	dB(A)		
1/3 Octave				1/3 Octave			
160	29,4	28,6	24,5	160	36,3	27,7	29,6
200	33,0	33,2	36,0	200	37,6	43,1	46,4
250	30,5	34,4	41,6	250	42,6	48,7	48,1
315	33,5	33,8	31,8	315	53,3	35,1	42,4
400	36,9	33,3	42,2	400	39,2	43,4	46,8
500	34,7	37,2	40,2	500	36,7	43,7	41,9
630	30,0	36,4	33,6	630	42,7	40,3	52,2
800	29,9	29,1	39,6	800	50,1	52,5	49,8
1000	34,0	32,6	45,9	1000	49,2	56,8	41,4
1250	30,3	38,0	32,1	1250	35,0	57,3	46,2
1600	27,7	35,0	31,6	1600	46,4	42,0	43,0
2000	30,7	30,0	33,5	2000	39,2	47,6	48,8
2500	33,3	41,6	43,9	2500	38,8	46,0	53,4
3150	32,9	33,0	31,8	3150	52,9	47,8	60,7
4000	33,9	34,8	31,7	4000	38,8	41,6	55,2
5000	33,6	32,6	37,1	5000	40,3	45,1	52,7
6300	32,4	33,9	36,3	6300	40,7	43,7	56,1
8000	30,4	37,5	51,9	8000	40,6	51,1	55,9
10000	27,0	32,4	34,5	10000	40,2	49,6	60,1

Table 19. 2000 rpm/100 bar—standard pump/ELIKA pump comparison—sound power

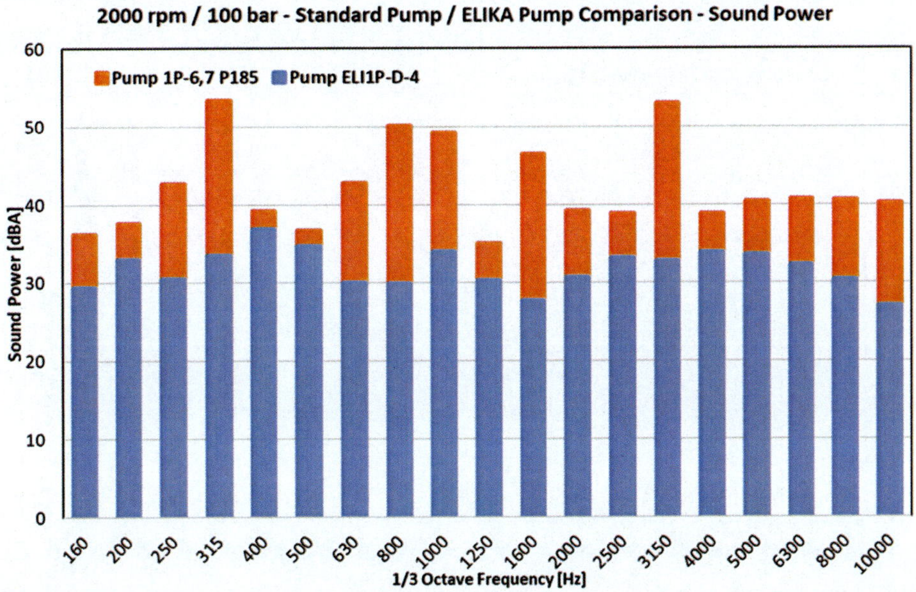

Table 20. 3000 rpm/100 bar—standard pump/ELIKA pump comparison—sound power

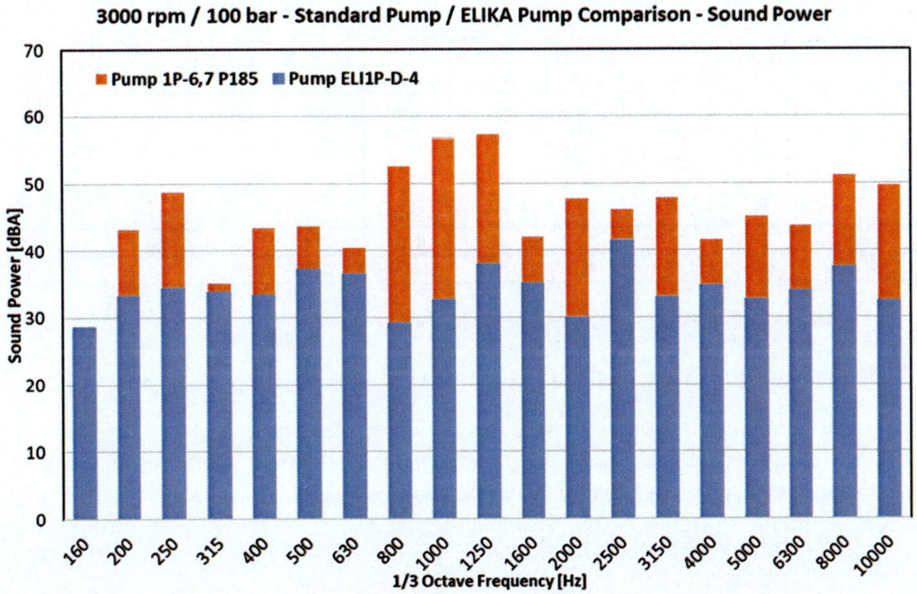

Table 21. 4000 rpm/100 bar—standard pump/ELIKA pump comparison—sound power

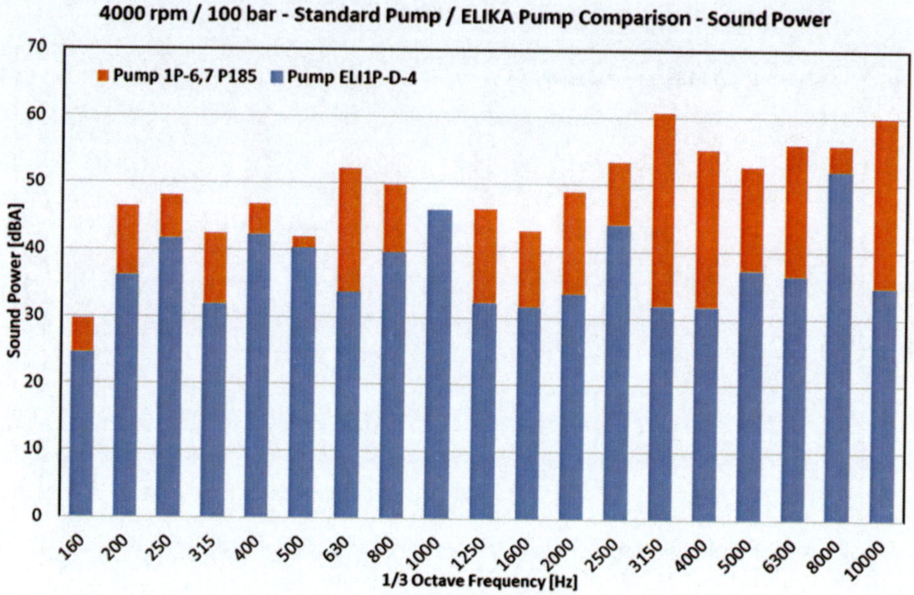

Table 22. Sound pressure at 1 m [dBA]

Sound Pressure at 1 m [dBA]									
Pump ELIKA ELI1P-D-4					Pump MARZOCCHI 1P-6,7 P185				
rpm / bar	10	50	100	200	rpm / bar	10	50	100	200
1500	40,7	41,2	42,3	43,2	1500	54,0	56,1	58,2	60,5
2000	42,5	41,7	42,5	46,3	2000	56,0	58,2	60,6	63,1
2500	43,2	44,3	44,2	47,8	2500	57,5	59,7	63,1	63,6
3000	45,4	47,0	45,9	49,2	3000	59,0	61,2	65,6	64,1
3500	48,1	47,8	48,7	52,1	3500	60,1	62,8	64,1	68,1
4000	52,5	51,0	55,7	57,7	4000	60,8	62,8	64,1	66,0
Average	45,4	45,5	46,6	49,4	Average	57,9	60,1	62,6	64,2

2.7 Pressure Ripple Comparison

Due to the elimination of the volume trapped between the sprockets, the amplitude of the pulsations is considerably reduced, up to 70%. As with noise, similar evaluations of frequencies can also be made for pressure pulsations (Table 23).

Table 23. Pressure ripple

Pressure ripple a 1500 rpm - Amplitude [bar]					
1P-6,7 P185	10	50	100	150	200
	1,55	2,33	2,40	2,66	2,77
ELI1P-D-4	10	50	100	150	200
	0,28	0,37	0,48	0,88	0,98

2.8 ELIKA1P in Automotive Production NEPLA

As explained, ELI1P pump was developed to supply the automotive market with a small displacement pump with superior performances in terms of efficiency, noise and pulsation.

Being aware of the automotive requirement, Marzocchi planned to produce the ELI1P pump for automotive Customer in the NEPLA facility, IATF certified, in order to fulfil all the necessary quality guidelines.

Park by Wire System for Current Electric Drive Units

Jan Nowack(✉), Gereon Hellenbroich, Arnab Ghosh, Valerij Shapovalov, and Ralph Fleuren

FEV Europe GmbH, Aachen, Germany
Nowack_j@fev.com

Abstract. Park lock systems are safety critical units. Due to intellectual property rights and different transmission configurations, a large variety of park lock design exists in the market. Furthermore, many park lock designs are still manually actuated. A new mechanical park lock system will be presented which is actuated by wire. The system has been optimized for rotary actuators and is extraordinarily compact. In addition, it minimizes the contact stresses due to optimized geometries. Furthermore, the functional safety and diagnostic concept for park-by-wire park lock systems, exemplarily for this system, will be presented. This will be done in accordance to current multi speed power-shift EDU concepts, including FMEA, legal, requirements summary, safety goal definition and error detection mechanism. The following chapters are divided according to the park lock development steps shown in Fig. 1.

Keywords: Functional safety and diagnostic concepts · Mechanical park lock concept · Park by wire concept

1 Generic Requirements Analysis

A requirement analysis has been performed to identify the major system requirements from a Park lock system for a passenger vehicle (Fig. 1). An overview of the identified requirements is shown in Fig. 2. Some sample OEM requirements were assumed, based on prior experiences of FEV. These requirements helped in creating a baseline design of the system, hardware and software. The system design has been, however, kept flexible to meet customer and market specific requirements. Intellectual Property restrictions were also considered while creating the baseline requirements.

2 Item Definition

A typical passenger vehicle has been considered for creating the item definition, see Fig. 3.

This helps identifying the key elements of the vehicle and the Park lock system. The key elements are described below:

Euroforum Deutschland GmbH (Ed.): *CTI SYMPOSIUM 2019*, pp. 398–406, 2021.
https://doi.org/10.1007/978-3-662-61515-7_35

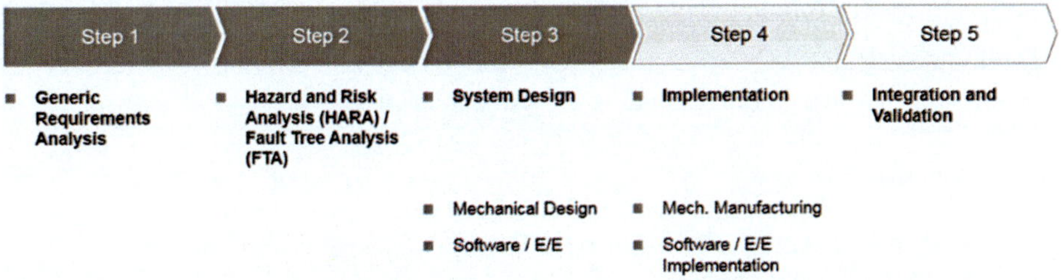

Fig. 1. Park lock system development steps

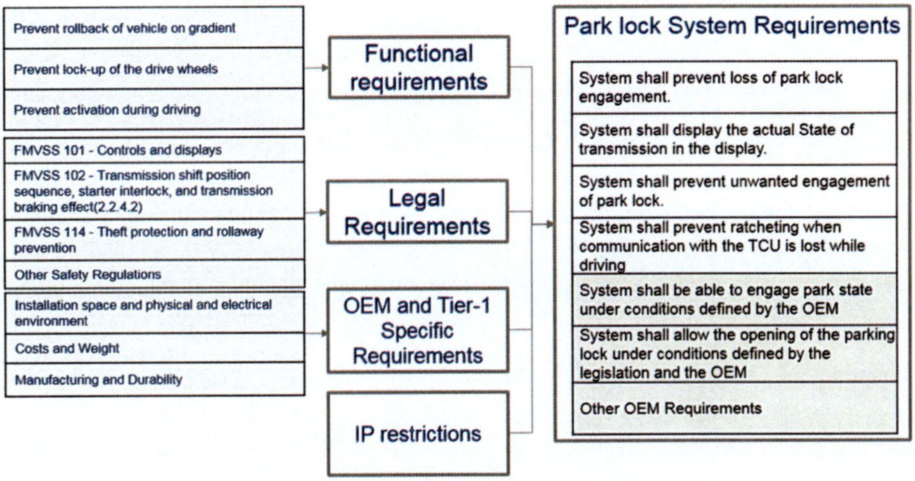

Fig. 2. Park lock requirements overview

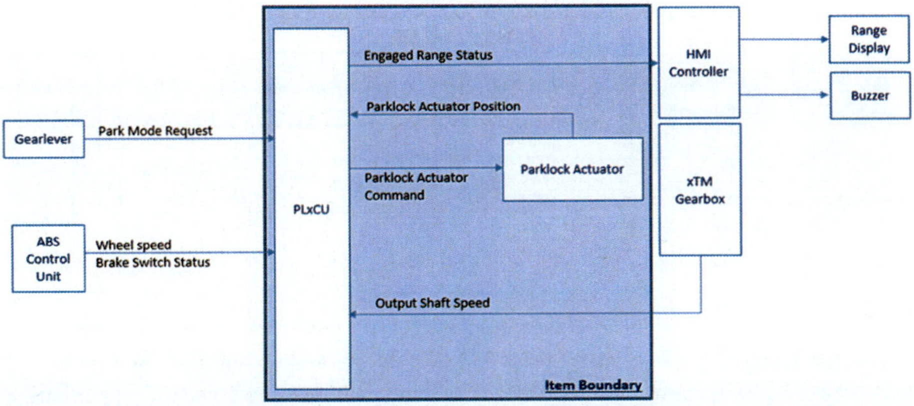

Fig. 3. Item definition for FEV Park lock system

- Gear Lever: Element in vehicle through which driver request for Park state is communicated to PlxCU
- ABS Control Unit: Antilock braking system control unit
- HMI Controller: Human Machine Interface Controller responsible for controlling display to driver
- xTM Gearbox: Transmission (AMT, DCT, EDU etc.)
- Range Display: Display of Actual Engaged range to driver
- Buzzer: Audio warning to driver.

The item definition has acted as a starting point for the control system and functional safety development.

3 Hazard and Risk Analysis (HARA)

Based on the item definition, a preliminary Hazard and Risk analysis (HARA) has been performed as described by ISO26262 Part 3. An overview of the process is shown exemplarily for a specific malfunction in the actuation control and resulting propulsion due to loss of Park range in Fig. 4.

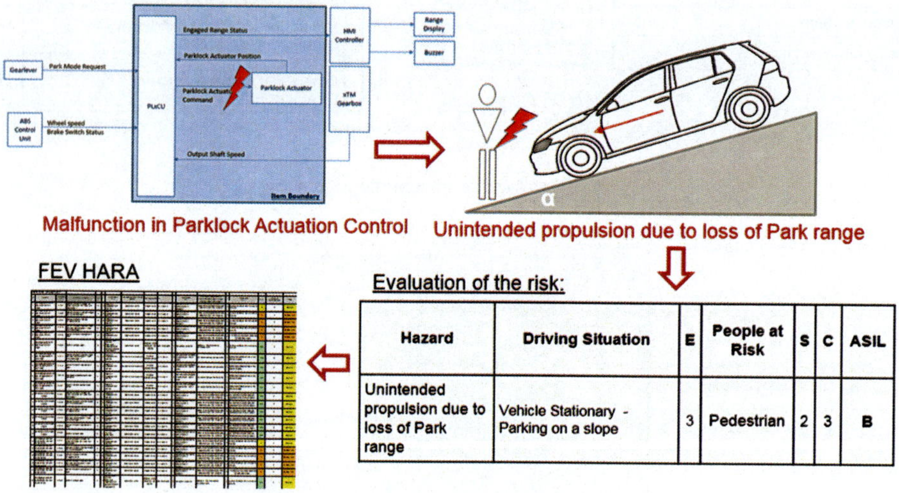

Hazard	Driving Situation	E	People at Risk	S	C	ASIL
Unintended propulsion due to loss of Park range	Vehicle Stationary - Parking on a slope	3	Pedestrian	2	3	B

Fig. 4. Overview of FEV HARA process

As step 1, the various hazards of the Park lock system has been identified.

As step 2, the hazards has been analyzed with various use scenario's to identify Severity (S), Exposure (E) and Controllability (C).

As step 3, the Safety goals and ASIL has been identified for the Park lock system.

Based on FEV preliminary HARA, the ASIL for the park lock system has been determined, as ASIL B. An overview of the safety goals including ASIL rating is given in

ID	Safety goal	ASIL	Safe state / fallback level (if applicable)
1	Prevention of Unintended Display control for Park Function	B	1. Range to Show '-' (Or Blank) 2. 'Press Brake lamp' to turn on 3. Buzzer to turn on until driver presses brake pedal (All are CAN Signals)
2	Prevention of Unintended Park Mode disengagement	B	1. Range to Show Actual Range with Blinking 2. 'Press Brake lamp' to turn on 3. Buzzer to turn on until driver presses brake pedal (All are CAN Signals)
3	Prevention of Unavailability of Park Lock Control	QM	
4	Prevention of Unintended Park Mode engagement	QM	

Fig. 5. Overview of safety goals

Fig. 5. These were used for current system development and design of safety measures. However, for the final product development, Safety Goals/ASIL coming from OEM HARA shall also be considered. Based on final Safety goals, OEM safety Concept and ASIL, adequate safety measures shall be ensured by FEV.

4 Park Lock Development and Optimization

FEV has broad experience in the development of park lock systems. Within more than 15 projects, FEV has developed an unique in-house tool chain, which includes powerful lay outing tools for quick concept development as well as advanced CAE methods including Finite-Element-Analysis and Multi-Body Simulation, see Fig. 6.

The range of services does include:

- Benchmarking and target setting (function, durability, NVH)
- Concept development incl. full system specification
- Design & CAE incl. tolerance analysis
- DFMEA and definition of critical characteristics
- DVP definition based on vehicle duty cycle incl. limiting sample testing
- Hardware validation on bench and in vehicle according to FEV and/or customer standards
- By-wire S/W development incl. Functional Safety.

4.1 Universally Applicable Park Lock Module for All Global Markets

Based on generic requirements and targeting applications in all global markets, FEV has developed a compact park lock module. Due to its compact size and flexible orientation,

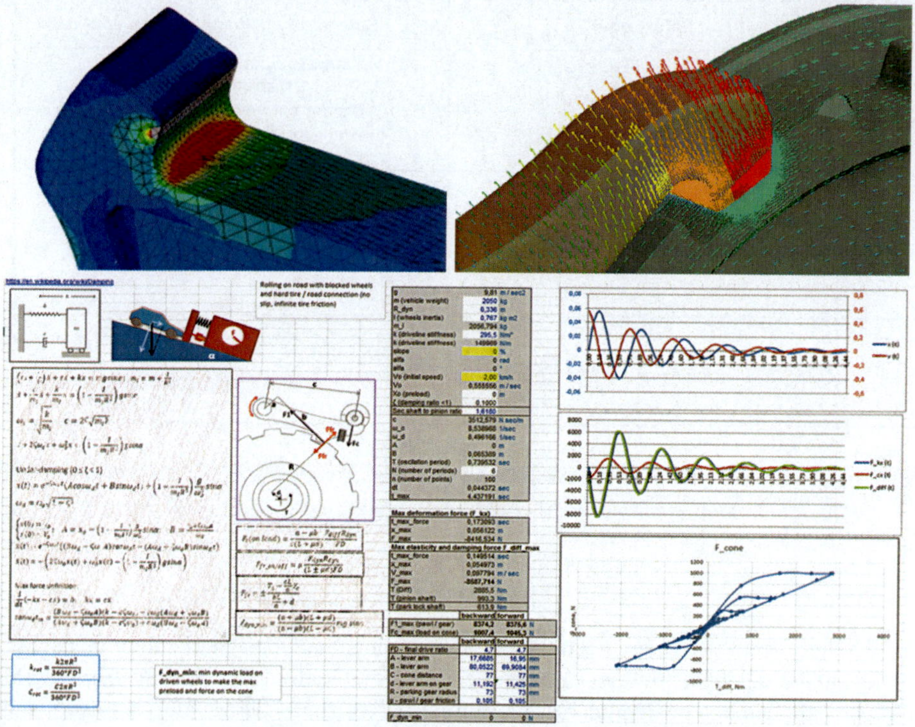

Fig. 6. Park lock development tools

it can easily be packaged in most transmissions. The module has been designed for rotary actuation and can be combined with a wide range of actuators and sensors. Paired with FEV's software and sensor concept, no smart actuator is needed. Manual cable actuation is also possible.

The current design has the following specification:

- Up to 2.400 kg gross vehicle weight
- Up to 4.000 kg with small modifications
- Maximum parking slope: 30%
- Ratio park lock gear to wheels: >3.

4.2 Compact, Modular Design

Figure 7 gives an overview of the different components of the park lock module.

During the concept development, a lot of attention has been paid to the optimization of the contact area between the actuation mechanism and the pawl. Conventional pull-cone designs inherently have to deal with high contact stresses due to point or line contacts. High contact stresses can lead to friction instabilities and high wear, eventually resulting in lower lifetime, risk of functional failures and increased actuator power requirements.

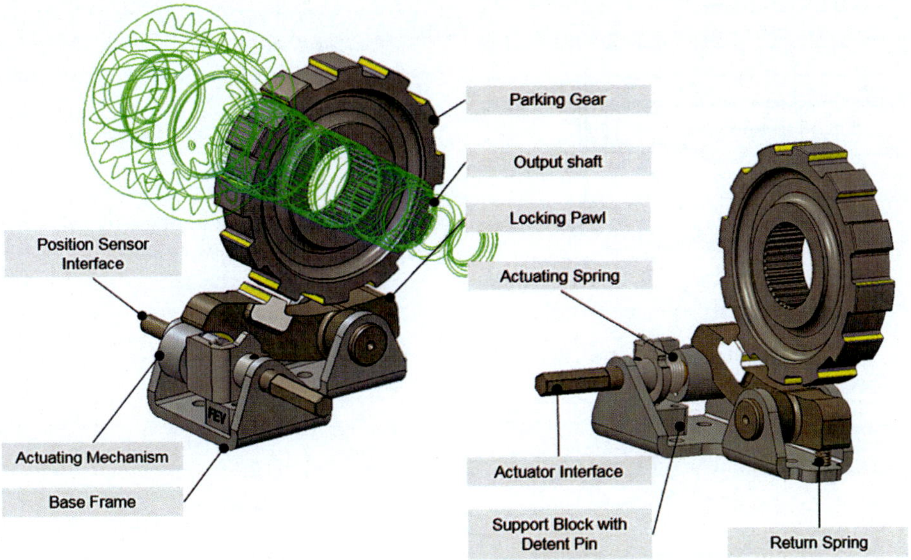

Fig. 7. Overview of park lock actuation system

The new design uses an eccentric actuation mechanism with a convex/concave contact zone, resulting in a much larger contact area and significantly reduced contact stresses. The advantages are higher reliability and less sensitivity to tolerances. In addition, the actuator torque and power requirements can be reduced.

Figure 8 shows the contact stress between pawl and actuation mechanism as well as the required actuator disengagement torque for two different designs. Both designs have to deal with the same reaction force introduced by the pawl. The new design exhibits significantly lower contact stresses and more robust disengagement behavior.

5 FEV Transmission Software Architecture

For actuation of park lock system, a ‘PARK LOCK MANAGER’ control software has been developed. The Control Software has been developed using AUTOSAR architecture approach on application software level. An overview of the software architecture is exemplarily shown in Fig. 9 for a multi speed electric drive unit.

Thus, it can fit in the preexisting customer SW Architecture for the xTM transmission – thereby creating a predeveloped PARK LOCK MANAGER’ for the OEM. The primary characteristics of the FEV ‘PARK LOCK MANAGER’ are shown below.

- Physical components based on hardware components
- Logical components with encapsulated functionalities
- Standardized physical interfaces between all components
- Park Lock functionality is integrated in the Composition Vehicle and will run on TCU or combined VCU.

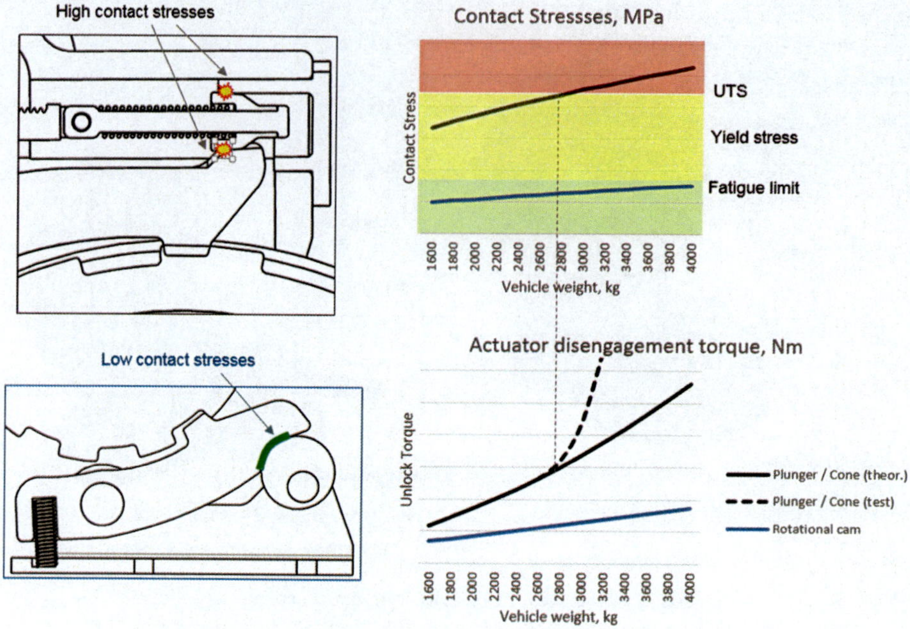

Fig. 8. Actuator disengagement torque over contact stress

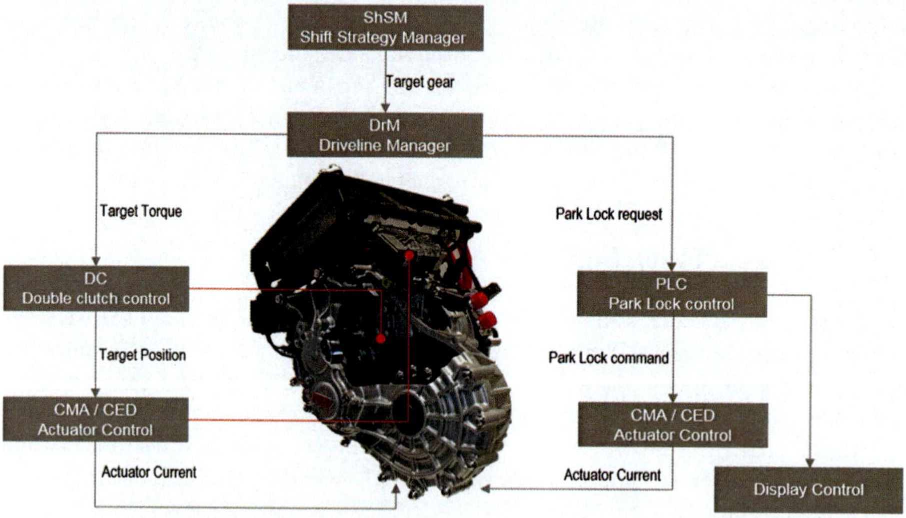

Fig. 9. Overview of FEV control software

6 Diagnosis Monitoring

Diagnostic coverage has been defined based on the ASIL and legal requirements. However, the coverage can be increased or decreased based on OEM requirements/Safety

Concept. Currently the FEV Park lock Control System satisfies Medium Diagnostic Coverage as specified by ISO26262 Part 5 and satisfies sample OEM requirements from FEV experience. An overview of the included diagnostic is shown in Fig. 10.

Hardware Component	"Low" diagnostic coverage:	"Medium" diagnostic coverage:	"High" diagnostic coverage:	Required by OEM? (Sample)	Safety Mechanism / Measures required to cover
Harnesses including Splice and Connectors	Open circuit;	Open circuit;	Open circuit;	Y	Failure detection by on-line monit
	Short circuit (to ground);	Short circuit to ground (d.c coupled)	Short circuit to ground (d.c coupled)	Y	
		Short circuit to vbat	Short circuit to vbat	Y	
		Short circuit between neighboring pins	Short circuit between neighboring pins	Y	
			Resistive drift between neighboring pins;	N	NA
			Contact Resistance	N	NA
Actuator (detailed analysis may be necessary for technology specific failure modes")	· Open circuit;	· Open circuit;	· Open circuit;	Y	H-Bridge internal diagnostic
	· Short circuit (to ground);	· Short circuit (to ground);	· Short circuit (to ground);	Y	
		· Short circuit (to ground, to power);	· Short circuit (to ground, to power);	Y	
			· Short circuit(to ground, to power, neighboring pins); Out-of-range;	Y	
	· Out-of-range;	· Out-of-range;	· Out-of-range;	Y	Failure detection by on-line monitoring'Functional diagnostic
	· Stuck (electrically, physically)	· Stuck (electrically, physically)	· Stuck (electrically, physically)	Y	
		· Frozen (physically or electrically stuck, locked);	· Frozen (physically or electrically stuck, locked);	Y	
			· Drift or oscillations or Offset in the valid range;	N	NA
Sensor/Switch (detailed analysis may be necessary for technology specific failure modes")	Open circuit;	Open circuit	· Open circuit	Y	Failure detection by on-line monit
	Short circuit (to ground);	Short circuit to ground (d.c coupled)	Short circuit to ground (d.c coupled)	Y	
		· Short circuit (to ground, to power);	· Short circuit (to ground, to power);	Y	
			· Short circuit(to ground, to power, neighboring pins); Out-of-range;	N	
	Out-of-range	Out-of-range	Out-of-range	Y	
	Stuck in range	Stuck in range	Stuck in range	Y	
		Offsets	Offsets	Y	
			Oscillations	N	

Fig. 10. Overview of diagnosis coverage achieved for FEV park lock system

7 Functional Safety Concept

The result of the development of the functional safety concept is shown in the following Fig. 11.

The system was been decomposed into

- Normal Control: QM(B)
- Functional safety Monitoring: B(B).

The Functional safety monitoring checks for unintended disengagement and Wrong Display Control during park lock control. It actuates the safe states based on detected Safety violation.

- Disable Display of range
- Disables actuator.

The functional safety concept is also scalable for higher ASILs. Based on OEM HARA, the appropriate and adequate safety mechanisms can be applied to the system.

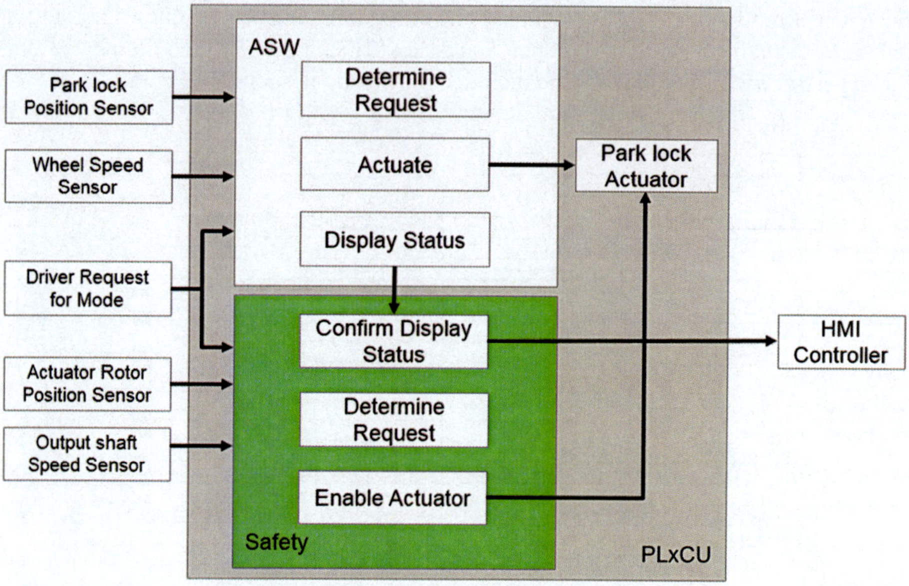

Fig. 11. Overview of functional safety concept

8 Conclusions and Outlook

FEV offers a one-stop solution for the complete park lock system including hardware, control software, diagnostics and functional safety, which was shown here exemplarily for a park by wire system, suitable for common EDUs. Such a synergistic approach reduces risk, cost and reworks as compared to distributed development. The predeveloped and pre-validated products help in accelerating projects. Further, the flexible software architecture allows plug and play of the product on an existing or a new control unit while managing implementation across multiple vehicle platforms. The architecture also allows flexibility to determine the failsafe operation dependent on vehicle or platform requirements. The functional safety software is scalable allowing the product to be compliant with ASIL B to D requirements.

The current development is in implementation phase and hence ready for series development project. A demonstration model of this Park lock system will be presented at the 18th CTI Symposium in Berlin from the 9th to 12th of December 2019 at the FEV Booth.

References

1. Standardized E-Gas Monitoring Concept for Gasoline and Diesel Engine Control Units
2. ISO/FDIS 26262 parts 1-9: 2018

Innovative and Highly Efficient Clutch System for Multispeed BEV with Highspeed Powertrains

Tackling the Efficiency and Drag Loss Challenges Through a Novel Latching and Actuation System

Sascha Ott[1](✉), Hüseyin Gürbüz[1,2], Falk Nickel[3], and Andreas Genesius[4]

[1] Institute of Product Engineering (IPEK), Karlsruhe, Germany
sascha.ott@kit.edu
[2] Mubea Tellerfedern GmbH, Daaden, Germany
[3] Miba Frictec GmbH, Roitham, Austria
[4] Kaco GmbH + Co. KG, Kirchardt, Germany

Abstract. According to the state of the art in research and technology, previous clutch systems based on multidisc design are not suitable for shiftable BEVs with highspeed transmissions up to 30,000 rpm in order to meet the higher requirements with regard to controllability at high clutch input speeds, efficiency and freedom from drag torque. The consortium project "Highspeed Clutch" was started with the IPEK—Institute of Product Engineering and transmission suppliers Mubea Tellerfedern GmbH, Miba Frictec GmbH and Kaco GmbH + Co. KG with the aim of developing and prototyping friction clutch concepts for electric vehicles with a multi speed transmission. The focus is on electric drives with high speeds of up to 30,000 rpm. This article presents the clutch concept for multispeed electric drives that was developed in the project. Two designs of friction system for high sliding speeds and combination and arrangement of the subsystems shift piston, separating spring and sealing system are presented in order to improve the system behaviour in the synchronization phase as well as in the open phase. To increase efficiency, the clutch concept is equipped with an innovative latching system that is based on the ballpoint pen mechanism and requires no energy when closed. This paper presents the first results of the functional verification. Furthermore, a novel release bearing in the form of a plain bearing was developed in order to realize actuation by a non-rotating shift piston at high clutch input speeds. In this lecture the test bench concept and the test specimen variants, test parameters and lubrication concepts are described also.

Keywords: Wet multidisc clutch for BEV · Mechanical latching system · Novel release bearing as plain bearing

Euroforum Deutschland GmbH (Ed.): *CTI SYMPOSIUM 2019*, pp. 407–416, 2021.
https://doi.org/10.1007/978-3-662-61515-7_36

1 Introduction

Most electric vehicles available on the market are equipped with a reduction gear. However, the 1-speed transmission only offers a compromise between the maximum achievable vehicle speed and the required starting torque. For vehicles in higher vehicle segments, this topology leads to an oversizing of electric motor or the use of several electric motors and thus to higher costs. A more efficient solution is to use a variable ratio for start-up and maximum speeds, so that the size of the electric machine can be reduced. Highspeed concepts in combination with a shiftable transmission allow an increase in power density of the drivetrain. In addition, the operation of the electric machine with optimum efficiency through multi-gear stages makes it possible to increase the overall efficiency of the electric drive train [1, 2].

Compared to conventional drive trains, an electric drive train with a highspeed electric machine has changed requirements and boundary conditions for the clutches. Due to the lower number of gears and higher gear spread, the maximum differential speed for clutches in electric drive trains is significantly higher than for clutches in internal combustion engine vehicles, which lead to high sliding speeds and heat input during the synchronization process [3].

In the open state of the clutch, the high differential speeds can lead to high drag losses due to the phenomenon of multi-disc tumbling. Further challenges arise in the area of material stress, since with increasing circumferential speed the centrifugal forces reach values that can lead to plastic deformations of the components. The frictional clutches used in powershift transmissions of conventional drives have not been developed for high-speed applications with speeds up to 30,000 rpm according to the state of the art in research and technology [4].

The "Highspeed Clutch" consortium project was launched to demonstrate the innovative components in the field of clutch technology for e-mobility. To transfer the system competence to supplier and to enhance development input for e-mobility, consortium was founded by suppliers who already supply components for conventional friction clutches. Consortium partners Kaco GmbH & Co. KG (Kaco), Miba Frictec GmbH (Miba) and Mubea Tellerfedern GmbH (Mubea) are responsible for the development and implementation of components. The project partner IPEK—Institute of Product Development at the Karlsruhe Institute of Technology (KIT) (Fig. 1), ensures the system-side validation of the clutch. In this project, Miba develops and produces the friction pairs and sliding coatings required for torque transmission. An innovative drive and sealing system will be developed and implemented by consortium partner Kaco to actuate the friction clutch. Besides the development and supply of technical springs to separate the friction pairings and the piston return, Mubea is also consortium leader in this project [4].

The IPEK contributes know-how from various completed and ongoing projects to the consortium project. In addition, they create simulation models and carry out test bench tests required for early validation of the clutch system. The goal of the consortium project is to carry out a feasibility study for multi-plate clutches in a shiftable high-speed drive with 30,000 rpm for electric vehicles. Furthermore, the aim is to maximize the input speed of the clutch and to identify today's limits. Additionally, the load transfer capability and control under high-speed conditions will be demonstrated in test bench tests.

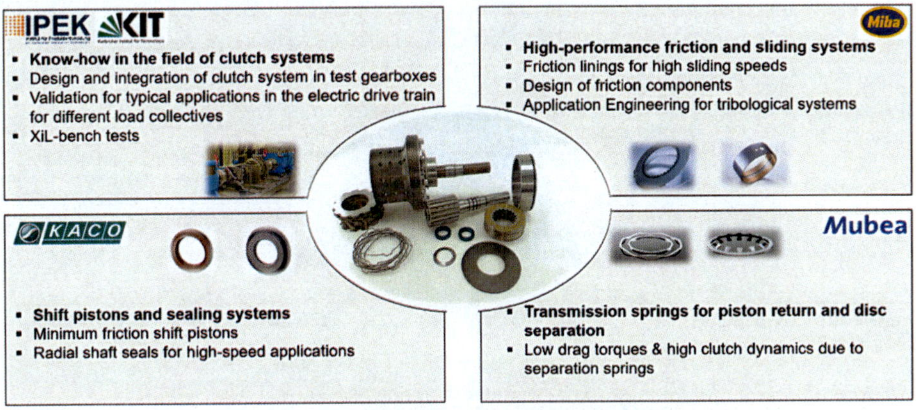

Fig. 1. Consortium highspeed clutch [4]

2 Wet Multi-disc Clutch for Shiftable BEV with 30,000 Rpm

For the initial design of the frictional clutch system, the requirements and boundary conditions from the research project Effect360° are used [1]. The clutch concept for shiftable highspeed drives in electric vehicles developed as part of the consortium project is shown in the Fig. 2. When designing the friction system for a clutch system with a input speed of 30,000 rpm, the permissible sliding speed dominates. The result is, for example, the design of average friction diameters which are below the values known in automotive applications. At a given engine torque, small friction diameters mean that the number of friction pairings must be increased. With an average friction diameter of 55 mm, 12 friction linings are required for fiber composites and 9 for friction discs with sintered material. Due to the higher thermal power density of sintered linings, less friction pairs are required here. A higher number of friction partners can occur higher drag torques in the open state and a higher loss of axial piston force due to tooth friction at engaging. The axial force losses due to the high number of discs are compensated by actuating from both sides of the disc pack. This also ensures that the total stroke is divided between two pistons and that there is a high dynamic response. A further increase in clutch dynamics is achieved by the separating springs, in which the clutch clearance relevant for drag losses is set as small as possible [4].

Due to their wave shape, waved springs lose spring force at high speeds. A new separating spring could be developed at Mubea through a variation in design and principle. The so-called torsional separating spring has a high speed stability up to 30,000 rpm and is used here.

To avoid dynamic pressures due to the rotation of the shift piston at the clutch input speed, the concept presented here uses release bearings. In contrast to conventional actuation, a plain bearing is used in this case. The requirement for the sealing lips is challenging due to the varying direction of rotation. The sliding speed ranges from 30 to 50 mm/s. Kaco has developed high-speed seals that are precisely matched to these requirements. [5]

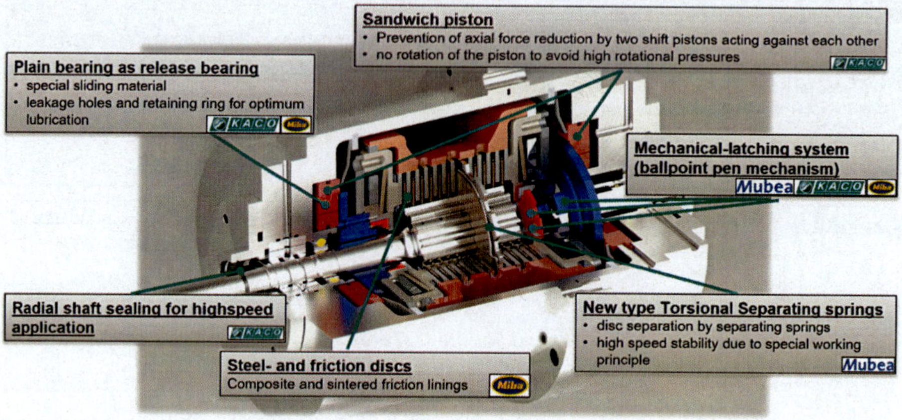

Fig. 2. Demonstrator concept

3 Mechanical-Latching System (Ballpoint Mechanism)

In wet multi-disc clutches with hydraulic shift pistons, torque transmission requires maintaining pressure to compress and hold the discs in the engaged position. However, because the operating pressure must be maintained permanently, energy is needed to keep the hydraulic piston closed or open in addition to the operating energy. Reducing the energy requirement is therefore one of the main development goals in actuation. One way is to use mechanical locking systems that keep the shift piston latched and allow the required pressure to be minimized.

A new latching mechanism was developed based on the ballpoint pen mechanism.

For this purpose, the three essential components latching crown, guidance crown and sliding sleeve are adapted for a clutch integration (Fig. 3) [5].

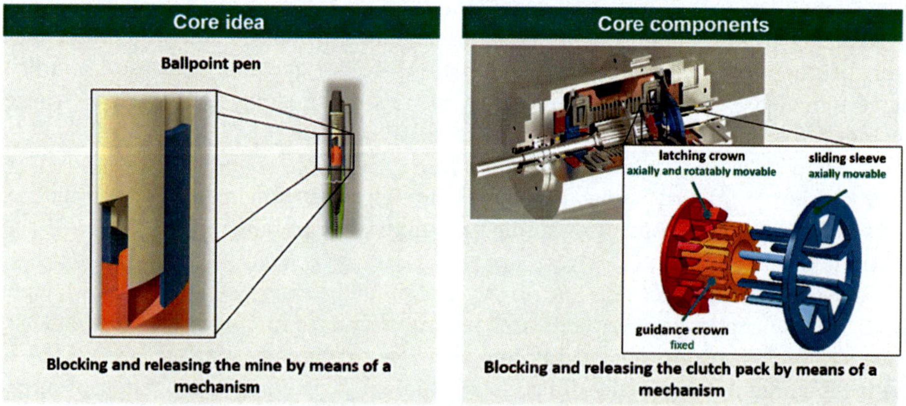

Fig. 3. Core idea latching system

The latching crown is axially movable and rotatable relative to the shaft. The base body has axially extended cams at one end. A fixed guide crown is used for axial positioning of the latching crown and sliding sleeve. Furthermore, the adapted tooth shape allows a defined rotation of the locking crown into the next tooth gap. In order to move the latching crown into the various latching positions, it is moved axially with the aid of the piston. The latching assembly forms a mechanical stop with which the required contact force on the discs can be held [4].

Figure 4 shows the latching sequence when closing the clutch. When the clutch is open, the piston is first moved back so far that there is no contact with the plain bearing. In the latching system, two disc springs are used due to their speed stability. The disc spring 1 is preloaded to the maximum required axial force, so that axial load required for synchronization and torque transmission can first be transmitted by the piston without influence of latching system. The second disc spring with smaller load and diameter is used to minimize the forces on the tooth flanks of the latching crown and guidance crown and to enable independent locking with lower forces than the axial force of the clutch. Increasing the pressure in the piston chamber continuously moves the shift piston axially and moves the complete latching unit in the direction of the clutch pack, so that first the clutch clearance is closed and then the speed and torque are adjusted (Fig. 4 left). Now the pressure is further increased to overpress the preloaded disc spring 1 (Fig. 4 right). Sliding sleeve presses the cam of the latching crown out of the grooves of the output shaft. The disc spring 2 and the specified tooth geometry of the latching crown, guide crown and sliding sleeve twist the locking crown and push it into the next tooth gap. In this position, the cams cannot engage in grooves or recesses in the output shaft [5].

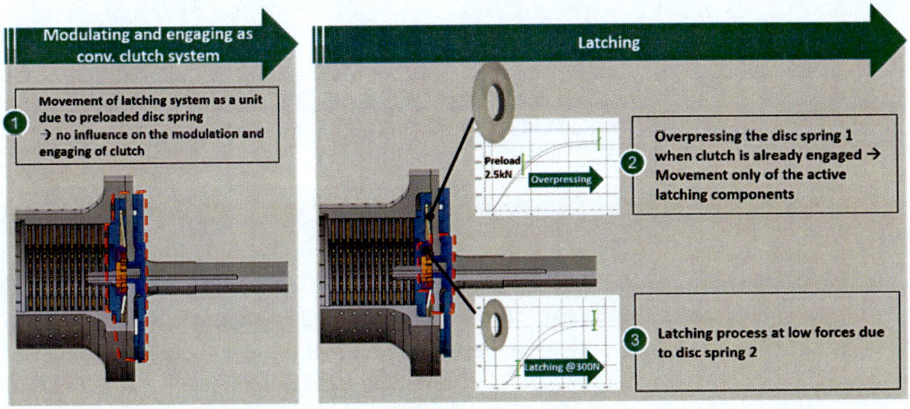

Fig. 4. Latching sequence

The springback of the disc spring 2 and the tooth contour rotate the latching crown and the cams now meet an end stop in the output shaft. When the piston pressure is reduced, the piston moves back through the disc spring 1 and the latching elements are supported on the output shaft. Now the axial force required for torque transmission is fully provided by the disc spring 1. A return spring can return the piston to its initial position so that there is no contact with the release bearing and it is not loaded. To disengage the clutch, a pressure must be built up and briefly released again. [5]

3.1 Functional Verification of Latching System

In order to carry out the first functional tests on the latching system, a test bench with lower speeds and torques was selected. The research focus here is on gaining knowledge in the interaction of the latching components in a clutch system under speed and torque. For this reason, the testing of latching system and plain bearings as release bearing are carried out on separate test benches. Therefore, a test bench was set up in which the actuation is carried out by means of a release bearing with rolling bearing (Fig. 5). The release bearing is designed also by ourselves and can reach speeds of up to 30,000 rpm with an actuating force of 5kN with spindle bearings. Furthermore, it is initially determined to press only from one side. The piston return is performed by separating springs and oiling is carried out by a rotary feedthrough. At the test bench it is possible to rotate the input and output side in clutch operation with max. 9,000 rpm and to build up a torque of max. 30.8Nm.

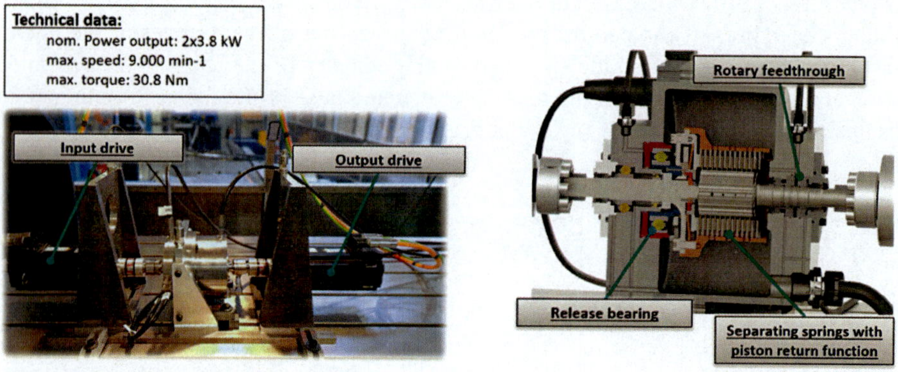

Fig. 5. Validation environment latching system: Image from test bench (left) sectional view CAD (right)

Figure 6 shows achieved test results in the first functional verification. To demonstrate the latching system, a differential speed between the input drive and output drive is set in phase 1. Here a maximum speed of 3,000 rpm and a differential speed of 1,000 rpm were set, since the maximum torque of the electric machine is in this speed range.

After a relative speed has been set, the pressure is increased in phase 2 to reach Kisspoint and synchronize the speeds. The total clutch clearance here corresponds to 2.4 mm (0.1 mm/friction surface).

It can be seen that input and output speed are equal and clutch is engaged. In phase 3, the pressure is further increased in order to close the latching mechanism. Here disc spring 1 (see Fig. 4) is pressed over and disc spring 2 can move the latching device. Pressure is then reduced to zero in phase 4. This completes the latching process. The drive speed is changed here in order to demonstrate the function of the latching system.

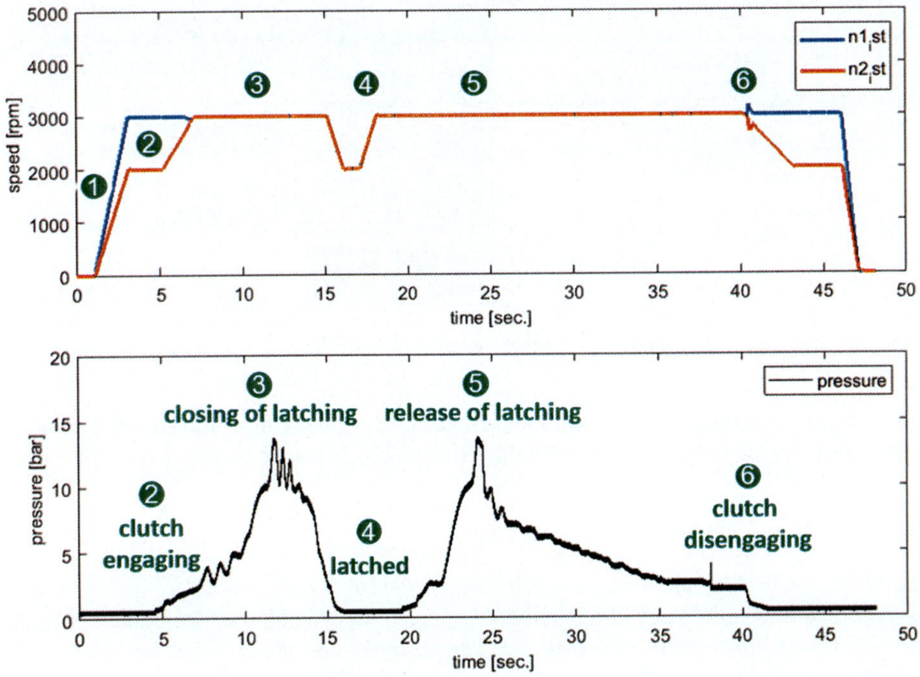

Fig. 6. Test results: Functional verification of latching system

It is noticeable that output speed changes synchronously with input speed change and clutch remains engaged despite zero bar. In order to release the latching, a pressure of 12 bar is built up and immediately released again in phase 5. Now the pressure is reduced by means of the springs in the clutch. In phase 6 the pressure is further reduced to reach the slip limit and to disengage clutch.

4 Novel Plain Bearing as Release Bearing for Cost-Efficient Clutch Actuation

For the demonstrator concept, we decided to develop a plain bearing for actuation, which we call sliding piston (Fig. 7). The clutch is actuated by two identical plain bearings at both ends of the housing. In order to avoid rotational pressure and the related countermeasures, the pressure chambers are located on the stationary side in the clutch housing. At a speed of 30,000 rpm, differential speeds of 30–110 m/s result in sliding contact.

The plain bearings each transmit a maximum force of 4–5 kN at surface pressures in the range of 1–5 N/mm^2. However, this peak load only occurs in the short phase of "overpressing" during clutch latching. In addition, there are the phases in which the force is built up or reduced and the longest of the phases in which the clutch is latched and there is no contact. In this phase, the lubricating gap is opened by means of a piston return spring, which avoids a drag torque.

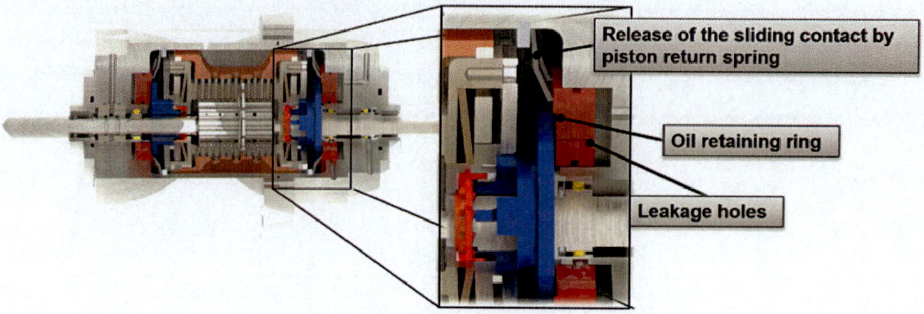

Fig. 7. Actuation concept sliding piston

All in all, this offers an economical, robust and space-saving solution compared to rolling bearings, which has the potential to meet the high requirements.

4.1 Validation Environment Sliding Piston

The test bench is designed for a release bearing according to the "hardware-in-the-loop" method. The degree of modeling is chosen in such a way that on the one hand the most important boundary conditions of the HSC are mapped and on the other hand external influences are reduced.

In a first step, the focus is on the validation of the plain bearing and the gathering of information to optimize it. For this purpose, the test bench is reduced to half of the HSC and thus one sliding pairing. The elasticity of the lamella pack is simulated using springs. One of the two sliding contact partners is driven via the input shaft, the other on the piston side is fixed. The suitability/quality/friction coefficient of the respective sliding pairing results from the measured braking torque. The temperature is also measured at several points in close distance to the sliding contact in order to narrow down problem areas and generate temperature characteristic maps.

For quick access to the test specimens, the test bench is designed to document possible wear between the test cycles and, if necessary, to change the sliding partners. At the same time, a wide range of variations can be mapped in the given time and supplemented almost as required. The tests are performed on the eDrive-in-the-Loop Test Bench (eDril) from IPEK, where a speed of up to 20,000 rpm is possible (Fig. 8).

Material pairing, geometric design and oiling strategy are varied in order to determine favourable designs of the plain bearing. Solid materials in the form of bronze/steel or brass/steel are initially used as the sliding material, as these are most suitable for the given conditions. Special coatings on steel plates will also be tested in the further course. The contact surface is defined by the geometric shape, mainly d_i and d_o, the contact partner. From this the variables surface pressure, sliding speed and shear rate do/di are defined. In addition to the geometry proposed by demonstrator, a design designed for a minimum inner radius and sliding speed as well as a design for low surface pressure at high sliding speed are also investigated in the present experiment. As oiling strategy three variants are investigated (Fig. 9). The first is designed as a simple surface, which is supplied with oil from the inner radius. The acceleration at the rotating contact partner creates

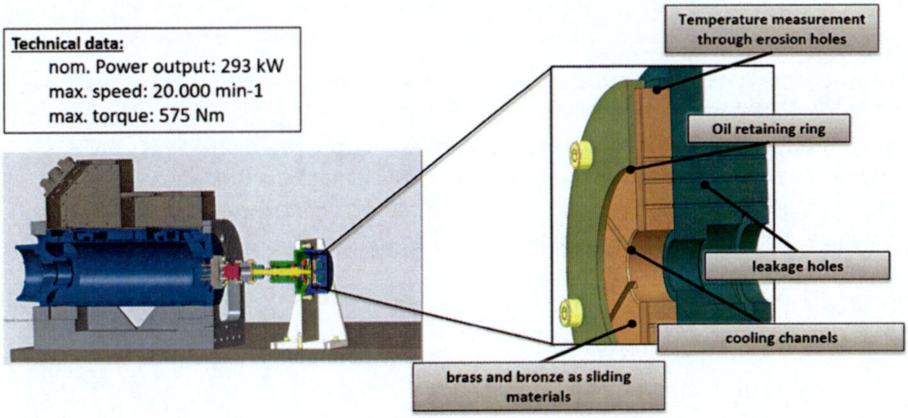

Fig. 8. IPEK eDril: Validation environment sliding piston

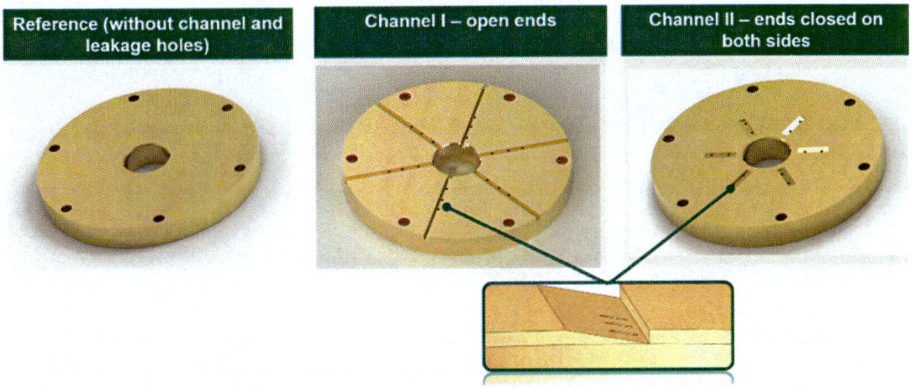

Fig. 9. Test samples and lubrication strategy

a centrifugal force which drives the oil through the gap and thus separates the contact partners from each other.

The second variant provides channels as shown in the figure above, which are supplied with lubricating oil from the piston pressure chamber via leakage holes. In addition, the oiling from the inner radius can also be adjusted here. The lubricating wedge converted in the channels leads to a hydrodynamic pressure build-up and finally to the separation of the contact partners. Due to the opening of the channels on both sides, a lubricating oil flow is guaranteed, which dissipates the resulting heat.

In the third variant, the channels are closed on both sides in the same way as hydrostatic plain bearings. A pressure is created in the channels and a lubricating oil flow is created in the gap to separate the contact partners from each other. A lubricating wedge can also be found in this version.

5 Summary and Conclusion

This article presents a new clutch system for electric drives with shiftable high-speed motor up to 30.000 rpm, which was developed as part of the Highspeed Clutch consortium project. It offers high potential for use in future drives in terms of efficiency and system behavior. The efficiency is maximized by the mechanical latching system so that no energy is required when closing the clutch. First tests have proven the functionality and potential. After latching, the clutch could be kept engaged at a pressure of 0 bar. Furthermore, an actuation via a release bearing in the form of a plain bearing was presented in this article. For this purpose, a new test bench was set up at the IPEK with 20,000 rpm max. speed and allows the variation of the test samples and oiling strategies. Leakage holes and retaining rings are specially attached to the piston in order to achieve optimum lubrication in the sliding contact. In the further process, the knowledge gained from the test bench tests on the subsystems latching and sliding piston will linked back to the overall highspeed clutch concept. Subsequently, the demonstrator concept with latching and actuation via sliding piston from both sides is prototypically constructed and tested on an IPEK test bench with 30,000 rpm. A validation with consideration of the interactions between driver, environment and residual system is carried out by means of the XiL method.

References

1. Reichert U, Bause K, Ott S (2017) Presentation of a multispeed gearbox for a BEV increasing the efficiency and power density. In: 16. Internationales CTI Symposium, Berlin
2. Gwinner P, Stahl C, Rupp S, Strube A (2017) Innovatives Hochdrehzahl-Antriebsstrangkonzept für hocheffiziente elektrische Fahrzeuge. ATZ – Automobiltechnische Zeitschrift, pp 72–75
3. Albers A, Reichert U, Bause K, Radimersky A, Ott S (2017) Entwicklung einer verlustoptimierten Kupplung für ein mehrgängiges Getriebe für ein Elektrofahrzeug. In: VDI Kupplungen und Kupplungssysteme in Antrieben, Ettlingen
4. Gürbüz H, Ott S, Albers A (2019) Entwicklungsansätze für innovative Hochdrehzahlkupplung in E-Fahrzeugen. In: Forschung im Ingenieurwesen
5. Gürbüz H, Ott S, Albers A (2019) Potentials for improving the system behaviour of highspeed clutches for BEV using optimal transmission springs. In: Forschung im Ingenieurwesen

Multipurpose Oil Filter Systems for Innovative Drivetrains and e-Axles

Marius Panzer[1](✉), Claudia Wagner[2], Anna-Lena Winkler[1], Alexander Wöll[3], and Richard Bernewitz[4]

[1] Lead Product Engineer, Oil Filter Element Development, MANN+HUMMEL GmbH, Ludwigsburg, Germany
marius.panzer@mann-hummel.com
[2] Senior Expert Transmission Oil Filters, Design Oil Filter Elements, MANN+HUMMEL GmbH, Ludwigsburg, Germany
[3] Principal Expert Oil Filter Systems, Oil Filter System Development, MANN+HUMMEL GmbH, Ludwigsburg, Germany
[4] Manager Oil Filter Element Development, MANN+HUMMEL GmbH, Ludwigsburg, Germany

Abstract. Changing drivetrain architectures and designs such as highly integrated e-Axles and hybrid transmissions require different filtration solutions. Not only the requirements towards oil cleanliness and differential pressure of the lubrication and cooling circuit become more challenging, but also the dielectric and electric oil properties and the insulating function, for instance, come more and more into focus. To protect all system components starting from the oil pump MANN+HUMMEL developed a new generation of filters using their depth filter media MULTIGRADE eM-CO. The new concept increases the degrees of freedom for the development of the lubrication circuit significantly. Offering an ultra-compact and flexible installation space, lowest differential pressure or highest filtration efficiencies for system reliability. The new MANN+HUMMEL filter media portfolio MULTIGRADE eM-CO offers filter media with excellent properties and a gradient structure that provides a high dust holding capacity at a low differential pressure. Customized numerical development tools are supporting the media development by analyzing and optimizing the structure and the filtration performance. To multiply these advantages and to add further benefits for the customers a new generation of pleated suction-side filters has been developed. With up to 60% higher filter area the new concept uses the given installation space to the maximum. A 3D structured drainage grid keeps the media performance at a maximum level over the whole filter life. The latest computational fluid dynamics (CFD) analysis tools are used to optimize the differential pressure of the housing even in challenging installation spaces. Special oil drying units are applied to minimize water induced increase of conductivity. This allows the dielectric properties of the oils to be kept constant and even water sensitive oils can be used for the applications. With this knowledge and expertise, MANN+HUMMEL developed a sensor-supported oil management system with intelligent cooling function and integrated oil drying for highest oil quality over the entire life of the system.

Euroforum Deutschland GmbH (Ed.): *CTI SYMPOSIUM 2019*, pp. 417–421, 2021.
https://doi.org/10.1007/978-3-662-61515-7_37

1 Introduction

Driving pleasure without emissions is a constant trend. With a rising number of congested cities, high levels of air pollution and stricter statutory requirements, companies are working towards even more sustainable solutions. And car buyers are on board: in Germany and other countries across Europe more and more vehicles are being registered with alternative drives that are no longer based entirely on internal combustion engines (ICE). Mobility related companies are focused on the electrification of the powertrain. In this respect the highly integrated e-Axle is a particularly promising concept. E-Axles are a system which combines the electric motor, gearbox and power electronics. This saves installation space, components and wiring due to the fact that the e-Axle is integrated directly on the drive axle. According to the system suppliers, this enables up to 20% less weight and between fife to ten percent greater efficiency. Furthermore, the easy scalability of the system makes it suitable for use in compact cars, sports cars up to commercial vehicles.

1.1 Lubrication and Cooling with the Same Oil

As all the drive components are integrated in an e-Axle, a common oil circuit is sufficient for cooling and lubrication. This advantage, however, has a drawback. Mechanical as well as electronic components are exposed to the same oil and metal particles and chips out of the transmission, which can penetrate the power electronics and electric motor and cause serious damage. As a result, these particles need to be efficiently, quickly and permanently removed from a reliable working system. In this respect, just as with conventional transmissions the focus is on the separation of particles, low space requirements and a maximum of filter surface area. The MULTIGRADE eM-CO filter media require less space as the engineers at MANN+HUMMEL were able to exploit the company's proven pleating technology. An installation space of just 20 mm is enough for the pleating. This allows the fitting of a substantial amount of filter media in even the smallest installation space. The result is an enormous increase in the filter surface area and therefore also the dust holding capacity by a factor of almost six. An advantage which is particularly evident in the compact design of the e-Axles that is also valid for the classic automatic transmissions. As a result, a significant increase in operating time and thus filter lifetime can be achieved. Under consideration of the differential pressure, which is particularly important for the suction side operation, the advantages of a pleated solution are also clear to see as there is a further significant decrease in the differential pressure due to the pleating of filter media. In comparison to a flat, non-pleated filter media, the pleated variant enables a differential pressure which is up to 83 percent lower already at an installation height of 20 mm (Fig. 1). This gives the customer a larger degree of freedom with regard to higher filter fineness or dimensioning of the pump, as the energy dissipation is minimized by the pleating technique.

The filter media always retains its shape and performance, even when exposed to coldness or high differential pressure. Stability is ensured by a drainage grid or adhesive lines which maintain the ideal gap between the pleats. This enables more long-term use of the complete filter surface area. Therefore, a filter can be integrated directly into the

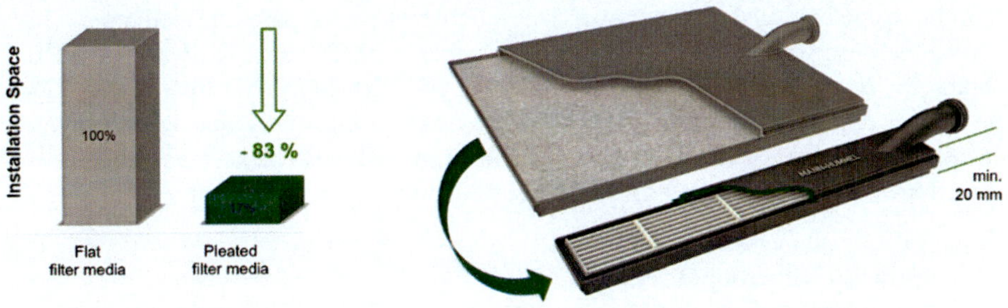

Fig. 1. The MANN+HUMMEL pleating technology enables highest filtration performances with up to 83% less installation space and simultaneously reduced energy dissipation

oil sump on the suction-side, upstream of the pump. This protects the complete system before the abrasive processes, including the pump.

2 Fully Synthetic Filter Media in MULTIGRADE eM-CO

The filtration experts at MANN+HUMMEL have taken on the challenges that come with alternative drivetrains and developed suitable transmission oil filters for e-Axle applications. The filters are equipped with the new MULTIGRADE eM-CO filter media. The filter media are specially designed to meet the requirements for the oil with regard to the lubrication and cooling of e-Axles. Here MANN+HUMMEL is able to transfer the advantages of fully synthetic filter media used in other oil filter applications.

The fully synthetic filter media is free of glass fibers and has a very high chemical resistance at considerably lower differential pressure in comparison to conventional transmission oil filters fitted on the suction-side (Fig. 2). MANN+HUMMEL utilizes consistent filter media layers without uncontrolled artificial media bypass, something which is required for conventional products due to their high loss of differential pressure.

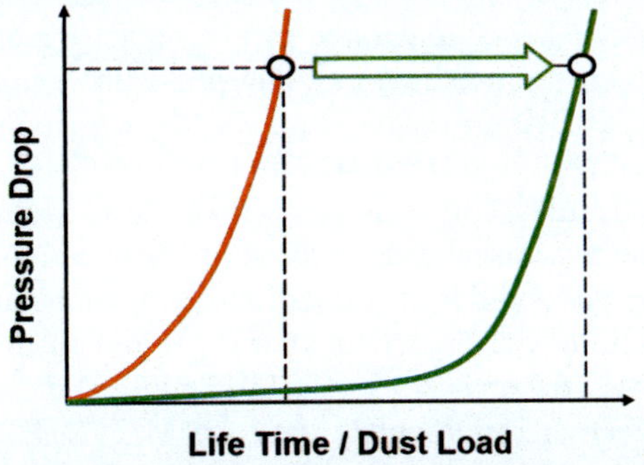

Fig. 2. Anonymized comparison of competitor's flat filter media (orange, left data curve) to MANN+HUMMEL MULTIGRADE eM-CO 180.1 pleated filter media (green, right data curve). The pleating technology enables a much longer lifetime at lower energy dissipation and installation spaces with very high chemical resistance

Vehicle owners can therefore depend on reliable particle separation at all times while keeping the energy dissipation at the lowest possible level. The use of a depth filter media enables the secure entrapment of the separated particles in the 3D fiber matrix. This prevents the reintroduction of particles or even long chips into the system, a danger which is common with conventional systems. The oil supply can also be permanently ensured without use of an uncontrolled filter bypass, which increases the system reliability.

2.1 Additional Filtration Tasks

Not only the abrasive risks of the oil contamination through particles are in the focus when it comes to e-Axles, also another important oil condition needs to be considered. On the one hand there is the foaming behavior of the oil. Due to the high motor speed air can introduced into the oil in such amounts, that foam is generated. The filtration system including an intelligent oil guiding is key for a successful foam reduction.

On the other hand the dielectric and electric properties of the oil becomes more and more important. In particular at highly integrated e-Axles, where the oil is used to cool the motor and the power electronics it is important to have low conductivity and a high breakdown voltage. Due to oil ageing, temperature and humidity, the oil can lose its strong insulating properties, leading to severe risks such as electric shortages and destruction of the e-Axle. The oil ageing and the temperature are correlated and can be influenced by the additives and the thermal management system. It is very important to remove the humidity as good as possible out of the system. By using a MANN+HUMMEL coolant dryer cartridge it is possible to keep the air induced humidity increase in the oil on a very low level, as the dissolved water is constantly removed from the oil. The coolant dryer cartridge can be easily and sensor-based serviced to ensure an always low humidity level without oil change.

2.2 Modular Thermal Management System

For highest integration, the newly developed filters can be arranged in a customer specific and flexible housing design. However, either as standalone filters within the customer given installation space or integrated in a compact modular thermal management system, all filters ensure maximum system protection against abrasive particles and water, at smallest installation spaces (Fig. 3). Many years of plastic expertise at MANN+HUMMEL and the use of modern simulation methods such as the finite element method (FEM) and mold filling simulations in the development process ensure that all components are able to withstand the highest mechanical requirements. The engineers also exploited the possibilities of simulations with computational fluid dynamics (CFD) to enable efficient flow characteristics for the filter housing in particularly tight installation spaces, to minimize energy losses by a minimal differential pressure with maximum system protection. Additionally the long MANN+HUMMEL expertise in acoustic design optimization can be exploited.

Fig. 3. Pressure, temperature and humidity sensor supported oil management system with intelligent cooling function and integrated oil drying for highest oil quality over the whole system lifetime

Together with an intelligent oil cooling, the MANN+HUMMEL oil management modules protect e-Axles and cooling oil circuits at a very high level, by including filters for the protection of bearings, gears and channels, for instance, as well as controlled and optimized oil flow and humidity sensor-supported oil drying. As a result, cleanest oil at constant high quality from the first second is ensured.

The Propulsion, Energy Storage and Charging System of the New Opel Corsa-e

Peter Ramminger(✉) and Hans-Georg Schade

Opel Automobile GmbH, 65423 Rüsselsheim, Germany
{peter.raminger,georg.schade}@opel-vauxhall.com

Abstract. Various forms of electric mobility contribute to the target of decarbonizing the traffic sector. Depending on the energy source for charging of Electric Vehicles (EVs), the CO_2 reduction contribution for Battery-EV (BEV) and Plugin-Hybrid EV (PHEV) cars differ, same as the emissions for electricity production differ by country. Once BEV and PHEV cars are stronger represented in the passenger car fleets, societies will take further benefit from progress in de-carbonization of the energy sector through this growing, coupled lever in the traffic sector. The new Corsa-e marks another milestone of EVs in the Opel product lineup. Other than the well-known, roomier Ampera-e, the Corsa-e is more clearly placed in the B-segment. It will make electric mobility more affordable for a wider customer range. Besides an enthusing exterior and interior design, great drivability and class-leading charging capability were key development targets. This paper describes how EV Functionality was maintained and improved subsequent to the Ampera-e. High Voltage (HV) architecture design simplifications were brought in place. Commonality leverage with the “Corsa” sister model as well as manufacturing insourcing within the corporation were employed to widen the affordability of this car to a broader customer area. With approx. 330 km in WLTP (preliminary figure), further optimized roadload for motorway driving and DCFC charging up to 100 kW peak, the car provides a perfect balance of Battery size even for longer distance usage. Details on the major HV Powertrain components are compared to the Ampera-e as reference. Key functions related to electrical driving and charging are explained, supplemented by performance data. The paper also compares charging options and delivers charge recommendations for different types of EVs, including the BEV vehicles, like the Corsa-e.

Keywords: BEV · DC fast charging · Electric vehicle · HVoltage architecture

1 CO_2 Emissions Attributed to EV Operation

Many societies on the globe are moving towards a sustainable energy management. This movement is not uniform, neither in speed nor in the individual strategy. Different countries or societies are moving with different speeds, powered by capability to invest financially for the change, lead time to implement new infrastructure—partially against the interest of individuals or groups—and progress in research and technical development, e.g. regarding generation of energy from renewable sources, distribution, storage and power-to-x coupling.

Euroforum Deutschland GmbH (Ed.): *CTI SYMPOSIUM 2019*, pp. 422–436, 2021.
https://doi.org/10.1007/978-3-662-61515-7_38

The agreement on CO_2 emission reduction achieved in 2015 under the authority of the UN marks an important milestone on the way to a future where mankind will continue to employ energy for many purposes, but will strive towards the ideal of complete usage of renewable energy sources. As a part of the UN agreement, European Union and its member states have set tough targets for de-carbonization of road traffic. Electric driving in all forms is a contribution to this CO_2 reduction, starting with mild and strong hybrids, over Plugin Hybrid cars with efficient combustion engine onboard all the way to either pure rechargeable Battery electric Vehicles or to Fuel Cell Vehicles which may or may not offer a plug to externally charge the Battery. To take maximum benefit for CO_2 avoidance with BEV cars, these need to be charged with CO_2-neutral, better even renewable energy. Otherwise, the CO_2 avoidance benefit for the environment will be lower at begin of operation and will rise every year with the progress in CO_2-reduction for electrical energy production (Fig. 1).

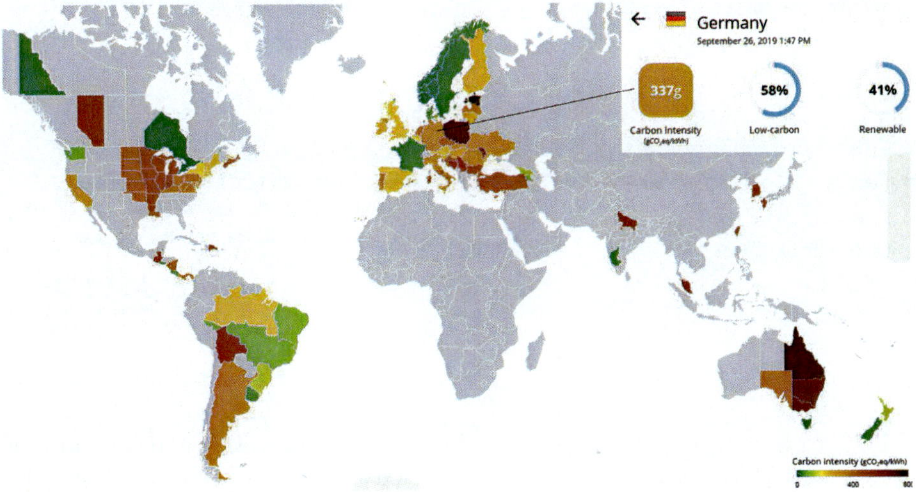

Fig. 1. Carbon intensity related to electricity production: www.electricitymap.org

Taking the arbitrary example of average electricity production in Germany on a cloudy autumn day (26th Sep 2019 13:47), one kWh electrical energy creates 325 g Co_2 [1]. Assuming 20% distribution and charge losses source until arrival in a BEV's energy storage system, and further assuming a BEV car with 15 kWh per 100 km, the operation consumption from the Battery results in 60 g CO_2 per km attributed to electricity production ("well to energy storage"), if the car is charged with electricity from this energy mix. This figure compares to a 2020 fleet average requirement of 95 g CO_2 per km, equaling 4 L gasoline per 100 km ("energy storage to wheel"). If the very same BEV car is charged from renewable sources, e.g. own photovoltaics or dedicated "green energy" contract, the emission in operation is zero grams CO_2. Similar beneficial figures are valid for countries with almost 100% of electricity production from renewables (as e.g. observed at same time in Western part of Norway with 34 g CO_2 per kWh, translating into 6,4 g remaining CO_2 per 100 km).

The sustainable conversion of the electricity sector to carbon-free operation remains a challenge for at least one entire generation. Multiple complex levers must be identified and employed to arrive at the zero-carbon target [2]. Figure 2 Offshore Wind Analysis Europe [2] shows an analysis related to offshore wind used for electricity production. The resulting cost of energy (LCOE) for the investigated locations is displayed in ct/kWh.

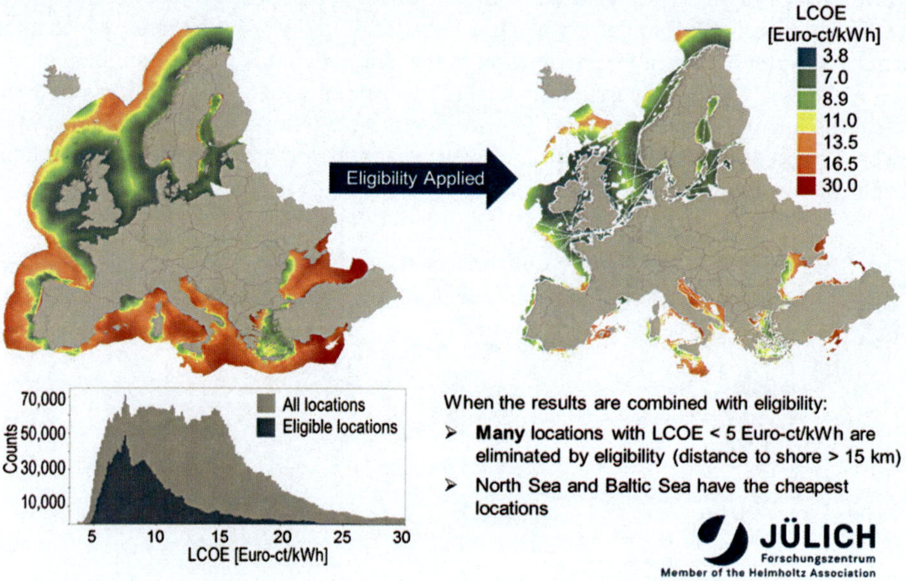

Fig. 2. Offshore wind analysis Europe [2]

Principally, enough regenerative energy is available worldwide in different forms to support a sustainable energy approach, but the suitable harvesting, conduction between source and sink areas, mid-term storage, power buffer capability to maintain reliable electricity grids and the final distribution including intelligent load managements are subject to technical and scientific progress (Fig. 3).

2 Electric Vehicle Experience at Opel

Opel has built, operated and sold a variety of Electric car series, being it Battery Electric Vehicles (BEV), Fuel Cell Electric Vehicles (FCEV), Extended Range Electric Vehicle (EREV). Some of the products were operated as test fleet including for public funded projects.

Especially EVs that went into mass production and were sold to the public generated a lot of learnings for the development teams in charge, for the manufacturing processes, for service operations and finally generated an overall knowledge about how these vehicles are used in a "real world" environment.

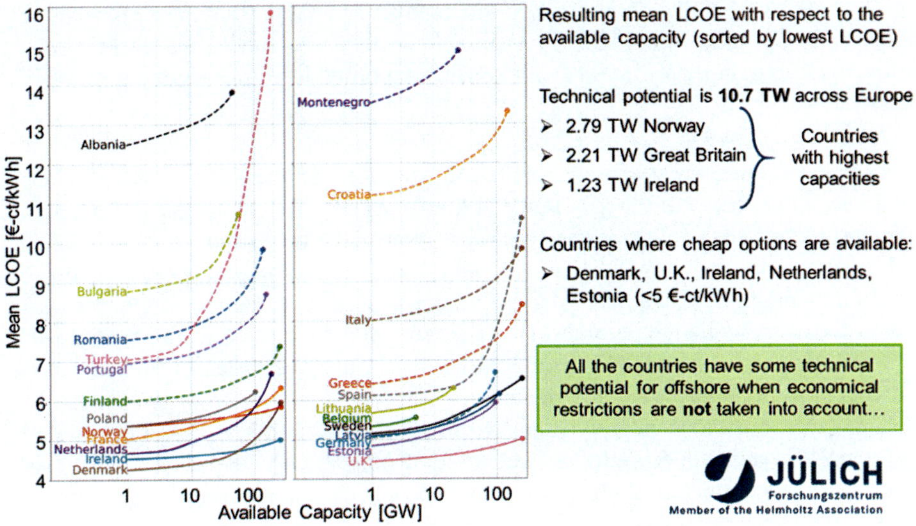

Fig. 3. Cost and availability of offshore wind energy for European countries [2]

2.1 Ampera (EREV Vehicle)

The Ampera is an EREV, hitting the market very early and setting metrics. It offers full Electric Drive Performance (max 150 kW), but the pure electric range is limited to 80–40 km (summer/winter). Public charge infrastructure for AC mode 3 charging was not well covered over all places at the time of launch, of course with different progress per sales country. Mode 4 (DC Fast Charging) was not applicable and not necessary, since a combustion engine with more than 70 kW power was onboard for power generation in range extension mode. Many users primarily relied on 220 V household sockets for EREV charging. This mode2 charging was overly used and limitations for the installations became apparent. Most important learnings from the Ampera operation were:

- Charging should occur through dedicated and orderly installed mode 3 equipment.
- Majority of customers could manage their daily distance within limited EV range, but autonomy range was substantially influenced in winter and to some extend in summer.

2.2 Ampera-e (BEV Vehicle)

The Ampera-e [3] did not hit the European market as the first BEV, but its performance was outstanding in the comparable class when it arrived. When it was developed, time to market was an imperative. The product was and still is working as catalyst to expose more customers to long distance EV driving and DC Fast Charging (DCFC). It was preparing the market, and defending the Opel position as an EV pioneer brand. The Ampera-e was launched with an overall manageable ratio of public charge points to EV cars in use, but still today there are many white spots on the landscape regarding public charge, especially in rural areas. The number of DCFC points was substantially increased in Germany in the last two years (after Ampera-e launch), supporting longer distance traffic with BEV cars in general. Most important learnings from the Ampera-e operation were:

- Especially for pure BEV operation, balancing interior roominess versus aero efficiency may lead to reduced autonomy range at higher speeds. This emphasizes the general importance of aero-optimized surfaces and optimized wheels/tires.
- Regarding charging, new BEV users tend to hold the car fully charged, just for the case they might spontaneously decide to take a longer trip. More experienced users hold the average charge level around 50%, charging once or twice per week, better factoring in the distance they would really expect to drive instead of preparing for drives they might exceptionally make.
- Range autonomy is negatively impacted in winter season due to cabin heating and entire HV system conditioning as well as some other friction and aero effects, especially for accumulated short distance drives, where energy used to heat-up will dissipate when car is parked.
- Even if a BEV battery has good DCFC capability, average outside temperatures and thermal mass of the Battery may frequently limit fast charging to lower than peak power, impeding a quick charge-up on travel.
- Reviewing the company's test fleet data, we recognize the usage of installed 150 kW peak power and related torque, same as driving at the max velocity of 154 km/h, were reserved for short to medium distances only. For mid to longer distances, customers learned how to drive efficiently and did not use installed power.
- A common experience for long distance driving with EV's is, that going slower can mean: arriving faster! This effect is not known from ICE propulsion, where refueling takes minutes only. While the gap between re-fueling time (ICE) to re-charging time (BEV) is being reduced by progress in quick-charge capability, higher power charge connections and automated authentication and payment, the effect mentioned remains generally valid, as illustrated in Fig. 4:

a. The propulsion power to maintain constant speed (also called roadload) includes a dependency from 2nd order of velocity. In a given example, when travelling with 120 km/h compared to 40 km/h, the required power rises by a factor of 10.5, requiring 3.5 times more energy to travel the same distance (blue curve).
b. With ICE propelled vehicles, this effect is partially compensated by the basic drivetrain not having reached its maximum efficiency at part-load low speeds (red curve, dashed).
c. Electric Propulsion systems typically have very good efficiency already at low speeds. With higher power (and for higher speeds) losses in the entire system rise in all parts of the high voltage system, which add to the roadload effect (red curve).
d. As a result, the driver of a BEV may achieve record ranges beyond the advertised range (WLTP based) when going constantly slow, while same driver would observe a further declining range with constant higher speeds e.g. on motorway (black curve). In contrast, a driver of conventional propulsion is used to higher energy consumption

at low speed (like in cities), an optimum point between 60 and 80 km/h on rural roads, and again rising consumption with higher speed (black curve, dashed).

e. To avoid an additional DCFC stop, driving slower may be a suitable method.

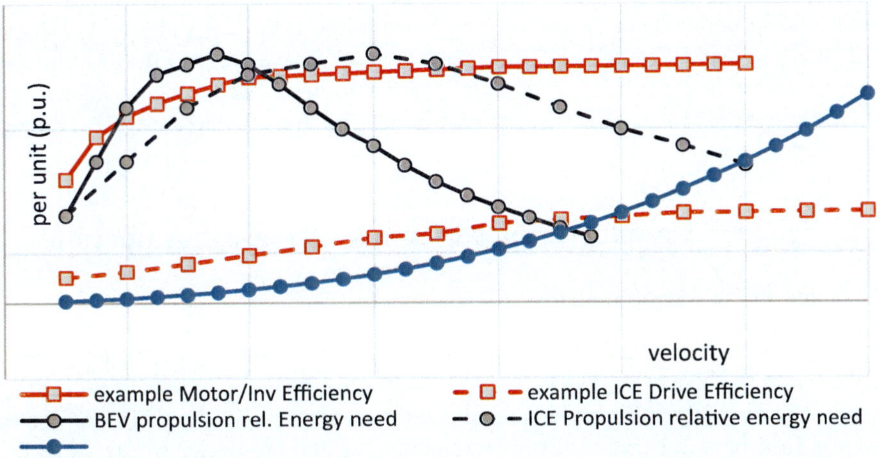

Fig. 4. Driving slower and arriving faster with a BEV

2.3 Grandland PHEV (Plugin-HV)

This PHEV is launching at fall 2019 and therefore only briefly covered. It features an electrically (or hybrid) driven front axle and an electrically propelled rear axle as 4WD option. The AWD version, equipped with a 13,2 kWh Battery, has less energy onboard for the pure electrical drive, but the PHEV offers better drive performance in the Hybrid operation combined with more interior flexibility.

3 Marketing Targets Corsa-e (Compared to Ampera-e)

The experience gathered with electric vehicles at Opel contributed to the marketing targets for the new Corsa-e. First of all, a compelling and enthusing exterior and interior design remain some of the main purchase criteria for vehicle customers. This aspect includes the expectation not only to have the "EV specifics" embedded into a car but as well to benefit from other leading-edge product characteristics, as e.g. exterior lighting, modern seat technology.

In order to increase the percentage of BEV sales of our group, the product must be affordable for a higher share of our customer base. Still EV technology as such is expensive, but a huge step has been made with the current product as will be explained below. The public subsidies for EV purchases as well expected penalties for CO_2 generation in operation of the vehicle factor into the equation for overall cost of ownership. But apart from a pure financial consideration, industry should expect customers with pronounced interest for sustainability to constitute a substantial part of the BEV sales.

The range autonomy for a given car is one of the major product characteristics for a BEV, it may become more important then what used to be the power figure of a combustion engine. A lighter car with better aero and drag behavior can afford a smaller capacity (and thereby lighter) battery. Rising fast charge capabilities for eventual long distance travel will balance against the requested initial autonomy range.

The Corsa-e and Grandland PHEV mark new milestones in the effort to provide a wide portfolio of electrically propelled vehicles in our lineup. Technologies and strategies used are supposed to take effect on future products, contributing to the wide EV lineup mentioned before.

4 Design and Manufacturing Difference Corsa-e vs Ampera-e

4.1 Evolution in Manufacturing

The Corsa-e is built on PSA's "Common Modular Platform (CMP)". This decision was taken in order to balance production volumes between conventional and EV powertrains and to maintain, where possible, identical parts (besides the drivetrain, HV and Battery system.) This main imperative also took effect on the Battery design as explained later on.

To some extent, also the logistics and production process makes use of this synergy: The powertrain plant produces an e-Powertrain assembly consisting of Reduction gear, e-Motor w/Inverter, OnBoard Charger with DC/DC and eVCU plus internal connections and ships this "Electric" Powertrain to the vehicle assembly plant (Fig. 5, Table 1).

Fig. 5. PSA e-Propulsion assembly delivered to the vehicle plant

Table 1. Main differences regarding components make or buy

Subsystem	Corsa-e	Ampera-e
Battery	Make assembly (e.g. from from purchased modules)	Buy
Reduction gear (RG)	Make. Design with side e-Motor, parts partially from MT machinery, c/o differential and housing strategy	Buy. Design with integrated e-Motor, hollow shaft, premium parts, oil pump on demand, Park Lock
e-Motor	Buy	Buy.
Inverter	Buy with motor (interconnected)	Buy (separate)
OnBoard Charger	Buy, 1Ph and 3Ph option	Buy, 1Phase
DC/DC	Integrated in OBC	Separate, buy
HPDM	Up-integrated in OBC	Separate, buy
PLCM	Up-integrated in OBC	Separate, buy
Connection points/connectors	Less (integrated)	More (split)

4.2 High Voltage Architecture Design

The HV Architecture for the Ampera-e was designed for a quick- to—market approach. Most of the suppliers were challenged to concentrate on developing dedicated modules with dedicated functionality, as e.g. a standalone APM (DC/DC converter), a standalone OnBoard Charger (OBCM) a external Battery heater or the dedicated connection module (HPDM) itself. Clearly defined interfaces helped to reduce interdependency and thereby the engineering complexity of every individual component.

The HV architecture of the Corsa-e design features multiple steps of up-integration, which deliver package, weight and cost advantage. As a tribute to this up-integration, the complexity of interdependent functionality concentrated in one device was risen. The most compelling example is the combined OBC and DCDC module, also providing a junction box function plus that it delivers the Powerline communication for DC Fast charging (according to the fast charge limits governed by the Battery Control System).

Also, the Battery heater is included in the Battery Housing. With every module not showing up separately, the Corsa-e saves a housing, mounting provisions, Housing-side HV connectors, cable-side HV connectors and the entire associated assembly time (Fig. 6).

4.3 High Voltage Battery Design

Groupe PSA is producing the HV Battery system assembly in own facilities. Due to a larger rollout of BEV vehicles in the group, PSA decided for the use of pre-manufactured modules. Same or similar modules like the ones used for the Corsa-e can be used in other Battery assemblies as energy storage for cars with other package requirements.

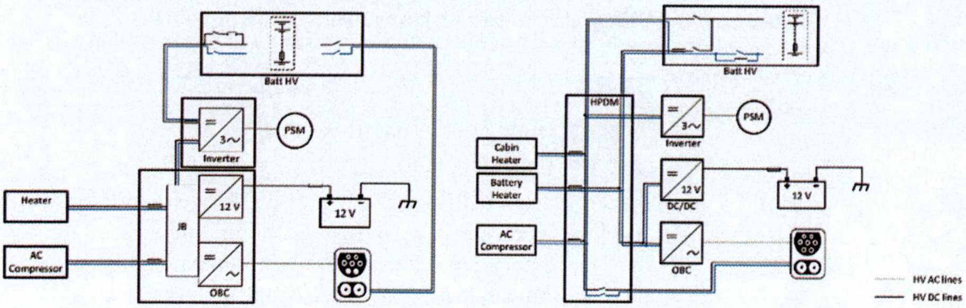

Fig. 6. High Voltage Architecture Corsa-e (left) versus Ampera-e (right)

The mentioned modules include 12 cells (2 in parallel, 6 pairs in series) and approximate dimensions 390 mm*150 mm*110 mm (Fig. 7, Table 2).

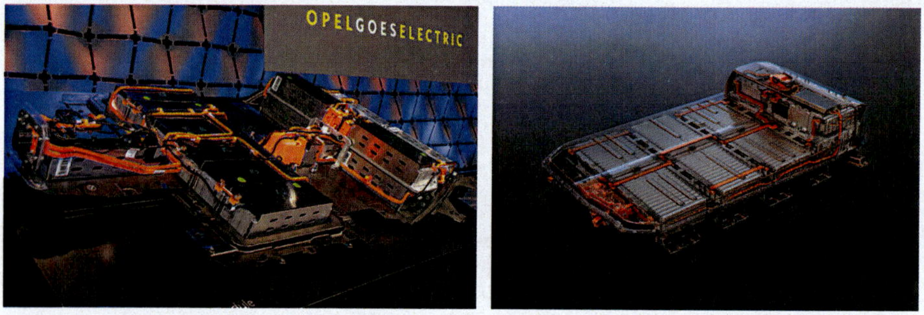

Fig. 7. High Voltage Battery of Corsa-e (left) versus Ampera-e (right)

Table 2. Comparison of high voltage battery properties

	Corsa-e	Ampera-e
Battery chemistry	Li Ion	Lithium Ion
Battery nom. energy	50 kWh	60 kWh
Battery mass	350 kg	430 kg
Max battery power (peak)	163 kW	160 kW
Battery nom. voltage	394.2 V	350 V
Peak power/energy	3,3	2,67
Spec. energy (pack) wh/kg	143	140
Cells, configuration	216 (2P, 108S)	288 (3P, 96S)
Sub-modules	18 identical modules (12 cells each)	5 mules assemblies, 3 with 60 cells, 2 with 54 cells

4.4 The e-Motor (Propulsion Motor)

In the Corsa-e, the electro motor is side-mounted to the RG and therefore has an own water cooling system. Also, the motor comes pre-assembled with the inverter power electronic on top including an internal controller, receiving the overall directions from a vehicle Controller (eVCU) (Fig. 8, Table 3).

Fig. 8. Propulsion Motor of Corsa-e (left, incl. Inverter) and of Ampera-e (right, Motor only)

Table 3. Comparison of e-Motor (propulsion motor) properties

	Corsa-e	Ampera-e
Peak torque	265 Nm	360 Nm
Peak power	100 kW	150 kW
Continuous power	57 kW	55 kW
Max motor speed	14565 rpm	8810 rpm
Stack length	175 mm	125 mm
Outer Ø	190 mm	204 mm
Pole pairs	4	4
Cooling	Water jacket	Oil. Dissipation within drive unit housing

4.5 The Reduction and Differential Gear

The respective reduction gears of Ampera-e and Corsa-e both are equipped with fixed ratio, include the differential drive and include a park lock system.

To support the cost target and to leverage existing production facilities for manual transmissions, the reduction gear for Corsa-e is manufactured in-house PSA and uses, where possible, carry-over parts from the manual transmission, e.g. the Differential drive with its ring gear. Same as for some manual transmissions, the housing is a 2-piece design and the gears are lubricated by splash. The motor is side-mounted to the RG. To realize the park-lock function, an intelligent actor is installed in the RG.

In the case of Ampera-e, the e-Motor is an integrated part of the Transmission assembly. Here, the stator of the e-motor leverages the thermal mass of the Transmission and is also cooled by the transmission oil, when necessary. In contrast, the e-Motor in the Corsa-e is side-mounted and therefore needs a separate cooling, in this case a water circuit shared with the power electronic components.

The reduction gear in the Corsa-e is adequately designed for the given power, torque and speed characteristics of the e-Motor, while it still maintains synergy with one of the group's manual transmissions.

On a propulsion system base, the overall weight made up from for reduction gear, e-motor and inverter is almost identical in the two cases (Table 4, Figs. 9 and 10).

Table 4. Comparison of reductions gears

	Corsa-e	Ampera-e
Peak axle torque (Nm)	2525	2500
Gear ratio	9.71	7.05
Mass (kg)	25 (excl. Motor/Inv)	76 (incl. Motor)
Total oil volume (L)	1,2	2,9

4.6 OnBoard Charger and DC/DC Converter

The Corsa-e comes with a standard OBC designed for 1-phase 32A charging, resulting in a maximum power intake of 7,4 kW. A similar 1Ph charger functionality but with different design can be found in the Ampera-e (Fig. 11).

As an option, the Corsa-e will offer a 3phase Charger for max. 11 kW Power Intake. This option is getting increasingly important due to widespread public charging opportunities (3Ph AC) and imbalance restrictions for the case of 1-phase charging, differing from country to country. The 3Ph charger of the Corsa-e will automatically reconfigure to a 1Ph 32A charger if paired with a 1Ph supply.

4.7 System Functions Comparison (Related to EV Drive and EV Charge)

See Table 5.

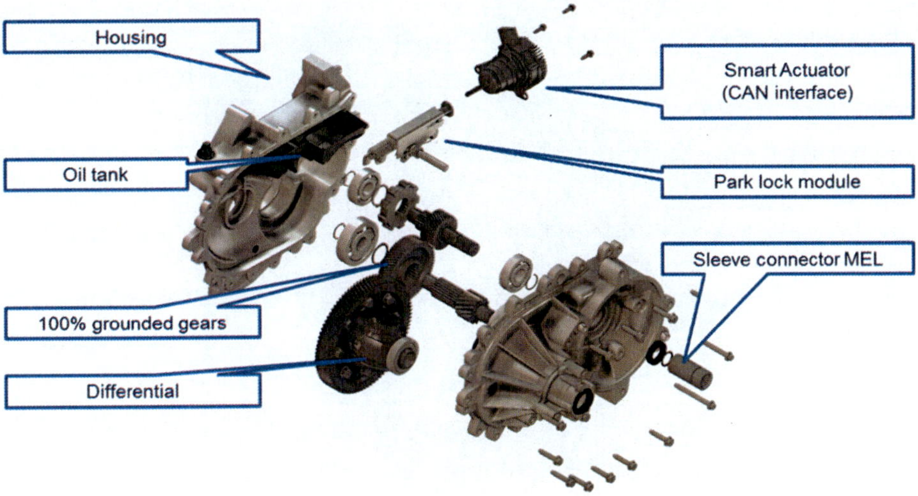

Fig. 9. Reduction gear of Corsa-e

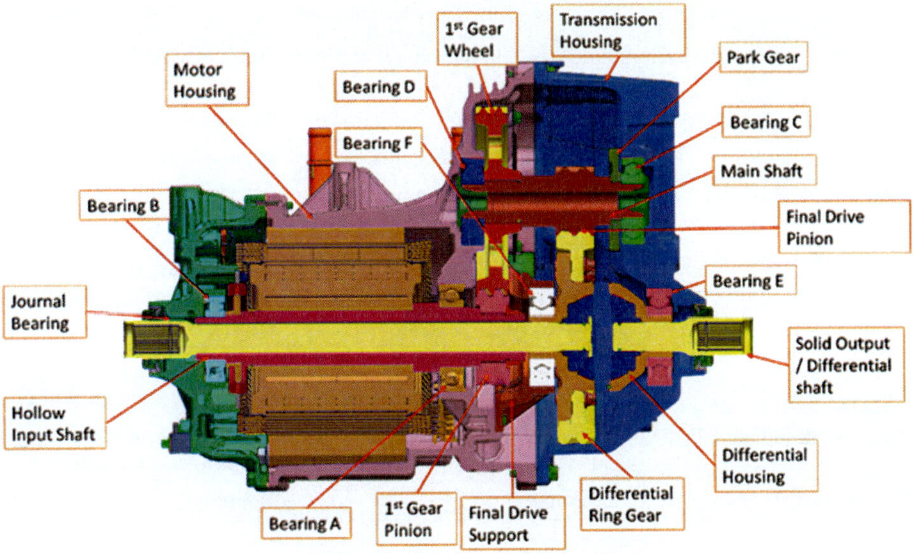

Fig. 10. Drive unit of Ampera-e

5 Preliminary Performance Data of the New Corsa-e

- Autonomy Range ~330 km
- Maximum speed 150 km/h
- Acceleration 0–50 km/h: 2,8 s 0–100 km/h: 8,1 s

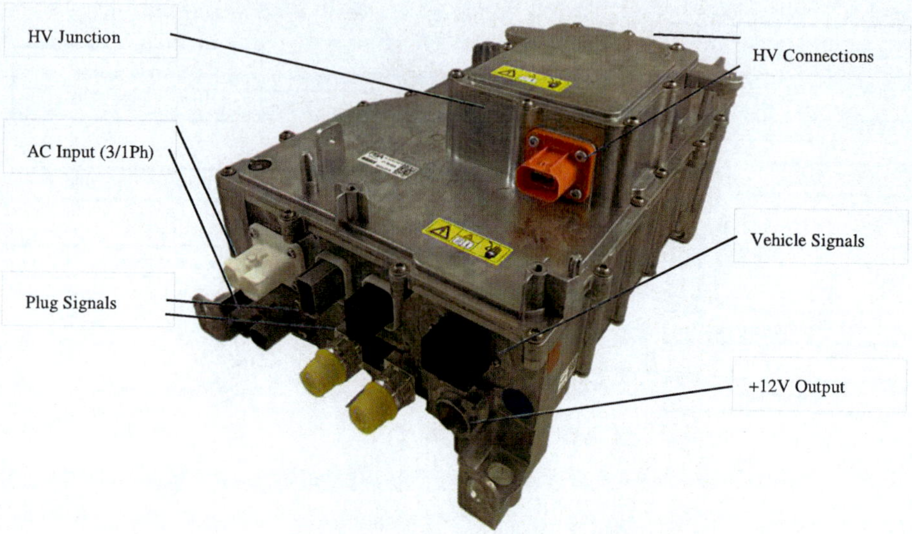

Fig. 11. 3Phase OnBoard Charger, DC/DC converter and Junction box integrated in one part

Table 5. System functions comparison related to drive and charge

	Corsa-e	Ampera-e
Dynamic modes	Eco, normal, sport	Normal, sport
Drive modes, recuperative braking	D and B (max ~50 kW)	D (max 40 kW), L (max 70 kW)
Charge modes	mode 4 with max 100 kW mode 3 AC 1PH 7,2 kW mode 3 AC 3Ph 11 kW	mode 4 with max 53 kW mode 3 AC 1Ph max 7,2 kW
	Mode 2, 8A (1,8 kW) to 12A (2,7 kW)	Mode 2, 6A (1,3 kW) or 10 A (2,3 kW)
Charge point find and navigate	Onboard data base and over the air service	Find on android phone navigate: android auto
Climate control	Remote	Remote

6 Using an Electric Vehicle: The Importance of Charging!

The following paragraph is dedicated to the importance of charging but also the different needs for EV charging, depending on EV type and charging modes. It is a generic statement not exclusively related to the Corsa-e.

Any recharge should occur with Electricity from renewable energy. The recommended charging method is mode3 charging, i.e. from a dedicated wallbox or public charge point, while mode 2 charging, connected to a household socket should be an exception on travel and reasonably suitable for BEVs.

6.1 Hybrid (HEV) and Mild Hybrid Electric Vehicles (MHEV)

Those are not designed for external charging. Hybrid batteries are installed to recuperate all or most of the energy during deceleration phases (or downhill phases) and to insert this energy again in the next applicable propulsion phase. With Hybrid vehicles, there is no need for the customer to change his/her fuel filling behavior, just the overall energy consumption is lower. Batteries in Hybrid Electric Vehicles are relatively small, typically <2 kWh usable Energy and partially even<0,5 kWh in cases of mild hybrids. (Also, those FCEV which do not offer a charge plug cannot be electrically recharged.)

6.2 PHEV, EREV and FCEV-Plugin Cars

This class of "plug-in cars" today come with typical Battery Energy of up to ~20 kWh (e.g. 16 kWh for Ampera-1). Most common in this vehicle group are 1Phase AC charge interfaces for mode-3 and mode-2 charging 16A/240 V maximum. The resulting charge power input of 3.7 kW would fill up a completely depleted 12 kWh Battery in less than 4 h, a duration which constitutes a common requirement e.g. for half-day parking. For the reason of reduced CO_2 emissions, it is important that PHEV drivers recharge their car regularly whenever they can. A full charge in shorter time then e.g. 4 h, is desirable to facilitate a multiple use of the installed electrical energy per day. Therefore, newer PHEV designs sometimes offer optional 1Ph/32A or 3Ph/16A charge interfaces. Regarding future developments we should expect to see wireless charging first in upper-segment PHEV, to make multiple and frequent recharging more convenient and thereby more likely to occur.

6.3 BEV Vehicles

Current BEV vehicles today are sold in capacity ranges between ~20 kWh (small) and over 100 kWh. For these vehicles, a closer look into charging options is needed. Any mode2 charging should be regarded a seldom exception, due to the duration, the low power level. These BEV vehicles are appropriately charged through a dedicated wallbox installation (mode3).

Private AC charging for owners of parking slots (shared garage/own house). In a part of European countries, the electricity supply to private houses is implemented in 1-phase technology, high current connection. Depending on the location, up to 32A/230 V Power intake (7,4 kW, 1-Phase) for EV charging is permitted for installation at households. A related power output will provide a 50 kWh Energy "refill" in ~7,5 h (a typical overnight charge requirement for 50 kWh Battery). In another part of European countries, private houses are supplied with 3Phase systems, and therefore offer 3Phase supply for EV charging (11 kW, 3Phase), resulting in delivery of 50 kWh Energy in 5 h if a BEV is equipped with such a 3Phase charger. In those counties where 3 Phase supplies are common, phase imbalance limits do exist. A vehicle equipped with 1-phase Onboard charger will only be allowed to draw e.g. 3,7 kW of power since throttled down to the permitted imbalance limit, resulting in 15 h to deliver 50 kWh energy.

Public AC charging. The growing majority of public 3ph charge posts in Europe provide 3Ph 22 kW power. Older installations provide 3Ph 11 kW only. In about half of the European countries, newly installed charge posts will be throttled down to 16A or 20A imbalance limit in case of a vehicle with a 1Ph charger is connected to the charge post. The corresponding charge up time for a full 50 KWh charge are the values mentioned in previous paragraph. More intelligent installations, as charge farms may level out the imbalances imposed by cars of multiple cars are charging on 1Phase each.

AC charging (general). In order to best cope with the different charge power needs for the different customers and in order to provide maximum versatility regarding the public charging installation, the Opel Corsa-e offers two OBC charger options. First, a standard OBC charger for 1Phase and 7,4 kW intake and second, a option of automatically reconfiguring OBC charger for 3 Phase and power up to 11 kW intake which can fully use both, a 16A 3Ph supply and a 32A 1Ph supply.

DC Fast Charging, High Power charging. The power flow for this charging method bypasses the OBC. Instead the voltage conversion occurs at the (bigger and more expensive) public charging station. The power directly flows into the Battery through the high-current connector. This method is suitable for charge-up on travel, especially along the motorways. Limited by Battery protection, the Corsa-e is designed to accept peak charging power of about 100 kW (up to 270A) resulting in a best recharge time of 30 min to reach 80% (a state of charge where usual Battery technologies require substantial derating). This duration changes depending on e.g. Temperatures of Outside environment and Battery [4].

HPC Charging stations for >100 kW are still the exception and will primarily be needed for Sport/Lux cars with very high Battery capacity (e.g. >80 kWh) and which target to provide fast-driving (e.g. 180 km/h and more) where high driving power is needed.

7 Summary

With the new Corsa-e, Opel launches a all-new BEV in the B-segment. A strongly improved DCFC capability and an optional 3phase charger will make it an ideal fit for many customers, while the mixed energy platform approach which it is built on, an up-integrated HV architecture and an increased amount of OEM make parts make it affordable for a wider customer base.

References

1. Electricitymap homepage. www.electricitymap.org. Accessed 26 Sept 2019
2. Stolten D, (FZ Jülich, IEK-3: Institute of Electrochemical Process Engineering)
3. Dr Ramminger P (2016) The propulsion system of the new Opel Ampera-e. In: 15th international CTI symposium automotive transmissions, HEV and EV Drives, Berlin
4. Matthé R (2017) EV battery technology and charging strategies. In: 16th international CTI symposium automotive transmissions, HEV and EV drives, Berlin

Fast and Accurate Road Interference Compensation for Objective Drivetrain Evaluation

Johann Rutz[1](✉), Thomas Ebner[2], and Ferit Küçükay[3]

[1] Institute of Automotive Engineering, TU Braunschweig, 38106 Brunswick, Germany
lncs@springer.com
[2] AVL List GmbH, 8020 Graz, Austria
[3] Director, Institute of Automotive Engineering, TU Braunschweig, 38106 Brunswick, Germany

Abstract. In cooperation with the Institute of Automotive Engineering of the Technische Universität Braunschweig and AVL List GmbH, a system to compensate the influence of the road surface on the objective driveability assessment is developed. Based on an existing simple function in AVL-DRIVE^{TM} (The Objective Tool for Driveability Assessment and Development) a system with two vertical-longitudinal axis wheel sensors is created. By this, road surface interferences on the left and right side as well as on the front and rear axle of the vehicle are considered, and the influence on the longitudinal acceleration signal, which is the main input for driveability evaluation, can be effectively compensated. In this work the influence of different road interferences on the longitudinal acceleration is measured and investigated in different driveability maneuvers like accelerator pedal tip-in or gearshifts. The principle of the road interference compensation based on multiple wheel acceleration input signals is shown, and a performance optimized algorithm is developed. While four sensors would be required to acquire the inputs of all four wheels a way to effectively compensate all wheels with only two sensors is integrated in a real time capable environment. The aim of this cooperation was to take a further step towards public road driveability assessment. Usually, this is done on test tracks with consistently good road surface conditions and exactly reproduced driving maneuvers. In contrast, during public road driving, all kinds of road surface conditions are found, significantly complicating the objective driveability assessment. With the road interference compensation, the influence of the road surface is decisively reduced.

Keywords: Influence of road disturbances in the longitudinal acceleration · Elimination of unwanted oscillations in case of road interferences · Virtual sensors for optimisation · Real-time capable method

1 Introduction

For the objective driveability assessment of a vehicle the longitudinal (chassis) acceleration is the most important input signal. In combination with other signals like accelerator pedal position, engine and electric motor speed, gear, brake and vehicle speed,

Euroforum Deutschland GmbH (Ed.): *CTI SYMPOSIUM 2019*, pp. 437–447, 2021.
https://doi.org/10.1007/978-3-662-61515-7_39

which provide information about the current driving state of the vehicle, the longitudinal acceleration is used to calculate objective ratings for all driveability relevant driving maneuvers.

1.1 Objective Driveability Attributes of the Powertrain

The input signals are processed within AVL-DRIVE™ (The Objective Tool for Driveability Assessment and Development) [1] in three steps (Fig. 1):

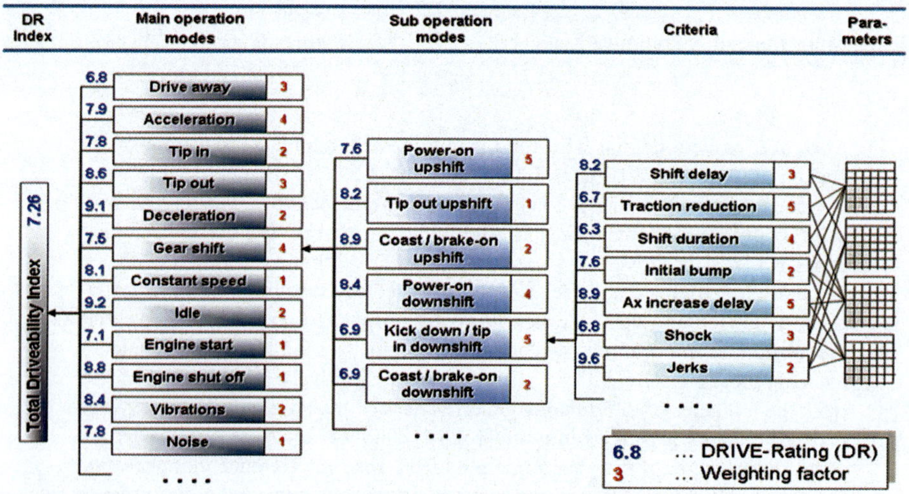

DR	Evaluation	Description
9 - 10	excellent	The driveability exceeds all customer's expectations
8 - 9	good	The driveability meets all customer's expectations
7 - 8	satisfying	The driveability meets most customer's expectations
6 - 7	acceptable	Driveability at basic level only, does not meet most customer's expectations
5 - 6	poor	Some customers complain about driveability
4 - 5	unacceptable	Most customers complain about the driveability
3 - 4	defective	All customers complain driving the vehicle
2 - 3	unsafe operation	Only limited or unsafe vehicle operation possible
1 - 2	no operation	Vehicle not operational

Fig. 1. Attribute assessment in AVL-DRIVE™ [2]. (© AVL List GmbH 2019. All Rights Reserved)

1. Detection of driveability operation modes
2. Calculation of physical parameters for driveability criteria
3. Rating of the criteria

In the first step driveability relevant operation modes like accelerator pedal tip-in or gearshifts are detected. A further categorization into sub-operation modes to characterize the current driving status completes this step. For example, the main operation mode *Gearshift* is further subdivided into *Power-on upshifts*, *Coast/brake-on downshifts*, *Kick-down downshifts* and so on. In the following step, for each sub-operation mode physical parameters characterizing specific driveability criteria are calculated. For a *Power-on upshift* such criteria are the *Shift duration*, the *Traction reduction* during the gearshift or the acceleration *Shock* amplitude at gear engagement for instance. Based on the physical parameters, finally the assessment for each criterion as well as sub- and main operation modes is done in form of unitless rating scores.

Many comfort relevant criteria with strong influence on the overall driveability score use the longitudinal acceleration signal as input. Disturbances in the road surface directly translate into disturbances in the longitudinal acceleration signal, and hence cause changes in the assessment of driveability attributes.

1.2 Influence of Road Disturbances on the Longitudinal Acceleration

In Fig. 2 the longitudinal acceleration during an identical drive away maneuver of an automatic transmission passenger car without and with road interference is shown. The brake (not shown) is released and the accelerator pedal (not shown) is pressed down to 100% in this maneuver. Close to the point in time where the engine speed is synchronized with the synchronous speed of gear 1 a bridge joint is crossed, which causes strong disturbances in the longitudinal acceleration signal (Fig. 2b). This results in deterioration of the drive away score. Without further information, it is not possible to distinguish between a bad synchronization shock event and the negative influence of the road surface. Additional wheel carrier mounted sensor information is required to enable compensation of this road interference.

2 Investigations

In order to develop a suitable method originating from an existing simple compensation of undesired interferences in the longitudinal acceleration signal, an extensive measurement campaign was carried out. Using acceleration sensors on the wheel carriers and chassis, a detailed examination of the influences of road interferences on the longitudinal acceleration signal was performed. A comprehensive measurement plan was used to investigate various maneuvers and road interferences. These investigations serve as a basis for the new compensation method.

2.1 Driveability Assessment and Prototyping Environment

As signal acquisition and assessment system AVL-DRIVE™ and the DRIVE Main Unit hardware were used (Fig. 3. Objective driveability assessment setup and wheel carrier acceleration sensor positions.), which connected to the acceleration sensors mounted on the wheel carriers as well as the passenger seat rail. For development purposes, each wheel carrier was equipped with a 2-axis acceleration sensor (longitudinal: x—vertical:

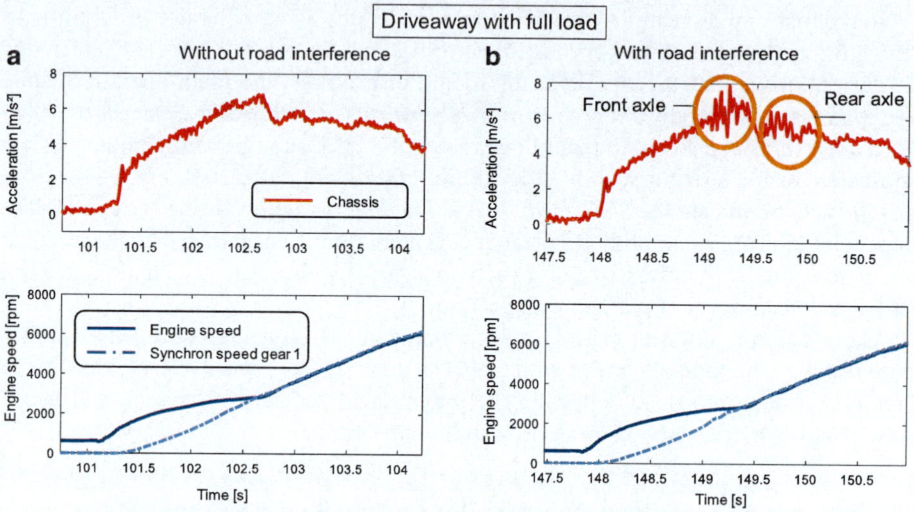

Fig. 2. **a**, **b** Influence of road disturbances on the longitudinal acceleration.

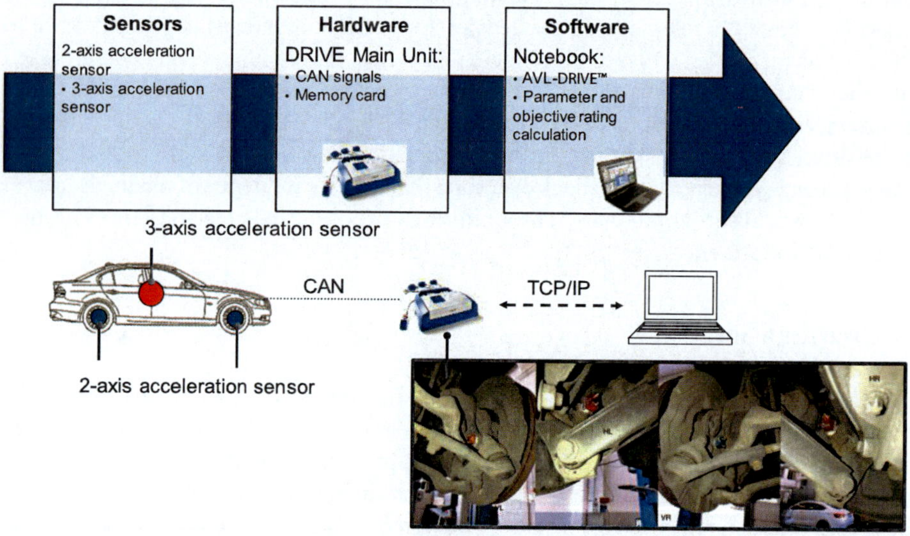

Fig. 3. Objective driveability assessment setup and wheel carrier acceleration sensor positions. (© AVL List GmbH 2019. All Rights Reserved)

z) and the seat rail with a 3-axis sensor. Vehicle CAN signals served as further inputs for the objective driveability assessment.

This system allows a real time evaluation of the driveability of a test vehicle. Using a MATLAB interface, rapid prototyping of new methods is possible, which was utilized to develop an enhanced compensation method for road interferences in the longitudinal acceleration.

2.2 Road Obstacles and Maneuvers

As basis for the development of the new method a comprehensive data set with typical public road obstacles, combined with many drivability relevant maneuvers, were measured with two different vehicles. The obstacles were divided into transient and stochastic road interferences (Fig. 4). In the case of transient disturbances, a bridge joint with negative impact on the wheel carrier was investigated.

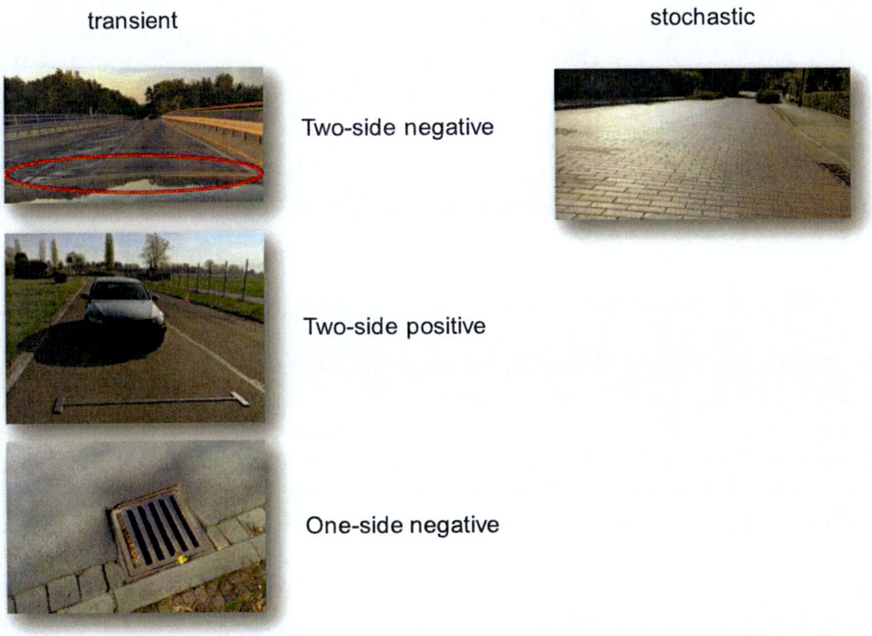

Fig. 4. Variants of road obstacles. (© AVL List GmbH 2019. All Rights Reserved)

To represent positive wheel carrier impacts, a wooden strip was placed on the test track. Positive impacts occur for example in concrete slabs. Other frequently occurring obstacles are manhole covers. In contrast to the first two transient road interferences, these occur only on one side. Cobblestone pavement was selected as a form of stochastic road interference. Driving on cobblestone causes increased and continuous noise in the signals which is not event-based. For later verification of the developed method, reference measurements were also carried out on a test track with very good road surface. The compensation is successful if the same maneuvers with and without road interference result in the same driveability ratings.

The driveability maneuvers were performed at different vehicle speeds, engine speeds and accelerator pedal positions (Table 1). Thus, an extensive measurement database to develop and test the new road interference compensation method was created.

Table 1. Driveability maneuvers.

Number	Maneuver
1	Deceleration without brake
2	Constant drive
3	Acceleration (constant pedal position)
4	Acceleration (increasing pedal position)
5	Power on upshift
6	Power on downshift
7	Coast/brake on downshift
8	Coast/brake on upshift
9	Driveaway
10	Accelerator pedal Tip-in

3 Method for Road Interference Compensation

The compensation method needs to fulfill several important requirements to be suitable for implementation in a driveability assessment system. Primary goal of the compensation is a longitudinal acceleration signal, which cannot be distinguished from the signal measured on a perfect road surface. Real-time capability plays a decisive role, as immediate feedback of the driveability assessment is expected by the test driver.

Independency from the vehicle speed as well as avoidance of over- or under- compensation are further required properties. An economical requirement is to use the lowest number of expensive acceleration sensors, while compensating interferences from both vehicle sides as well as the front and rear axle.

To fulfill the real-time requirement, the filtered acceleration signal is calculated iteratively. This is achieved with an algorithm based on an FFT analysis (Fig. 5). As an illustration, an FFT analysis of the longitudinal acceleration signal of the chassis and the vertical accelerations of the wheel carriers from the front left (Z_FL) and rear left (Z_RL) are considered. When the vehicle is crossing a road interference, the front axle is hit first followed by the rear axle. The FFT analysis when the front axle passes the obstacle is shown on the right side. The interference occurs in a similar form at the front axle signals and within the chassis acceleration signal. Corresponding spectral components of the chassis longitudinal acceleration are damped considering the frequencies.

This compensation is designed in a way that it works independent from vehicle speed, and under- or overcompensation is avoided. The spectral components are then transformed back into the time domain using the inverse FFT analysis. Thus, the acceleration signal is filtered by the interferences caused by the road surface.

3.1 Comparison of Used Sensors

Disturbances caused by road interferences can be transmitted through all wheels of the vehicle, depending where they occur on the road. Based on the measurements containing

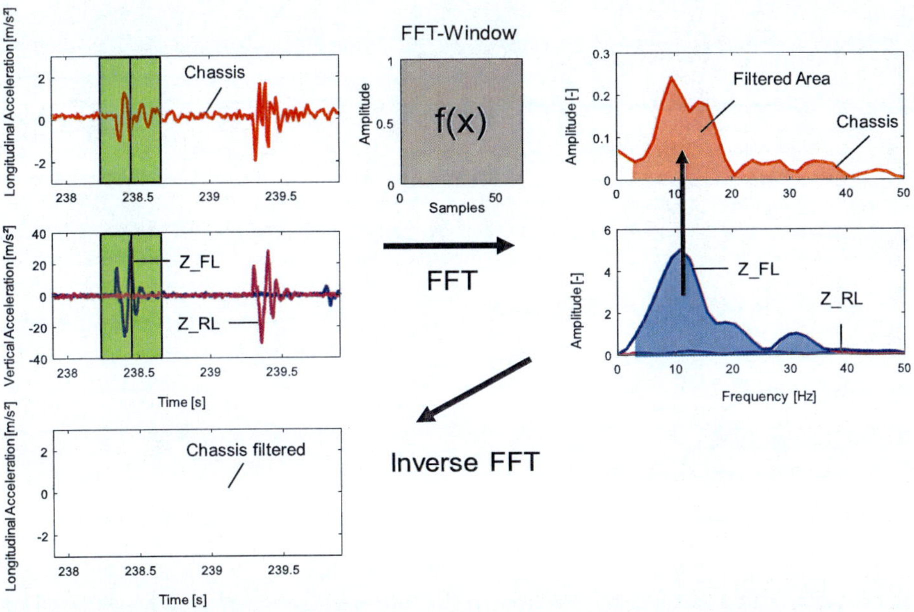

Fig. 5. Iterative FFT-Method (© AVL List GmbH 2019. All Rights Reserved)

wheel carrier x and z signals of all four wheels, the compensation quality is now examined, depending on the number and positions of sensors used. Exemplary, the crossing of a bridge joint is used. This excitation causes disturbances on all wheels, which are transferred to the longitudinal acceleration of the chassis. Figure 6 shows the uncompensated and the compensated chassis acceleration signals of the event. Figure 6a is the compensated signal if only the vertical part of the front left wheel carrier acceleration sensor is used for filtering. In this example, the acceleration signal is well filtered when the front axle crosses the obstacle. The disturbance caused by the crossing of the rear axle is not compensated at all. If the vertical component of the acceleration sensor of the rear right axle is now considered (Fig. 6b), a slight compensation follows. However, in this example the compensation is only satisfactory if the longitudinal component of the acceleration sensor is used as well (Fig. 6c, front left x/z and rear right x/z sensors).

With the addition of the remaining two 2-axis acceleration sensors, the compensation quality is often not considerably improved (Fig. 6d). In conclusion, two 2-axis acceleration sensors are sufficient for two-sided disturbances like the bridge joint. In the case of one-sided disturbances, such as a manhole crossing, the two 2-axis sensors are no longer sufficient.

3.2 Virtual Sensor Signals

Ideally, test vehicles are equipped with four 2-axis acceleration sensors to achieve the most accurate compensation result. However, this also increases the instrumentation effort and the costs for the sensors. As shown before, two 2-axis accelerometers provide good compensation for two-sided disturbances. In case of a man-hole on the right side

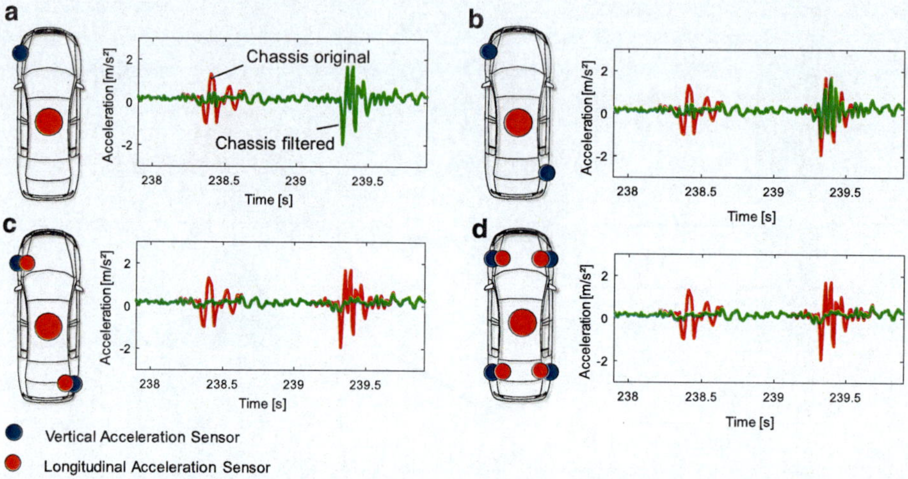

Fig. 6. **a–d** Comparison of used sensors. (© AVL List GmbH 2019. All Rights Reserved)

the obstacle is first crossed by the front right followed by the rear right wheel. The disturbance transferred by the front right wheel is not covered by a sensor and thus will not be compensated sufficiently.

For this reason, a method is developed to generate virtual sensor signals from the real measured sensors for the remaining wheels (Fig. 7). The virtual sensor signals are generated by combining the measured signals with the vehicle speed and wheelbase information, and then further used for FFT analysis.

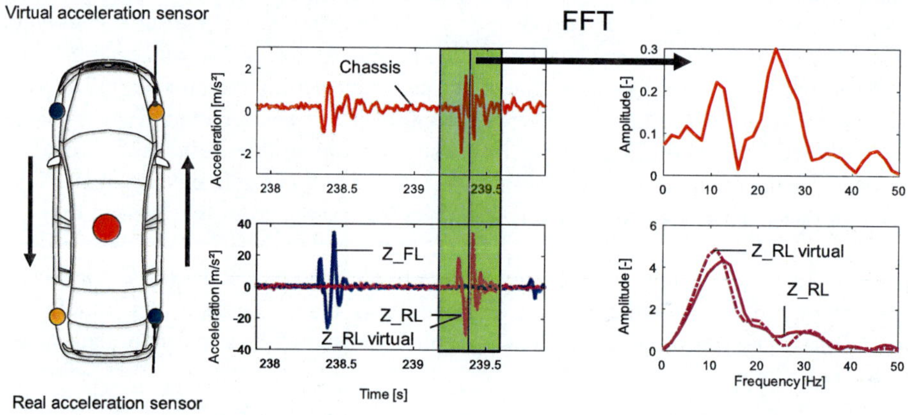

Fig. 7. Virtual sensor signals. (© AVL List GmbH 2019. All Rights Reserved)

In Fig. 8a the compensation quality using four real 2-axis sensors against two real 2-axis sensors plus two virtual 2-axis sensors is compared (Fig. 8b). The considered driveability maneuver is constant driving over a manhole on the left side. The example demonstrates an almost identical compensation quality. In Fig. 8b the lack of compensation without the virtual sensors can be seen.

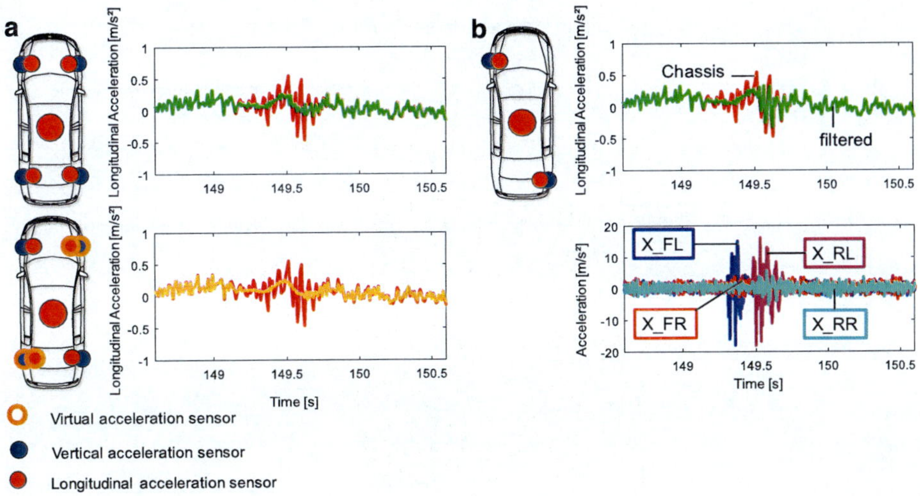

Fig. 8. **a, b** Virtual sensor signals—manhole event. (© AVL List GmbH 2019. All Rights Reserved)

Using the virtual sensor approach, the required sensors and instrumentation time can be cut into half. The demand for calculation performance, which is an important factor for real-time capability, is significantly reduced.

3.3 Real-Time Capable Method and Compensation Examples

A further optimization of the performance of the road interference compensation to enable online driveability assessment in AVL-DRIVE^{TM} is required. The algorithm and the window function of the FFT analysis are modified so that the calculation can run with 10 Hz instead of 100 Hz. As a result, the required calculation time can be considerably reduced without significantly losing compensation quality (Fig. 9).

The final developed compensation method was tested with all gathered measurement data. A comparison between identical events without and with road interference, as well as using two 2-axis sensors versus 1 z sensor on the front left wheel only, was carried out. In Fig. 10 a full load drive away maneuver and a gear 2–1 power-on downshift maneuver with 50% accelerator pedal are shown as examples.

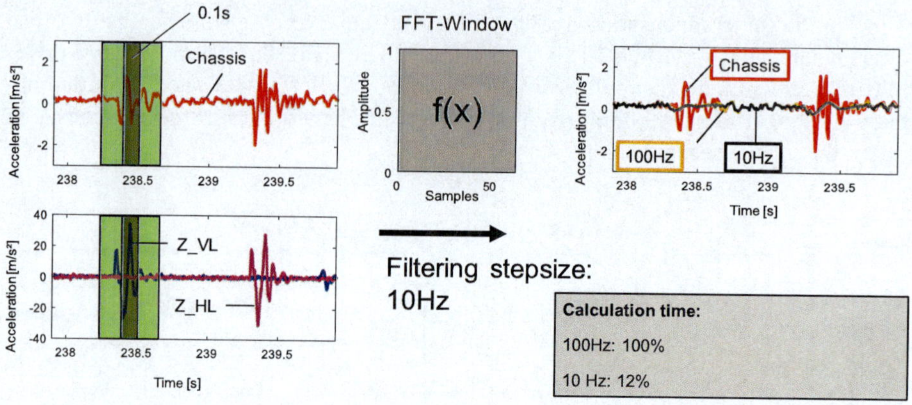

Fig. 9. Real-time capable method (© AVL List GmbH 2019. All Rights Reserved)

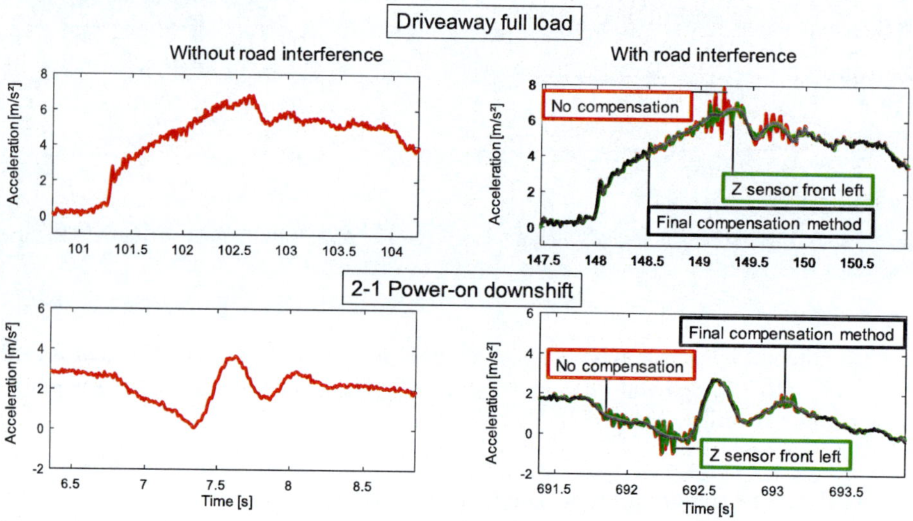

Fig. 10. Comparison for different events (© AVL List GmbH 2019. All Rights Reserved)

4 Summary

For objective driveability assessment the road surface has considerable influence on the final scores, which cannot be neglected. In this paper a method to compensate interferences on both vehicle sides as well as on the front and rear axle, combined with lowest possible expenditure of sensors and real time capability, is shown. The described compensation method presents an important step towards driveability assessment capability during public road driving.

A current limitation of the method is present, if disturbances in the wheel carrier sensors are induced by the powertrain itself. Further development will be undertaken to investigate the possibility to distinguish road- from powertrain originated disturbances.

The target is a selective compensation by integrating additional powertrain signals and vehicle information.

References

1. Schoeggl P, Ramschak E (2000) Vehicle driveability assessment using neural networks for development, calibration and quality tests. SAE Technical Paper 2000-01-0702. https://doi.org/10.4271/2000-01-0702
2. AVL-DRIVE™ 4.2.0 Product guide

BEV AWD EDU Gear Ratio Selection for Efficiency

Yogesh Mehta, Cong Liao, and Michael Schulte(✉)

AAM, Detroit, MI, USA
Michael.Schulte@aam.com

Abstract. The single speed gearbox is a largely preferred solution for electric drive units (EDU) in hybrid vehicle P4 and Battery Electric Vehicles (BEV). A proper gear ratio selection is required to meet the power and speed performance requirements, and also to minimize the parasitic losses over the life of the vehicle. Conventional wisdom holds that a higher ratio power dense gear box design for an EDU is lighter, more packaging friendly, and more efficient. This study examines the validity of this assumption for an auxiliary EDU installed in the rear of a BEV CUV providing AWD functionality. This paper outlines the power loss analysis for a single speed EDU propelled by an induction electric machine, and provides a useful approach for gear ratio selection at the concept phase of a design. Specifically, this paper determines the range, mass and cost tradeoffs for two auxiliary EDU design options without disconnect function, with gear ratios of 7 and 10 for a given duty cycle, and for two different induction motor material choices (aluminum and copper). We show that the lower gear ratio design option with aluminum rotor saves approximately 450 kWh energy compared to the higher gear ratio over the life of the vehicle. For an assumed battery miles/kWh, this saving translates into approximately $30 lower overall cost (battery+EDU). Alternately, for a given battery pack capacity, the ratio 7 design with aluminum rotor is expected have a longer range (+2.13 miles) for an estimated $31 cost and 6.5 kg mass increase.

Keywords: Efficiency · Gear ratio · Edrive

1 Introduction

Determining the gear ratio of a single speed EDU gear box is critical during the concept design phase of the development. The gear ratio directly affects the acceleration, range and overall system cost. A properly selected gear ratio will not only satisfy the performance requirements, but also maximize system efficiency and minimize the overall cost of ownership for the end user.

Conventional wisdom in the industry is that a higher gear ratio leads to a lighter, more packaging friendly and more efficient EDU. However, it will be argued here that general understanding of efficiency cannot be used as a rule. The conclusion at the end shows that efficiency results are highly dependent on duty cycle and usage assumptions. As the

Euroforum Deutschland GmbH (Ed.): *CTI SYMPOSIUM 2019*, pp. 448–458, 2021.
https://doi.org/10.1007/978-3-662-61515-7_40

EDU propels the vehicle (or is being dragged when passive), losses are accumulated due to frictional dissipation of energy in gears, bearings and seals (mechanical losses), oil churning in the gearbox, motor electromagnetic inefficiency, rotor windage losses, and the losses incurred in speeding up and slowing down the weight and rotational inertia of the EDU. These losses are dependent on speed, loads, and temperature, but the exact nature of dependency is different for different frictional interfaces. The choice of gear ratio changes the speed and torque values of the various components and interfaces. Thus, the system efficiency will change with gear ratio and with the content and distribution of the energy in the duty cycle.

A comprehensive power loss analysis for a variety of gear box ratio options needs to be considered in order to determine the best proposal. The analysis object of this paper is a passively cooled, induction motor driven auxiliary EDU without a disconnect function. Because of the nature of this application (auxiliary), most of the time the EDU is passive (being dragged). This paper compares the power loss, mass and cost tradeoffs of two EDU design options with gear ratios of ~7 and ~10 using a customer provided duty cycle for two different rotor material options. Estimated battery and rotor costs are considered in the analysis. The ratio 10 aluminum rotor design is the lightest and least expensive choice if the battery capacity is assumed fixed. However, this study will show that the lower ratio 7 design dissipates less energy, which allows reducing the size and cost of the battery up front, or extending the range of the vehicle.

2 Power Losses

The power loss of an EDU is an indicator of the amount of available battery energy, which is not used for propelling the vehicle (i.e. irrecoverably lost as heat). The smaller the parasitic losses, the higher the efficiency of the energy usage from the battery. Figure 1 shows the various mechanisms that contribute to the losses that are considered in the analysis presented here. They include the mechanical (gear/bearing/seal) losses, oil churning losses in the gearbox, motor propulsion and rotor windage losses, as well as the weight and inertia-related losses incurred during acceleration and deceleration of the EDU.

Figure 2 shows the layouts of the two EDUs with different gear ratios; both are parallel shaft offset designs. There are two stages of gears and seven ball bearings in the two layouts. The input gear is mounted directly on the rotor shaft and is cantilevered at the drive side bearing. The gears and bearings in both designs satisfy the durability and strength requirements.

Gear ratio 7 (7.15 actual ratio) design has a motor stack length of 95 mm, while the gear ratio 10 (9.88 actual ratio) design has a motor stack length of 70 mm. Additional specifications are presented in Table 1 below. Both design options were analyzed for two different rotor material options (aluminum and copper).

2.1 Mechanical Loss

The mechanical losses come from the gear mesh, bearing and seal losses. In this EDU design, there are two stages of helical gears; seven bearings; and two seals, which contribute to the mechanical power loss.

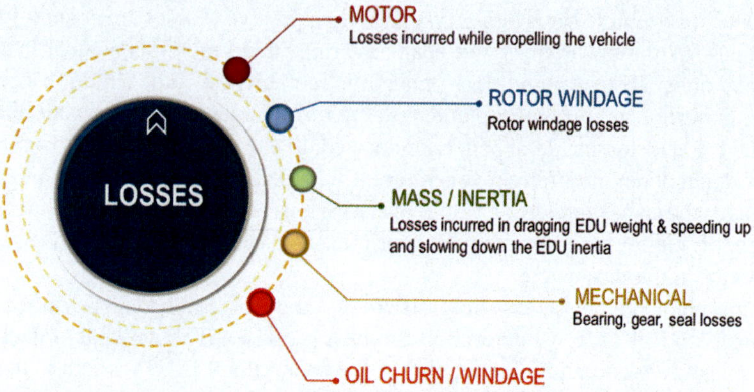

Fig. 1. The sources of the power losses

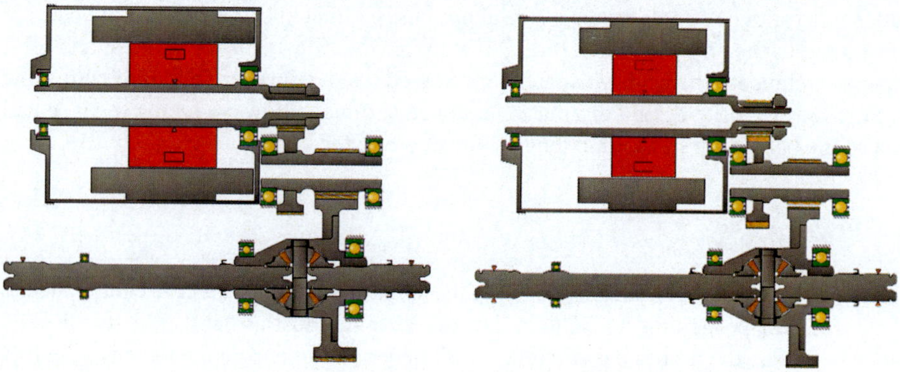

Fig. 2. The layouts of the two designs. Left—gear ratio 7. Right—gear ratio 10

Table 1. High-level parameters for the two design options

Parameter	Ratio 7	Ratio 10
Motor OD, mm	212	212
Height, mm	95	70
Ratio	**7.15**	**9.88**
Max torque, Al (350 V, 322 Arms), Nm	258.5	189.5
Peak power, Al (350 V, 322 Arms), kW	93	86

The gear mesh power loss is a function of sliding ratios and the coefficient of friction, as well as the torque, rotational speed, and operating helix angle. The sliding ratios at the start of approach and the end of recess, along with the transverse working pressure angle, determine the mechanical advantage. The mechanical advantage is gear geometry dependent, and the mesh coefficient of friction is determined by kinematic oil viscosity at sump temperature; load intensity; and tangential pitch line velocity, among other factors.

The bearing power losses consist of load dependent loss, and bearing churning loss. The load dependent power loss is calculated based on bearing designation, equivalent bearing load, basic static load rating, the axial and radial components of dynamic bearing load, and the bearing geometry. The bearing churning loss is influenced by the oil kinematic viscosity at the bearing temperature; rotational speed; bearing geometry; and dip factor. For the sealed bearing (e.g. the bearing on the jackshaft outboard), the power loss caused by the friction moment of seals is also included.

2.2 Oil Churning Loss

Both the gears and shafts churn the oil in the gearbox, which results in power losses. Bearings also create oil churning losses, which are included as a part of the ISO/TR 14179 -1: 2001 calculations.

For the gear churning losses, the gear teeth and sides of the blank both contribute to the total. The gear tip diameter, face width, tooth surface roughness, and helix angle are involved in the tooth churning. The gear root diameter is used in calculating the churning losses from the side faces. In both churning calculations, the dip factor, oil kinematic viscosity, and rotation speed are parameters that affect the calculations.

Shaft churning loss is determined in a similar way as gears. Two contributions are obtained, one from the sides of the shafts and another from the outer diameter and length.

2.3 Motor Propelling Loss

Motor propelling losses occur when the motor is energized. The high-frequency current in the winding not only induces the current in the squirrel cage for generating the torque, but also eddy currents in the lamination of the rotor and stator, which create the "iron losses". The current passes through the winding and squirrel cage and generates heat, and these losses are called "copper losses".

2.4 Rotor Windage Loss

The rotor spins and churns the air in the motor chamber. The drag from the air causes rotor windage losses, and mainly comes from shearing of the air gap. The loss is a function of length and speed of the rotor. Speed has a stronger influence on the losses than the length.

2.5 Inertial Loss

When the vehicle is moving, part of the electric energy of the motor is converted into the translational and rotational kinetic energy of components in the vehicle. During braking, this kinetic energy can be recycled by the regeneration of the motor. For an auxiliary EDU (which does not regen by itself), the energy that is stored can still be recycled to some extent by the EDU on the other (front) axle. This is possible because the two EDUs are connected through the vehicle body. The translational and rotational kinetic energy of the auxiliary EDU can be transferred to translational kinetic energy of the vehicle, and in turn recycled by the front EDU, which is engaged in the regen process. However, the portion of the energy restored will be lower than if the auxiliary EDU itself were used in directly regenerating, as the energy now must go through more mechanical components before being harnessed in the front. In this analysis, a round-trip efficiency of 50% was assumed for the EDU. This also means that the higher the initial inertial energy stored in the rear rotor for any given vehicle event, the more of it will be wasted during the deceleration.

3 Assumptions and Calculation Strategy

For this paper, the mechanical and churning losses were calculated using ISO/TR 14179 -1: 2001. It was assumed that the oil level does not change regardless of whether the component dipped in oil is spinning or not. This assumption reduces the fidelity of the churning loss assessment but was considered appropriate for this case study.

To calculate the motor losses incurred while the auxiliary EDU is propelling the vehicle, efficiency maps were obtained from MotorCAD for aluminum and copper rotors for both ratio 7 and ratio 10 designs. These maps provide loss estimates in Watt as a function of motor rotation speed and torque, assuming working temperature of 125 °C.

Windage losses were obtained from MotorCAD for both ratio 7 (95 mm) and ratio 10 (70 mm) rotors as a function of speed. With this it was possible to determine the windage losses expected to be incurred for each cell in the JTIB duty cycle file. The total losses expected for the entire duty cycle were obtained by simply summing up the losses in all cells.

The calculation of losses incurred during accelerating and decelerating the vehicle requires a time domain schedule to be available. The LA92 schedule and 200 k miles of total driving were assumed in this analysis. The two EDU designs have different total mass and reflected inertia. When copper or aluminum are used in the rotor, the mass and inertia of the rotor are different and considered in the calculation. As the vehicle speeds up and slows down, a portion of the kinetic energy stored in the mass and inertia is absorbed through regenerative braking with the active (front) EDU. The remaining energy, which is not absorbed back into the battery is dissipated as heat at various frictional interfaces in the regen load path.

Figure 3 summarizes the approach used in power losses calculations and subsequent comparison between designs. The JTIB duty cycle was used for mechanical, oil churning, motor propelling, and windage losses. The LA92 schedule was used for assessing inertial losses.

4 Results and Comparison

The high-level duty cycle analysis is shown below. As shown in Fig. 4, the auxiliary EDU is dragged passively more than 90% of the time, without propelling the vehicle or regenerating any electric power.

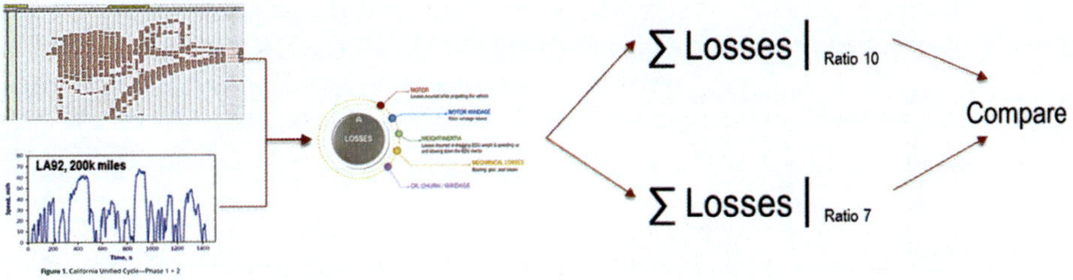

Fig. 3. Calculation strategy for power losses

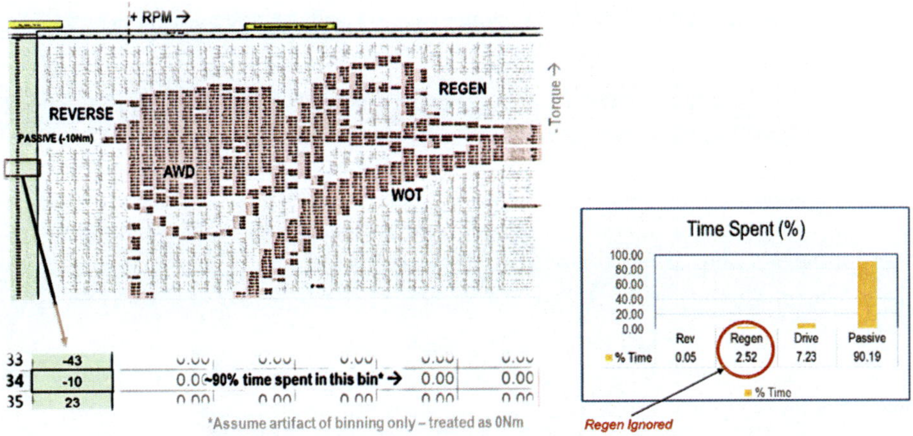

Fig. 4. The composition of the duty cycle

When the EDU is spinning, the mechanical, oil churning, and rotor windage losses are unavoidable, as the system architecture does not have an auxiliary EDU disconnect function. Figures 5 and 6 show the mechanical and oil churning losses, assuming various average oil sump temperature (40 °C, 60 °C and 80 °C) throughout the vehicle life (duty cycle). It becomes clear that an auxiliary EDU with a higher gear ratio has higher mechanical and churning losses than one with a lower gear ratio.

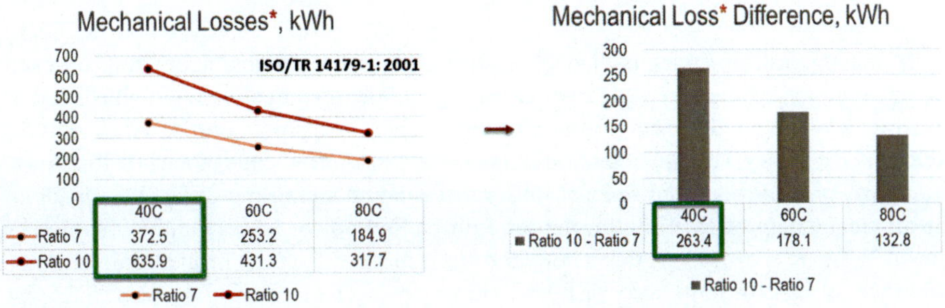

Fig. 5. Mechanical losses of the EDUs of ratio 10 and 7, and the difference of the losses

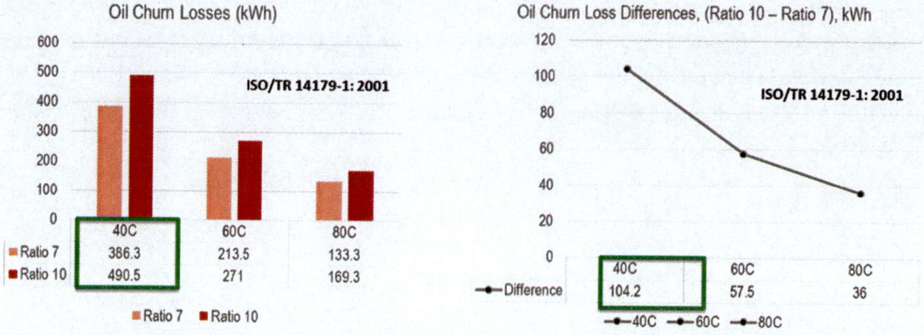

Fig. 6. Churning losses of the EDUs of ratio 10 and 7, and the difference of the losses

Rotor windage losses of the two design options are related only to the rotor speed and stack length. The squirrel cage is made of either aluminum or copper, but the rotors are of the same length and diameter for a given gear ratio. Though the design of ratio 10 has a shorter rotor (70 mm) than the design of ratio 7 (95 mm), the overall windage power loss incurred with the ratio 10 shorter rotor is higher. Figure 7 shows the windage losses of the two ratio design options over the duty cycle.

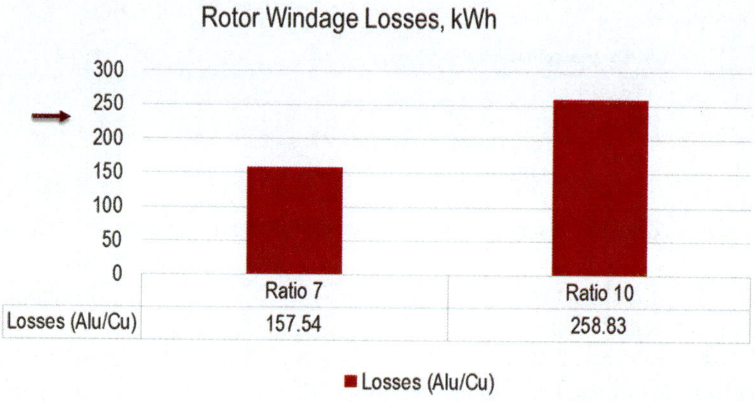

Fig. 7. Comparison of rotor windage losses of auxiliary EDUs with ratio 7 and 10

Motor propelling losses only happen during the drive cycles, depending on each motor's load and efficiency map. For a given output (axle) duty cycle, EDUs with different gear ratios will have different speed and torque operating points on the motor curves. Additionally, for a given EDU gear ratio, the choice of squirrel cage material (aluminum or copper) also introduces efficiency differences. By mapping the drive cycle loads to the efficiency map of each motor, the total propelling power losses can be obtained as shown in Fig. 8. It is obvious that a copper rotor is more efficient than an aluminum rotor while propelling, for both ratios (EDUs). However, when comparing the two EDUs for a given rotor type, the power loss differences between ratios are small.

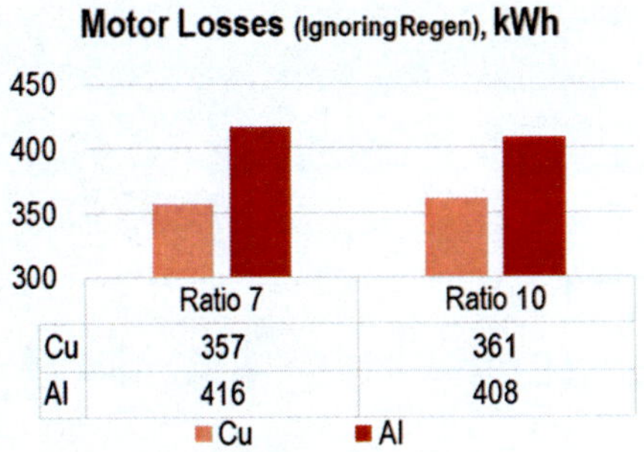

Fig. 8. Motor propelling losses for the EDUs with ratio 10 and 7 for copper and aluminum rotor

Between aluminum and copper rotors, the former is lighter and has less inertia than the latter. This works in favor of aluminum for inducing smaller inertial power losses (for a given ratio). The two EDU design options have a different total mass and reflected inertia because of differences in squirrel cages and gear ratios. Table 2 summarizes the mass and inertia specifications for the two rotor designs with aluminum and copper.

Table 2. Mass and inertia information of the two designs.

Attribute *	Rotor	Ratio 7	Ratio 10
Mass (Kg)	Alu	27.35	20.93
Inertia (Kg x m^2)	Alu	0.024964	0.019076
Mass (Kg)	Cu	30.35	23.83
Inertia (Kg x m^2)	Cu	0.033345	0.026626

* Includes rotor shaft
* Housing mass difference between R10 and R7 (~0.235kg)

As the vehicle accelerates and decelerates, a portion of the kinetic energy stored in the auxiliary EDU is absorbed through regenerative braking by the prime mover (front) EDU. Assuming 50% of the kinetic energy is recycled as stated above, the rotor reflected inertia losses were calculated based on LA92 drive cycles. The results are shown in Fig. 9. The lighter aluminum rotor has lower losses than the heavier copper rotor for both designs. Even though the ratio 10 EDU uses a shorter motor, it incurs higher losses than the ratio 7 EDU as the motor spins faster (gear ratio effect) and speed has a larger influence on losses compared to length.

5 Conclusions

In this section, all losses calculated above are summarized and compared. 40 °C was selected as the assumed average temperature for all transmission side temperature dependent losses.

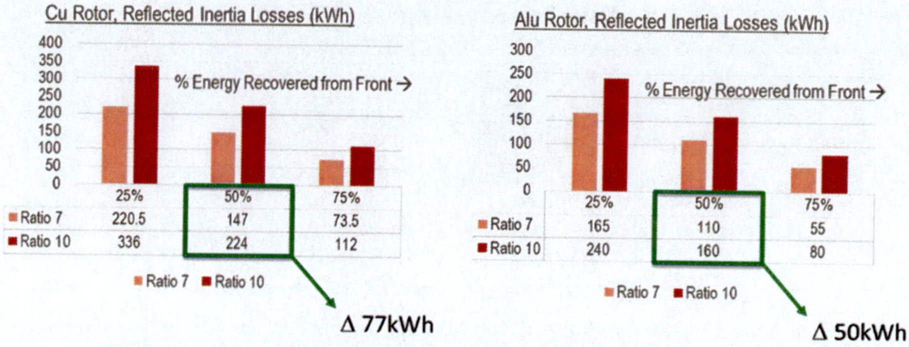

Fig. 9. Rotor reflected inertia losses assuming 50% energy recycle rate

Table 3 shows the power loss differences between ratio 10 and ratio 7 EDU design options for both copper and aluminum rotors per the analyzed duty cycle. The ratio 10 design with copper rotor has 489 kWh higher losses than the ratio 7 design. With an aluminum rotor, the ratio 10 design has 451 kWh higher losses than the Ratio 7 design.

Table 3. Total power loss differences between EDU designs for gear ratios 10 and 7 with copper and aluminum rotors

Losses	Al Rotor, kWh (Ratio 10 – Ratio 7)	Cu Rotor, kWh (Ratio 10 – Ratio 7)	Assumptions
Motor	-8	4	No rear regen
Rotor Windage	101	101	
Mass	-61	-62	LA92, 50% recovered,0.3275m tire
Inertia	50	77	LA92, 50% recovered, 0.3275m tire
Mechanical	265	265	40C, ISO 14179-1: 2001
Oil Churn	104	104	40C, ISO 14179-1: 2001
Total	**451**	**489**	

****Positive number = Ratio 7 more efficient*

Although a copper rotor is more efficient than an aluminum rotor while propelling, since the motor is passive most of the time, the efficiency advantages of copper are almost fully negated by the added mass/inertia of the ratio 7 EDU design. The result is shown in Table 4.

For completeness, please note that for the gear ratio 10 EDU design option, the impact of mass/inertia becomes greater due to its higher reflected inertia across the higher gear ratio. The net result for the ratio 10 design option is that the lighter aluminum rotor incurs lower net losses even though it is less efficient while propelling the vehicle.

Table 5 shows the impact to vehicle range for the two EDU gear box ratio options with aluminum or copper rotor material choices. The assumed vehicle life was calculated from the duty cycles and tire rolling diameters provided by a customer. The battery capacity,

Table 4. Total power losses difference between EDU designs with copper and aluminum rotors (shown for ratio 7 only)

Losses	Cu – Alu, kWh (Ratio 7)	Assumption
Motor	-59	No rear regen
Rotor Windage	0	
Mass	27	LA92, 50% recovered,0.3275m tire
Inertia	37	LA92, 50% recovered, 0.3275m tire
Mechanical	0	40C, Masta ISO 14179-1: 2001
Oil Churn	0	40C, Masta ISO 14179-1: 2001
Total	**5 kWh**	

****Positive number = Aluminum more efficient*

weight, and vehicle range per charge are reasonable estimates based on the background program information available at the time. The deep charge cycles are calculated based on assumed life and range per charge. Since the energy loss difference of the two design is available from the previous analysis, the energy saved per deep charge for the ratio 7 design over the ratio 10 design can be calculated.

Table 5. Impact of gear box ratio 7 on vehicle battery range

Attribute	Value (Cu)	Value (Alu)	Units	Source
Vehicle Life	182k – 200k	182k – 200k	Miles	Estimate
Battery Size	60	60	kWh	Assumption
Battery Mass	960	960	lb	Assumption
Vehcile Range	238	238	miles	Assumption
Loss Difference	489	451	kWh	Calculated
Deep Charges	770 - 840	770 - 840	-	Calculated
Charge Saved/Deep Charge	0.58 - 0.64	0.54 - 0.59	kWh	R7-R10
Additional Range	**2.31-2.51**	**2.13 – 2.32**	**Miles**	**R7-R10**

Assuming a fixed battery capacity and 4 miles/kWh, the additional range for ratio 7 design was calculated to be about 2 miles.

Alternately, if designing to a fixed range requirement, the ratio 7 design would require about 0.5 kWh smaller battery pack.

For a given battery capacity, the ratio 7 EDU design option saves more energy, and is expected to extend the range of the vehicle compared to the ratio 10 EDU design. However, this design requires a longer motor and thus incurs more cost in the upfront motor, housing and shaft materials.

Table 6 shows that the ratio 7 EDU design is expected to cost 35 or 31 USD more than ratio 10 design with copper and aluminum rotors respectively.

Table 6. Cost estimation for EDU gear box design options (ratio, rotor material)

Attribute	Guidance Cost (Cu) (R7-R10)	Guidance Cost (Al) (R7-R10)	Units
Motor Cost (estimate)	33.5	29.5	USD
Housing Cost (estimate)	1.5	1.5	USD
Total	**35**	**31**	**USD**

In conclusion, Table 7 summarizes the cost-range-mass trade-off comparisons for the two EDU design options with two rotor material choices.

Table 7. Cost-range-mass trade-off matrix

Rotor Type	EDU Cost Impact Estimate, Range & Mass	
	Ratio 7	Ratio 10
Aluminum	*Base + $31* *Range + 2.13* *Mass + 6.65*	*Base (USD)* *Range (miles)* *Mass (kg)***
Copper	*Base + $63* *Range + 2.10* *Mass + 9.65*	*Base + $28* *Range – 0.21* *Mass + 2.90*

For a fixed battery capacity, the ratio 10 aluminum rotor EDU design is the lightest and least expensive choice in terms of upfront cost.

However, the ratio 7 aluminum rotor EDU design can extend the range by 2.13 miles for an additional $31 & 6.65 kg added mass (range impact of added mass is included in these calculations).

An Efficient and Automated Design Strategy for Multiphysics E-Motor Development

Torben Schulze[1](✉), Jean-Baptiste Mouillet[2], Thomas Lehmann[3], and Lars Fredriksson[3]

[1] Mercedes-AMG GmbH, Affalterbach, Germany
torben.schulze@daimler.com
[2] Altair Engineering France, Antony, France
[3] Altair Engineering GmbH, Böblingen, Germany

Abstract. Standard design processes for e-motors foresee siloed development of different attributes depending on nature and type of requirements. Thus, electromagnetic, thermal, mechanical and NVH attributes are developed individually which leads to unfavorable compromises and/or late costly firefighting. Mercedes-AMG GmbH are developing high-performance e-motors that deliver maximum performance under severe driving conditions. This requires high attention to durability and thermal effects and although serving high-performance cars, the motors must fulfill highest NVH standards. Altair Engineering has introduced Simulation-Driven Design (SDD) to support holistic, multiphysics development of e-motors. SDD processes are introduced to enhance classical development by accounting for essential development requirements simultaneously using optimization and by continuously providing a maximum of information to support the design process. This paper introduces a pilot application of a unique, highly automatic, multiphysics design process for e-motors, based on a current e-motor program at AMG. The process is developed to consider essential development requirements including electromagnetics and thermal requirements, NVH, stress and durability. It accommodates for DOE, multi-objective optimization and design exploration methods to be used to explore and find feasible motor designs. The paper will show how the strategy adds efficiency to the e-motor development process and how it can impact the total costs of development. The uniqueness of this development lies in the automation of the multiphysics design process and the creation of an environment to democratize multiphysics design optimization within the e-motor design departments.

Keywords: E-Motor design · Optimization · Multiphysics

1 Introduction and Motivation

The rate of which electric hybrid and full-electric vehicles are reaching the market and are being developed has increased significantly over the last years and associated requirements are increasing rapidly as well. The goal is to develop better vehicles including better e-drivelines within tighter time and cost schedules. Simultaneously, the technical

Euroforum Deutschland GmbH (Ed.): *CTI SYMPOSIUM 2019*, pp. 459–472, 2021.
https://doi.org/10.1007/978-3-662-61515-7_41

requirements on the drivelines being developed are increasing rapidly, both in terms of level and bandwidth of requirements. FE and other simulation methodologies have traditionally been used very successfully to verify designs and design directions.

2 Simulation-Driven Design

Today, FE and especially numerical optimization is increasingly used to drive the design process, i.e. optimization is used to help design teams finding best alternatives, executing sensitivity studies, performing trade-offs between different design alternatives, etc. This design strategy is often denoted *"Simulation-Driven Design"*. Simulation-driven design is especially beneficial, where the design is less intuitive because of high design complexity and/or complexity of loads and targets for the design.

Multiphysics design optimization methodologies make it possible to design e-drivelines and e-motors for multiple, completely different design requirements simultaneously, thus avoiding a siloed development strategy, where a larger number of design iterations are necessary to fulfill all requirements and unfavorable design compromises need to be accepted.

Multiphysics and multi-disciplinary optimization, however, need efficient processes to be executed within the narrow technical and time constraints of a live product development. The processes need to be integrated with the different departments involved in the e-motor development. Simulation models for different physics need to be made available and need to be up to date in order to successfully perform holistic multiphysics design studies.

Design input and restrictions from design, packaging, production etc. need to be considered when setting up the design optimization problem. Otherwise, the results will not be feasible, nor relevant. Finally, all input and restrictions need to be brought into one single optimization process as shown in Fig. 1.

3 The Reference E-Motor Optimization Environment

3.1 Description of Reference Process

A pilot study conducted together with an industry partner is used as a reference for the development presented in this paper, see Luthardt et al. [1]. A simplified view of the reference multiphysics design optimization environment is depicted in Fig. 2. Altair HyperStudy serves as the optimization engine and is used to drive all other subprocesses and tools necessary to execute a complete optimization study.

The addressed design optimization problem is depicted in Fig. 3. Three working points were considered, and a total of 10 responses were extracted. In order to execute all simulations covering all working points, a sequence of solver execution scripts was defined for HyperStudy. The main effort was spent to create the design variables for the shapes in Flux, to setup the process manually in HyperStudy incl. linking the shape DV's of all models and linking the output responses from one model (e.g. current, speed, control angle) to input of next model. Using the established reference process, optimization and design exploration could be conducted efficiently and a final design could be found that fulfilled targets at improved objective function values.

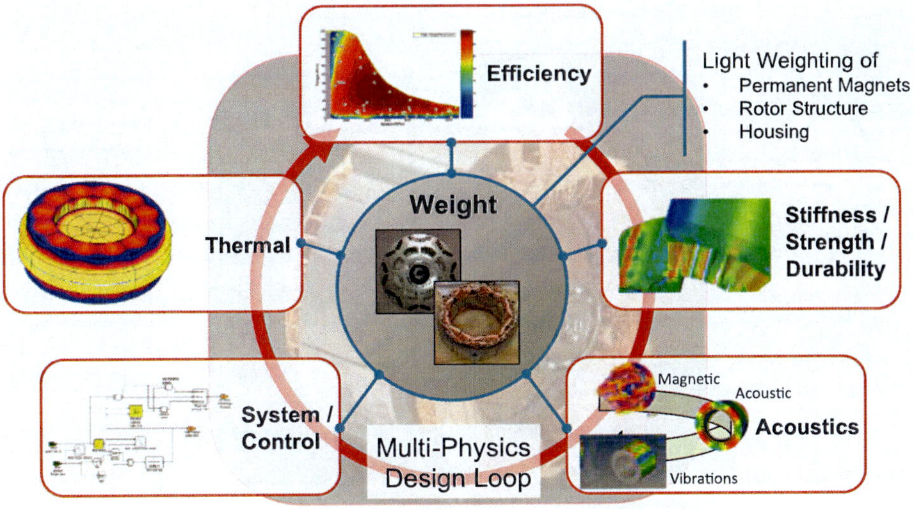

Fig. 1. Generic e-motor multiphysics optimization environment

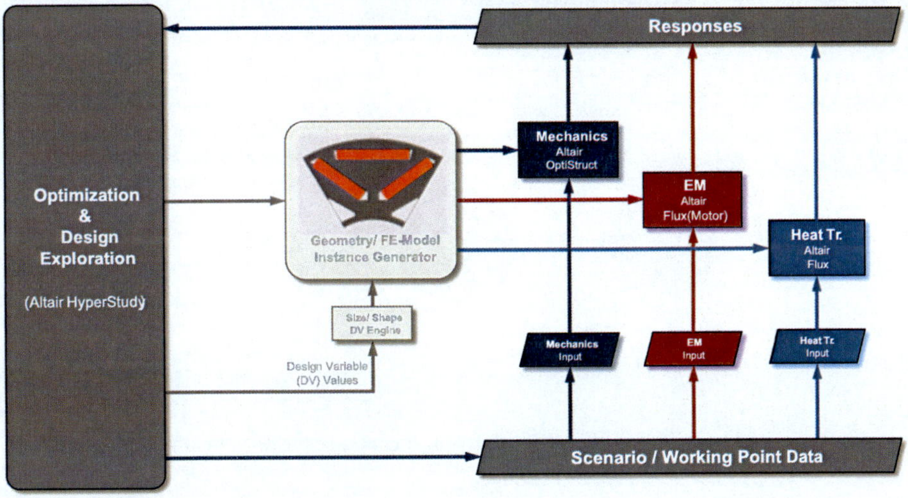

Fig. 2. Physics considered for the reference optimization environment

3.2 Benefit of Reference Process

With the reference process described in this section, it is possible to efficiently find a design optimum considering all selected constraints and additionally reduce the mass of the magnets with about 40%, see Fig. 4.

The process allows further to change the geometry and execute the same study if the setup of the considered working points and responses stay the same. The reference

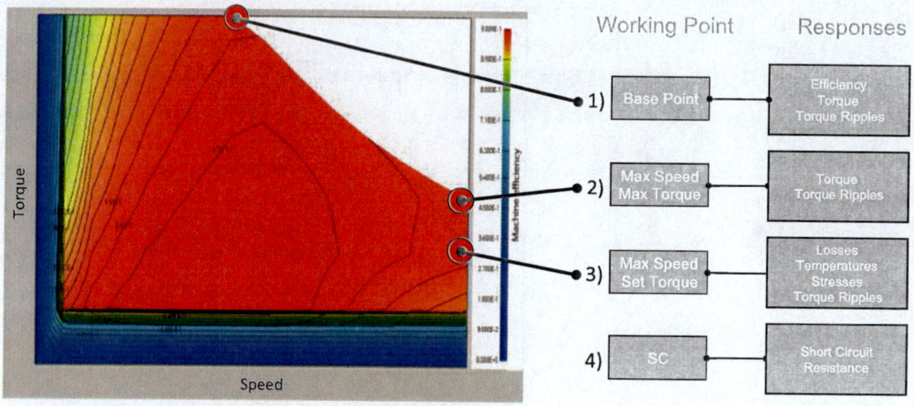

Fig. 3. Test study executed using the reference process

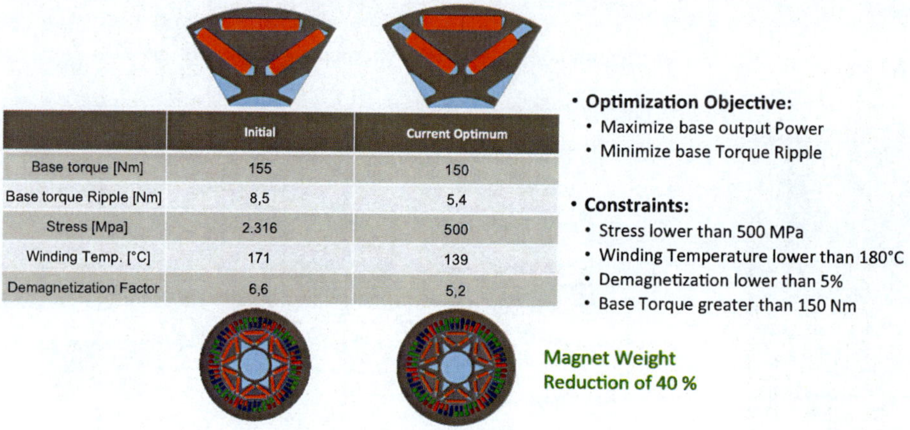

	Initial	Current Optimum
Base torque [Nm]	155	150
Base torque Ripple [Nm]	8,5	5,4
Stress [Mpa]	2.316	500
Winding Temp. [°C]	171	139
Demagnetization Factor	6,6	5,2

Fig. 4. Complete e-motor optimization problem—All constraints fulfilled within acceptable tolerances

process allows to be extended if further working points, responses or physics need to be added.

3.3 Improvement Potential—Baseline for the Proposed Process

Although the described process provides a general environment to efficiently evaluate and optimize an e-motor design against chosen criteria, it has room for improvements concerning flexibility, simplicity and speed of setting up a design study.

- The process to execute all solvers in the right order and to extract all responses for all working points has been setup manually by experienced engineers and optimization experts.

- The setup of the study lacks visibility, especially when associating working points, physics simulations, and responses.
- In the current process there is no consistent geometry for all solvers. FluxMotor is used with predefined rotor shapes. The inclusion of local details like noses etc. must be done in Flux.
- Data handling is specific for each individual situation. Thus, manual linking of design variables and responses is used to provide output from solver A as input to solver B in a solution chain.

4 Requirements on an Improved Design Strategy/Process

Using the reference process as baseline, several requirements were defined for an updated process for e-motor design. The application described in this paper aims at being a pilot to test solutions associated with these requirements and serve as a basis for future software developments.

As stated in previous section, the requirements are associated with visibility of the setup, automation, the consistent definition of geometry for all involved solvers and data handling. The following subsections elaborate more in details on the requirements for the new process.

4.1 Visible Setup of the Design Study

It should be possible to define the study, i.e. the choice of responses to be evaluated and working points to be considered, using a graphical user interface. The interface should provide clear view on which working points to consider, which responses to be evaluated for each working point and which solvers are necessary to extract and evaluate the chosen responses.

4.2 Consistent Definition of Shape for All Involved Physics/Solvers

Consistency of geometry for all solvers and all simulations to generate the chosen responses is key. Thus, an approach must be found for which a unique geometry description can be fed to all solvers, thus that associated FE-models/analytical models are all based on this unique geometry description.

4.3 Process Automation

The new process to setup a study must be highly automated and simplified. This is mandatory for the creation of the solver execution scripts for the optimization engine (Altair HyperStudy). In addition, the automatic creation of the setup file for HyperStudy (session file), that mostly eliminates the need for manual setup in HyperStudy, simplifies the use of the tool further. A democratization of a process to use design optimization as a part of the e-motor development will require that such setup processes can be handled by a broader range of engineering employees.

In the reference process, experts must be involved every time when something in the process is changed: e.g. new responses are needed or some solver parameters like max current or max speed changed. This is time intensive, since the experts must open their tools, update the model and export the HyperStudy interface. After each change, the HyperStudy session file must be recreated manually.

In the new process the important solver parameters and all available responses must be provided such that the MDO-user (not the experts) is able to change these parameters easily and the HyperStudy session must be built automatically based on these changes. The experts may only advice the MDO-user, but no time-consuming interactions of the experts are needed after they delivered the baseline models.

4.4 Data Flow and Data Handling

Extensive automation of the design optimization process will require that data can be handled easily throughout the process (Fig. 5). The evaluation of different working points and the usage of different solvers within the multiphysics design process will require a standardization of how data is handled throughout the process, see Fig. 6. In certain scenarios, solvers must be executed sequentially using the output of the previous solver as input. Such scenarios must be handled easily and with a maximum of automation.

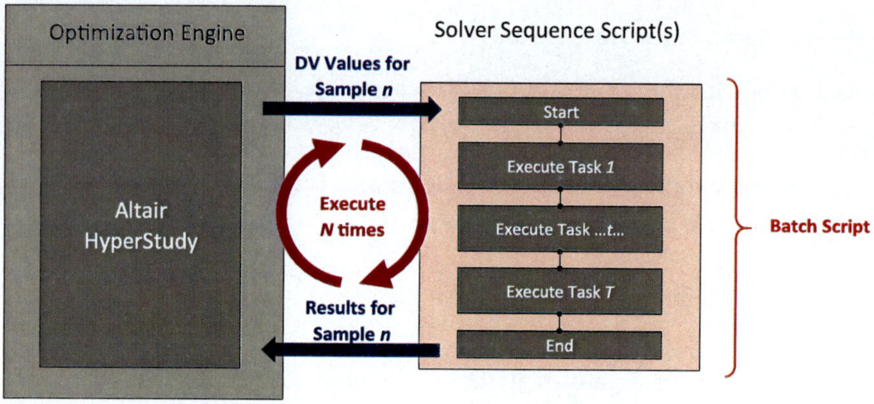

Fig. 5. Interaction between the optimization engine and the solver scripts

5 Process Implementation

As described in previous sections, the major challenge of creating an automated, simple to use design optimization process is to support the creation of the solver execution scripts, see Fig. 7.

The task of creating a complete batch-executable solver script for HyperStudy requires several sub-tasks to be executed automatically. The study which is to be executed by the solver execution script must be defined in a simple, user-friendly manner.

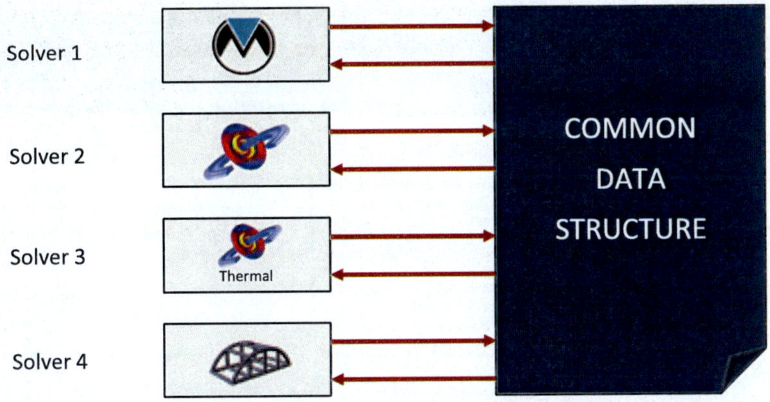

Fig. 6. Consistent data handling by using a common data structure

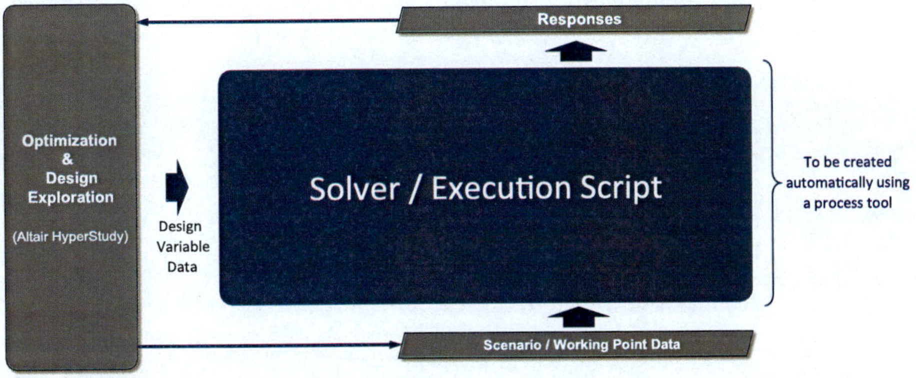

Fig. 7. Main challenge is to create the solver execution scripts automatically

The data used as input for the study and the data transferred between simulations must be handled and solver models need to be provided into the study that allow for parametric shape changes.

Figure 8 shows a principle sketch of the master application for the "e-Motor Director" pilot application including the most important functionalities.

Two functionalities are addressed more in detail in this paper since they are essential for fulfilling the requirements set in Sect. 4:

- *"Create/Modify"*—Setup of a multiphysics study
- *"Run"*—Execution of a DOE study.

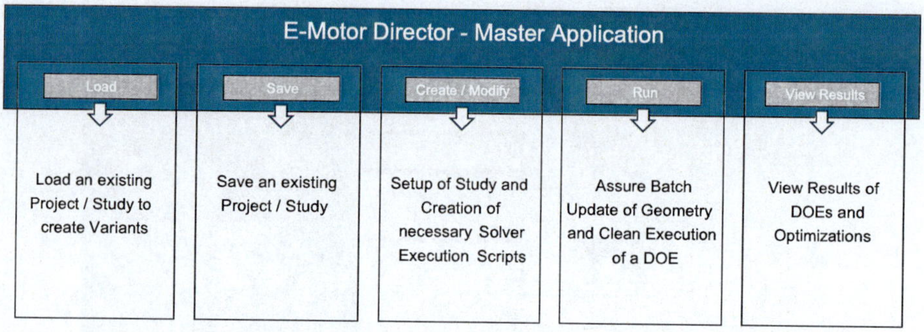

Fig. 8. Essential functionalities of an "e-Motor Director" master application

5.1 Setup of a Multiphysics E-Motor Design Study

The *"Create/Modify"* function of a study includes several steps to be executed before the study finally can be launched using the *"Run"* function of the master application. Figure 9 provides an overview of the complete process to setup a study.

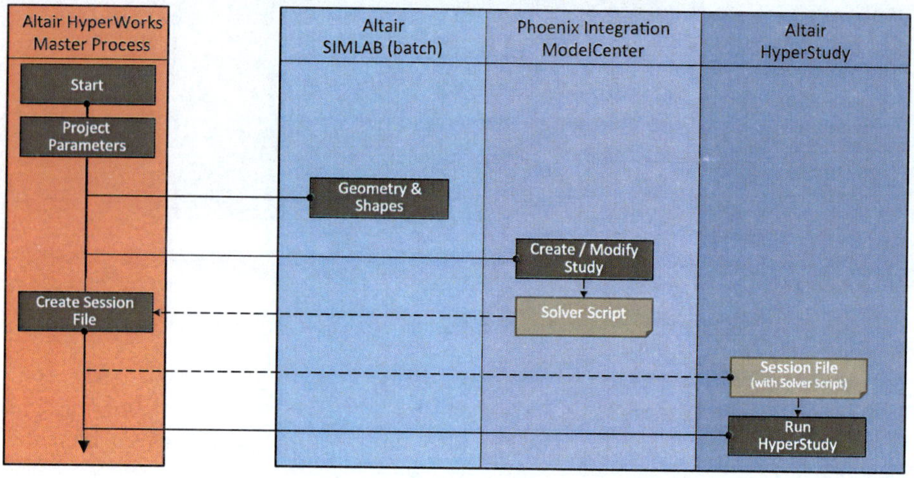

Fig. 9. The process to define and setup a new study using the master application

Start and setting project parameters.

The master application provides input masks for the user to define project and study directories as well as target directories for DOE and other output data. Physics related input, such as max current and max voltage for electromagnetics solvers and material information for structural analysis can be defined using dedicated input panels.

Geometry and shapes

Using a parameterized CAD file containing parts of the motor subjected to design optimization, Altair SIMLAB supports the creation of a list of design variables associated with the CAD parameters and proposes lower and upper bounds for each design variable. The geometry, the design variables and the bounds can be inspected in the master application and the design variable bounds can be modified as necessary.

Create/modify study

The visual setup of the design study including definition of relationships between load cases, physics and output responses is aided by the software tool ModelCenter from Phoenix Integration [2]. The user chooses tasks from a predefined tasks library which each represent the execution of a sub-problem for a specific physics such as thermal or structural calculations. Each sub-problem/task requires certain input and output. Tasks are added one by one and ordered in branches to represent different load cases (working points) and physics. Parallel branches could also potentially be executed in parallel in order to save physical time necessary to complete the DOE, see Figs. 10 and 11.

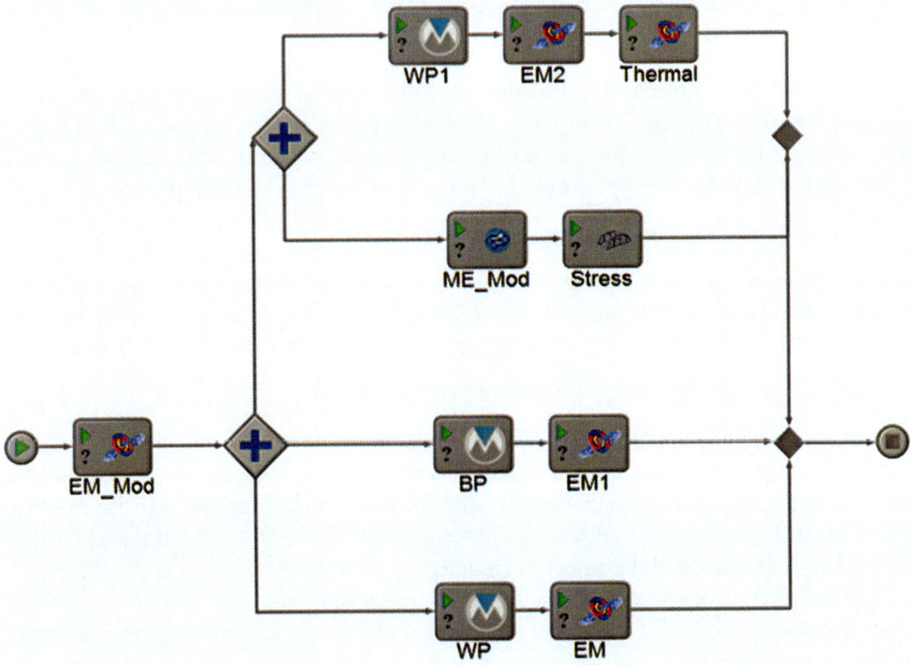

Fig. 10. Visual Setup of working points and simulation chains to get desired output parameters

In this setup phase no simulations are executed, the study is defined in order to create the necessary solver execution scripts for HyperStudy.

Working point 1 (WP1) contains two sub-branches in order to execute both structural analyses to extract tension stresses in the rotor and electromagnetics/thermal analysis to extract motor temperatures. WP2 determines the basepoint and the efficiency and torque ripples at this point". WP3 is a low torque, low speed working point to evaluate efficiency at moderate motor power.

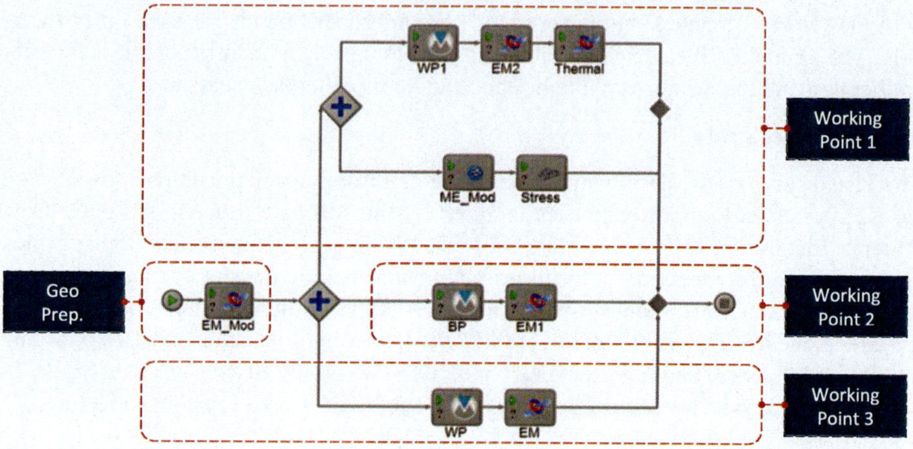

Fig. 11. Association of working points to the study definition in Fig. 10.

The complete definition of tasks to be executed in order to generate all responses is exported into a batch script. This script file is executable from a command line and/or from other applications. In the context of e-motor director, the script is registered as a solver execution script in HyperStudy to be executed within the DOE process.

Create session file for HyperStudy

Based on all input given, the geometry including design variables and the solver execution script created using ModelCenter, a HyperStudy session file is created which contains necessary settings and definitions to execute a DOE study based upon the given data.

Launch of HyperStudy using the created session file

After the session file for HyperStudy has been created, HyperStudy is launched to finalize the setting and start executing the desired DOE study. The use of the automatically created session file for HyperStudy significantly simplifies and speeds up the process to setup the study. Most parameters and settings have already been defined and the user merely must confirm the data and the selected choices.

The one section where the user will need to intervene manually is the selection of constraints based on the list of available responses and the definition of associated target values for all responses.

5.2 Execution of a Defined E-Motor Design Study

The execution of a design study corresponds to the execution of a DOE by varying all design variable values according to the chosen DOE scheme and then extracting all responses from all working points for all the different DOE samples. In the context of this study, the general execution procedure depicted in Fig. 5 reduces to the execution of two distinct solver execution scripts:

- Creation of a geometry instance for DOE sample n

- Execution of all study tasks for sample ***n***

This is repeated for 1 <= ***n*** <= N samples defined for the study. The choice of scheme how the samples are selected can be left as chosen by the system or be changed manually by the user. All samples are written to a predefined directory structure and are executed from HyperStudy. It is the aim to be able to execute the complete study outside HyperStudy, thus using available HPC submission procedures. Figure 12 shows the interaction between the optimization engine and the solver execution script sequence.

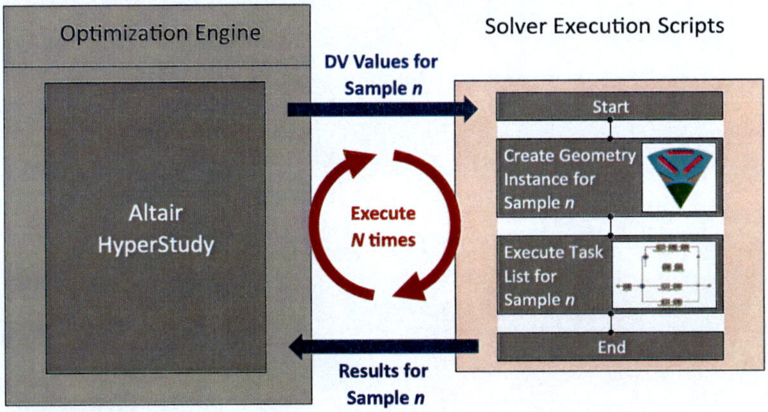

Fig. 12. Execution of a DOE

The creation of the geometry instances for each of the DOE samples is managed by using Altair SIMLAB as a translation tool between a CAD tool of choice and the solver execution tasks defined in ModelCenter. The main advantage of using SIMLAB in the process is, that a single interface to different CAD-Systems can be used. Only small modifications are needed if the CAD-system change. In addition, SIMLAB has a direct interface to HyperStudy which simplifies the setup.

The creation of the solver execution script for SIMLAB to create the geometry instance could be broken out of the study setup and was done once. This current solver execution script is valid for all geometries and can be used together with different CAD tools provided that certain modeling and naming rules are followed, see Fig. 13.

6 Test Case for the e-Motor Director Pilot

The "e-Motor Director" pilot application was developed using a test case which cover a typical scope and set of challenges for an e-motor design project but still be easily possible to handle and fast in execution. A three working point example was selected as depicted in Fig. 14.

A representative set of design variables was chosen both to cover global changes of a magnet configuration as well as local changes of radii to capture stress peaks see Fig. 15.

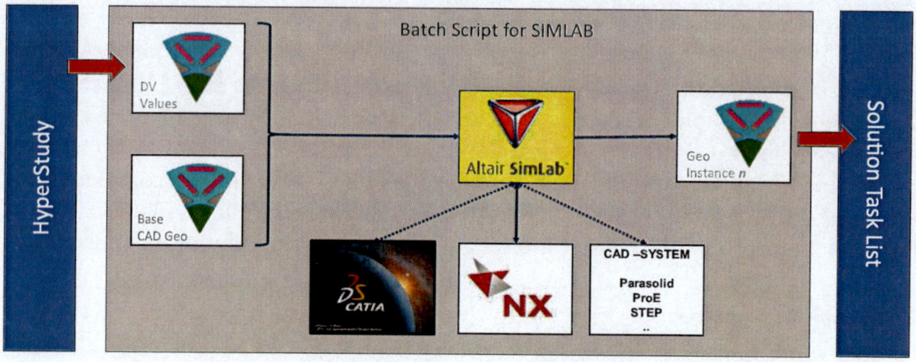

Fig. 13. Creation of geometry instance for the DOE

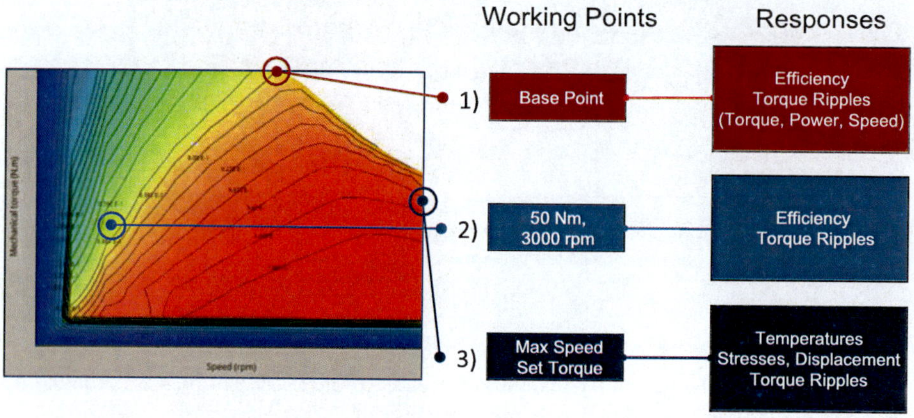

Fig. 14. Test case or the E-Motor director pilot application

The study was setup using the "e-motor Director" pilot application and a DOE was executed using the MELS scheme including 51 valid samples. By applying typical targets for all included responses, the valid list of samples reduced to 6, see Fig. 16. The responses used for this study was torque, power, torque ripples, efficiency, max stresses, temperatures

No final optimization was carried out from the DOE but a "best run" was selected from the valid runs. A qualitative comparison of efficiency map, motor temperatures and rotor stresses are depicted in Fig. 17.

Efficiency at working points increased by 15%, torque and power by 50–60%. Other constraints like torque ripples and stresses were inside the defined limits.

7 Summary

This paper summarizes the development of a pilot application for multiphysics e-motor design. The application has been developed to address and manage four specific challenges in the area of multiphysics e-motor design.

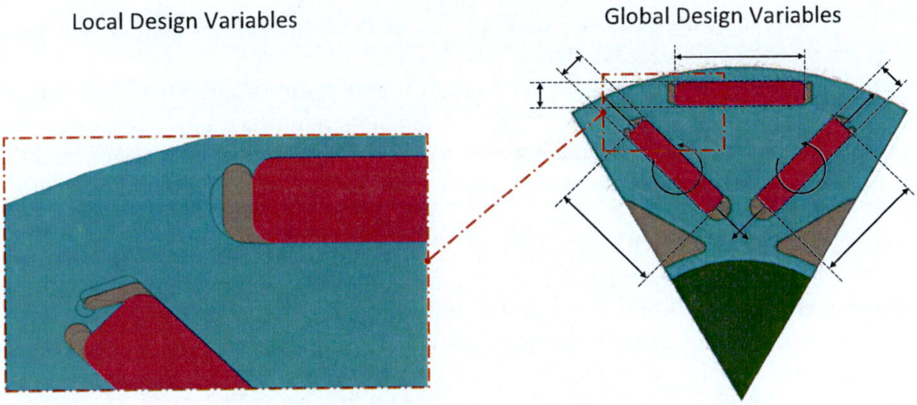

Fig. 15. Shape design variables for the test case

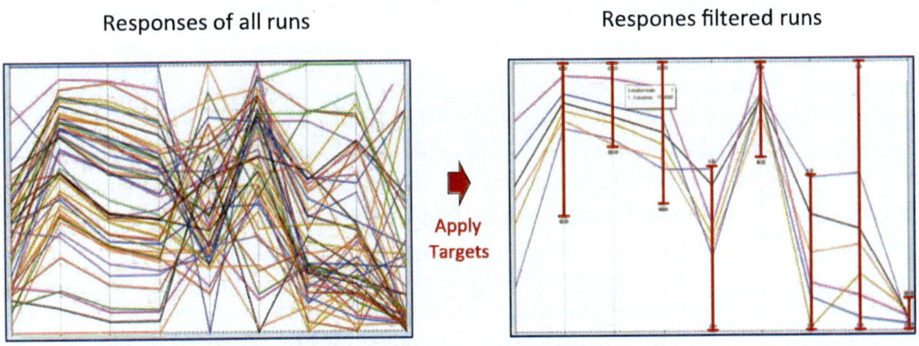

Fig. 16. Snake plot of all samples (left) and samples fulfilling the design targets (right

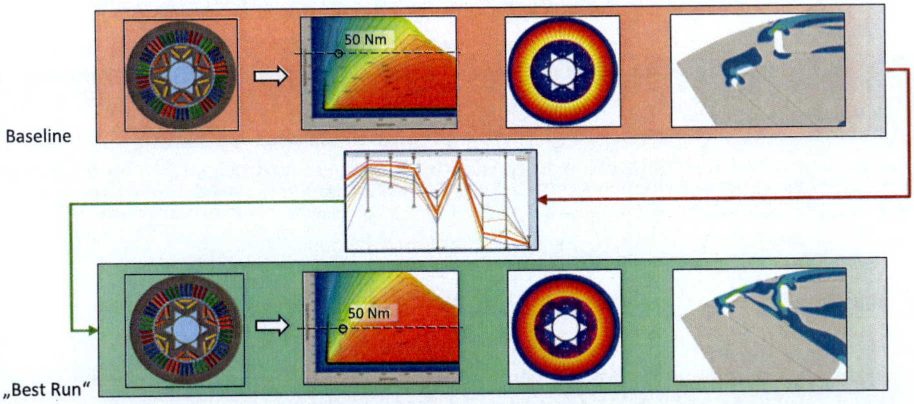

Fig. 17. Finding best run from valid samples

1. The use of consistent model description/model geometries throughout different physics in the optimization process.
2. Visual and simple definition of a design study supporting to establish the relationships between responses, load cases, physics and input/output data.
3. Automation of the process to allow engineers to setup and execute a design study.
4. Consistent handling of data to be passed between physics simulations and between load cases

The pilot application has been developed using the Altair HyperWorks framework. The open APIs and the programming capabilities of HyperWorks have allowed to create the new software pilot using the new HyperWorks Unity interface. All development challenges were successfully managed to create a working pilot software able to setup and execute the test case described in this paper. Although the pilot application of the e-motor director has limited functionality and has not undergone longer debugging periods at this time, it clearly shows the potential of such application and shows the potential of Altair HyperWorks framework as a host for such an application.

8 Future Work

The current pilot application of the "e-Motor Director" can handle a restricted range of e-motor design situations. The framework behind the e-Motor Director is however general. Thus, the e-Motor Director application can be extended without changing the general framework. High priority extensions include:

- Additional standardized tasks to be included in the tasks library for handling additional events and physics in an e-motor design study. High priority tasks to be added include:
  - Noise and vibration
  - Systems simulation of inverter and possibly other parts of the e-powertrain
- Improved handling of responses. Generalization of the "Common Data Structure" concept.
- Improvement the robustness of geometry and design variable handling
- Improvement of task definitions to be able to re-use standard tasks for a wider range of simulation situations.

References

1. Luthardt S, Fredriksson L, Dyer A, Lombard P (2018) Ganzheitliche E-Motor-Entwicklung mittels "Multi-Physics"-Optimierung. SIMVEC Baden-Baden
2. www.phoenix-int.com

How to Ensure Safety of EV or FCV Without Reliable, Helpful Saving Functions Made by ICE

Takashi Shibayama(✉)

Shin Nippon Tokki Co., Ltd., 1349 Ohkubo-Cho Nishi-Ku, Hamamatsu-City 432-8006, Japan
t_shibayama@shinnippontokki.co.jp

Abstract. The power train electrification has been remarkably evolving and HEVs or EVs are increasing in world-wide markets. Especially recently, pure electric driving system such as EVs or FCVs without any ICEs are highly desired as better solutions for environmental protection. However, we have to recognize the important role of engine for conventional ICE vehicle. Engine has reliable, helpful saving function on long and steep downhill road. It provides everlasting reliable engine brake. On the contrary, EVs and FCVs do not have engine. Instead of engine brake, regenerative brake is usually utilized for EVs and FCVs. However there remains emergent situation where battery SOC is full on long and steep downhill road driving. Thus, we have to develop emergency brake for those pure electric driving systems. In this paper, the background of this issue is explained in detail and several effective solutions to fix this issue are proposed.

Keywords: EV · FCV · Brake · Battery · SOC · Full SOC · Downhill · Safety · Engine brake

1 Background

Following descriptions are the more detail background why such emergency brake is inevitable.

(1) Engine braking and exhaust braking are everlasting reliable, saving function made by conventional power-train. There are many caution signs on long and steep down hill road recommending or requiring to use low gear to save mechanical friction brake. In low gear range, engine braking and exhaust braking are preventing vital fade of mechanical friction brake system in conventional ICE vehicle especially for heavy weight trucks and buses. And the saving is everlasting during long and steep downhill road driving.

(2) Only temporary alternative saving function by regenerative control is supplied by today's power-train for EVs or FCVs. Regenerative control is surely one alternative technology to obtain sufficient braking force for EVs or FCVs if it is active. However, the regenerative brake control cannot be last for long duration. Battery SOC will be soon or later "Full". Especially FCV is critical, because the capacity of battery is much smaller than battery EV. So, on long and steep down hill road,

Euroforum Deutschland GmbH (Ed.): *CTI SYMPOSIUM 2019*, pp. 473–487, 2021.
https://doi.org/10.1007/978-3-662-61515-7_42

SOC of FCV will become soon "Full" and such FCV is braked only by mechanical friction brake that leads to fade trouble. Off course battery EV has also similar risk if battery is charged on high altitude charge station and EV is driven on long and steep down hill road just after battery is charged [1, 2].

(3) What is happening in the market? On SNS, descriptions by customers about the strange behavior of regenerative brake of electrified vehicles are gradually increasing year by year. Some customers are mentioning frightened issue just after full SOC. They had felt sudden lack of braking force. Today, general design trend is adopting no emergency brake for EVs and FCVs in case of such full SOC situation. Only caution and guidance are written on owner's manuals for full SOC situation. It is allowable for light weight pure electrical vehicles. However, the era of heavy e-vehicle is coming now. The increasing descriptions on SNS are important signs and warnings for us. We have to prepare some everlasting saving function for EVs and FCVs. Especially for heavy weight vehicles, emergency brake for long and steep downhill road driving with full SOC situation is inevitable.

In this paper, the importance of such emergency brake is explained in detail. And new emergency brake concept named Down Hill Brake (DHB) is proposed with several concrete example which can provide good solution for the important issue for EVs and FCVs.

2 The Key Functions

There are three key functions to be considered. One is function for obtaining everlasting braking, and the second is rapid converting function. Conversion from position energy into heating energy. And the third is rapid exhausting function. Exhausting the heating energy from inside of the vehicle to outside of the vehicle. For conventional ICE vehicles, ICE itself provides these three functions. Adequate alternative apparatus which can work everlastingly is required for EVs and FCVs.

3 Everlasting Brake Function and Rapid Transaction Function Obtained by ICE

Figure 1 shows an example of calculation estimating heating energy during typical steep downhill constant vehicle speed driving. Typical passenger vehicle specification is used for estimation and typical steep downhill road condition is used for assumption. As shown in this figure, large amount of heating energy is generated during steep downhill driving. For conventional ICE vehicle, engine provides everlasting engine brake with very rapid transaction. Very rapid conversion from position energy into heating energy. And very rapid exhausting the heating energy from inside of vehicle to outside of vehicle. The engine acts as an everlasting useful air pump for cooling, exhausting heating energy without depending on any heat conduction or heat transfer such as cooling fan or radiator. For example, 2000 rpm speed engine is equal to huge capacity air pump with 4000 L/min output flow capacity. It is really a huge exhausting air pump.

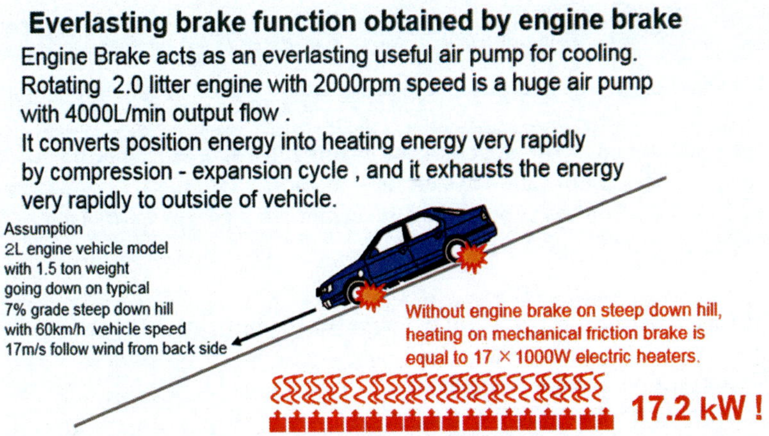

Fig. 1. Large amount of heating energy during typical steep downhill driving

In Fig. 2, similar estimation result has been executed for 10 times heavier vehicle such as heavy-duty trucks. Reasonably, in the case of 10 times heavier vehicle, 10 times larger amount of heating energy will be generated. Huge amount of heating energy as shown, 172 kW heating will be generated. Because Shin Nippon Tokki is a tester supplier, we have a lot of experience for cooling system of testers. Usually, 172 kW power absorber dynamometer tester is cooled by 80 L/min water jacket cooler. Cooling is often a big issue. If such a heavy truck is driven on long and steep downhill road only by using mechanical friction brake without any engine brake, any exhaust brake, nor any retarder, the mechanical friction brake will suffer from enormous amount of heating energy. Therefore, in the case of heavy e-vehicle, driving on long and steep downhill road without engine nor retarder, is very risky. Risk of brake fade becomes much larger.

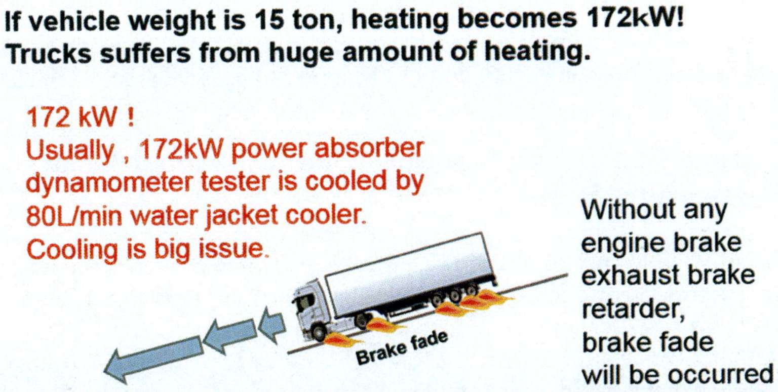

Fig. 2. Huge amount of heating energy during steep downhill driving by a heavy weight truck

4 Temporary Brake Until Full Charge by Regenerative System

Figure 3 shows how e-trucks is saved by regenerative brake. Regenerative brake of an e-truck is surely effective if it is activated. As shown in Fig. 3, regenerative brake acts similar as a generator on top of mountain pulling life line rope. Mechanical friction brake is saved by that. The combination system of generator and battery provides important converting function, from position energy to accumulated energy in the battery through electric energy.

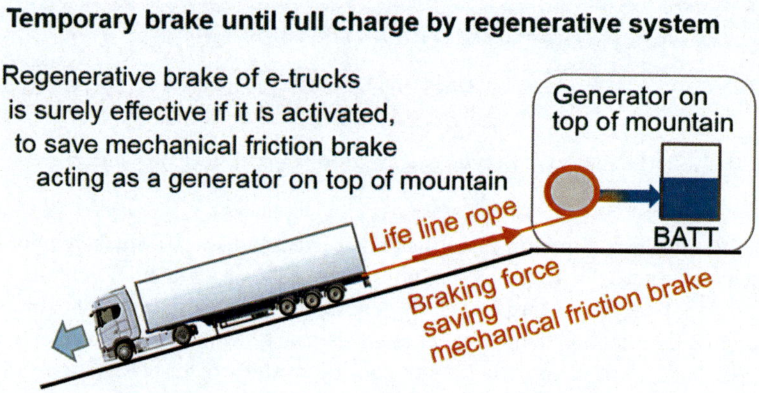

Fig. 3. The important role of regenerative system of e-trucks on steep downhill road

However, the useful function by generator and battery does not last everlastingly. During driving on long and steep downhill road, soon or later, the battery SOC becomes to be full. The generator must stop electric generation and the position energy cannot be converted to electric energy. Then, regenerative brake will be stopped. In this view point, the regenerative brake is only a temporary brake until full battery SOC. As shown in Fig. 4, very awful situation as if life line rope is broken off will happen. In this way, just after battery SOC becomes full, e-trucks becomes to be a dangerous vehicle.

5 Increasing Customer's Description on SNS About Behavior of Regenerative Brake

Figure 5 shows the history of electric based vehicles which had been introduced into the market. Many kind of electric based vehicle has been introduced one by one. Author of this paper has investigated descriptions on SNS written by customers. Many owners of electric based vehicle have written about behavior of their own car. Among those descriptions, the number of descriptions about behavior of regenerative brake is increasing year by year. For example, some owner mentions about strange brake feeling, and another owner begins analysis by experiment using his own car. The report of the experiment calls another discussion. Such long and many discussions are recorded on SNS. By analyzation of such descriptions on SNS, important trend has been found. For heavy

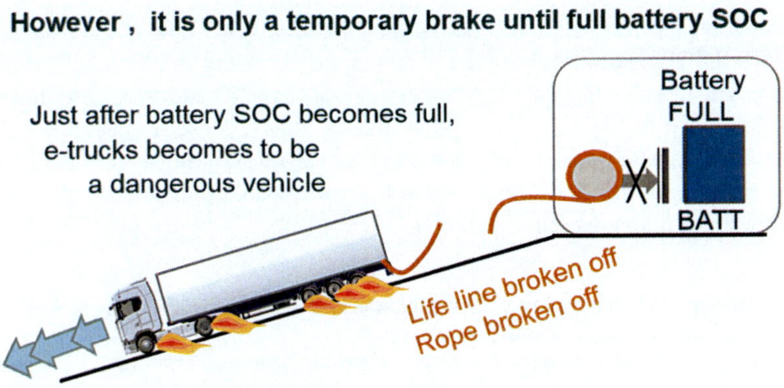

Fig. 4. Dangerous situation for heavy e-trucks in the case of full SOC of battery

weight vehicle, more severe descriptions are written more frequently. For example, "I was frightened when brake force became suddenly to be almost zero just after SOC becomes full" is written for heavy weight wagon. It is very important sign and warning for us. As shown in Fig. 5, today, heavy weight EV, FCV has just started to be utilized or tested for urban use. FCV buses is already used in many cities in the world, however, those buses are for public ride in the city. So, long and steep downhill issue is not noticed yet. We think that the big challenge will come soon. Many OEMs have already announced their introducing strategy about pure electrified vehicle such as EVs and FCVs. We have already also heard many announcements mentioning about introduction of heavy EVs or FCVs into the market in very near future. After 2020, heavy EVs and FCVs will increase rapidly. Heavy e-vehicle era is now coming. Many heavy e-vehicles will be used for mountain road in very near future. If so, similar scenes with danger as descriptions on SNS today, will increase.

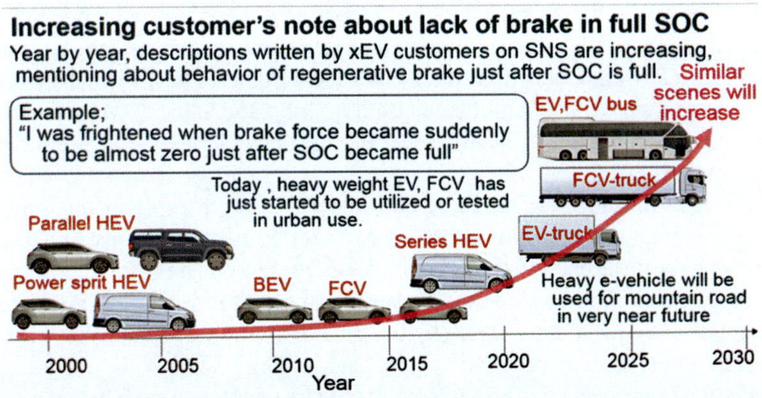

Fig. 5. Increasing customer's description on SNS about behavior of regenerative brake

Figure 6 shows another important fact today. Several OEMs selling e-vehicles already know the full SOC issue. On several owner's manuals of EV or FCV, we can find cautions about the behavior of vehicle in the case of SOC becomes full. One manual says that "In the case of full SOC, regenerative brake will become to be very small". And it is followed "Please use foot brake in that case". Another manual says "Please drive downhill by slow speed in the case of full SOC". And another manual says "Please stop vehicle, if brake is not enough. And please take a rest for a while". Those OEMs knows very well what will happen in the case of full SOC. Such cautions are allowable for light weight e-vehicles, but dangerous for heavy weight e-vehicles.

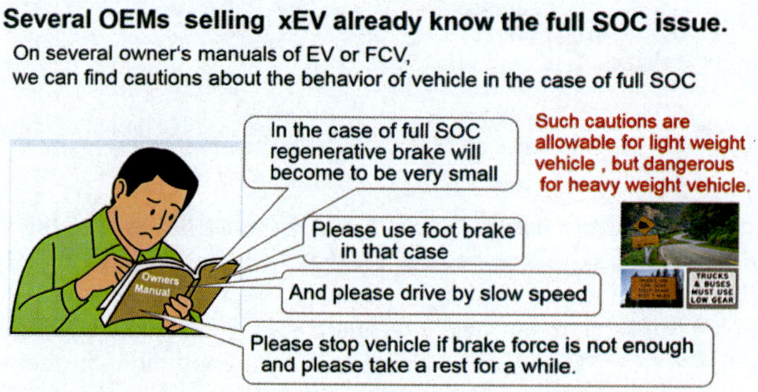

Fig. 6. Well known behavior of regenerative brake for several OEMs who is selling e-vehicles

Figure 7 explains why the heavy weight e-vehicle is so dangerous. Heavy weight e-vehicle accumulates huge position energy during climbing mountain. Off course, heating energy is also generated during climbing because of the loss of drivetrain. However, large portion of input energy is converted into position energy. The big issue will happen on down hill road. During down hill road driving, all position energy accumulated in vehicle at high altitude point have to be converted into heating energy and have to be exhausted to outside of the e-vehicle. As shown in Fig. 7, heating energy caused by position energy during downhill is much larger than heating energy generated during uphill driving. The heavier the vehicle weight becomes, the greater the amount of heating energy becomes. Thus, if only mechanical friction brake is used on long and steep downhill road during full SOP, enormous amount of heating energy will be accumulated at the mechanical friction brake.

The influence of wind also has to be considered. Figure 8 explains the influence of wind. If follow wind speeds from the back side becomes large, air resistance decreases, so, the energy to be absorbed by retarder or brake increases. On the other hand, cooling air flow decreases that will reduce cooling capability of radiator or air cooling for retarder. So, heating issue becomes more severe than normal condition without follow wind from the back side. Thus, some good brake system and cooling system without depending on heat transfer to air nor heat conduction to air should be adopted to such heavy e-vehicles.

Here, we would like to make proposal. Figure 9 shows our proposal. Emergency brake is inevitable for heavy EV, FCV. Our proposal is Down Hill Brake(Here after DHB) by compression—expansion cycle. By the emergency brake, position energy will be converted into heated air very rapidly, and the warmed up air will be exhausted to

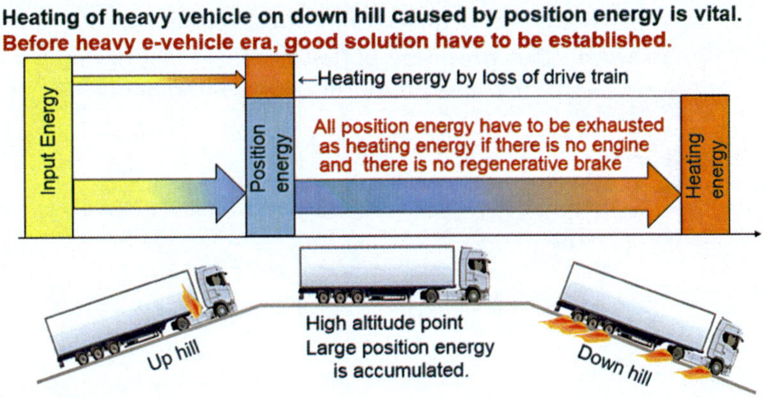

Fig. 7. Much greater amount of heating energy generation during downhill driving

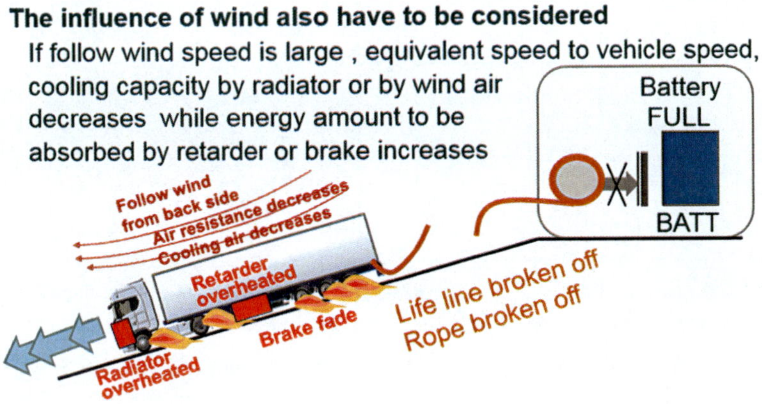

Fig. 8. The influence of wind to be considered

outside of vehicle very rapidly, while making useful braking force to save mechanical friction brake.

6 Proposal of Smart Everlasting Brake

Figure 10 shows our proposal in detail. Compression—expansion chambers are installed in a transmission case housing. Inside of those chambers, pistons with air passage will be moved by the crank system. The crank shaft is connected to a DHB on-ff clutch. And the clutch will be controlled by DHB controller at adequate situation. For example, if the SOC sensor detects full charge situation or driver select DHB selector on steep downhill road, clutch for DHB will be connected and by that, DHB will be active.

Figure 11 explains how to work the proposed DHB in detail. As shown in the figure, first, cool air comes into chamber, second, the air is compressed and brake torque is generated, simultaneously, air is warmed up by the compression, and then, the warmed

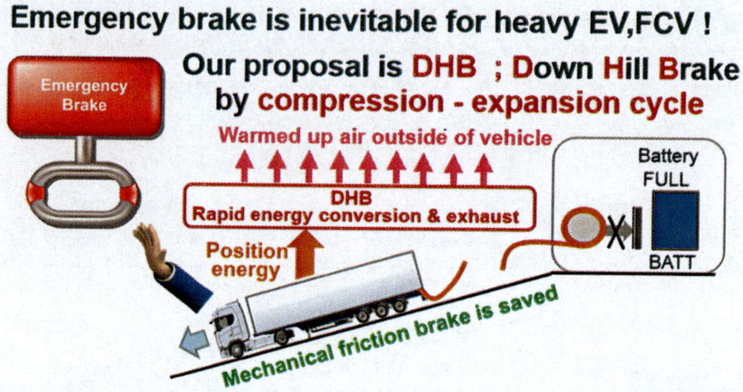

Fig. 9. Inevitable emergency brake DHB for heavy weight EVs and FCVs

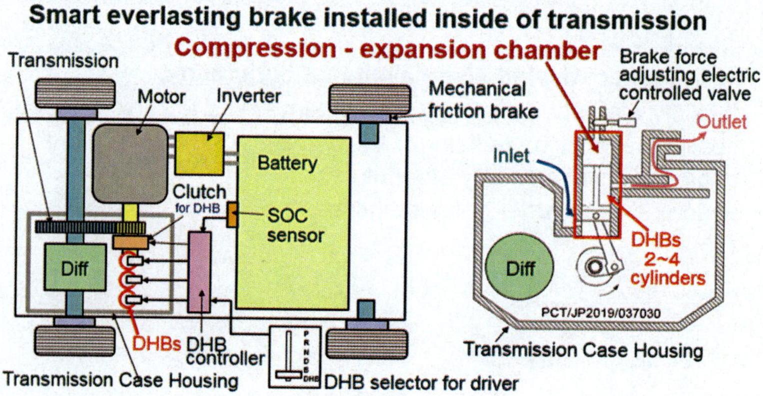

Fig. 10. Compression-expansion chamber for DHB built in the transmission of e-vehicle

up air is very rapidly exhausted by expansion, and then next, the chamber volume will be enlarged making vacuum air. The resistance against vacuum makes another brake torque. In this way, this system can very rapidly convert position energy into heating energy and exhaust the heating energy without depending on heat transfer such as cooling fans or radiators.

To confirm fundamental characteristics of proposed DHB, we have established a testing system as shown in Fig. 12. Piston of DHB is moved by the motor-crank system, and brake torque has been measured by cradle system as illustrated in the figure. Temperature and pressure in the chamber have been also measured by sensors. In order to investigate the influence of side leakage, two kind of DHB piston are prepared. One is spool type piston without o-ring by which side leakage is large. Another is plastic piston with o-ring by which side leakage is very small.

Figure 13 shows the difference between proposed DHB and engine brake by conventional 4 cycle engine. Proposed DHB works every cycle, 2 cycle system and additional remarkable feature is obtained during expansion part of the cycle. Air is expanded during expansion and resistance against expansion also make good braking torque. As shown

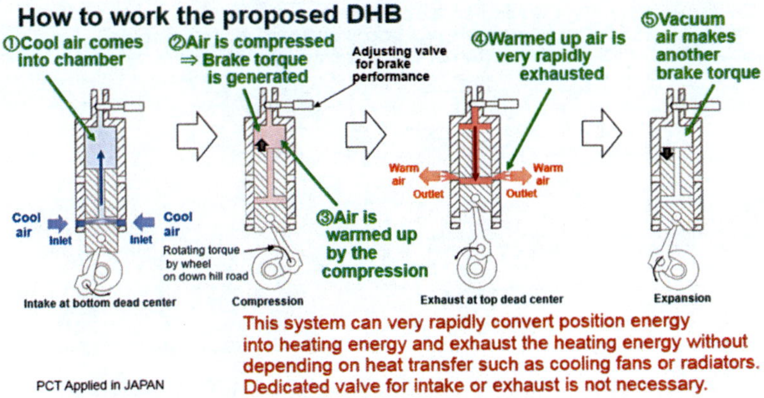

Fig. 11. Explanation about how to work the proposed DHB

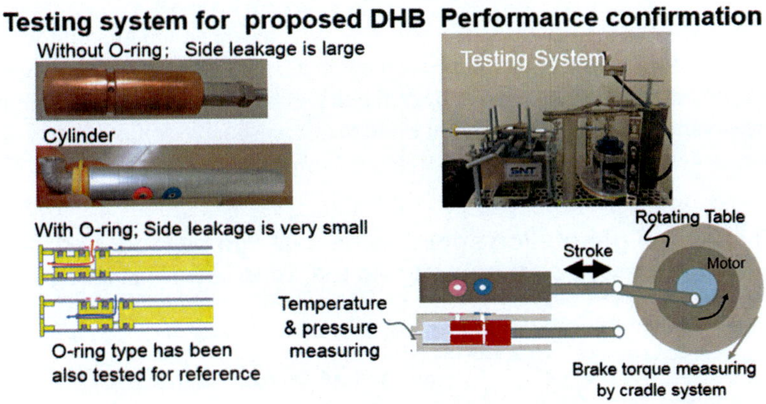

Fig. 12. Testing system for proposed DHB performance confirmation

in the figure, resistance force in both direction generates good braking force. On the contrary, conventional 4 cycle engine generates braking torque during short duration and driving torque just after compression i.e. just after top dead center is generated. Therefore, by proposed DHB, stronger brake by 4 times can be obtained compared with the engine brake of conventional 4 cycle engine. That means, 500CC DHB can provide equal brake force same as 2000 cc engine brake.

The influence of side leakage has been investigated by the testing system shown in Fig. 12. Figure 14 shows the test results. Red line indicate the braking torque in the case of o-ring type. Blue line indicate without o-ring type. Without o-ring, because of side leakage, maximum pressure in the DHB chamber becomes low at low rotation speed. However, at 500 rpm, maximum pressure becomes large enough. As indicated in the Fig. 14. Over 500 rpm, both braking torques have become almost same level. It means that in high rotation speed area, compression is rapidly executed before side leakage

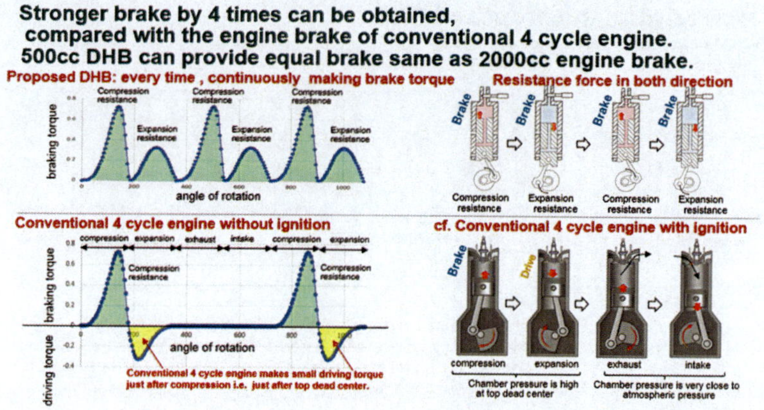

Fig. 13. Stronger brake by 4 times than 4 cycle ICE engine brake obtained by proposed DHB

happen. So, almost same performance as o-ring type has been obtained by without o-ring type. Therefore, using high rotation speed, side seal can be loosened, then, DHB can be manufactured without precise manufacturing machine. It can be recommended to make reasonable suitable side clearance and to use reasonable suitable rotation speed. O-ring less system is available by such way.

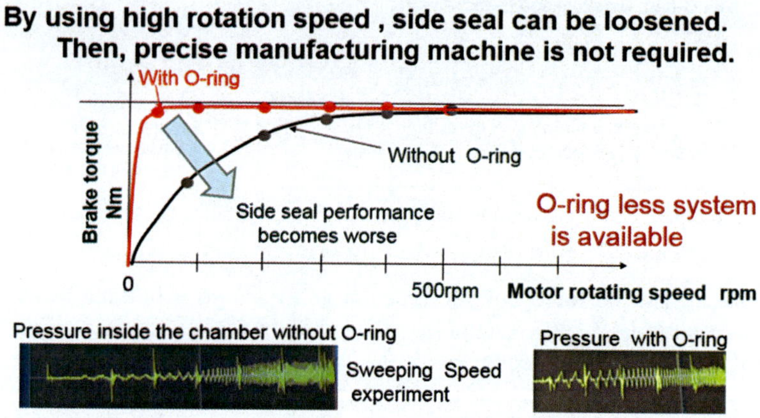

Fig. 14. Compensation effect by high rotation speed which helps enhancement of DHB

Figure 15 shows the necessity of DHB for heavy weight trailer towing usage of EVs or FCVs. For small wight EVs or FCVs, DHB is not inevitable because its position energy on high altitude position is not so large. However, if they are used for heavy weight trailer towing usage, summed up position energy on high altitude position becomes dramatically enlarged, so in that case, DHB is also innevitable. Attachment type DHB is required and it seems to be better to facilitate it into the trailed vehicle. For example, DHB attachment to adopt to rotating axis of camping car or rotating axis of motorboat carrier. In those case, to adopt attachment type DHB with speed increaser gearbox may be desireble.

Fig. 15. Necessity of DHB for heavy weight trailer towing usage of EVs or FCVs

For such requirement of DHB attachment for aftermarket, a DHB unit shown in Fig. 16 is one solution. It has dry multi-plate clutch for DHB on-off control which works by air pressure. And it has multi-piston systems. Multiple pistons act similar way as to above described single piston type DHB. From one piston to another piston, pistons move in order and provide brake torque one after another.

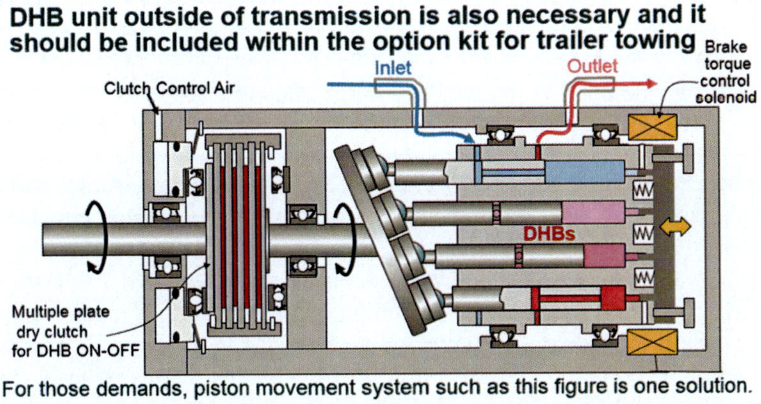

Fig. 16. Concept of DHB unit for adopting DHB as an option parts for after markets

Here, we have to consider about issues to be solved. There are several issues to be solved for the proposed DHB. Figure 17 shows major issues. Size down, noise and vibration, and to assure durability are off course important issues. However, for proposed DHB, cost reduction and weight reduction is also very important issues. Because DHB is emergency brake for safety, it does not work always. It works in rare case for short duration. So, cost reduction and weight reduction is very important for better vehicle planning. In this view point, there are several ideas for cost reduction and weight reduction. As shown in the figure, rectangular piston with gutter on side surface of the piston for air passage is one idea. Expensive air hole tooling can be avoided by that shape. Multi-layer structure manufactured by press machine is another idea. Air passage can be very easily composed by combining different shape plates as shown in the figure. Here, those pictures indicate just image of the shape. In order to assure reliability and durability, adequate material should be applied.

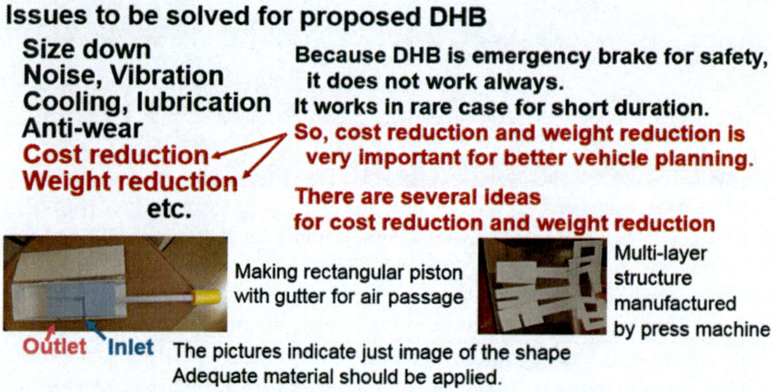

Fig. 17. Issues to be solved and several ideas to make cost reduction for proposed DHB

Figure 18 shows Rotary type DHB concept to improve size, noise, vibration, and durability. Fundamental principal is the same as spool piston type DHB. It has one way valve. The oneway valve opens during exhausting and closes during expansion. As shown in the figure, vacuum pressure is generated during expansion and the vacuum pressure provides brake torque. Compressed high pressure air also provides brake torque and high pressure air is exhausted just after compression. Electric valves is utilized for brake torque adjustment. The rotor is directly connected to rotating axis, so, crank mechanism is not neccessary. So, it is more simple and compact than spool piston type DHB with crank mechanism.

In Fig. 19, The concept of dual cycle compression rotary DHB is shown. In this concept, compression is executed at two compression chamber positions. As shown in the figure, both at left side in the figure and at right side in the figure, compression is executed. Compression ratio may become a little bit smaller than previously described rotary type, however this system provide good balance and symmetric action, and air flow way looks better, inlet and outlet are clearly separated and air flow is not counter flow. The air flow is cross flow passing though compression chamber. And large brake performance by dual compression is hopely expected.

In Fig. 20, Double cavity with balanced rotor DHB concept is shown. In this concept, the rotor does not have unbalance. And compression-expansion is more frequently executed. Probably, there may be many other concepts for obtaining further improvement. Some more excellent DHBs is expected for safe EV, FCV era.

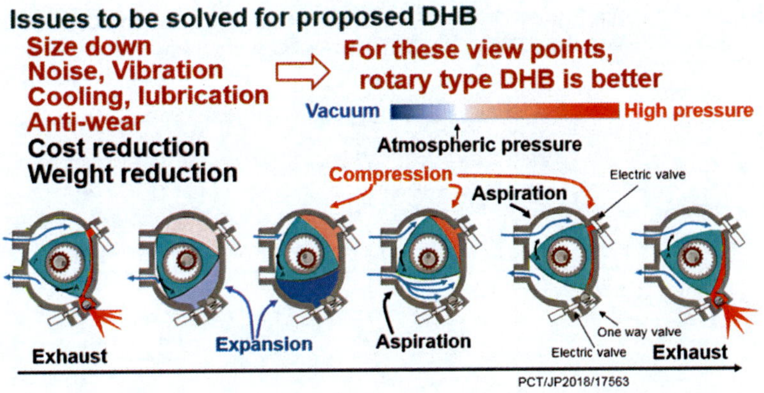

Fig. 18. Rotary type DHB concept to improve size, noise, vibration, and durability

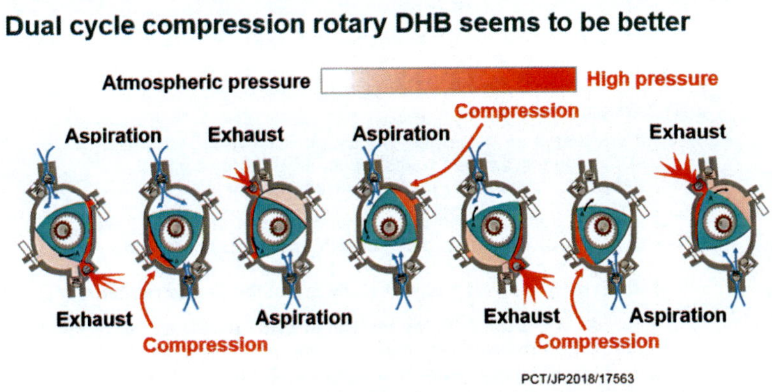

Fig. 19. Dual cycle compression rotary DHB concept

Now, Fig. 21 shows the summary of this paper. Before heavy e-vehicle era comes, we have to develop good DHB.

We believe that proposed compression-expansion type DHB will surely contribute to establishment of safe e-vehicle era.

And finally, We would like to make call for our partners, as shown in Fig. 22. Please contact us if the readers of this paper can help us.

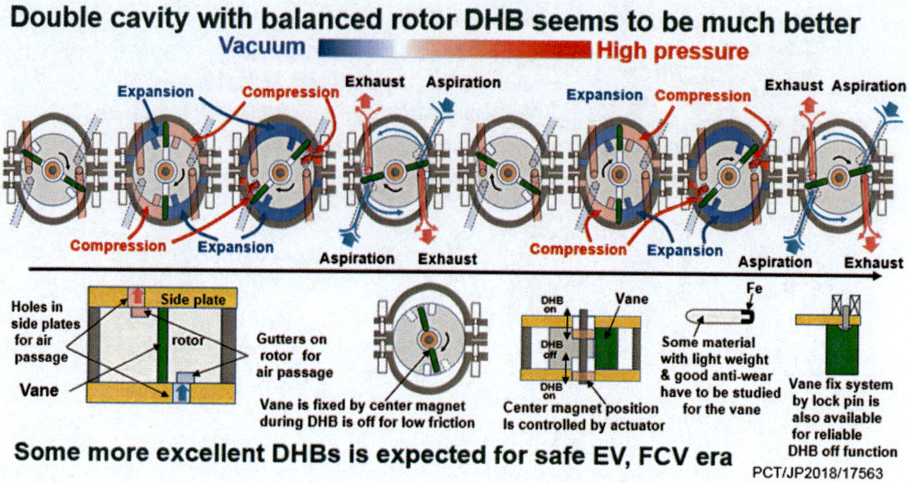

Fig. 20. Double cavity with balanced rotor DHB concept

Summary

1. For steep and long down hill EV or FCV driving with full SOC , down hill brake (DHB) as an emergency brake is inevitable, especially for heavy weight vehicle.
2. DHB by compression-expansion cycle system is the best , because it can very rapidly transfer position energy into heating energy and exhaust the heat energy without depending on heat transfer such as cooling fan or radiator.
3. Increasing customer's description in SNS about lack of brake under full SOC are important signs and warnings for us . DHB is highly required for coming EV and FCV era.
4. Several DHB concept has been proposed, which have to be developed before mountain use heavy EV and FCV increase.

Fig. 21. Summary of this paper

Fig. 22. Call for our partners

References

1. SNT: SHIN NIPPON TOKKI Web Site. http://www.shinnippontokki.co.jp/en/
2. Shibayama T, Shin Nippon Tokki Co. Ltd. Critical safety issues for EV, FCV and useful testing machines to solve them. CTI Berlin presentation and paper, Dec 2018

P2i, a Family of Modular, Scalable, and Integrated Hybrid Drive Modules

Heiko Jausel[1] and Wolfgang Wenzel[2(✉)]

[1] BorgWarner Transmission Systems, Ketsch, Germany
[2] BorgWarner Corporate Advanced Engineering, Ludwigsburg, Germany
wwenzel@borgwarner.com
http://www.borgwarner.com

Abstract. Among the many hybrid architectures, the P2 layout is a cost-effective solution that provides one of the best combinations to achieve CO_2 reductions especially within a given ICE vehicle architecture. The P2 layout allows the engine to be disconnected from the driveline, and the electric motor leverages the ratios of the gearbox to drive the wheels. However, packaging the motor and disconnect clutch without increasing the overall length of the powertrain is very challenging and needs detailed knowledge in electric motor and inverter development and integration. By combining a unique method for winding the stator of the electric motor, a scalable stack length and years of experience designing compact clutch systems, BorgWarner has developed a family of new drive modules including a scalable and modular custom engineered hydraulic actuation systems suitable for MHEV up to PHEV applications. P2 On-Axis modules for Dual Clutch Transmissions (DCT) and Automatic Transmissions are in various stages of development. For the DCT, the disconnect clutch and dual launch clutches are integrated inside the electrical machine rotor. Currently in pre-production phase at multiple customers, this P2 and P2i (with integrated inverter) Triple Clutch Module has a very compact design that minimizes the impact to overall powertrain length. For planetary automatic transmissions and CVTs, modules are being developed that are compatible with both torque converter and launch clutch configurations. Various rotor and stator lengths are available to match vehicle torque and power requirements, and these motors are compatible with both high voltage and 48-voltage architectures. A common lamination stack is used along with a unique winding pattern to realize a motor for the desired voltage. In addition, the three-phase inverter interface is standardized to have a modular P2i concept for both 48 V and high voltage. The P2i family also includes a matching inverter that can be integrated in the housing of the module. This inverter integration leads to significant benefits in overall system efficiency and EMC behavior. A custom engineered hydraulic module is also integrated into the housing to actuate the disconnect clutch and provide cooling flow to the clutch and motor. This presentation will describe the technical innovations that were necessary for successfully developing this family of P2i Hybrid Drive Modules.

Keywords: Parallel hybrid · P2 architecture · Hybrid modules

Euroforum Deutschland GmbH (Ed.): *CTI SYMPOSIUM 2019*, pp. 488–497, 2021.
https://doi.org/10.1007/978-3-662-61515-7_43

1 Introduction

The upcoming limits for fleet CO_2 emissions, especially in Europe and China, are significantly pushing electrification of vehicles. To achieve a balance between powertrain costs and reduction in CO_2 emissions, different levels of hybridization were developed and will exist in parallel on the market for a certain time. Besides very low electrification of an entry level of 12 V dual storage hybrids (also called micro hybrid vehicles) [1], which already allow a small amount of brake energy recuperation and engine off coasting, so-called mild hybrids with 48 V will occupy a significant portion of the market and are expected to replace conventional vehicles as a new baseline in some segments. Especially in Europe and China, a big market could be expected.

48 V hybrids can improve efficiency and driving performance at moderate costs. A 48 V board net, on the one hand, could be an enabler for electric climate compressor, electric air boosting devices [2] and supports electric systems like an anti-roll system. On the other hand, 48 V offer already a significant potential of CO_2 reduction. However, this potential is strongly linked to the architecture of the hybrid system and the power of the electric motor.

The standard architecture for the first mild hybrid vehicles is the P0 architecture with maximum power during recuperation of up to 12 kW (mainly limited by the belt) and usually less power for electric boosting on the crankshaft. P0 applications are simple to install, however, the losses on the recuperation path from the wheels to the motor generator unit (MGU) are relatively high [3, 4], which limits the CO_2 benefits of this approach. On the other hand, the belt starter alternator allows a quick engine restart, which enables extended start-stop and engine off coasting. The latter provides benefits which might not be fully covered in the test cycle but can be addressed by eco-innovation credits.

However, as the need to reduce CO_2 emissions increases so does the need to recuperate a higher share of the available kinetic energy. An alternative to achieve this is to move to a different hybrid architecture which allows for higher electric power and a closer path to the wheels. In this context, a P2 architecture represents a good compromise between added functionality, cost and potential for CO_2 reduction.

One of the biggest benefits of a P2 hybrid architecture is its potential for scalability both in terms of both power and voltage, which makes it a suitable solution for a variety of electrification needs. An option to further increase the performance in a transversal P2 configuration for front wheel drive is the combination with an electric rear axle, which allows a temporarily all-wheel drive in addition [5]. Moreover, it can be combined with other Px architectures to produce a so-called dedicated hybrid powertrain including a multi-mode dedicated transmission (DHT). This approach allows both series and parallel hybrid operation while significantly reducing the mechanical complexity of the transmission. The basic principle of these approaches is, that the base load is covered by the combustion engine and the driving dynamics are covered by the electric subsystem.

Due to these advantages regarding scalability and packaging, as well as the potential to reduce the CO_2 emissions, the market for P2 hybrids is expected to grow in the upcoming years. In this context, Fig. 1 presents the market outlook for 2024 which shows that P2 hybrids will reach a market share of about 15%. Moreover, it is projected that the dominant P2 layout will be on-axis with a fairly even split between 48 V and high voltage.

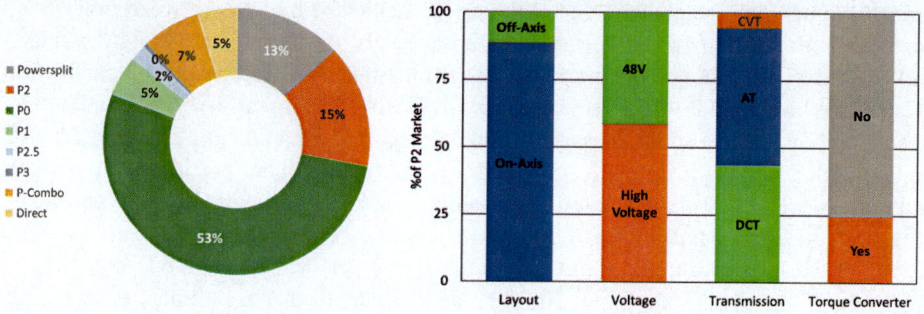

Fig. 1. 2024 hybrid architectures market outlook

In the following section the potential and limitations of P2 architecture in a 48 V level are discussed. Additionally, the technical innovations developed at BorgWarner to address the challenges and create a family of P2 modules for on-axis applications with and without integrated electronics are described in detail.

2 P2 Hybrid Architecture

Advanced mild hybrid powertrains and powertrains with high level of electrification are equipped with more powerful electric Motors (EM) compared to a P0 architecture. The most common solution is the P2 topology where the EM is located on-axis between the internal combustion engine (ICE) and the transmission. The fixed link between the combustion engine and EM is broken up by the inclusion of an additional disconnect clutch. In this configuration the EM operates at engine speeds and leverages the ratios of the gearbox to drive the wheels. Alternatively, a P2 configuration can be realized with the EM located off-axis and connected via a gear drive or chain to the crankshaft. In that case, an additional degree of freedom is given by the gear ratio which defines the speed ratio between combustion engine and EM. The gear ratio is limited by the max speed of the EM. Figure 2 shows both configurations. The application depends on the axial package, off axis parallel modules can be offered with a very short axial length, this is important for transversal powertrains. In this paper, the focus in on the on-axis layout, however, BorgWarner offers both versions.

One of the main advantages of the hybrid P2 architecture is the scalability regarding electric power. P2 modules can be standardized to a certain extend for 48 V and high voltage applications like PHEV. This offers flexibility for the OEM to balance the market share between different levels of hybridization with a relative low effort.

2.1 P2 Hybrid Architecture Characteristics

In general, the CO_2 benefit of a hybrid powertrains is mainly influenced by the level of brake energy recuperation and the opportunity to operate the combustion engine as close to the efficiency sweet spot as possible.

Fig. 2. P2 Architecture for hybrid powertrains: Electric motor configuration on and off axes

Considering recuperation from a vehicle perspective, the brake energy increases with vehicle mass, but on the other hand is lowered by increasing driving resistance. From a system perspective, the energy portion which can be harvested in the battery is influenced by the losses in the recuperation path from wheel to the battery and the maximum power of the generator. For most vehicles, 48 V system power levels allow the recuperation of a high portion of the brake energy available in the standard homologation cycles [6]. However, the exact level depends on the architecture and the power of the electric motor. For example, the common P0 architectures electric power level is limited by the torque transfer capabilities of the belt. The overall recuperation path including the engine and belt losses also reduce the overall efficiency in recuperation significantly.

The P2 hybrid architecture offers significant advantages compared to the P0. For example:

- The possibility of decoupling the engine from the EM significantly reduces the losses during recuperation.
- The EM power can be significantly increased as it is not connected to the engine through the belt, this allows to recover a higher share of the available kinetic energy on a cycle.
- The increase in EM power together with the possibility of disconnecting the engine enables P2 hybrids to provide a limited amount of electric driving.

In order to illustrate some of these advantages, a comparison of the recuperation potential of a P0 with 10 kW BSG vs. a P2 MHEV with 20 kW MGU is shown in Fig. 3. Both cases were simulated for a mid-size SUV in a WLTP cycle with an ECMS energy management controller [7]. The share of brake energy which can be stored in the battery is approx. 66% for the P2 system compared to 40% for the P0 system.

In a mild hybrid vehicle, the combustion engine is the component that mostly determines overall powertrain efficiency. Load point shift and engine charge can be used to optimize engine operation by shifting operating points to a more efficient area of the operating map. But with the chosen voltage and power level and the chosen battery size there is a limitation to what can be achieved.

Comparing these two hybrid architectures, the P2 provides the flexibility to run the engine in a high efficiency range. The operation areas of the combustion engine in a P0 and P2 architecture are shown in Fig. 4. It can be observed that the highly inefficient

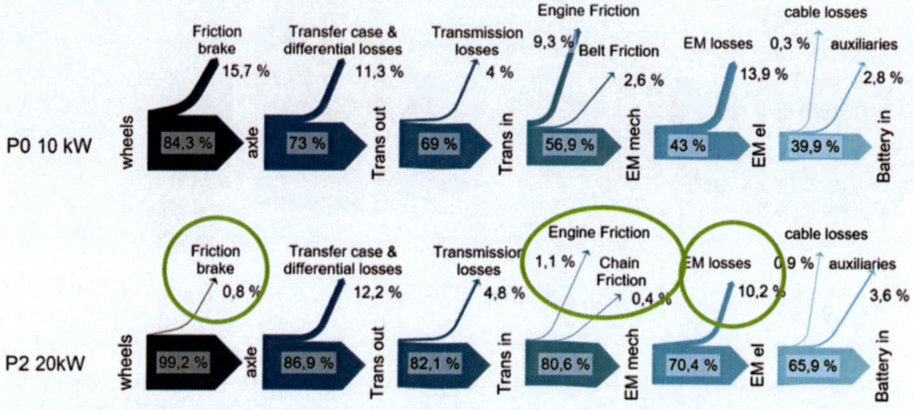

Fig. 3. Energy flow for the brake energy recuperation of a mid-size SUV in a WLTC

area at low propulsion power below 10 kW could be significantly reduced which leads to a more efficient operation of the combustion engine.

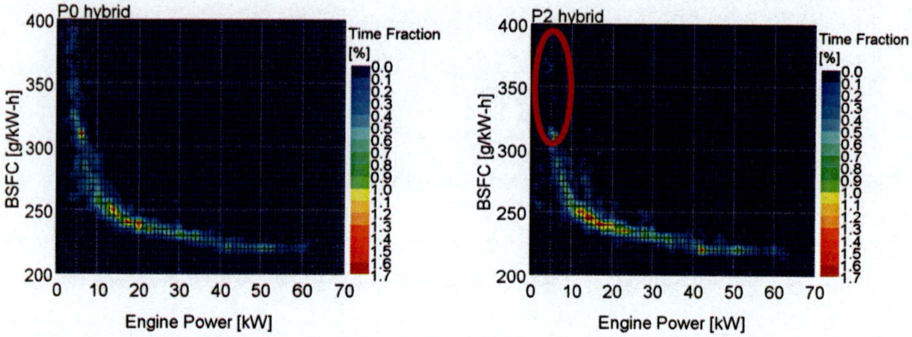

Fig. 4. Engine Operation of a 2.0 l GTDI engine in a mid-size SUV MHEV in the WLTC

Based on these benefits, the fuel consumption results for the modeled vehicle are summarized as in Table 1. The P2 architecture offers a significantly higher reduction in fuel consumption with respect to a P0. It should be noted that the presented results are for a SUV, and that higher improvements in fuel economy could be expected for a smaller vehicle with a lower driving resistance.

Table 1. Fuel consumption results of a mid-size SUV MHEV in a WLTC

	P0 10 kW	P2 20 kW
Fuel consumption benefit (compared to a conventional vehicle)	6%	13%

The main difference between 48 V MHEV and high voltage HEV is the ability for electric driving. Electric driving in a hybrid vehicle requires the option of a quick engine restart, if the battery SOC is reaching the lower limit or high torque is required by the driver. This restart must happen very fast and be comfortable to the passengers. In P2 architectures without engine starter, the engine restart is usually carried out by the EM which in parallel drives the vehicle. This means that a torque reserve for the engine restart limits electric driving power to a relative low value. Due to this limitation, so called electric coasting is possible, when the electric motor keeps the speed during engine off phases (compare [8, 9]). However, no significant electric operation is allowed due to the power limitation. This power limitation in electric operation also limits the CO_2 benefit in the homologation cycles.

2.2 P2 Modules

P2 modules can be combined with dual clutch transmissions (DCT) and automatic transmissions for transversal and longitudinal powertrain applications. Compared to a conventional powertrain, a P2 hybrid is equipped with an additional disconnect clutch and an electric motor between combustion engine and the transmission. In dependency of the transmission type, additional elements can be combined in a module to reduce the required package.

In the combination with a DCT, the disconnect clutch and dual launch clutches are integrated inside the rotor of the EM. This P2 and P2i (with integrated inverter) Triple Clutch Module has a very compact design that minimizes the impact to overall powertrain length. This is especially important for transversal applications. A possible assembly of the components two versions of a Triple Clutch Module can be seen in Fig. 5. The scheme shows the high integration of the three clutches and the electric motor. The example shows the benefit of standardization for P2 modules. Both versions have the same outer contour and connection interfaces which is ideal for a powertrain design, that should offer different electrification levels depending on the market needs.

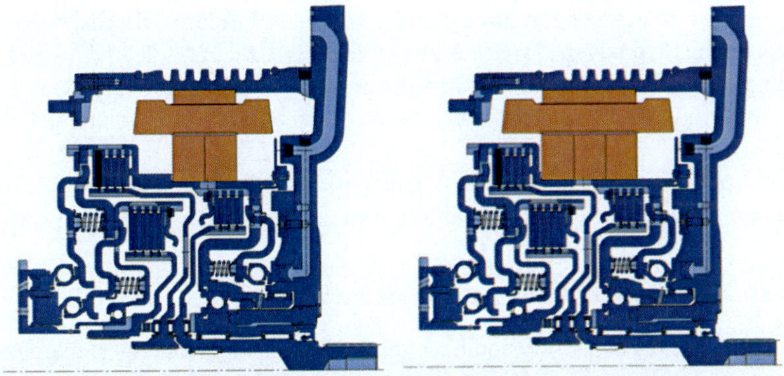

Fig. 5. Triple Clutch Module for a 48 V P2 and a HV P2

The motor is designed to operate in the same speed range as the ICE, having a maximum speed of about 7krpm. The stator uses BorgWarner's S-wind PACE award winning technology with rectangular wires, as it can be seen in Fig. 6, to provide a high torque and power density, and a peak efficiency of 95%. The torque and power can be varied by changing the stack length and winding configuration (selected depending on the target voltage level).

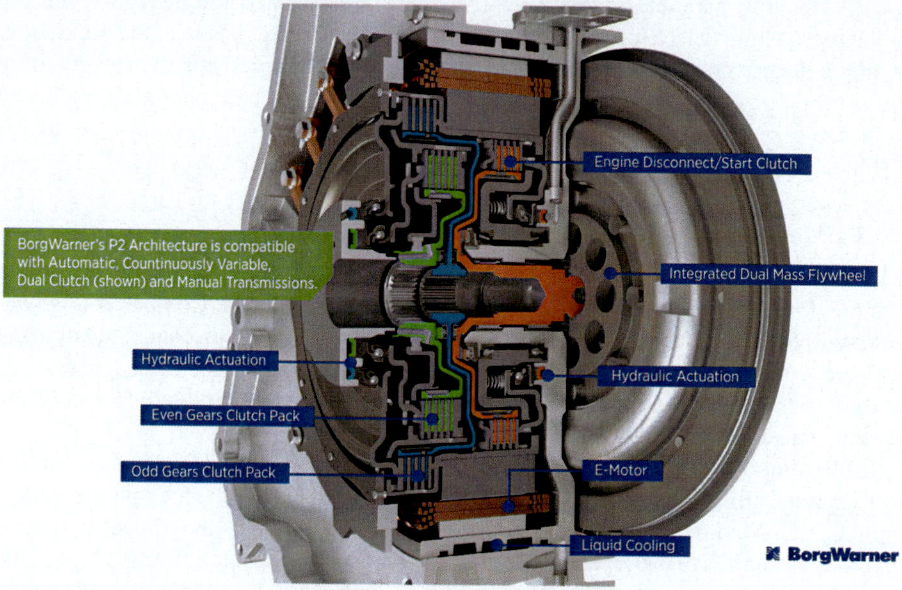

Fig. 6. P2 module for DCT

For planetary automatic transmissions and CVTs, modules are being developed that are compatible with both torque converter and launch clutch configurations.

I addition to the electric motor stator and rotor, the P2 module consists of the housing with the cooling circuit, the hydraulic control module for the clutch, the disconnect clutch and the connector. Moreover, if space is available around the module, the inverter can be integrated into it as well. Alternatively, if the available packaging space does not favor a mechanically integrated inverter, it can be placed separately and connected to the P2 module.

Figure 7 presents the different components assembled inside the P2 module during the production phase. The complete module is then used for the end of line testing.

2.3 Alternative Starter System for Hybrid Powertrains

To further improve the capability and benefit of a P2 hybrid architecture, there are a few alternatives. A simple option is further increasing the power, however, after around 30-35 kW the high currents force a move towards high voltage applications. Another option is the separation of the engine starting function from the hybrid module. Due to the fact that standard engine starters can only be used with standing combustion engine, they are not suited for hybrid powertrains. Adding a belt starter alternator is an option, which

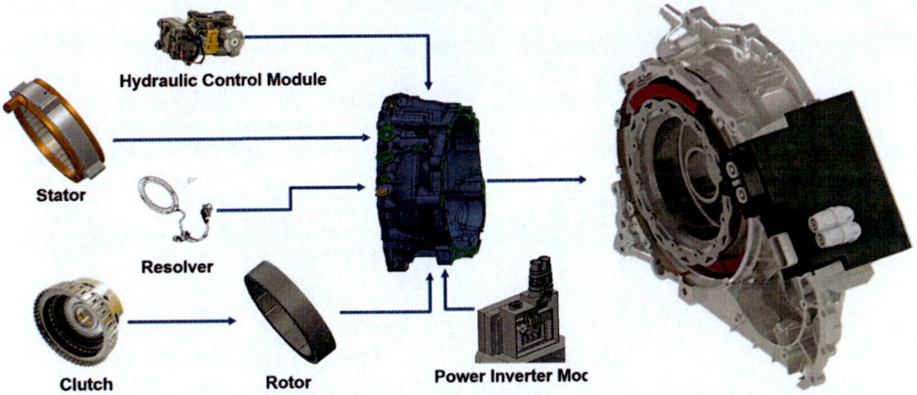

Fig. 7. Triple Clutch Module with integrated Inverter

would lead to a P0-P2 configuration. However, a BorgWarner's permanently engaged starter (PES), which originally targeted improved Start/Stop systems, can be used to provide a simpler and more cost effective solution than a P0-P2 combination. The main goals during the development of BorgWarner's PES were:

- Faster system response and better repeatability.
- No additional drag torque and no wear at engine speeds above idle.
- Improvement of NVH behavior during engine start.
- Durability for higher amount of engine stop/start cycles.

BorgWarner's dry PES system enables lot of potentials for hybrid technologies in automotive powertrains with combustion engines:

- Advanced sailing and coasting functionality with early engine shut off.
- Hybrid applications with a high number of engine restarts.
- No additional system required for cold start.
- Reduction of crankshaft torque load during engine cranking.

With an almost unnoticeable shut off and restart of the combustion engine, this technology offers the ideal technical features for the combination with a P2 48 V architecture to handle the start of the combustion engine. The system layout for the PES is shown in Fig. 8.

The PES system is designed as a Plug & Play module that is positioned in the dry environment between the combustion engine and transmission. It replaces the original starter ring gear and uses existing interfaces with only minor modifications on the powertrain.

The basic requirements for the PES design are the torque which the system has to transmit to the crank shaft during engine cranking at start and the total number of engines starts. In the current design, 1 million engine starts can be achieved which can be considered sufficient for 48 V mild hybrid operation.

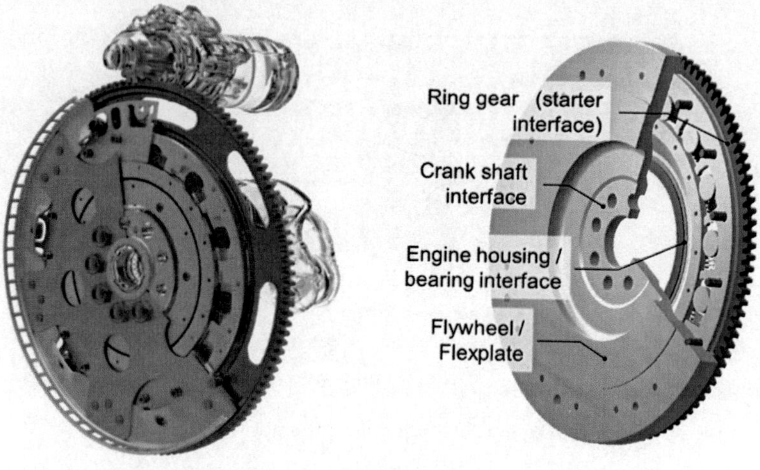

Fig. 8. Permanently Engaged Starter (PES) with one way clutch

2.4 System Advantage of P2 with Separate Starter

Adding a PES to a P2 MHEV architecture provides further benefits. The electric drive mode area can be enlarged by eliminating the engine start torque reserve. Thus, low efficient operation of the combustion engine at low load, and therefore CO_2 emission can be further reduced. However, the energy for the engine restart is still required, which limits the overall impact.

The efficiency gap between 48 V hybrids and HV hybrids is closing, which leads to a favorable ratio between cost and CO_2 benefit. Figure 9 shows the functionality and the CO_2 emission ranges for a combination of a P2 MHEV+beltless starter in comparison with P0 and P2 MHEV and a P2 HEV. The combination of a P2 hybrid with up to 25 kW in combination with a beltless starter brings the 48 V P2 architecture closer to a high voltage hybrid regarding CO_2 emissions while doing so at a lower cost.

3 Summary

BorgWarner has developed a family of hybrid drive modules including a scalable and modular custom engineered hydraulic actuation systems for P2 hybrid architecture. By high integration, modules are offered with low package requirements for all transmission combinations. A three clutch module as well as the combination with a torque converter is possible. The integrated high performance motor delivers enough power to fully achieve the possible benefits of a P2 architecture for most vehicle segments. In combination with a permanently engaged starter, the CO_2 potential can be further increased due to the elimination of the engine start power reserve. Simulation and test will be carried out to fully analyze the changes in operation strategy and the related fuel efficiency gains.

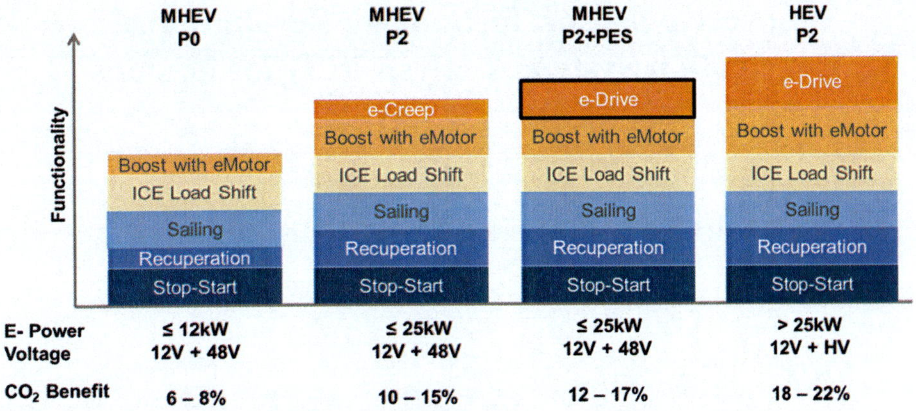

Fig. 9. Benefit of a P2 + PES combination compared to other MHEV and HEV configurations

References

1. Eichler F, Bennewith K, Helbing C, Philipp, K, Lück P, Weiß N, Felsch C (2017) Volkswagen electrifies the new golf. In: 38th International Vienna motor symposium
2. Breitbach H, Jausel H, Garrard T, Metz D (2018) Elektrisch unterstützte Aufladung - Status und Ausblick. MTZ - Motortechnische Zeitschrift 79:28–35
3. Bondards A, Mohon S, Semenov D, Wenzel W (2019) Comparing 48 V mild hybrid concepts using a hybrid simulation toolkit. In: Stuttgart symposium
4. Blumenröder K, Bennewitz K, Zillmer M, Mann A, Voeltz T, Dick V (2019) Das 48 V Mild-Hybrid Antriebssystem des Volkswagen Golf: Auslegung und Steuerung. In: 40th international Vienna motor symposium
5. Eglinger M, Hengst J, Kücükay F, Li M, Sieg C, Wenzel W (2019) Next Generation 48 V MHEV. In: Int. VDI Congress *Dritev - Drivetrain for Vehicles*. Bonn
6. Sorger H (2018) The tailored engine & powertrain for 48 V. Future powertrain conference, Birmingham, 28 Feb 2018
7. Bongards, Anita et al (2019) Comparing 48 V mild hybrid concepts using a hybrid-simulation-toolkit. 19. Internationales Stuttgarter Symposium. Springer Vieweg, Wiesbaden
8. Eckenfels T, Kolb F, Lehmann S, Neugebauer W, Calero M (2018) Eine intelligente Aufwertung des Antriebsstrangs; 18. Schaeffler Colloquium
9. Timmann M, Inderka R, Eder T (2018) Development of 48 V powertrain systems at Mercedes-Ben. Int. Stuttgart engine symposium
10. Bäumler R, Moser A, Schäfer M (2014) Dry running, Permanently Engaged Starter System (PES) for a comfortable start of combustion engines providing further potential to reduce fuel consumption. Motortechnische Zeitschrift, Dec 2014

Innovative Sensor Technology Revolutionizes Lubrication System Analysis in Transmissions

Mario Theissl(✉), Hannes Hick, and Peter Neger-Loibner

IME, Graz University of Technology, Inffeldgasse 21BII, Graz, Austria
{mario.theissl,hannes.hick,peter.neger-loibner}@tugraz.at

Abstract. With increasing requirements for transmissions in terms of rotational speed and geometrical constraints, the topic of lubrication gains in importance. In the future, more accurate and sophisticated lubrication system analysis methods will be needed to ensure functionality in series applications. Therefore, it is necessary to reconsider current testing methods like tilt tests based on visual inspections with transparent housings and pursue new and innovative approaches for test and verification as well as for condition monitoring. One of those approaches was developed at the Graz University of Technology and comes in the form of a unique fluid sensor. The innovative sensor was designed for the quantification of multiphase flows and is ideal for analyzing the lubrication in transmissions. By integrating several measurement principles into one miniaturized sensor, the flow and the temperature of the oil are measured. Furthermore, the sensor is able to detect gas bubbles and oil splashes. The combination of the obtained measurement data from several of these innovative sensors at different positions is used to determine the exact lubrication condition dependent on rotational speed, tilt angle and oil temperature. Different sensor designs like a sensor probe or an adhesive sensor foil allow measurements at individual test points inside of the transmission. Together with the objective of implementing a wireless data transfer, this sensor application provides a new way of analyzing lubrication conditions.

Keywords: Lubrication transmission sysem analysis · Lubrication sensor · Multiphase flow sensor

1 Introduction

The increasing power density and speed of electrified of powertrains brings up many new challenges for the field of lubrication. Especially for the lubrication system analysis, current test methods are facing the limit of their capabilities and need to be reconsidered for coming development projects. Overall, the fundamental knowledge about the behaviour of the lubrication becomes more and more a key factor in understanding the system from a tribological point of view and in increasing efficiency. Traditional testing methods are often not accurate enough for quantifying the oil supply of specific machine components inside of the powertrain. Therefore, the Institute of Machine Components and Methods of Development at the Graz University of Technology set up a new standard for verifying

Euroforum Deutschland GmbH (Ed.): *CTI SYMPOSIUM 2019*, pp. 498–508, 2021.
https://doi.org/10.1007/978-3-662-61515-7_44

the multiphase lubricant flow inside of a transmission with a patented sensor system. The innovative sensor technology combines two measurement principles at the size of a *match stick*, which allows new insights into multiphase flow analysis. The implemented functions of the sensor are gas and liquid detection, as well as a measurement of the flow velocity and the temperature of the fluid. The compact size of the sensor allows minimal invasive flow metering inside of a transmission.

The purpose of this paper is to give new insights into lubrication system analysis and to explore new technologies for that purpose. First, the targets of the research are explained, followed by an introduction of the measurement principle. Furthermore, a pilot project at the AVL-TU Graz Transmission Center is used to explain the technology, the application, and the benefits of its usage. In addition, a new procedure for lubrication system analysis implemented with this technology is explained and compared to traditional methods.

2 Targets

The Institute of Machine Components and Methods of Development at the Graz University of Technology operates a Transmission Center (TC) in cooperation with AVL List GmbH, where basic research on powertrains is performed. One research focus of powertrain development is tribology, especially the lubrication system analysis of gearboxes. A particular type of test cell, the dynamic tilt rig is used to evaluate the lubrication condition of the transmission depending on the rotational speed and tilt angle. The tilt angle represents the acceleration forces caused by lateral acceleration, longitudinal acceleration, or climbing conditions that act on the transmission at real drive operation and therefore cause dynamic lubricant flow.

Additionally, several methods are used to validate the oil distribution. Most often, these methods include visual analyses with camera systems through transparent housings. However, due to high-speed tests and complex structures of the gearbox, usually, there exists no optical path to the analysis point of interest. In Fig. 1, a concrete example of such a visual oil supply analysis of a bearing inside of the transmission is illustrated.

A test matrix defines a set of a specified tilt angles and the input shaft speeds of the gearbox, and is the starting point of every lubrication system analysis procedure. Based on this test matrix, the transmission is tested, whereas points of interest are being recorded by cameras simultaneously. Via the video stream, the engineering team assess, if the lubrication of the bearing seems is "*sufficient*". This binary assessment of "*ok*" or "*not ok*" is experience based and difficult to qualify decision. The assessment, whether the lubrication is "*sufficient*" or not, needs to be added to these currently *experience based decision*. The development of a more detailed solution to the visual inspection was the objective of this research project.

Due to geometrical constraints, the installation of conventional process flow meters inside of transmissions is usually not possible. Often a bypass of a lubricant supply line is established with transparent Teflon tubes to make the fluid flow visible. Such modification could dramatically influence the original lubrication condition and is not representative. Therefore, any measurement system to quantify the fluid condition has to be minimally invasive and should not disturb the initial fluid flow in a gearbox.

Fig. 1. Visual inspection of oil distribution inside a transmission through a transparent housing

Furthermore, the harsh environmental conditions inside require thermal stability over a broad temperature range and also chemical resistance against aggressive lubricants to any measurement probe. Observations through transparent housings in currently used gearboxes show, that the lubricant distribution is not consistent. Depending on the speed, and the acceleration in lateral or longitudinal direction, oil splashing and for example aeration in the oil sump increases. This aspect makes the differentiation between gaseous and liquid fluid conditions an essential requirement for a flow meter. Based on the listed requirements, the development of a new sensor system for a precise lubrication analysis is required.

3 Technology

The acronym FMwMD stands for **F**low **M**eter **w**ith **M**edium **D**etection and is the name of the patented sensor [1]. The sensor combines a thermal and an optical measurement principle and is capable of measuring the flow velocity and the temperature of a fluid. Furthermore, the sensor can detect fluid splashes as well as gas bubbles.

The flow meter is realized by using the calorimetric measurement principle, which is shown in Fig. 2. If the cross-section of the application e.g. a pipe is known and the flow is laminar, also the flow rate in liter per second [lps] can be calculated. The calorimeter front end consists of two PTC thermistor elements. One element acts as reference element and the other element as heater. Based on the temperature of the reference element, the heater element gets heated up by a controller. A control loop manipulates the temperature of the heater to establish a constant temperature difference between the heater and the reference element. The heat power of the heater is measured and represents the output signal of the calorimeter. When the flow equals zero, the heat power $\dot{Q}$ is at its minimum. With increasing flow velocity, more heat dissipates and the heat power also increases. Therefore, the heat power $\dot{Q}$ is proportional to the flow velocity v_{fluid} of the fluid. The physical law behind this behaviour is the general heat transfer equation [2]. To get an understanding of the relations, the quasi-one-dimensional steady conditions heat transfer equation is mentioned:

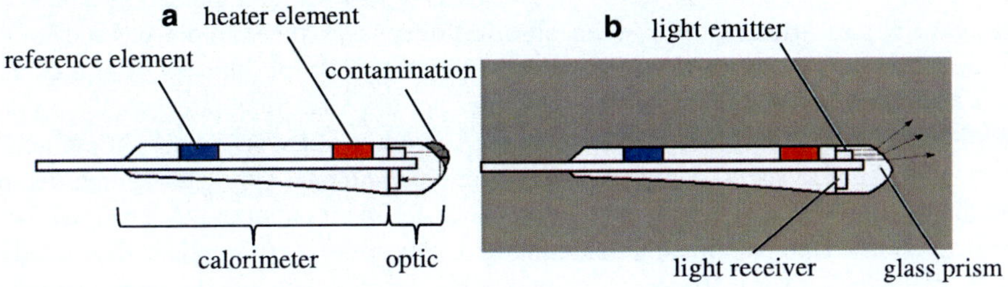

Fig. 2. Sensor measurement principle. (a) sensor in gas with a liquid contamination on the optic (b) sensor entirely in liquid

$$\dot{Q}(x) = -k(v_{fluid})\left(A\frac{dT}{dx}\right) \tag{1}$$

$\dot{Q}$ Heat power in W
x Direction of temperature gradient m
k Heat transfer coefficient $\frac{W}{mK}$
v_{fluid} Flow velocity of fluid $\frac{m}{s}$
A Area of heat dissipation m^2
T Temperature K

A controller holds the term $\left(A\frac{dT}{dx}\right)$ constant, so it can be replaced by the Constant factor C.

$$C = -\left(A\frac{dT}{dx}\right) \tag{2}$$

C Constant parameter
Therefore, the heat transfer equation simplifies as follows:

$$\dot{Q}(x) = k(v_{fluid}) \cdot C \tag{3}$$

Due to the fact, that the heat transfer coefficient is proportional to the flow velocity of the fluid at a forced convection boundary condition, the following simplification can be introduced:

$$\dot{Q}(x) \propto v_{fluid} \tag{4}$$

In simple terms: the higher the flow velocity gets, the higher the heat power has to be. The observed proportion between the flow velocity and the heat power is valid for one type of fluid. If there is a change in the fluid phase, for example, from liquid to gaseous, the heat power is no longer in *explicit* relation to the fluid velocity. For example, the heat transfer coefficient for stationary oil is similar to the heat transfer coefficient of high airflow. The differentiation between the fluid phase in the post-processing is not *distinct* any more. This observation leads to the conclusion that just a thermal measurement principle for multiphase flow metering is not sufficient.

For that reason, the calorimeter of the FMwMD sensor is supplemented by an optical sensor, which can differentiate between multiple fluid phases, for example, gas and liquid. The optical sensor uses the character of the different refraction indexes of liquids and gases. Contaminations or aging of the fluid does not have a critical effect on the optical measurement principle. The main elements of the optical sensor are a light emitter, a prism, and a light receiver. A light beam gets emitted and enters the glass prism. At the boundary surface of the prism to the fluid, the measurement effect of the sensor element comes into play. The fundamental function of the measurement principle is illustrated in Fig. 2. When the sensor is in air, the emitted light gets reflected, and the output signal is high because total refraction occurs at the boundary surface. In contrast, when the optics are in liquid, all emitted light gets decoupled, and the output signal is zero.

The sensor provides following raw data signals, which are then used for post-processing:

(1) Temperature from the PTC element of the calorimeter
(2) Heat power of the calorimeter which is depending on the flow velocity
(3) The signal from the photodetector measuring the fluid phase gas or liquid

To get the flow velocity out of the raw measurement data, a look-up table is used, where the heat power and the temperature are the two inputs (Fig. 3).

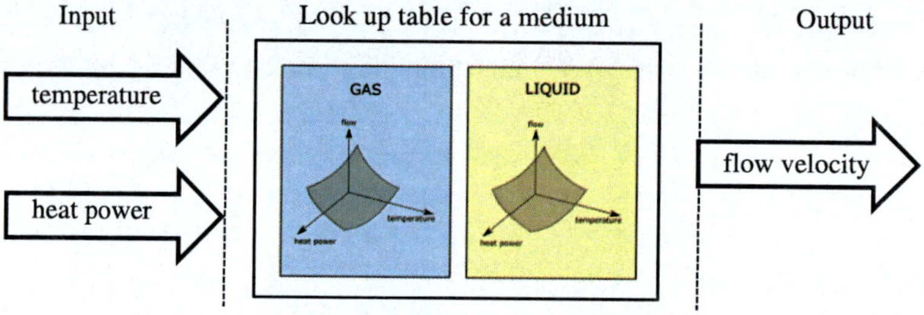

Fig. 3. Lookup table for data processing

The look-up table is a 3D map, which has to be calibrated for each lubricant and sensor separately. The calibration is necessary to match the fluid properties like the usually unknown thermal and viscosity parameter to the measurement system.

For any flow rate measurements, the sensor has to be fully under oil. If the measurement is performed in an open channel, the flow velocity is used to quantify the flow instead of the volume flow rate. Due to the fact, that the sensor measures the flow velocity, the pipe characteristics are needed to calculate the flow rate in out of the flow velocity. At a laminar flow profile the flow velocity is not constant over the diameter of the pipe (see Fig. 4). *Therefore, the distortion of the flow profile over the diameter has to be considered for the installation point of the sensor.* The easiest way of matching the flow velocity to a the flow rate in a pipe is the calibration of the measurement setup.

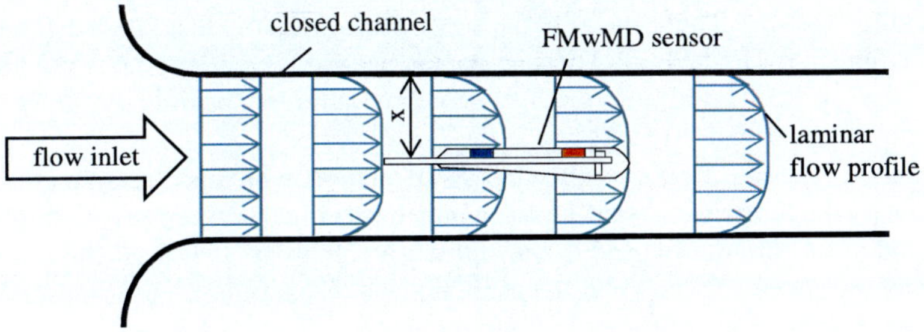

Fig. 4. Mounting position of the sensor in a pipe and dependency of the flow profile

4 Field of Application

The matrix in Fig. 5 shows possible applications, where the sensor could be used to validate different types of lubrication systems in the automotive powertrain development. In lubrication systems, different conditions of the lubricant are present. These conditions vary from pure oil to oil mist. Depending on the type of the lubrication system, the required, as well as the critical conditions, could vary and according to that, the sensor mounting position has to be chosen. This relation can be seen in Fig. 5. The examples in the red frames represent positions for monitoring critical operation conditions, whereas the green frames show the positions for evaluating the required conditions.

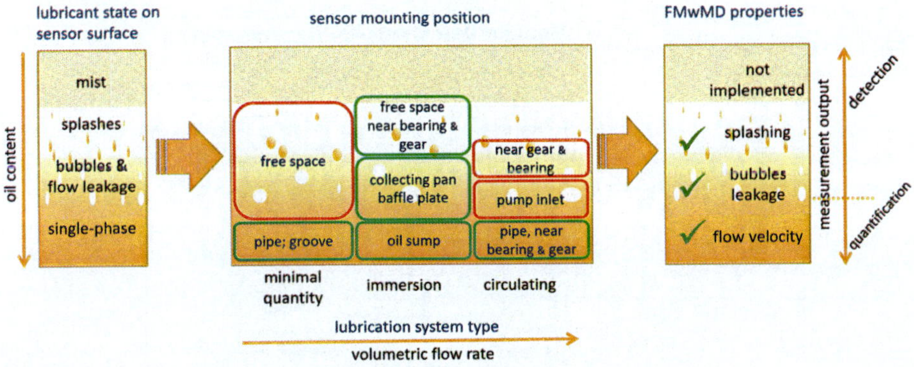

Fig. 5. Matrix of sensor measurement positions depended on the oil content and the lubrication system type

The choice of the lubrication system is based on the circumferential speed of the machine component. The speed of the bearing or gear relates to the flow rate of the oil supply to the bearings or gears [3]. Therefore, a minimal quantity of lubrication has a minimum flow rate and the oil sump lubrication has the highest flow rates of the lubricant.

The design of the powertrain and the lubrication system is done by specific experience. This preliminary design should fulfill the requirements of the theoretically needed oil amount [reference]. Therefore, simulations are used to provide a rough estimation of the oil distribution. The FMwMD sensor system is used to validate the simulation and measures the lubricant conditions precisely. The measurement position has to be chosen with caution to get the best results out of the sensor system.

The matrix (see Fig. 5.) is classified by the oil content, which is present on the sensor surface and by the lubrication type. The oil sump lubrication is cheap, easy to implement, reliable and most commonly used in the automotive industry. It is used for a moderate circumferential speed <25 m/s [4]. To ensure that the oil captures all bearings and gears, the FMwMD sensor could be used in different ways.

First, the sensor can measure the flow velocity in the oil sump of the powertrain to detect if smoothing zones for degassing the lubricant are reasonably designed. The gas/oil detector is also able to measure if there are air bubbles in the lubricant. In the region of the collecting pan, measurements after the startup could be performed to monitor lubrication in the early operating phase. In holes, grooves or drillings, the flow velocity is representative for the oil supply of the bearing. Depending on longitudinal and lateral acceleration as well as on the input shaft speed, the lubricant conditions can be quantified with the FMwMD sensor. Furthermore, the presence of oil splashing near bearings in free space can be detected with the sensor.

Figure 6 shows how the signals are used for medium detection. The blue signal represents heavy oil splashing. The red curve shows few bigger air bubbles in an otherwise laminar oil flow, indicated by the peaks. A starting lubrication, so the transition from air to oil is illustrated in the yellow signal.

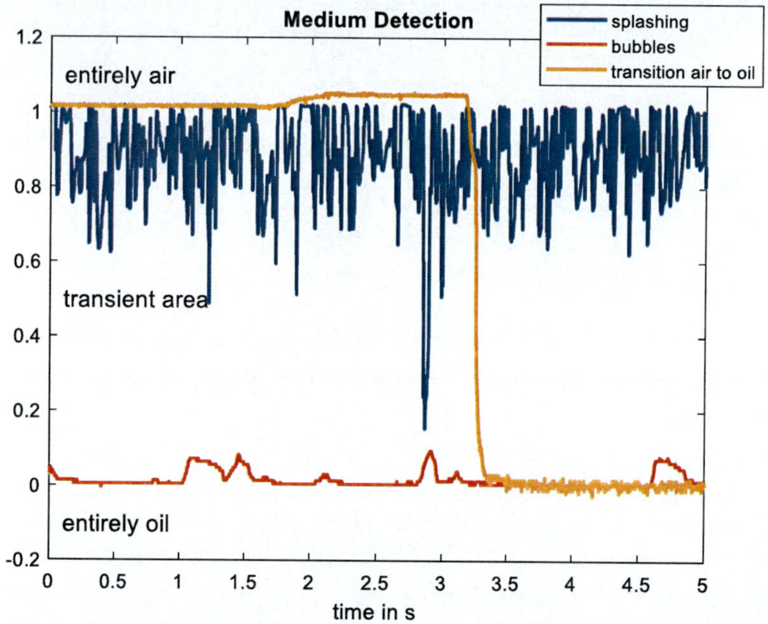

Fig. 6. Three exemplary FMwMD sensor measurement signals of the optical medium detection

In case of a higher input shaft speed, often a dedicated oil pump lubrication is implemented. For that type, the sensor could be used to measure either in the supply pipes of the machine components or directly near the machine components. The fluid conditions in the supply pipes is most often a one-phase laminar oil flow. However, due to any failure, this condition could be changed by the formation of bubbles because of cavitation. The bubbles can be detected by the gas/liquid detector of the sensor. Also, the detection of a leakage in the supply line is possible. Near bearings, the time until the first lubricant is present, if the bearing is flooded or if the lubrication flow stops due to any operating condition could be detected.

In high-speed gearboxes, minimal quantity lubrication has to be applied. The detection of oil mist or especially the measurement of the oil content of the mist is not implemented in the sensor. However, the calorimeter is theoretically able to measure the velocity of the single-phase oil film, which forms inside of pipe walls in a minimal quantity dosing unit. Therefore, the detection of too much or to less oil can be used as a condition monitoring parameter and trigger an emergency stop at any critical parameter deviation. Near bearings, splashing or the accumulation of lubricants could lead to over lubrication, which could damage the gearbox within seconds.

5 Potentially Interesting Application Areas

Besides using the sensor as a development and analysis tool for lubrication system design in the automotive sector, a potentially exciting use-case is condition monitoring. At testing prototypes, the sensor can prevent the device from form-critical lubrication parameters depending on the lubrication system types. The primary information, whether there is "*too much*" or "*insufficient*" oil present at the bearing, could enable early detection of damage. Therefore, the FMwMD-sensor can be a helpful device to save costs, to prevent prototype failures, and thus accelerate the development. In general, the sensor could be used in any application to monitor operating conditions and detect critical states. Another possible use case is the monitoring of coolant cycles. Current systems are using the thermal balance to monitor its condition. By implementing the measurement of the fluid flow, a much faster reaction to changes would be possible.

6 Comparison to Traditional Lubricant Analysis Methods

In this section, traditional methods are compared to the innovation of the IME and possible other substitute sensor methods. The evaluation criteria are composed of the in Fig. 7 stated aspects.

The first criterion is the required competency level for data interpretation and the ongoing evaluation of ultimate simulation results. The most commonly used traditional method for lubrication system analysis is a visual analysis. Either a transparent housing or sight glasses in combination with a translucent bypass tube and riser, which are installed in aluminum housings, are used. Engineers are using camera systems to record oil distribution. Based on the film material, the lubricant condition assessed by the engineering team with "*sufficient*" or "*critical*". This type of evaluation requires a high competence level. The comparison of the results is difficult, especially where the

	traditional methods					sensors alternatives		
	visual analysis transparent housing; riser, tubes	**dye test** with luminescent additive	**endurance tests**	**temperature measurement**	**new FMwMD**	**thermal flow sensors**	**ultrasonic flow measurement**	**capacitive measurement of fluid phase**
competence level × high ≈ medium ✓ low	×	×	≈	×	✓	✓	✓	✓
real time measurement & detection	×	×	×	✓	✓	✓	✓	✓
detection of bubbles; splashes; single phase flow	≈	×	×	×	✓	×	×	✓
flow measurement at single phase	×	×	×	×	✓	≈	✓	×
minimally invasive	×	✓	✓	✓	✓	✓	×	✓
system view × poor ≈ limited ✓ good	✓	≈	✓	×	✓	×	×	×

Fig. 7. Matrix to compare traditional analysis systems and possible substitute sensor systems to the FMwMD sensor

threshold value for this assessment between "*sufficient*" and critical lies. Furthermore, an optical path to the evaluation point of interest has to exist. Otherwise, the unknown product behaviour remains in these not properly visible areas. An alternative solution to get an estimate about the not observable regions is a dye test. In this procedure, a *fluorescent* colour is added to the lubricant while the test is running. At the end of the trial, where a specific tilt angle and input shaft speed was used, all the oil has to be drained and the transmission to be disassembled. All machine components are inspected under ultraviolet light to examine if the fluorescence colour coats it. Out of the image form the ultraviolet inspection, the contact of oil with the machine component is evaluated.

Endurance tests require high efforts in the form of time and cost until the first results are available. In case the device under test does not fail, and no annealing colours are noticeable, the lubrication seems to be "*sufficient*" for the applied test specifications. Nevertheless, real drive operating conditions can be different from the ideal test condition on an endurance testbed.

The temperature measurements near bearings are an indicator of the operating condition of a bearing. However, the high temperature on a bearing can be caused by under- or overlubrication. In the case of overlubrication, the bearing can get damaged within a few seconds. As the temperature measurement is most often located at a small distance from to the bearing, a dead time of a few seconds is present. Therefore, the damage could already have happened without an increasing temperature at the measurement point.

The FMwMD-sensor does require no special requirements to the competence level in data interpretation when the measurement location is chosen correctly. When the test setup is installed, the threshold between "*sufficient*" and "*insufficient*" oil can be strictly defined and does not drift over time. Therefore, the FMwMD sensor provides a tremendous additional measurement data of the lubricant flow to the engineering team, which did just visual analysis up to now. Thus, the measurements are more comfortable to compare.

If a single-phase fluid flow is present in a pipe, a conventional flowmeter could be used to measure the flow rate. But to ensure a proper function, the condition that the whole cross-section of the tube is entirely under oil has to be fulfilled.

The FMwMD measures the flow velocity in meter per second [mps] because the sensor system can be used for any pipe diameter. Therefore a quantification of fluid flow in an open channel is possible.

If a measurement system is minimally invasive, its influence on the original test setup is minimal. By using a plexiglass housing, the mechanical stiffness and the thermal conductivity is reduced in comparison to aluminum. This restricts the transmission from being tested at high speed, and the temperature range is limited to less than 70 °C. The dye test can be performed at the real transmission and is not limited to speed or temperature

The most reliable tests are endurance tests, where the original parts can be tested at high load and a wide temperature range. Additionally, the installation of thermocouples for temperature measurement does not influence the device under test noticeable.

Due to the compact size of the FMwMD sensor with a diameter of 3 mm, the disturbance of the lubricant flow is minimal.

Compared to sensors with ultrasonic or capacitive technologies, the FMwMD-sensor is significantly more compact in size.

For that reason, the FMwMD-sensor can either be placed near the housing insight of the transmission or with an extension in free space near a gear. The next generation of the FMwMD sensor will be realized in a foil-design, which could universally be installed on housings or a rotating shaft, which is not possible with conventional sensors.

7 Summary and Outlook

The technology discussed in this paper is capable of measuring the flow velocity and the temperature of the oil and can detect phase shifts in the fluid. This allows the FMwMD-sensor to differentiate between gas, splashes, bubbles, and oil. With the current implementation of the measurement principle in a compact measurement probe with a diameter of 3 mm, the sensor can be universally applied in a transmission. For linking more sensors together at one databus system, a CAN-bus interface for data transfer will be included in the next sensor generation, which will be available in 2020. For a more flexible application, an adhesive flexible foil sensor probe is in the design phase. This provides the advantage of installing the sensor on a curved surface as well as in holes. Furthermore, the foil is even less invasive than the probe. For measurements at rotating parts, a wireless solution including an energy harvesting system is also in development.

Based on the gained knowledge through the sensor usage at the AVL-TU Transmission Center, a standardized lubrication system analysis procedure will be established. The future of the lubrication analysis is based on practical knowledge and simulation in the early development phase. Nevertheless, the simulation of the transmission lubrication has to be validated with the FMwMD sensor system in different operating conditions on different testbeds. Therefore, this validation method reduces unknown product behaviour and therefore supports not only to improve the quality but also the safety of transmissions. The procedure for the application of the FMwMD sensor will be published free of charge on the homepage of the IME next year.

References

1. EP18190129.9.; PCT/EP2019/069315
2. Hering Martin Stohrer.: Physik für Ingenieure, 11. Auflage
3. Niemann, G Winter, H: Maschinenelemente, Band 2, 2. Auflage
4. Schaeffler Technologies AG & Co. Kg: Wälzlagerpraxis 4. Auflage 2015

Proposal of Application of Magnetostrictive Torque Sensor in EV—Seamless 2-Speed Shifting with Torque Feedback Control

Seigo Urakami(✉), Kota Fukuda, Junji Ono, Tomoyuki Miyazaki, and Shinji Okada

NSK Ltd., Tokyo, Japan
urakami@nsk.com

Abstract. This paper shows the high potential of magnetostrictive Torque Sensor applied in electric vehicles (EVs). Although most of the current EVs are equipped with fixed-ratio gearboxes, the demand for 2-speed transmission will be increased in the near future for improving cruising range and driving performance. The torque feedback control of clutches and an e-motor with Torque Sensor can realize seamless shifting even with large gear step ratio which causes uncomfortable shifting shock. In order to evaluate the advantages of Torque Sensor applied in 2-speed EV, preliminary evaluation in the powertrain system of an ICE vehicle was conducted. Moreover, seamless 2-speed eAxle concept is proposed for long cruising range and comfortable ride.

Keywords: Torque sensor · Torque detection · Feedback control · Transmission · eAxle · 2-speed

1 Introduction

Vehicle electrification is accelerating for worldwide requirement of CO_2 reduction and better fuel economy. The number of pure electric vehicles (EVs) is increasing due to above background. However, most of EVs have insufficient driving performance and cruising range compared with the internal-combustion engine (ICE) vehicles in the same segment.

Our group is focusing on 2-speed transmission with large step ratio for the EVs' powertrain called eAxle in order to improve the driving performance and cruising range. However, 2-speed transmission with large step ratio causes uncomfortable shifting shock and requires high clutch capacity. Feedback control of e-motor and clutches with Torque Sensor is one of the effective solutions for the issue.

This paper shows advantages of applying magnetostrictive Torque Sensor to EV powertrain by exhibiting the high sensing performance. Developed Torque Sensor is implemented in the ICE powertrain, and the achieved experimental results validate the effectiveness of torque sensing in powertrain.

Euroforum Deutschland GmbH (Ed.): *CTI SYMPOSIUM 2019*, pp. 509–522, 2021.
https://doi.org/10.1007/978-3-662-61515-7_45

2 Principle of Magnetostrictive Torque Sensor

The principle of torque detection with magnetostrictive Torque Sensor[1] is shown in Fig. 1. The sensor uses the inverse magnetostrictive effect. Permeability of ferromagnetic materials is changed by the strain on the crystal lattice. The permeability of the shaft surface is changed in proportion to the applied stress. The sensor equipped on the shaft with clearance detects the inductance fluctuation of its coil generated by the permeability variation of the shaft.

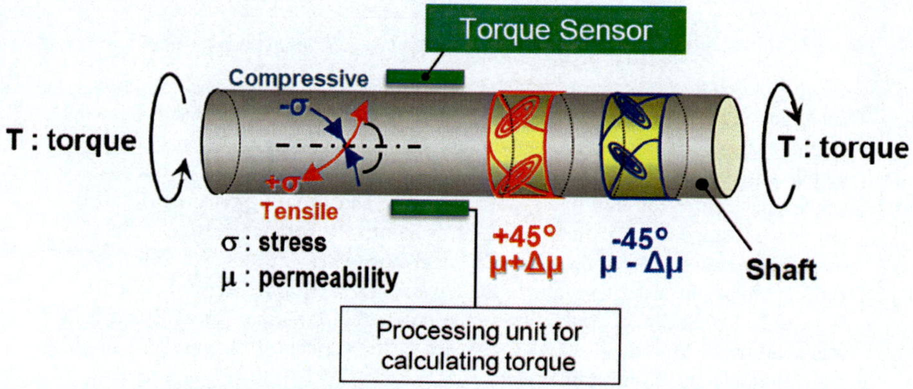

Fig. 1. Principle of torque measurement.

When the torque (T) is applied to the shaft, the tensile and compressive stresses with the direction of ±45° to the axis are caused on the shaft surface. Then, the permeability is changed from the original (μ) to $\mu + \Delta\mu$ at +45° and $\mu - \Delta\mu$ at −45° [2]. In order to maximize the inductance change by the torsional strain (= torque), the sensor coil is mounted with the direction of ±45° to the shaft axis. The bridge circuits are implemented to detect the generated voltage in the coils with high accuracy. Finally, the torque is calculated by the generated voltage on the sensor via the processing unit.

3 Shaft Properties Applicable to Magnetostrictive Torque Sensor

To evaluate the sensor performance, sensitivity and hysteresis error are defined as the following equations and Fig. 2.

Sensitivity = $\Delta V_S / \Delta T$ [mV/Nm]

Hysteresis error = $V_h / \Delta V_S \times 100$ [% full scale]

T: Torsional torque on shaft, V_S: Output voltage of sensor

V_h: Maximum deviation of output voltage of sensor

Δ: Difference between the maximum and the minimum values

The sensitivity and the hysteresis error of Torque Sensor are influenced by shaft properties (e.g. material, heat treatment, surface finish). In order to verify the feasibility of the shaft as a sensor in the powertrain, the following shafts were evaluated, which are used in common transmissions.

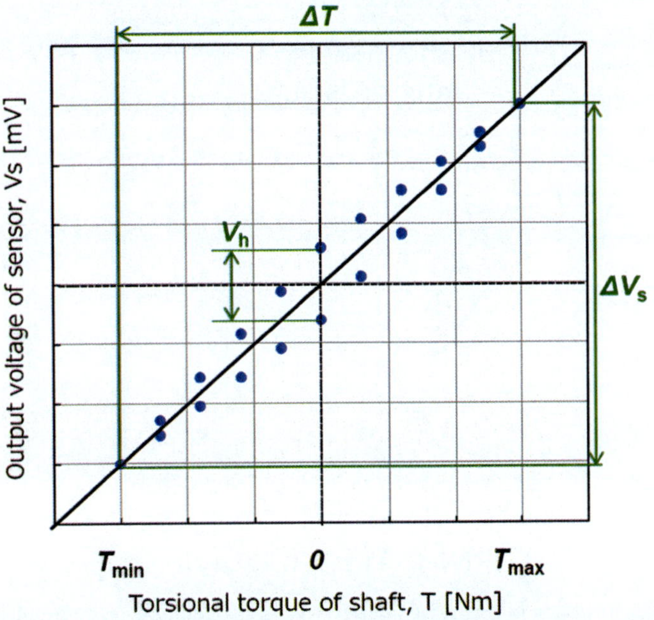

Fig. 2. Definition of sensitivity and hysteresis error of torque sensor.

■ Steel grade: JIS SCr420H (Equivalent to ASTM 5120H, ISO 20Cr4)
JIS SCM420H (Equivalent to ASTM 4118 with modified Cr composition)
JIS SNCM420H (Equivalent to ASTM 4320H)
■ Heat treatment: Carburized, quenched, and tempered
■ Surface finish: (a) Polishing and (b) Shot peening after polishing

The measurement results are shown in Fig. 3. The carburized shaft with polished surface showed small sensitivity and large hysteresis, which is unsuitable for precise torque measurements. Then, the further improvement of magnetic properties on shaft surface was implemented. It is widely known that coating of ferromagnetic material or shot peening on the surface is effective for improving the sensitivity[3]. In consideration of the manufacturing process of transmission shafts, shot peening was additionally treated on the polished surface. As a result, the shaft properties were significantly improved with 5 times of sensitivity and less hysteresis than 1% full scale.

Shot peening improves 5 times of sensitivity and less hysteresis than 1% full scale.

4 Sensor Performance

Schematic diagram of test equipment to evaluate Torque Sensor performance is shown in Fig. 4. Magnetostrictive Torque Sensor and the commercially available torque meter of flanged type are equipped between drive and driven servo motors. The measured torque by the torque meter is used for the reference value.

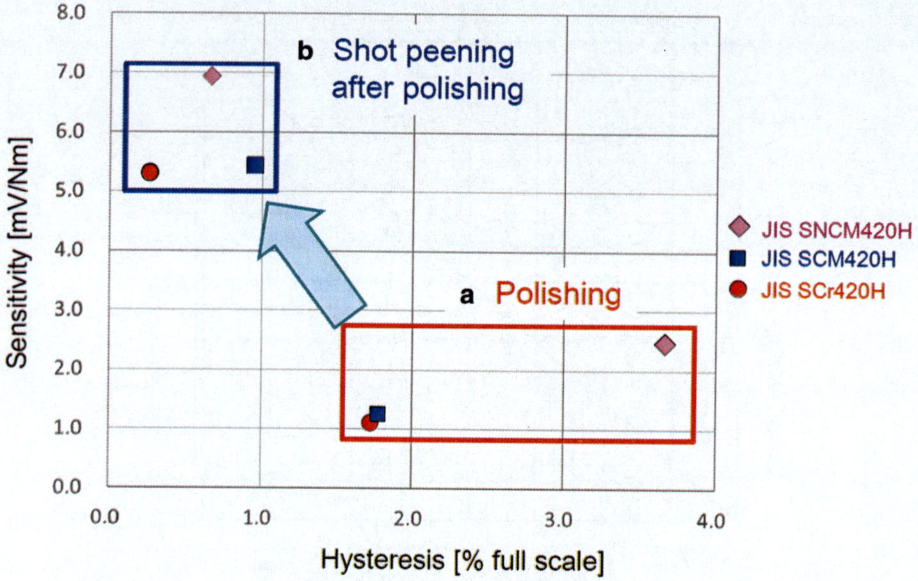

Fig. 3. Comparison of sensitivity and hysteresis error among the different shafts.

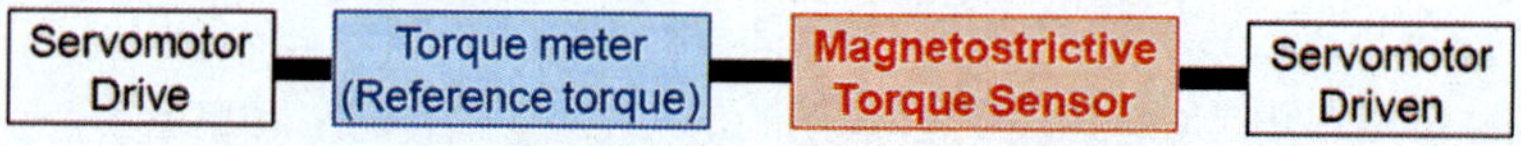

Fig. 4. Schematic diagram of the test equipment.

The measurements were conducted with 2000 min^{-1} of rotation speed and various input torque between −100 Nm and +100 Nm with 20 Nm of steps. The shaft specification for the test is shown below.

Carburized shaft (Grade: JIS SCr420H), Surface finish: Shot peeing after polishing Outer diameter: Φ18 mm

Figure 5 shows the result of static performance test with 15 times repetition. Figure 6 is the result of dynamic performance test to confirm the responsiveness to the rapid torque change. The obtained results validated high repeatability, linearity, and responsiveness of Torque Sensor.

Torque Sensor showed high repeatability and linearity.

Torque Sensor showed high responsiveness.

5 Applicability of Magnetostrictive Torque Sensor to the Powertrain

5.1 Installation of Torque Sensor on the Turbine Shaft of Belt Type CVT

The high performance of Torque Sensor is validated through the measurements in the test equipment. To verify the applicability of Torque Sensor to the powertrain system in an experimental vehicle, Torque Sensor was installed on the turbine shaft of belt type

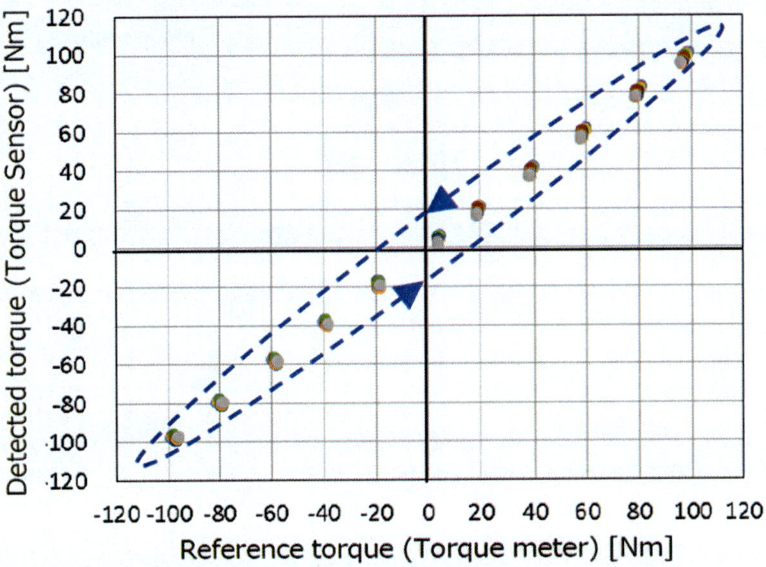

Fig. 5. Static performance of Torque Sensor with 15 times repetitive measurements.

continuously variable transmission (B-CVT) for monitoring the torque transmitted from the engine via the torque convertor. The installation space of the sensor on the turbine shaft was extremely limited. Especially, as shown in Fig. 7, the cross section between rotatable turbine shaft and stator shaft was significantly narrow like the space for the needle roller bearing. Therefore, the sensor was downsized to 3 mm of overall thickness by replacing wound coil with flexible printed circuits.

5.2 Responsiveness of Torque Sensor on the Turbine Shaft

The measurement results in driving condition are shown in Figs. 8 and 9. Figure 8a shows the overall result from the starting phase to stopping phase, while Fig. 8b, c are the magnified view of Fig. 8a in 100 ms scale. As shown in Fig. 8b, c, the torque deviation by cylinder explosion of engine could be measured precisely. Figure 8c shows the torque deviation at ON/OFF of torque convertor clutch (TCC). The torque deviation is large at engaged condition of TCC and small at disengaged condition of TCC.

Figure 9 shows the result of starting in a stall state with 100% of accelerator position. The torque could be detected even without the rotation of the shaft before the starting. The high responsiveness of Torque Sensor in the powertrain is fully confirmed by the experimental results in the vehicle powertrain.

6 Feedback Control with Magnetostrictive Torque Sensor

In general, the drivers' feeling is highly influenced by the change of the acceleration of the vehicle, which is subject to the change of the output torque of powertrain. The

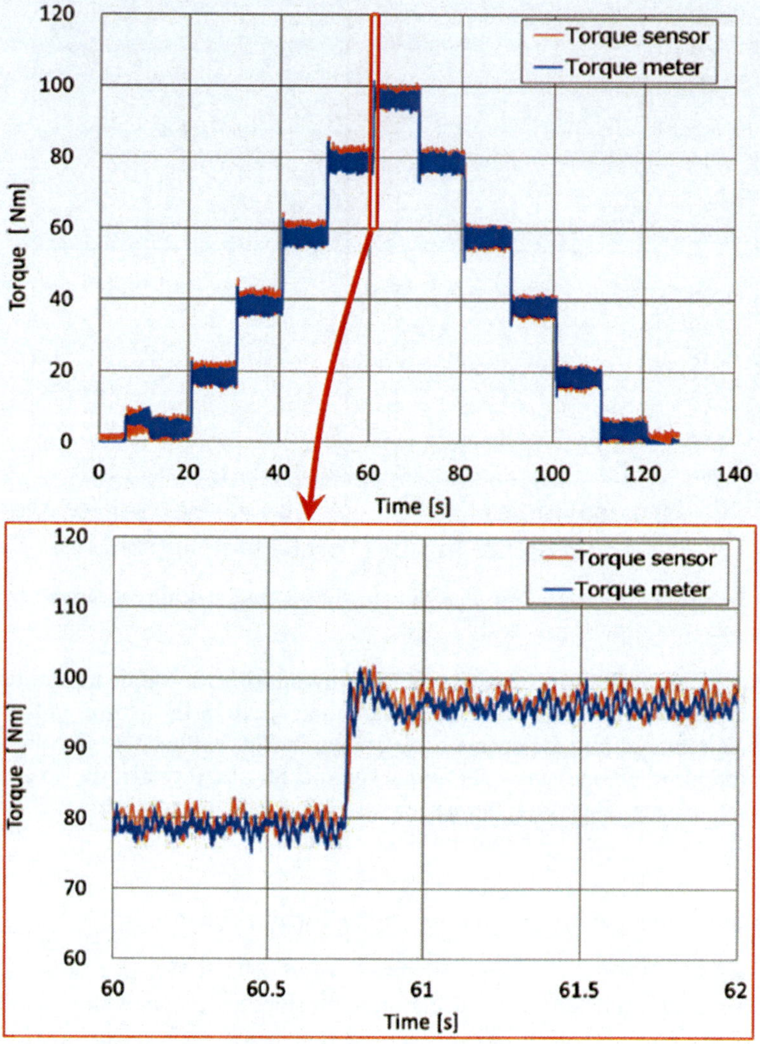

Fig. 6. Dynamic performance of Torque Sensor.

torque sensing of powertrain enables feedback control for comfortable driving. Therefore, the performance of the feedback control with Torque Sensor was evaluated by clutch actuation.

6.1 E-Clutch System with Magnetostrictive Torque Sensor

The tested system is shown in Fig. 10. The e-Clutch with magnetostrictive Torque Sensor is equipped into a manual transmission, whose release function is automated by an actuator. The transmitted torque from the engine to the input shaft is controlled by the e-Clutch. The real-time torque information is detected by Torque Sensor on the input

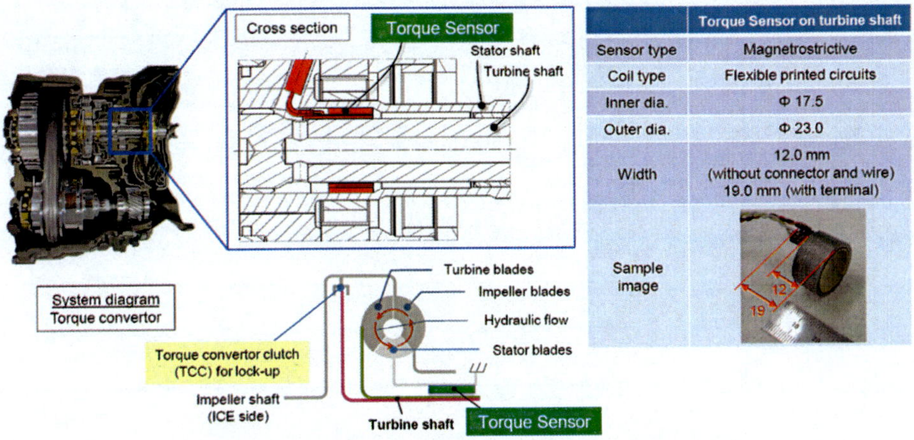

Fig. 7. Installation of Torque Sensor into belt CVT.

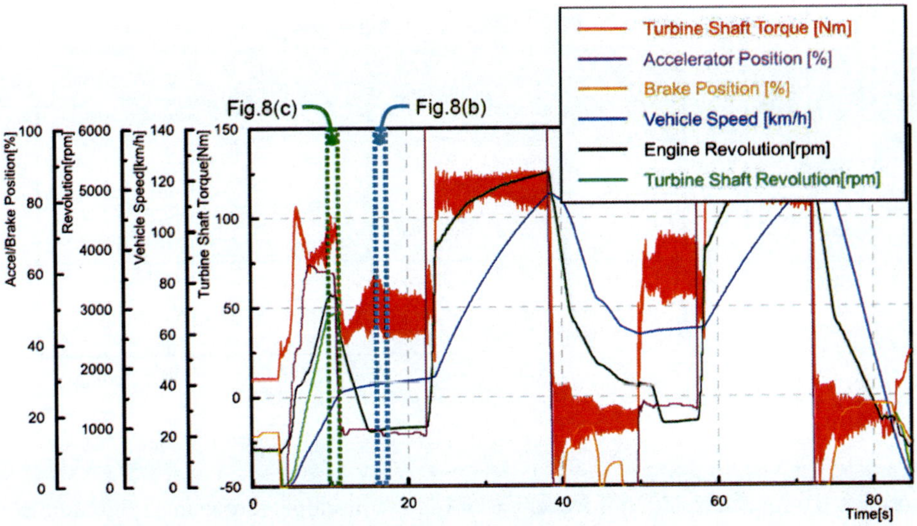

Fig. 8. **a** Torque measurement in the vehicle. **b** Magnified view of Fig. 8a (100 ms scale): TCC ON. **c** Magnified view of Fig. 8a (100 ms scale): TCC OFF/ON.

shaft. Then, the e-Clutch is controlled to follow the target torque determined by the detected torque, accelerator position, and shift information.

6.2 Test Result of E-Clutch Feedback Control

The e-Clutch test result at launching with 100% of acceleration position is shown in Fig. 11. Figure 11a shows the result when the target torque is generated aiming at smooth acceleration change during the clutch engagement. On the other hand, Fig. 11b shows the result when the target torque is generated aiming at rapid acceleration change. The

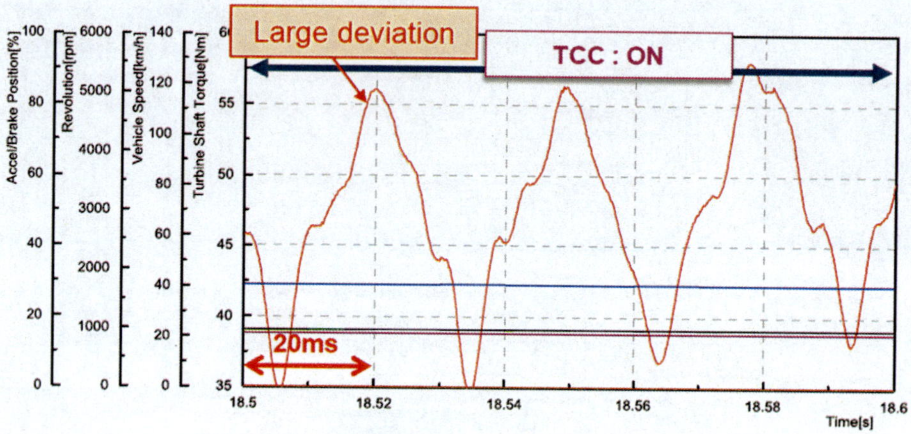

Fig. 8. (*continued*)

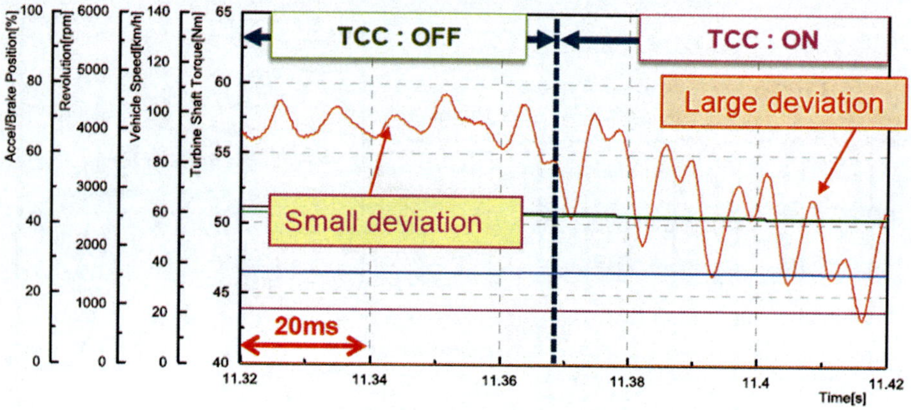

Fig. 8. (*continued*)

transmitted torque detected by Torque Sensor follows the target torque in any conditions by the feedback control of e-Clutch. The effectiveness of feedback control with Torque Sensor in the powertrain is validated experimentally.

7 Adoption of Torque Sensor into EV

7.1 Necessity of 2-Speed Transmission for EV

Most of the current EVs have the gearbox with fixed ratio combined with e-motor. It is difficult that EVs only with the fixed-ratio gear reducer satisfy the required driving performance (i.e. maximum driving torque and vehicle speed) and the cruising range compared with the ICE vehicles in the same segment. The cruising range of EVs is improved by increasing the use of efficient area in the vehicle performance curve, which is converted from the e-motor characteristic map and the gear ratio. Effective application

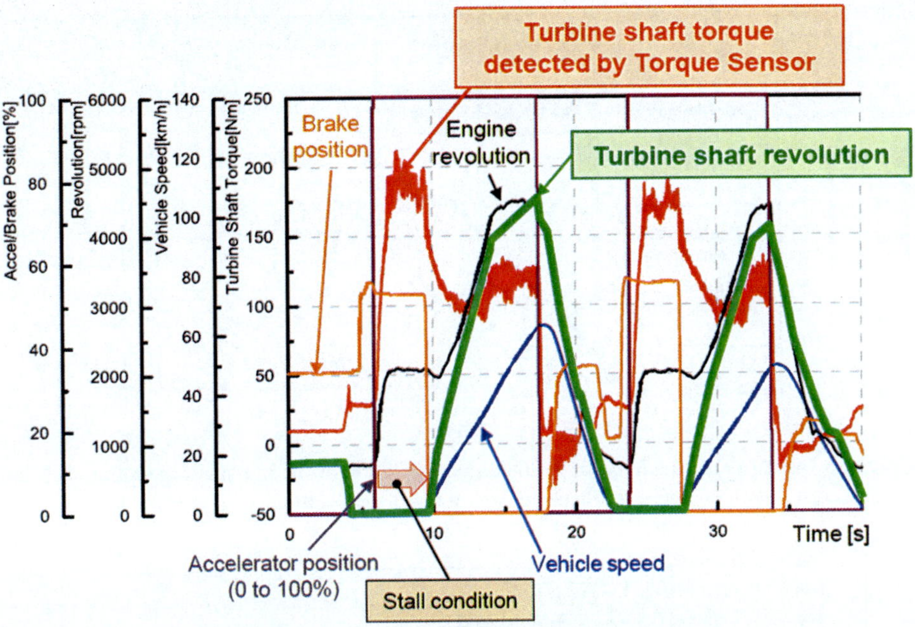

Fig. 9. Torque measurement in starting in a stall state.

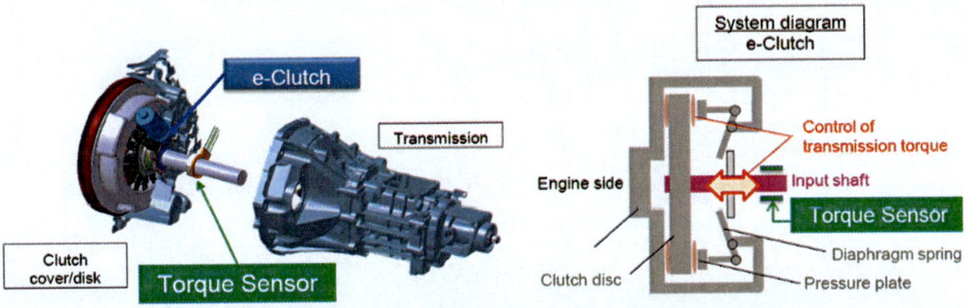

Fig. 10. Experimental powertrain system with Torque Sensor.

of 2-speed transmission to EVs can enhance the driving performance and the cruising range by the expansion of the maximum driving torque, the vehicle speed, and the high efficient area as shown in Fig. 12.

The gear step ratio between the low and high modes determines the cruising range of EVs. In case of the small step ratio, the expansion of the high efficient area in the vehicle performance curve is limited because of the large overlap between the low and high modes.

The efficiency calculation with various step ratio based on Worldwide-harmonized Light vehicles Test Cycle (WLTC) mode reveals that the step ratio of approximately 2.5 can maximize the improvement of cruising range as shown in Fig. 13. However, 2-speed

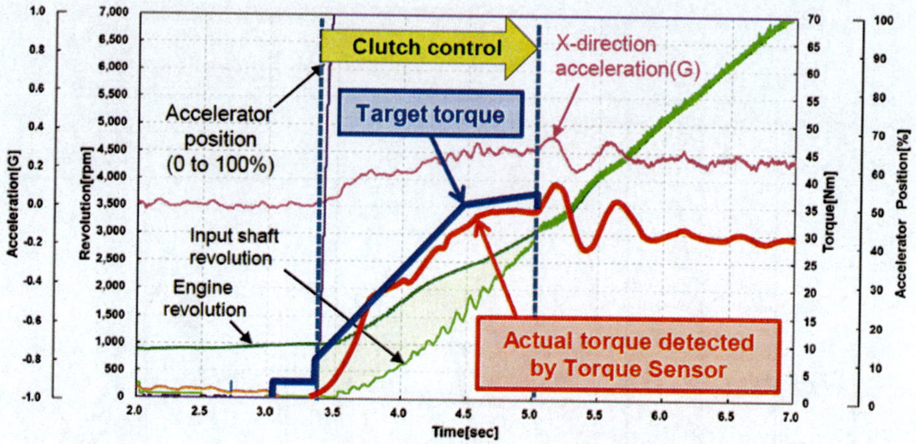

Fig. 11. **a** Measurement result with e-Clutch feedback control in smooth acceleration. **b** Measurement result with e-Clutch feedback control in rapid acceleration.

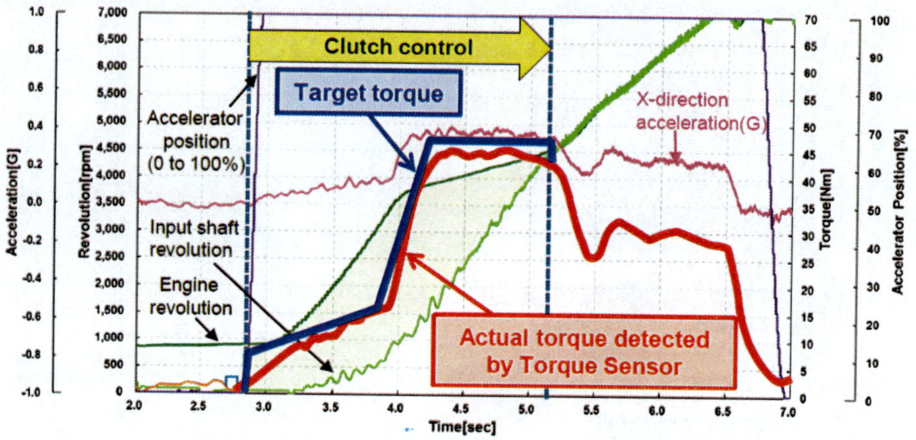

Fig. 11. (*continued*)

transmission with large step ratio of 2.5 could cause uncomfortable shifting shock and increase of absorbed energy of clutches.

7.2 Two-Speed eAxle with Torque Sensor

In order to avoid uncomfortable shifting shock and increase of absorbed energy of clutches, following system configuration and operational diagram are developed. Torque Sensor is equipped on the output shaft of 2-speed transmission with the clutch, and the brake in the eAxle system as shown in Fig. 14.

The operational diagram of low-high shifting is shown in Fig. 15. The e-motor torque is increased during the disengagement of the brake to avoid the fluctuation of the output torque at torque phase (Time slot: A–B). Then, the shifting period and absorbing energy

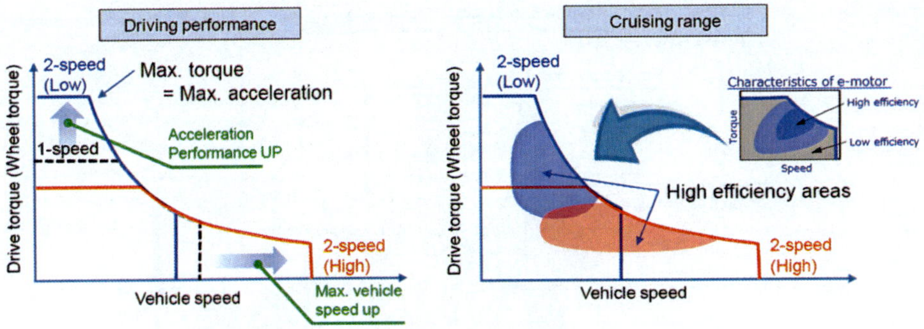

Fig. 12. Advantages of 2-speed transmission in EVs.

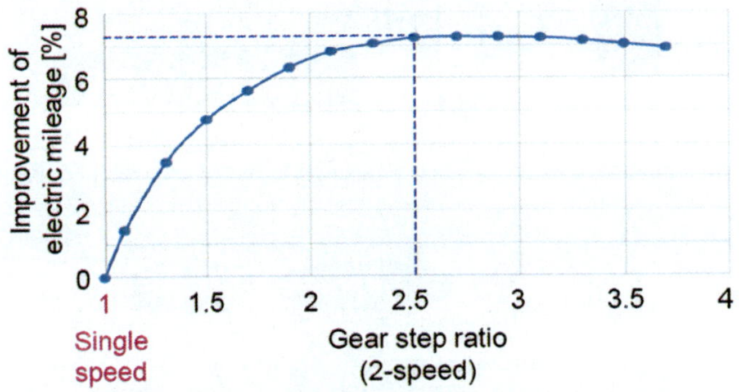

Fig. 13. Calculated optimum step ratio (2-speed).

of clutch are reduced by cancelling the inertia torque with regenerative operation of the e-motor at inertia phase (Time slot: B–C).

The feedback control of the e-motor, brake and clutch with the equipped Torque Sensor enables to minimize shifting shock by keeping the output torque constant and clutches' degradation even with large gear step ratio.

7.3 Seamless 2-Speed eAxle System Concept

The specification and the structure of seamless 2-speed eAxle concept is shown in Fig. 16 and Table 1. The power of downsized e-motor with 30000 min^{-1} of maximum speed is transmitted to a drive shaft via Traction Drive Speed Reducer, 2-speed transmission with 2.5 of gear step ratio, and a final gear. The brake and the clutch for 2-speed transmission are controlled by an electric shift actuator. Torque Sensor is equipped beside the final gear on the output shaft for the torque feedback control.

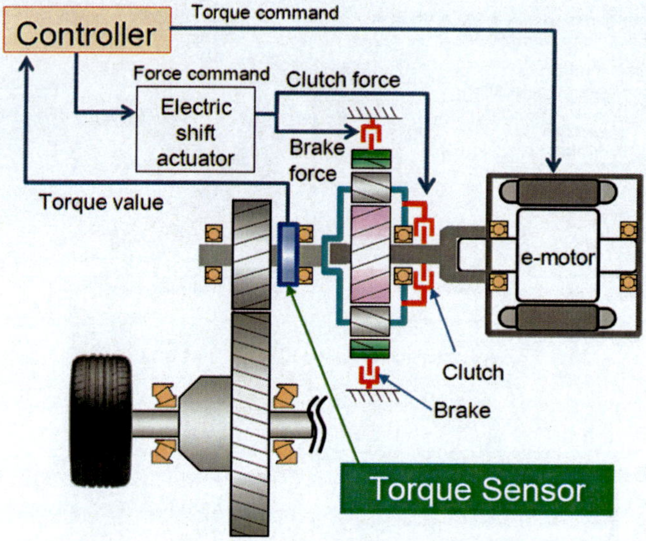

Fig. 14. System configuration of 2-speed transmission with the developed torque sensor.

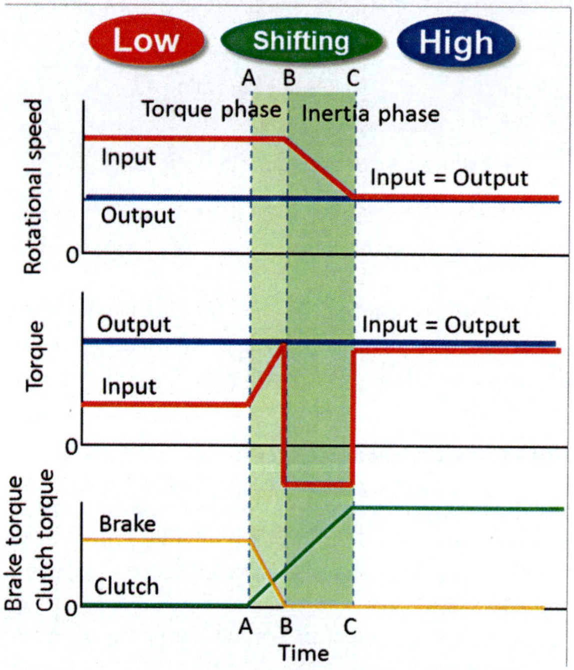

Fig. 15. Operational diagram of seamless 2-speed shifting.

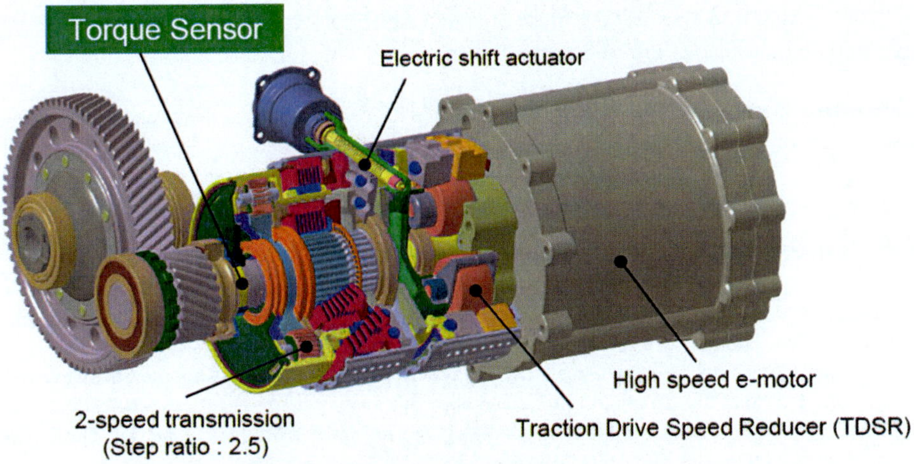

Fig. 16. Seamless 2-speed eAxle concept.

Table 1. Specifications.

Technical specification	
Max. e-motor power	150 kW
Max. e-motor torque	130 Nm
Max. e-motor speed	30,000 min^{-1}
Reduction ratio of TDSR	5.0
Reduction ratio of 2-speed transmission	Low: 2.5 High:1.0
Max. drive torque	4,000 Nm
Max. vehicle speed	250 km/h

8 Summary

Developed magnetostrictive Torque Sensor showed high responsiveness, accuracy, and applicability to the powertrain systems. The feedback control with clutch actuator and Torque Sensor was successfully implemented with high followability.

Two-speed transmissions with large gear step ratio are increasingly required for eAxles in order to compete against the driving performance and cruising range of ICE vehicles as the expansion of EVs' market. This paper validates the high potential of seamless 2-speed eAxle for realizing comfortable ride and long cruising range. Seamless shifting is achieved by the feedback control of e-motor and clutches with Torque Sensor.

The verification test of seamless 2-speed shifting with Torque Sensor is being prepared and the results are to be reported soon.

Acknowledgments. We are deeply grateful to Hitachi Metals, Ltd. for the cooperation of the development of magnetostrictive Torque Sensor.

References

1. Nakamura T, Sugiyama Y, Shimizu H (2019) Magnetostrictive torque sensor for in-vehicle transmission. Hitachi Met Tech Rev 35:18–23
2. Technical Committee on Magnetics (2013) Magnetics-fundamentals and applications-(revised version). Corona Publishing Co., Ltd., Tokyo, p 257
3. Kaneko H, Homma M (1995) Metallurgical engineering series 8 Magnetic material. The Japan Institute of Metals and Materials, Tokyo, p 212

Compact Pushbelt Variator Module to Improve Energy Economy in Electrified Powertrains

Gert-Jan van Spijk[1(✉)], Ingmar Hupkes[2], Mattijs Tweehuysen[2], Arno Klaassen[2], Jordi Meegdes[2], and Rokus van Iperen[2]

[1] Director Engineering Transmissions, Bosch Transmission Technology B.V., Tilburg, Netherlands
Gert-Jan.vanSpijk@bosch.com
[2] Bosch Transmission Technology B.V., Tilburg, Netherlands

Abstract. A lot has happened since the introduction of the Variomatic with its rubber belts based continuously variable transmission (CVT) in the DAF passenger cars in 1958. The rubber belt evolved into a compact steel belt with a far higher torque capacity, and the transmission is now built into the vehicle instead of a vehicle built around the transmission. Since then, the CVT is known as an excellent system regarding fuel economy and offers outstanding NVH qualities. It is these qualities that EV developers look for, when designing electrified powertrains and reducing the ecological footprint of mobility. CVT is ready for its next evolutionary step. Previously, studies have been published which show that CVT—just as it did for internal combustion engines (ICE) based powertrains—offers the potential to reach better energy economy in electrified powertrains. In this paper these results will be further explained. The CVT will look different from what we know from ICE applications; it is smaller and has a better energy efficiency. The solution consists of a small variator combined with an energy optimal actuation system as the core elements for an electrified powertrain.

Keywords: CVT · Electric vehicles · Motor efficiency · Vehicle performance · Modularity

1 Introduction

Electric vehicles are a fast growing segment in the automotive landscape. Though the first electric driven vehicles were already on the road over 130 years ago, the development went to the background until recently, due to the massive focus on combustion engines. Therefore electric vehicles are still a relative young technology in which still some challenges need to be solved. One of those is the compromise that in some cases has to be made to be able to drive off with maximum payload at a steep hill as well as to reach the desired maximum speed, as in Fig. 1. The majority of electric vehicles on the market at this moment are equipped with a fixed gear ratio between the e-motor and the wheels, by which often a compromise between these two drive conditions must be made.

Euroforum Deutschland GmbH (Ed.): *CTI SYMPOSIUM 2019*, pp. 523–535, 2021.
https://doi.org/10.1007/978-3-662-61515-7_46

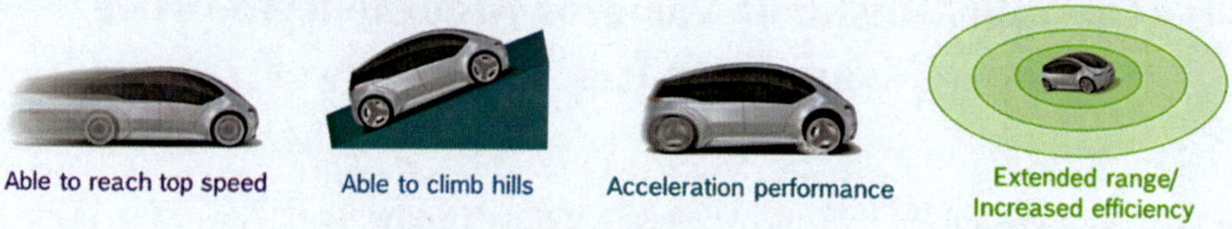

Fig. 1. Base requirements and benefits with CVT

A solution for this issue is to install a multispeed transmission, which brings a ratio spread. With the freedom to choose a ratio, also the choice for more optimal operational conditions of the e-motor becomes available, and with that better overall energy economy.

When for this purpose a stepped transmission is installed, a new challenge needs to be solved: when shifting from one to the other ratio a discontinuity in speed and torque is generated. As with conventional ICE, the solution can be found in powershift solutions, but these introduce additional losses and cost. Such disadvantages can be overcome with another proven technology from ICE powertrains: CVT. With CVT a smooth variation of ratio as well as a wider ratio coverage than e.g. a two speed offers will be obtained. Figure 2 shows an evaluation of different KPIs between CVT, DCT and AMT.

	EM spec.	Comfort	Transm. cost	Efficiency
2-speed AMT	o	–	+	+
2-speed DCT	o	o	o	o
CVT	+	+	o	+

Fig. 2. Evaluation of different multispeed transmissions in electric vehicles

With the wider ratio coverage of CVT—approximately 3,5 compared to 1,8 for two speed—further benefits with respect to single speed can be explored, such as a wider coverage of optimal efficiency operating conditions and keeping the maximum or continuous power operation of the e-motor. With this lower e-motor losses result as well as higher top speed and acceleration elasticity. Figures 3a and 3b offer a graphic explanation thereof.

2 CVT4EV Design Considerations

The process that translates the requirements for a powertrain into the right definition of the components, requires a holistic view. If the powertrain design is part of a product family, an even more extensive and complex challenge has to be fulfilled. A successful outcome demands a careful definition of the priority of the key performance indicators (KPI's), between which in many cases a tradeoff has to be made. In such a complex situation it is very convenient if one of the components offers flexibility which—as an example—enables modularity in component definition or less compromises.

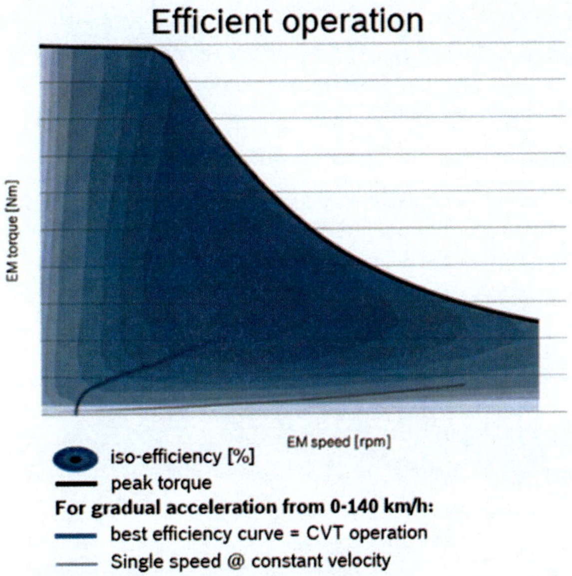

Fig. 3. **a** E-motor operation of single speed and CVT. **b** E-motor operation of acceleration of single speed and CVT

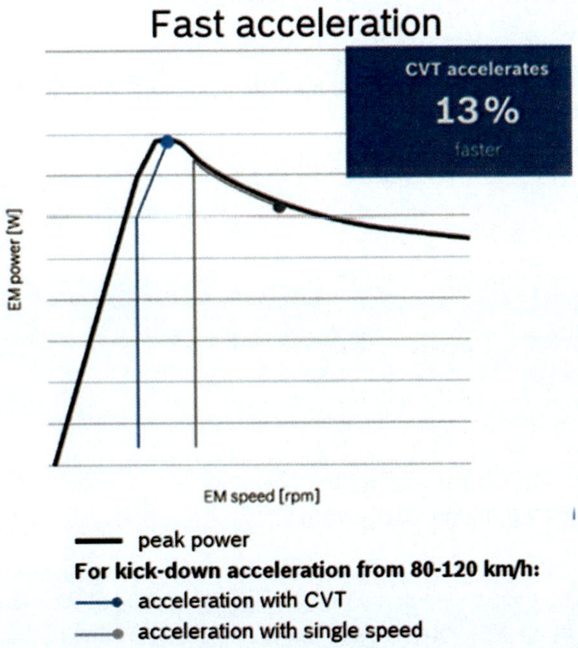

Fig. 3. (*continued*)

In this chapter by a concrete design example some clarification on such a process is given, which also shows the convenient flexibility of a CVT based powertrain.

The use-case is a C-class passenger car with specifications according Table 1. Specifications

Table 1. Vehicle and performance specifications and KPI prioritization

* Curb weight	1550 [kg]
* GVW (loaded weight)	2100 [kg]
* Gradeability	35 [%]
* acceleration 0–100 km/h	9.2 [sec]
* Top speed	170 [km/h]
* Motor power	94 [kW]
* Driving range	400 [km]
KPI priorization	
1. Cost	
2. Efficiency	
3. Packaging	

2.1 Holistic View on System Efficiency

The various components that influence the efficiency of the powertrain are characterized by their specific efficiency mappings. For each operating state of the vehicle the product of the efficiencies of all components determines the overall value. In the case of a direct link between e-motor speed and vehicle speed - fixed ratio transmission - this is a given condition. But when the powertrain contains a (continuously) variable gear ratio, the required wheel speed and torque can be obtained by different combinations of e-motor and transmission operating conditions. Especially in the case of a CVT this gives an almost infinite flexibility for optimization, which can lead to the recognition of e-motor operating conditions that are superfluous. This relates to the low speed - high torque conditions and the high speed - low torque conditions.

Such a change of e-motor operating conditions, towards a limited set within its specification, opens the consideration for a dedicated e-motor for application with a CVT. This offers a cost reducing effect by decreasing mass and volume of active motor parts as well as bearing and balancing related cost.

2.2 Variator Design Parameter Definition

The CVT variator, being a key towards the definition of the other components, has its own design variables. The variator is characterized by its lowest and highest speed ratio, by which also the overall ratio coverage is defined. A further parameter is the center distance between the pulleys, which has an effect on the variator efficiency and it's actuation, and of course the system packaging. Related to the center distance the pushbelt load condition is affected which can lead to the selection of a different belt.

For the specification of this case study, based upon powertrain simulations, the best choice for ratio coverage is 2,1 [-]. Figure 4 shows this relationship. For the concept 3,5 has been chosen to create a unit that is flexible to be applied in a variation of powertrains.

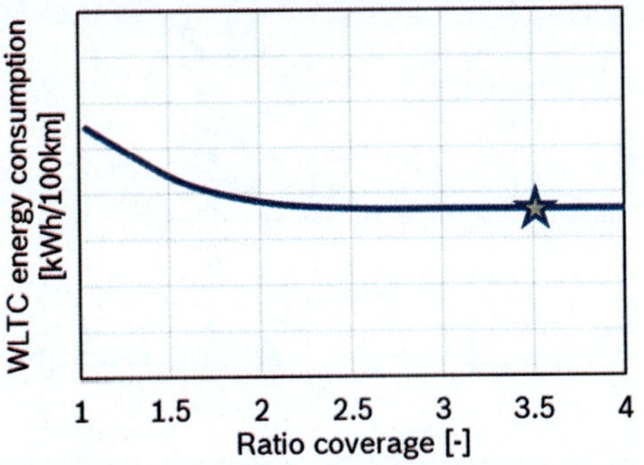

Fig. 4. Efficiency/system losses over ratio coverage

For the definition of the center distance another set of simulations was done, leading to the relationship as shown in Fig. 5.

The decision for this concept was not to go for the minimal possible center distance. The chosen dimensions offer a better overall efficiency and with that cost savings on smaller battery capacity for an equal range prevailed. Together with center distance, the torque capacity of the variator increases, which brings the opportunity to reduce the final drive ratio. The consequence of that again is a reduction of the variator speed, and thereby another reduction of losses.

2.3 Hydraulic Actuation Design

A principle difference between transmissions based on fixed gears and CVT exists in gear versus friction based power transfer. For this friction based transfer a clamping force

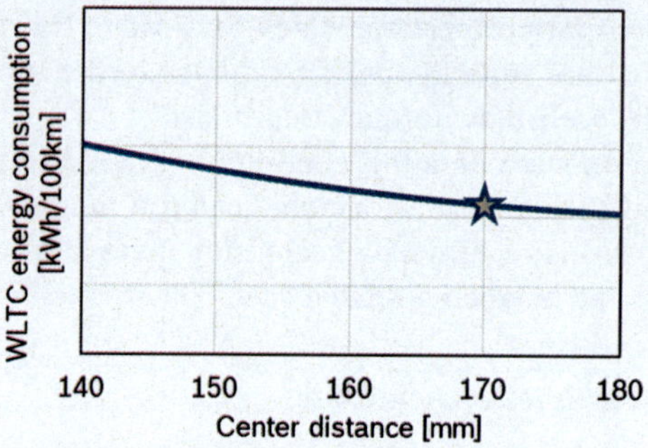

Fig. 5. Efficiency/system losses over center distance

is required, which normally is generated by hydraulic pressure. The system that provides and controls such hydraulic actuation plays a key role in the success of a powertrain. The challenge exists in the minimization of the actuation energy and at the same time to be able to deliver the responses that are required. Figure 6. shows the hydraulic scheme of the proposed control system.

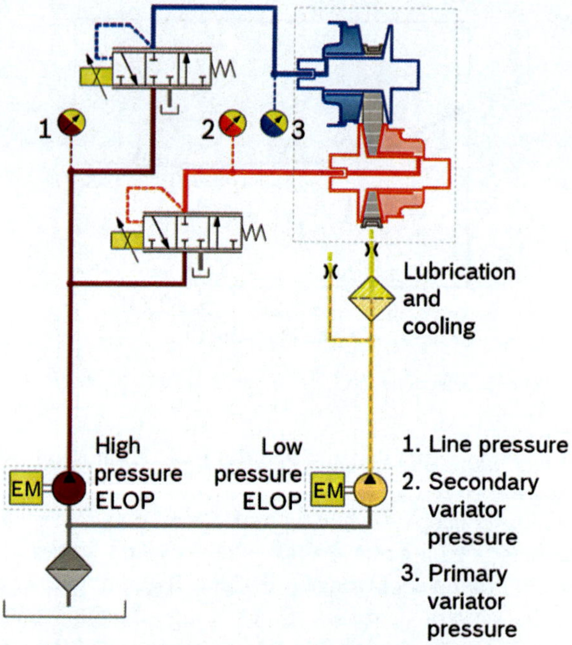

Fig. 6. Scheme of hydraulic actuation system

It is characterized by a minimal number of components, though the choice is made to separate the high and low pressure needs. With this "energy on demand" concept minimal energy consumption is enabled an important consideration in the definition of the component parameters is the responsiveness to throttle commands. As a base for the target setting, the full throttle response, from a steady state condition between ICE and electric powertrains, was compared. Figure 7. shows an example for a given kick-down action. The graph shows a distinctive difference between the two power sources. These could be characterized from comfortable to sportive.

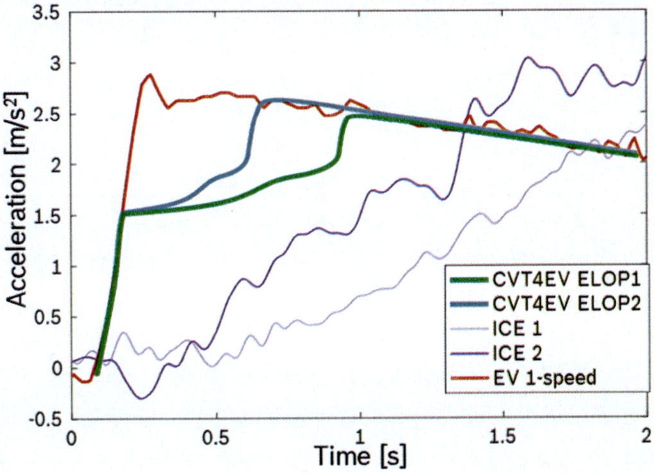

Fig. 7. ICE and EV full throttle response measurements, and simulated CVT4EV response

The response measurements of the electric vehicle are executed with a fixed gear powertrain. For the CVT-hydraulic system the actuation power was taken as a variable in order to evaluate the optimum choice between cost, performance and efficiency. The earlier mentioned KPI priority serves as a guide for this choice. Figure 7 demonstrates that the initial response as well as the maximum acceleration level is equal for a fixed gear and a CVT based powertrain. The difference exists in the transition towards the maximum acceleration which is smoothened in the CVT-case. The magnitude thereof being a function of the available clamping energy. In this case the choice is made for the smaller hydraulic power supply because of the direct relation to lower cost and power consumption. The penalty, like shown in the graph, is a relative small disadvantage which only occurs during full throttle acceleration, but still far more aggressive than the measured ICE responses.

2.4 CVT Control Strategy

The degree of freedom that is offered by the variability of the CVT needs an intelligence to make the right use of this. Therefore a transmission control unit (TCU) is programmed to decide for the right CVT set points. The right set point in this case can be defined for minimal energy consumption, maximum performance or comfort, based upon given priorities (programmed or commanded by a manual or analytic control). For the case of minimal losses the efficiency characteristic of each component is taken into account as well as their interaction. As an example for a given required wheel torque at a certain speed (so power need), various e-motor speed and torque combinations can be chosen by changing the CVT ratio. At the same time the efficiency of the battery, inverter, variator and reduction gears changes: *η-battery* * *η-inverter* * *η-E-motor* * *η-variator* * *η-reduction gears* = *η-system*. Based set upon the consideration of maximum efficiency the best operation can be decided, but if desired also a consideration for readiness for an upcoming request, such as maximum torque reserve for a kick-down, can be taken into account. Figure 8 demonstrates this process.

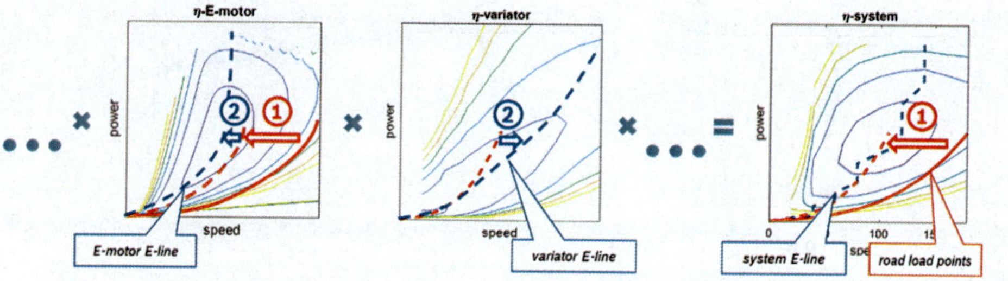

Fig. 8. Determination of system efficiency set-points

With reference to the earlier mentioned development from Variomatic till CVT4EV, Fig. 9 shows the more than 50% reduction of transmission energy losses from the current state of the art ICE CVT to CVT4EV. Figure 10 shows a comparison between the hardware of the ICE and EV application. The clear reduction in number of components and size demonstrates that the expectations about these characteristics for EV can be set to far more flexible integration and a lower cost.

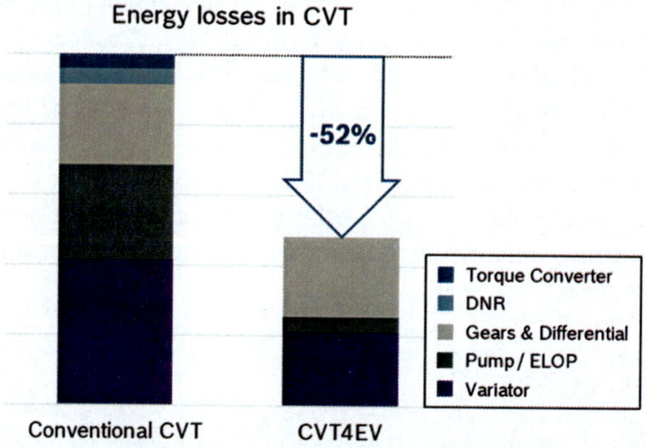

Fig. 9. CVT losses in conventional and CVT4EV powertrain

Besides that, on EV powertrain level, CVT brings a reduction of over 20% on e-motor and inverter losses, as Fig. 10 shows.

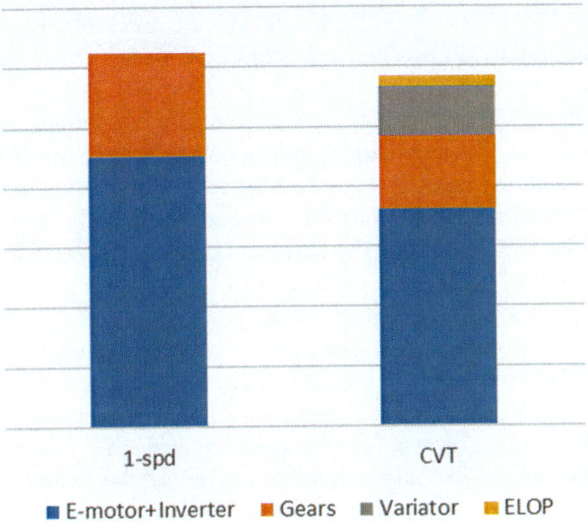

Fig. 10. Powertrain losses 1-speed versus CVT

From this it can be concluded that with the variability of CVT the electric vehicle can become your 'thinking along' partner in traffic (Fig. 11).

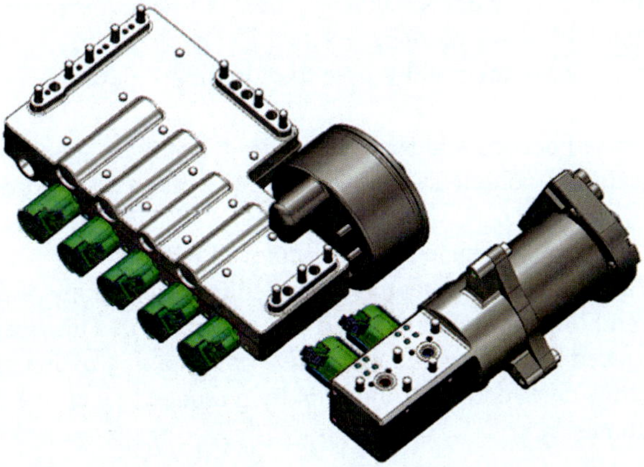

Fig. 11. hydraulic control unit for respectively conventional CVT and CVT4EV

3 CVT4EV in Hardware

The conclusion can be drawn out of the theoretical analysis of CVT in electric powertrains, that there is a potential on different KPI's. As ever, such an insight only convinces when it can be experienced. Therefore in parallel to this study a first hardware for

CVT4EV has been set up. The motivations for this are, next to demonstrating the performance, comfort and drivability, in the development of skills regarding hardware and controls of such a powertrain.

Within this activity the performance requirements were defined by benchmarking with the base vehicle which is an electric vehicle with a fixed ratio transmission. Concrete target is to prove the ability of a CVT to at least match the drivability of the base vehicle, but with an e-motor that has a lower maximum torque. Further objective is the evaluation of the changes in the operating points with respect to speed (NVH) and responsiveness and to develop the controls based on the findings.

For this first hardware study, as mentioned, an existing fixed gear electric vehicle was selected and rebuilt with a CVT, making use of ready available or slightly adapted components. For this purpose an ICE based CVT has been selected in which the mechanical driven oil pump is substituted by an electric oil pump (ELOP), the torque converter was removed and the drive, reverse function disabled. This means that no optimization with respect to gear ratio—like an input reduction before the variator—was done and a—for the function of an electric vehicle—bulky CVT unit was installed. With respect to the controls of the vehicle, a dedicated e-drive CVT control software was developed, by reverse engineering of the base vehicle's controls. Figure 12. shows the software architecture scheme for that.

3.1 First Results

From the graphs in Fig. 13, which show the results of a wide open throttle acceleration it can be learned that:

* with a reduced maximum motor torque the similar acceleration from standstill can be achieved

* the e-motor speed becomes independent from the vehicle speed

* a higher maximum vehicle speed is possible, without a compromise on launch performance

or the similar maximum vehicle speed is achieved with a lower e-motor speed

These results show that the potential for a dedicated e-motor, designed for lower maximum torque and lower maximum speed can be explored. This opens the option to a design with a reduced size and volume of active parts, saving on magnets and cupper. Reduced speed brings cost reduction potential by a reduced level of bearing and shaft balancing requirements.

4 What CVT4EV Can Bring

As addressed in the previous paragraphs, the flexibility that CVT brings to the powertrain can be exploited in different ways. The choice for that should be motivated from the priority setting between the KPI's as well as the use-case for which the powertrain will be applied. But some of the benefits are inherent. One of these is the reduction of e-motor speed at higher vehicle speeds, by which the vehicle performance is brought on a higher level.

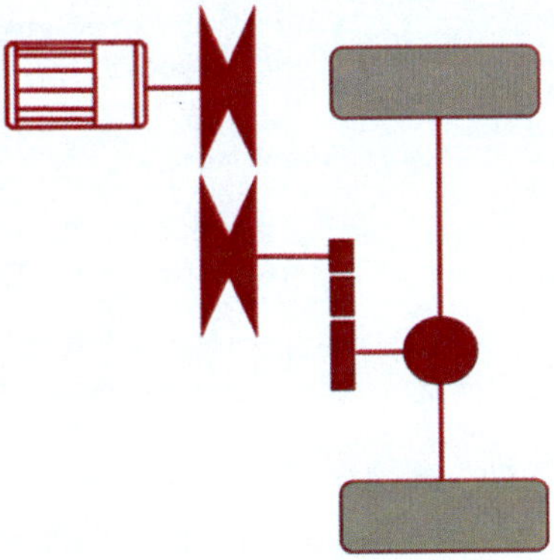

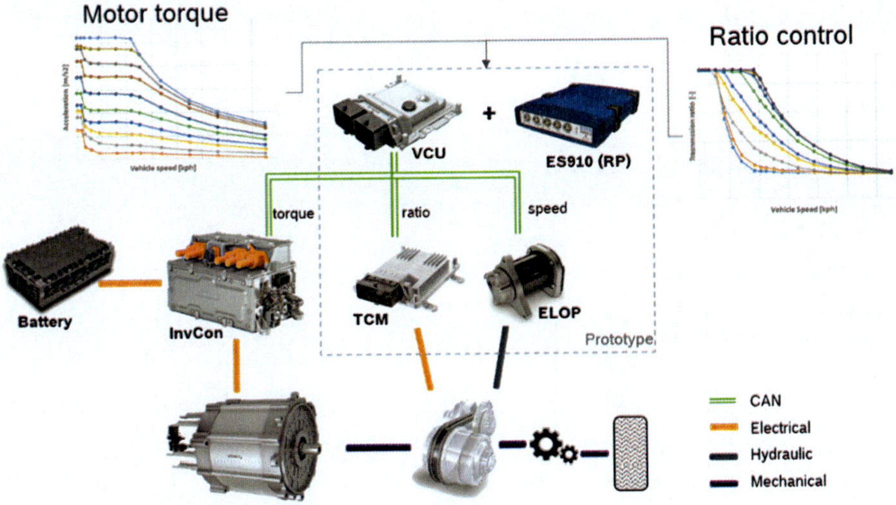

Fig. 12. Software architecture CVT4EV demonstrator

* for intermediate accelerations, like 80–120 km/h a better performance is delivered because the CVT will keep the e-motor in its peak power operation.

* a higher maximum speed results from a similar operation by keeping the e-motor in its peak continuous power operation.

The reasoning that concluded chapter 3.1 can also be valid for the flexibility to match e-drive components to a wider variety of requirements. This can be supportive to a modular line-up approach. Figure 14 shows how CVT can offer both an increased low ratio enabling to drive off with higher loads from standstill and uphill as well as a decreased overdrive gear ratio that prevents limitation of top speed performance.

With this insight the exploration of different use cases with the CVT as the enabling element can be challenged. Bosch is ready to support with its components and system know-how to create successful new solutions.

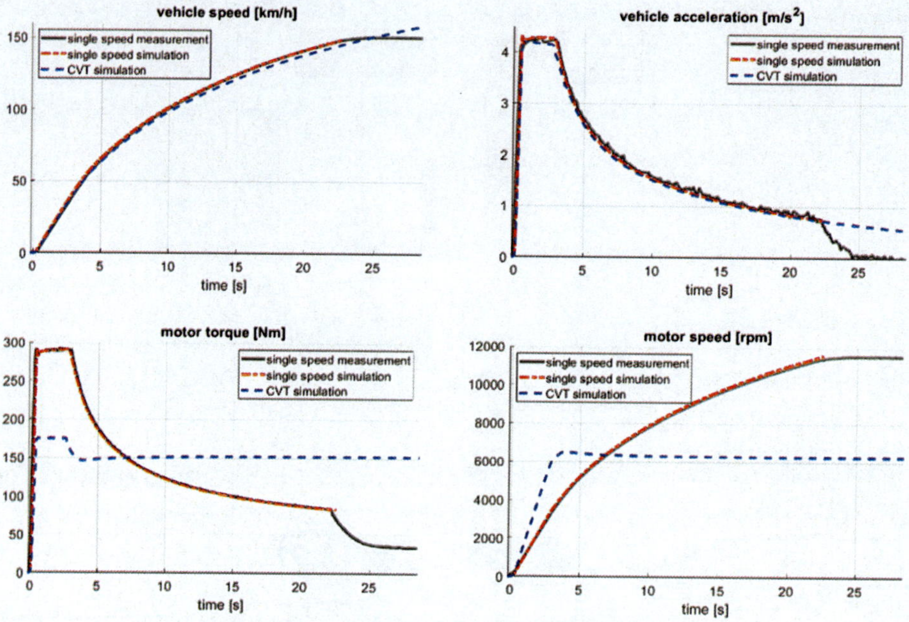

Fig. 13. Acceleration performance evaluation

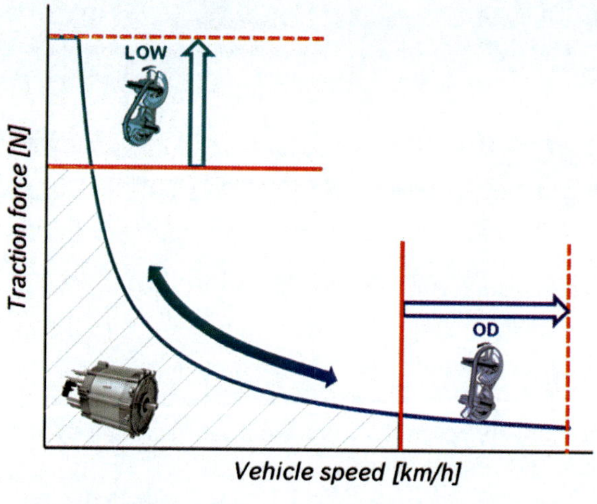

Fig. 14. Potentials for e-motor application with CVT

5 Conclusion

- CVT is ready to be applied in electric powertrains and can be the key for the improvement of different customer values
- A design study is presented that focuses on cost and energy economy

- The design considerations for the variator and the hydraulic control system are explained
- Based upon the understanding of the design parameters the control set points for the desired KPI performance are generated
- The flexibility that is provided by CVT enables downsizing of the e-motor or application of a given e-motor into new use-cases
- The study results have been included into a first hardware development and demonstration platform, with initial results available
- For the future, further optimization of this concept is envisaged that will further enhance the benefit of the CVT equipped electric powertrain.

Literature

1. Roth, Hans (March 2011) Das erste vierrädrige Elektroauto der Welt [The first four-wheeled electric car in the world]
2. van der Sluis F, Römers L, van Spijk G, Kunze M (2018) A synergy between latest CVT and hybrid technology. CTI conference, Nov 2018
3. Kunze, CVT pushes hybrid and electric vehicles forward. SAE conference, Shanghai, 2018
4. Nakazawa T, Traushold S, Lauinger C (2018) Development of a Chain type CVT with compact variator system. JSAE annual congress, no. 20186191, Nagoya 2018
5. Lauinger C (2019) Chain CVT highlights for new energy vehicles, VDI-CVT, Baden-Baden 2019
6. Yamamoto K (2019) CVT challenges in the electrification age, VDI-CVT, Baden-Baden 2019
7. Demmerer, Efficiency of electric axle drive systems. The potential of multispeed drives. DRITEV converence, Bonn 2019
8. ZF, Paradigm shift for electromobility – ZF presents new electric 2-speed drive for passenger cars, Press release 2019
9. https://www.wired.com/story/electric-car-two-speed-transmission-gearbox/
10. Satyanarayana, Development of 2-speed Automatic Transmission for battery electric two wheelers. SAE conference paper, SAE 2019-26-010

Supply Chain Implications of Increasingly Integrated EV Drivelines

Claudio Vittori[1] (✉), Wen Gao[2], and Graham Evans[3]

[1] Senior Research Analyst, Supply Chain and Technology, Automotive, Milan, Italy
claudio.vittori@ihsmarkit.com
[2] Senior Research Analyst, Supply Chain and Technology, Automotive, London, UK
wen.gao@ihsmarkit.com
[3] Associate Director, Supply Chain and Technology, Automotive, Frankfurt, Germany
graham.evans@ihsmarkit.com

Abstract. IHS Markit undertakes component-level research in the automotive industry through a combination of peer-level interviews and discussions with contacts and thought leaders. This is underpinned by their industry leading vehicle and powertrain-level forecasts. The approach generates a unique level of varied and comprehensive insight, which helps drive the business lines and investment decisions of our clients. With the move to electrified powertrain technology, the automotive supply chain is undergoing a transformational shift towards more compact configurations that will liberate interior space in the vehicle. Whilst this 'end state' might appear somewhat simplistic and uninspiring, the interim period will generate a huge number of creative technical evolutions. IHS Markit has undertaken component-level research focusing on the extent to which power electronics, transmission and e-motor technology as well as the thermal management will become increasingly integrated, analyzing the approaches of OEMs and interpolating the supply chain dynamics to understand how solutions will evolve with time.

Keywords: Electric vehicles · Vehicle architecture · Integrated drivelines

1 Alternative Propulsion Vehicle Market

The global market of passenger vehicles in 2018 was still predominantly represented by internal combustion vehicles (ICE) and stop-start technology covering between 80 and 95% of the market share according to the different regions, with a minor role played by the alternative propulsion vehicles.

1.1 Production Volumes

Alternative propulsion (AP) vehicles have been a relative niche market to date, with a penetration rate ranging from 3 to 12% worldwide with significant differences across the various regions. Nevertheless, the evolution over the years has been relatively rapid

Euroforum Deutschland GmbH (Ed.): *CTI SYMPOSIUM 2019*, pp. 536–544, 2021.
https://doi.org/10.1007/978-3-662-61515-7_47

from 2012 onwards. This, of course, has been reflected in production from about 500,000 units in 2012 to 1.8-million vehicles by Q2 2019.

Design was previously dominated by hybrid technology (HEV), led by vehicles such as the first-generation Toyota Prius, but 2018 marked a change with ~50% of increase largely driven by plug-in vehicles (PHEVs) and battery electric vehicles (BEVs) (Fig. 1).

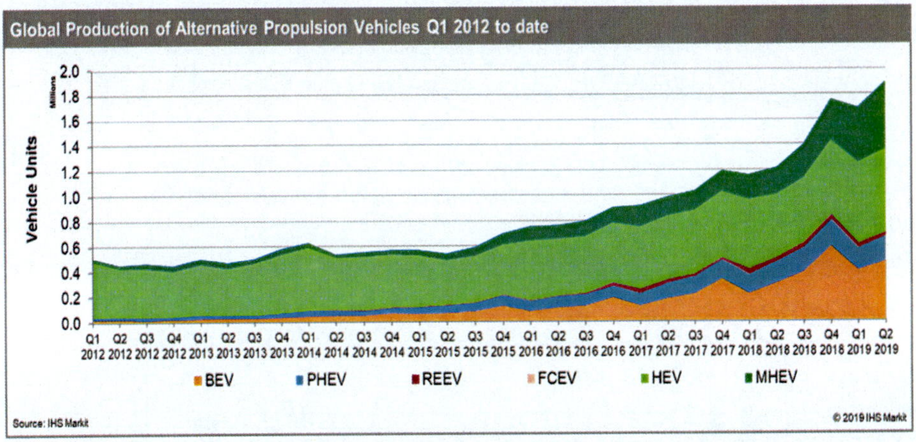

Fig. 1. Global production of alternative propulsion vehicles from Q1 2012 to date.

This growth rate is expected to continue according to IHS Markit data, with AP vehicles forecast to hold half of the passenger vehicle market by 2025.

Within this electrification trend, mild hybrid technology (MHEV) is expected to displace the internal combustion engine (ICE), stop-start. MHEVs represent a cost-effective means by which OEMs can hybridize a large portion of their ICE fleet to achieve a circa 8–12% CO_2 saving. This will most typically be delivered through use of a belt starter generator in lieu of an alternator and introduction of a 48 V electrical architecture.

Meanwhile, PHEVs and BEVs will represent around 10% of the total market share in 2025, growing up to 15–20% by 2031 (Fig. 2).

1.2 EV Component Demand

As shown by the Figures below, AP growth will lead to exponential growth of xEV-related components, with innovation focused on the various e-motor sub-system designs. While IHS Markit expects BEVs and hybrids (including PHEVs) to experience a near-equal growth trend up to 2025, mild hybrids will present a more accelerated rate of adoption for the reasons mentioned previously and fuel cell electric vehicles (FCEVs) are forecast to remain extremely niche for light passenger vehicle applications in the short to medium term (Fig. 3).

The same trend can be observed with DC-DC converter, with global production set to almost grow tenfold between 2018-25 from about 5 million to more than 45 million

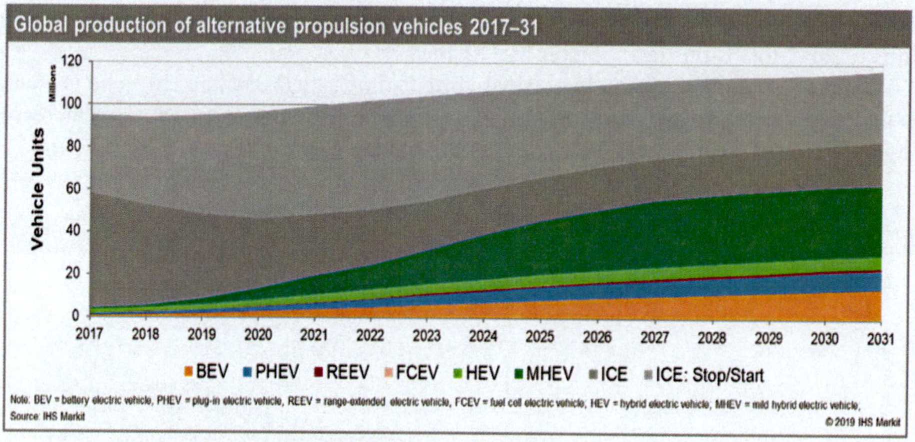

Fig. 2. Global vehicle production by propulsion design from 2017 to 2031.

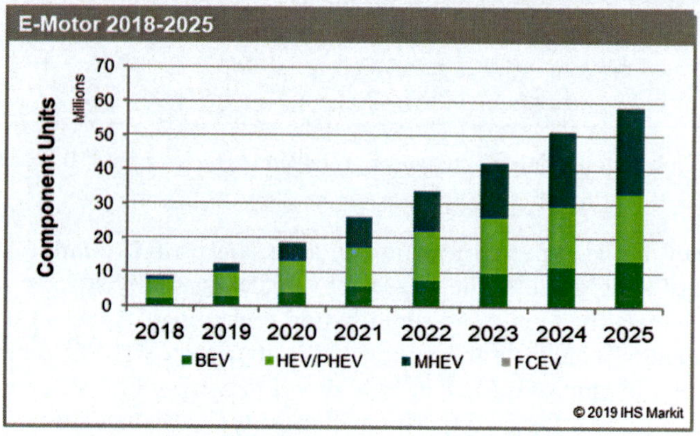

Fig. 3. E-motor global production from 2018 to 2025.

units. Market share will again be skewed to MHEVs, with the main electrification driver claiming circa 60% of DC-DC converter production (Fig. 4).

Due to the higher battery capacities required to support electric only propulsion, BEVs will be the main driver of battery cell demand. Global production will grow from around 2 billion units in 2018 to more than 6 billion by 2025. In comparison, cell demand for full hybrid applications is forecast to reach circa 1 billion that same year (Fig. 5).

Thermal components are also expected to follow a similar trend, albeit with more modest volumes given Fig. 6 shows only full hybrid, electric and fuel cell applications. More than 12 million electrically driven water pumps are forecast to be produced annually by 2025, growing from 2 million units in 2018. Again, the primary driver will be BEVs accounting for around 60% of the market demand.

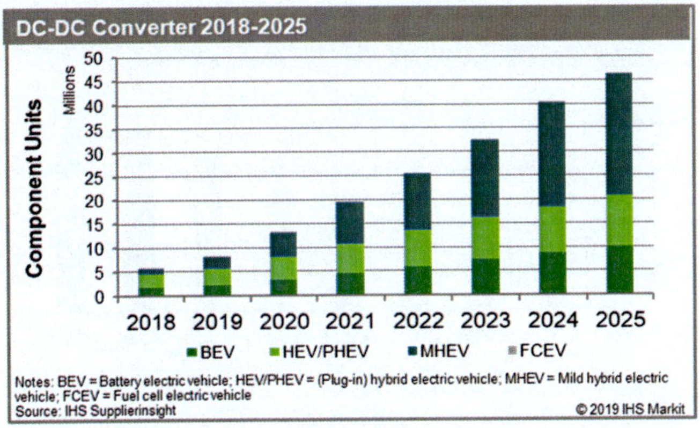

Fig. 4. DC-DC converter global production from 2018 to 2025.

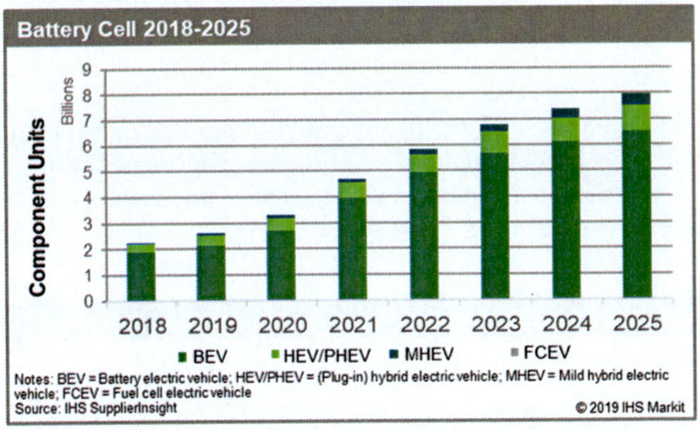

Fig. 5. Battery cell global production from 2018 to 2025.

2 Component Integration

With the move to electrified powertrain technology, OEMs are looking for new, more simplistic integration solutions, which ultimately leads to a smaller number of components required. Paired with this is reduction of component volume, cost and weight are optimized whilst the vehicle's interior space is liberated from intrusion by propulsion system components, such as the transmission. In turn, this is expected to open up greater revenue generation streams and opportunities from feature content and experiences within a car.

Whilst the 'end state' might appear somewhat already defined and moving towards full integration, the interim period is generating a huge number of creative technical solutions for e-motor and power electronics.

2.1 E-Motor Integration

Electric motors are evolving towards integrated solutions incorporating transmissions and/or inverters in so-called '2-in-1' and '3-in-1' solutions. According to IHS Markit

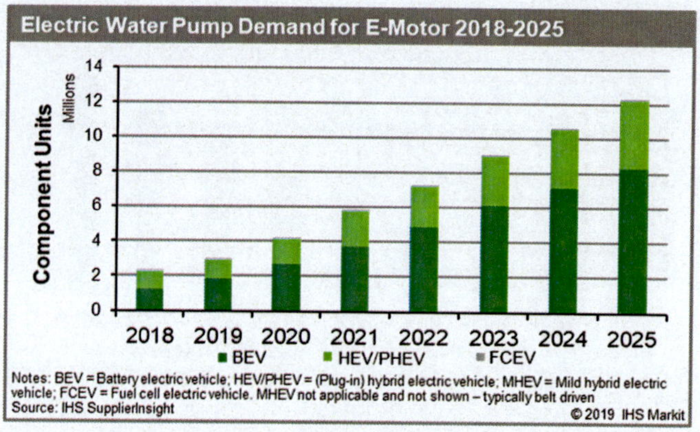

Fig. 6. Electric water pump global production from 2018 to 2025.

forecasts, integrated e-motor applications will be mainly driven by a motor inverter solution, which is expected to be a commonplace solution for MHEV applications incorporating a belt starter generator (BSG).

Motor transmission, in contrast, is forecasted to be the typical solution for HEV applications. E-motors in BEV applications currently stand alone for the most part, and this is expected to continue on a steady rate. However, various integration levels can be found dependent on the OEM and Tier-1 supplier (Fig. 7).

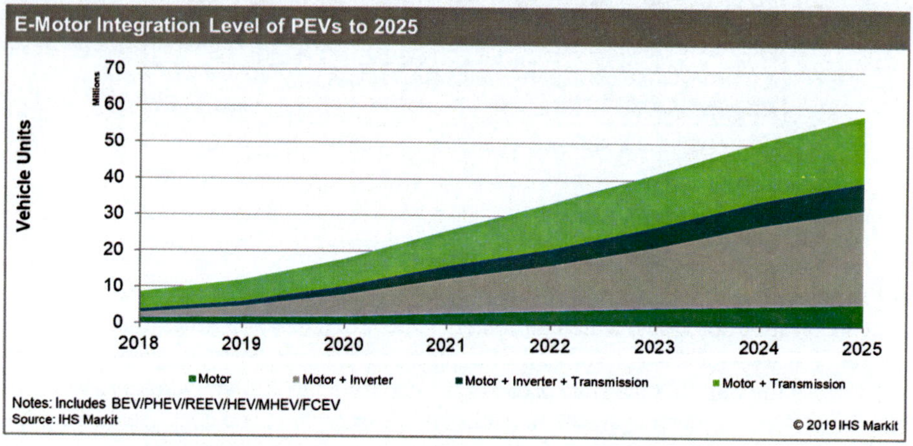

Fig. 7. E-motor integration level of PHEVs from 2018 to 25.

2.2 Power Electronics

While IHS Markit foresees a growing demand of integrated solutions for power electronics that offer higher efficiency, as well as cost and weight synergies, this will be restrained by existing architecture limitations.

Further to this, the market is forecast to be quite fragmented across the different solutions, excluding that of the inverter and motor, which remain a logical combination in electric drive units. Nevertheless, a higher level of integration is expected in future BEVs, as opposed to the current standalone solution. This can be seen in the Tesla Model 3 where the On-Board Charger (OBC) and DC-DC converter are integrated in a unit integral with the battery under the rear seat. The OEM's previously released Model S, on the other hand, placed the OBC in the cabin under the rear seat and the DC-DC converter in the front trunk (Fig. 8).

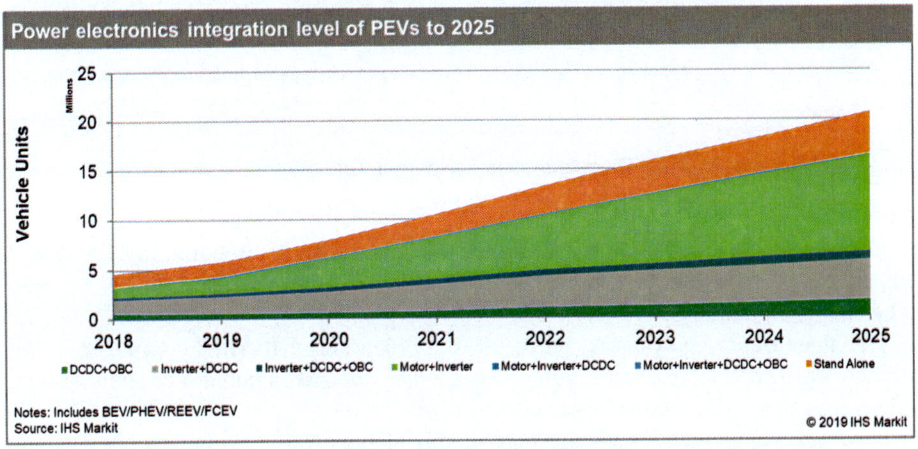

Fig. 8. Power electronics integration level of PHEVs from 2018 to 25.

2.3 Thermal Systems

Electronic devices gain heat during their operation, as well as that endured during cold ambient conditions. In order to operate efficiently and safely, the e-motor and power electronics require specific cooling architectures and sub-systems to keep the components in their workable and optimized temperature range and to avoid short-term damage or non-acceptable degradation over their lifetime. Weight, cost, and the complexity of the vehicle—especially taking the battery thermal management into consideration—all add to cooling efforts.

IHS Markit sees thermal system integration driven by requirements of high efficiency as well as a compact package (Fig. 9).

Electric motors are typically cooled by coolant and oil, while some lower performance e-motors are still cooled by air. Commonly made up of a water-ethylene glycol mixture, coolant simply transports heat away through a cooling jacket. Another method uses the oil to transfer the heat by directly being in contact with the main thermal components. The dielectric characteristic of the oil enables this approach in cooling. Unlike electric motors, power electronics are only cooled by coolant or air in automotive components.

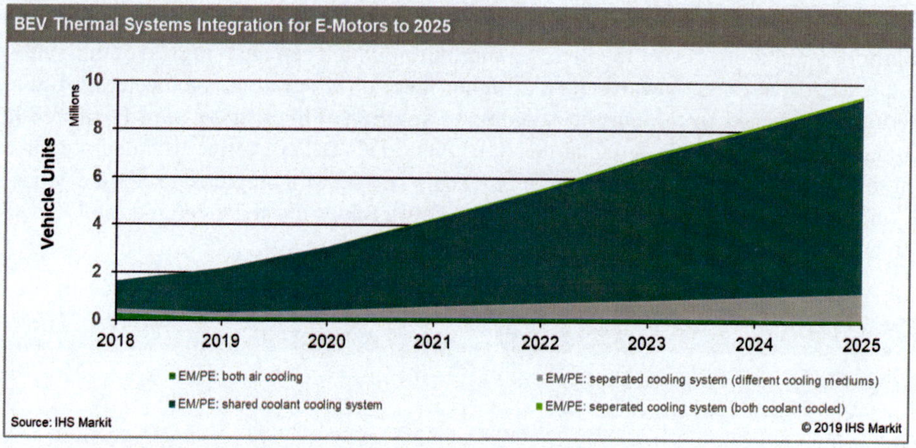

Fig. 9. BEV Thermal system integration for E-Motors to 2025

Considering the trend of power electronics and electric motor integration, IHS Markit foresees a huge growth in the market share of coolant cooling systems shared commonly by electric motors and power electronics for battery electric vehicles, shown in Fig. 7.

The shared cooling systems can result in a reduction of thermal components. For instance, only one electric water pump is needed in the shared coolant cooling system, which lends to a compact and lower-cost system.

However, some vehicles on the market still feature separate cooling circuits and therefore systems. For example, the e-motor is oil cooled while the power electronics are coolant cooled. Air-cooled systems for both components are very rare due to their low heat transfer capabilities.

This picture differs for plug-in or full hybrid electric vehicles. Many of these vehicles' propulsion systems are modified from engine-based powertrain platforms, often leading to oil-cooled motors (Fig. 10).

The development of the e-axle also influences cooling systems. The component is a compact unit for BEVs and hybrid vehicles, in which the e-motor, power electronics and transmission may be combined, as discussed earlier. Due to the diverse range of power levels versus a relatively low production volume, e-motors are currently not fully integrated into the transmission. This helps e-axle suppliers to adapt the size of their e-motors to meet the requirements of performance for different OEMs and for different vehicle models. In this type of e-axle, the electric motors are coolant cooled.

However, with the growth of the electrified vehicle market, mass volume production of e-axles can be expected alongside more motors being fully integrated into the transmission, which will also result in more oil-cooled e-motors.

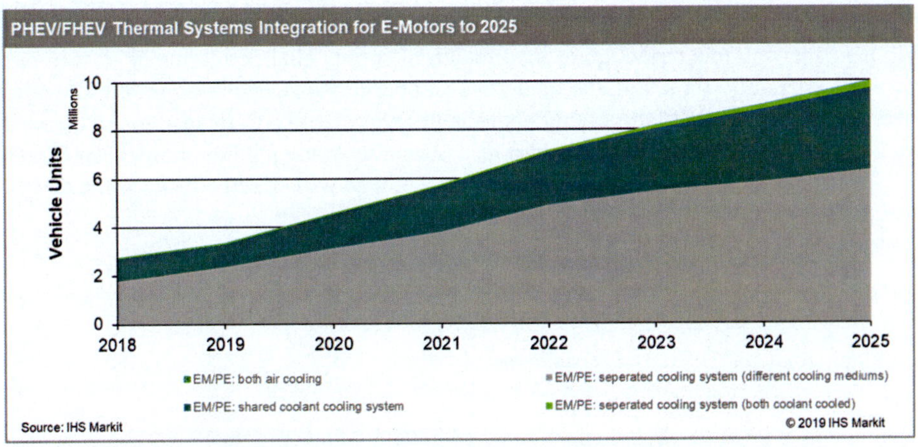

Fig. 10. PHEV/HEV e-motor thermal systems integration from 2018 to 25.

3 Supply Base Implications

3.1 E-Motor and Power Electronics

EV programs have been relatively low volume to date, but this will change in the coming years leading to higher component integration as one step in the pursuit of higher profitability.

Such capacity efficiency will be achieved on larger EV platforms, such as the Volkswagen Group's MEB architecture, since ramping up scale is difficult for all but the biggest Tier-1 suppliers. A lack of integration in the short to medium term provides opportunities for smaller market players to compete for programs.

3.2 Thermal Systems

The main components for e-motor and power electronics cooling are the radiators and electric water pumps. Only a handful of suppliers like Denso and Hanon can provide the full system, and such cases are forecast to be in the minority. This is because the thermal supply chain is not a system business, rarely providing integrated systems or modules, but rather a component business made up of specific suppliers with a narrow product remit. This is not forecast to change. Unlike the battery thermal systems, which in many cases are designed by battery suppliers, cooling systems of electric motor and power electronics are and will continue to be designed by OEMs, selecting component technologies from a Tier 1 supplier portfolio. This also means that the conventional OEM-Tier 1 dependencies are expected to remain.

4 Conclusion

A more rigorous integration of the driveline and its associated components will occur as scale increases in the E-Mobility market, requiring power electronics and thermal management functionality to be more flexible and to support modular approaches. This will pave the way to a reduction of the breadth of supply chain opportunity as volumes increase, especially for small and medium sized participants. This also implies a wider involvement of the OEMs in component manufacture thanks to licensing agreements, where OEMs produce components in-house using the strategic know-how of the Tier-1 supplier or a strategic technology partner/provider, who will effectively act as the design parent.

Reference

1. Multiple Authors, IHS Markit Data (2018)

Fuel-Saving Potential of Hybrid Electric Vehicles Using Surroundings Sensor System Information

Dominic Waldenmayer[1,2](✉), Johannes Buyer[2], Katharina Bause[1], Nikolas Andriessen[2], and Hermann Koch-Gröber[2]

[1] Department of Product Engineering, Karlsruhe Institute of Technology, Kaiserstr. 10, 76131 Karlsruhe, Germany
dominic.waldenmayer@hs-heilbronn.de, katharina.bause@kit.edu

[2] Automotive Systems Engineering, Heilbronn University of Applied Sciences, Max-Plank-Str. 39, 74081 Heilbronn, Germany
{johannes.buyer,hermann.koch-groeber}@hs-heilbronn.de, nikolas.andriessen@student.kit.edu

Abstract. Operation strategies used in hybrid electric vehicles need a traffic speed profile for an initial operation strategy planning, which can be obtained from real-time traffic information services. However, microscopic traffic flow is unrepresentable in this initial planning which leads to deviations in the batteries' state of charge. Therefore, the presented method plans partially automated micro maneuvers to correct the deviations during the drive using local prediction based on the surroundings sensor system, which is assumed to offer exact data of the future speed profile within its visual range. The method is applied to various driving cycles such as the WLTP and to real speed profiles in cities and on highways. Comparing driving cycle runs with and without the use of local prediction shows a fuel-saving potential of up to 11% in emission test cycles. Moreover, validation with real data confirms the results and indicates a higher potential in urban areas.

Keywords: Hybrid electric vehicle · Energy management · Surroundings sensor system

1 Introduction

New car registrations of hybrid electric vehicles (HEVs) in Germany increased from 8.374 in 2009 to 130.258 in 2018 as seen in [1]. Regarding current political discussions and a tightened exhaust emission legislation, hybrid electric vehicles are one step in the development towards an emission-free traffic. The biggest potential of HEVs is their energy efficiency due to the variable combination of internal combustion engine (ICE) and electric machine (EM) in the powertrain.

In recent years, many publications dealt with energy management of hybrid electric powertrain structures. Besides further improvement of single powertrain components such as the ICE or EM, researchers addressed the problem of fuel-efficient operation strategies in HEVs. Advanced operation strategies aim to optimize the hybrid electric

Euroforum Deutschland GmbH (Ed.): *CTI SYMPOSIUM 2019*, pp. 545–561, 2021.
https://doi.org/10.1007/978-3-662-61515-7_48

powertrain's energy efficiency by choosing the optimal torque-split of ICE and EM at any given time during the driving cycle. Global optimal torque-split is achieved if the driving cycle is completely known a priori. Information about current traffic flow on the planned route, color phases of traffic lights and traffic jams can be checked prior to the drive and considered in the planning of the operation strategy but will most likely change during the actual drive. Therefore, real-life traffic is not accurately representable in an a priori planning. Moreover, microscopic traffic flow like the interaction between driver and vehicle or other road users is neither. As a result, the initial speed profile prediction and the actual speed profile deviate as seen in Fig. 1. Thereby, the deviation between the speed profiles depends on the amount and type of occuring unplanned traffic situations.

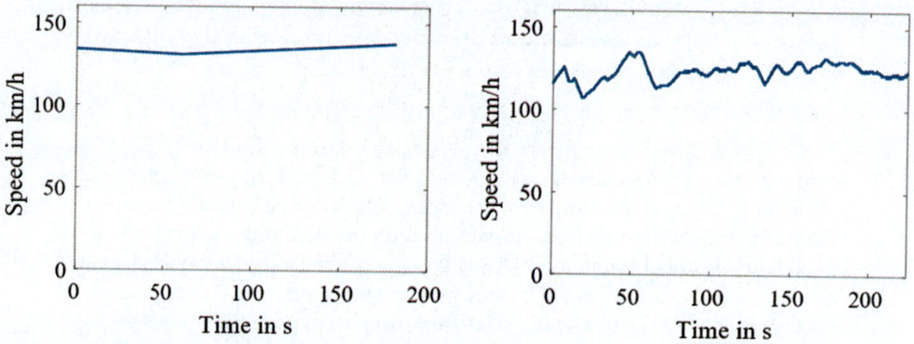

Fig. 1. Speed profiles of the German highway A81 between Untergruppenbach and Pleidelsheim retrieved from traffic information services (left) and measured with a test vehicle (right).

Deviations in the speed profiles lead to sub-optimal energy efficiency by influencing the actual State-of-Charge- (SoC) trajectory during the drive.

This paper presents a way to increase fuel efficiency during the drive with the usage of a surroundings sensor system to predict future deviations of the initial speed profile prediction and actual speed profile. Figure 2 exemplarily shows the surroundings sensor system built on the test vehicle's roof.

2 Background and Related Work

2.1 The Equivalent Consumption Minimization Strategy

The most well-known and researched causal operation strategy is the Equivalent Consumption Minimization Strategy (ECMS) firstly published in [2]. As described in [3], the total equivalent fuel consumption $\dot{m}_{eq}(t)$ is a sum of the ICE fuel consumption $\dot{m}_f(t)$ and the energy from the electric storage expressed as a fuel consumption $\dot{m}_{el}(t)$, weighted with an equivalence factor $s(t)$:

$$\dot{m}_{eq}(t) = \dot{m}_f(t) + s(t) \cdot \dot{m}_{el}(t). \quad (1)$$

Fig. 2. The surroundings sensor system of the Audi A3 e-tron test vehicle at Heilbronn University of Applied Sciences. The system consists of a combination of four LIDAR (Light Detection And Ranging) sensors, two cameras and one GPS (Global Positioning System) antenna.

As further seen in [3] the ECMS determines the optimal torque-split

$$u(t) = \frac{M_{EM}}{M_{TOTAL}} \tag{2}$$

by minimizing the cost function

$$\min_{u}\left\{ J(u) = \int_{0}^{t_n} \dot{m}_{eq}(u(t), t)\mathrm{d}t \right\}, \tag{3}$$

with M_{EM} representing the EM torque, M_{TOTAL} the total torque and J the cost functional.

The actual optimization only uses information about the current system state. Information about future depletion or recharging of the battery is considered via the equivalence factor. Given the equivalence factor is ideal due to a priori knowledge of the driving cycle, the ECMS can deliver a global optimal solutions as proven in [4]. Since a priori knowledge is not available in online operation, the equivalence factor has to be determined in another way. Many publications show different approaches to estimate the equivalence factor. Some of them are presented in the following section.

2.2 Related Work

The paper [5] proposes two methods for equivalence factor estimation. Firstly, a shrinking dartboard to adapt the equivalence factor online. Secondly, a weighted fractional, which compounds a torque-split factor from multiple weighted ECMS calculations performed simultaneously.

On the other hand [6] uses an approach, which adjusts the equivalence factor based on pattern recognition. The adjustment uses speed and elevation estimations from a model of the expected route created from previous drives.

An algorithm, which adapts the equivalence factor online at regular intervals based on the difference between actual and reference SoC is presented in [7].

The paper [8] presents a way to optimize operation strategies in HEVs by the usage of predictive data of vehicle driving states from surround sensors, GPS and car-2-x systems. The possible fuel savings for different prediction horizons, simulated with the FTP-75 cycle, indicate a higher potential for longer prediction horizons.

3 Research Method

This section presents the research method used to determine the fuel-saving potential of HEVs using predictive speed profile information of a surroundings sensor system. The simulation is based on the DPM toolbox provided by [9, 10] which has been extended with an ECMS function derived from [11]. The simulation uses a model of a hybrid electric vehicle with P2 configuration, whose parameters are shown in Table 1. The EM is located on the transmission input shaft and can be operated independently from the ICE. The torque-split $u(t)$ of ICE and EM is determined with the ECMS.

Table 1. Vehicle model parameters [9, 10]

Vehicle mass	1800 kg
wheel radius	0.3 m
rolling friction	144 N
aerodynamic coefficient	0.48 Ns^2/m^2
gearbox efficiency	0.95
Gear ratios	17 9.6 6.3 4.6 3.7 3.5
engine inertia	0.14 $kg \bullet m^2$
motor inertia	0.03 $kg \bullet m^2$
idle speed	100 rad/s
SoC minimum	40%
SoC maximum	70%
SoC start value	55%
SoC target end value	55%
Battery capacity	6 Ah
Max charging current	125 A
Max discharging current	100 A

An extended energy management system is added to the HEV model (see Fig. 3), offering an online equivalence factor adaptation based on prediction data. The prediction input parameters are the predicted speed profile v_{pred}, acceleration a_{pred}, gear selection g_{pred} and gradient of the road c_{pred}.

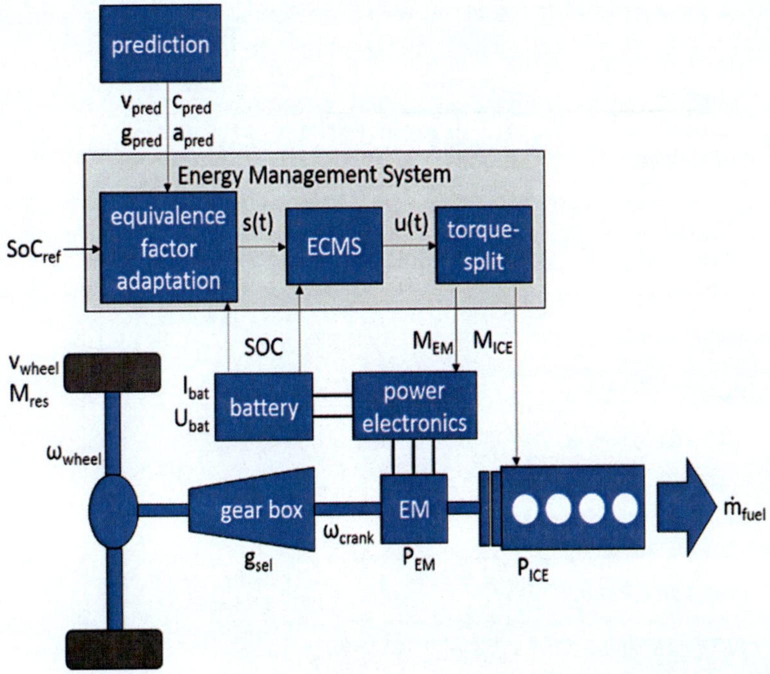

Fig. 3. Schematic model of the P2 hybrid electric vehicle model. The prediction mirrors the surroundings sensor system information.

Figure 4 gives an overview of the applied algorithm in the research method. It is divided in two phases: The initial global plan and the local online adaptation during the drive.

Initial Global Plan. Before the start of the drive an initial equivalence factor $s_{init}(d)$ and according torque distribution $u_{init}(d)$ are calculated using a global prediction of the speed $v_{init}(d)$ and (if available) altitude profiles $h_{init}(d)$. The resulting SoC-trajectory $SoC_{init}(d)$ serves as a reference for the local online adaptation of the equivalence factor during the drive.

Local Online Adaptation. The local online adaptation path in the algorithm uses local prediction to calculate an adapted equivalence factor $s(d)$ to correct deviations from the initial global SoC-trajectory $SoC_{init}(d)$. The adaptation is based on the driving, view and planning horizons displayed in Fig. 5.

Driving Horizon $\boldsymbol{H_{drive}}$. The driving horizon is the distance after which a new evaluation is executed.

View Horizon $\boldsymbol{H_{view}}$. The view horizon is the range of the local prediction of the future speed profile and equals the view range of the surroundings sensor system. The local speed prediction $v_{view}(d)$ is assumed to be accurate within the view horizon. The vehicle drives the next section of the route $\boldsymbol{H_{drive}}$ using the adapted equivalence factor before the next adaptation.

Planning Horizon $\boldsymbol{H_{plan}}$. The planning horizon indicates how far ahead from the current position d_i the equivalence factor is planned during adaptation. The SoC-value

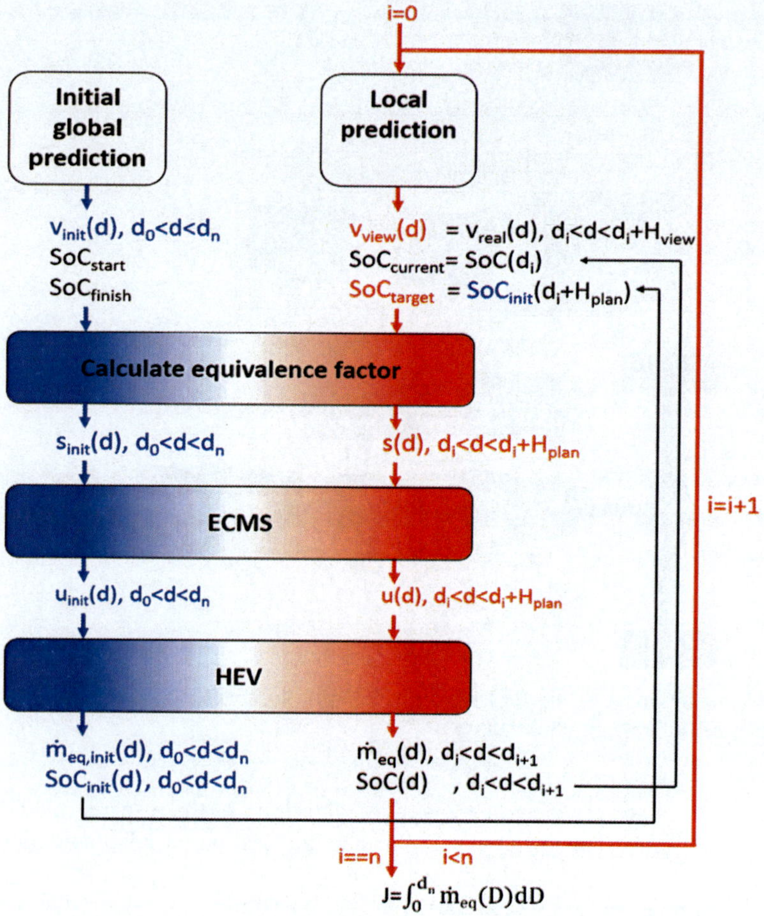

Fig. 4. Schematic principle of the algorithm used to identify fuel-saving potentials by using surroundings sensor system information in HEV. The blue side displays the initial global plan prior to the drive, the red side describes the local online adaptation during the drive.

from the initial plan at the end of the planning horizon $SoC_{init}(d_i + H_{plan})$ is used as a target when readjusting the equivalence factor. Using a more precise local speed prediction $v_{view}(d)$, the new equivalence factor is determined in such way that the SoC-trajectory converges to the reference SoC-trajectory within the planning horizon.

Horizon Variation. In this work, the length of these horizons is varied. The view and planning horizon have been set at equal length for this study, because this offers the possibility to plan the adapted equivalence factor for the maximum view range of the surroundings sensor system. The length is set to 75 m, 135 m and 200 m, based on typical view ranges of surround sensors. The driving horizon is set to 50 m, which is a compromise between quick response to deviations and a manageable demand of computational effort.

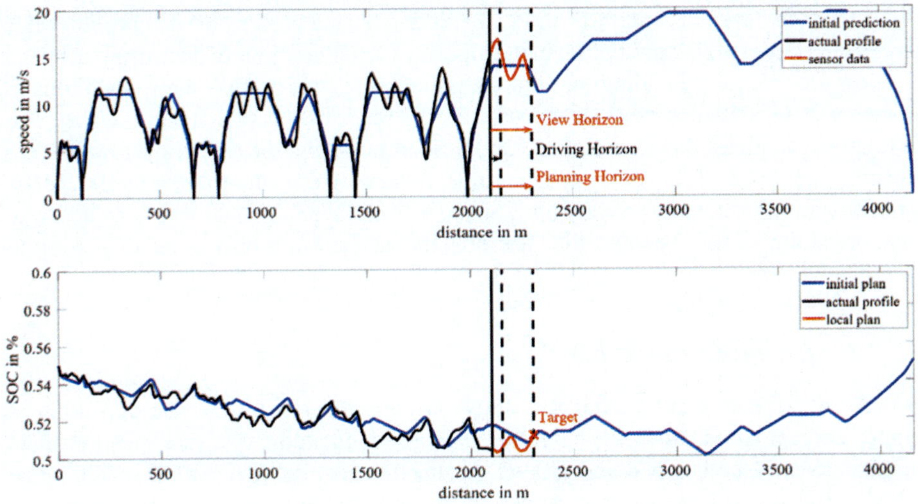

Fig. 5. Application of algorithm. In the upper diagram, the blue trajectory displays the initial speed profile prediction based on an emission test cycle speed profile, whereas the black trajectory describes the actual speed profile. The lower diagram analogically displays both SoC-trajectories. Within the view and planning horizon, the local online adaptation corrects the deviations between the SoC-trajectories, which is periodically executed after the driving horizon.

3.1 Study on Emission Test Cycles

At first, the algorithm is applied on emission test cycles. In order to have a great variety paired with relevance for the automotive market the authors chose five cycles from European, American and Japanese legislation, which offer a wide range of driving characteristics. These cycles are the Japanese 10–15 Mode (1015), the New European Driving Cycle (NEDC), the Worldwide Harmonized Light Vehicles Test Procedure (WLTP), the American Urban Dynamometer Driving Schedule (UDDS, also known as FTP-72) and the Highway Fuel Economy Test Cycle (HWFET). All cycles are classified into urban, highway and mixed profiles in Table 2.

Table 2. Classification of emission test cycles.

Type Profile	Emission test cycle
Urban	UDDS
Highway	HWFET
Mixed	1015 NEDC WLTP

In the study on emission test cycles, the default speed profiles of the chosen cycles serve as the initial global prediction in the algorithm. 21 alterations of each original speed profile were created by adding random Gaussian noise, which represents deviations in an actual drive. These alterations are used as the respective actual speed profiles. The simulation calculates the equivalence factor, the torque-split, the SoC-trajectory and the fuel consumption for every noisy speed profile. Afterwards the algorithm is applied to the same alterations but with the view horizon set to 0 m, meaning that there is no accurate local prediction of the speed profile. Thereby, the adaptation is only based on the initial global prediction.

3.2 Study on Real Speed Profiles

In order to validate results from the study on emission test cycles, the algorithm is applied on real speed profiles (see Table 3) in a second study. The real speed profiles are actually measured speed and altitude profiles on real roads recorded with the test vehicle. The first driving cycle is a section of the German highway A81 between the cities Untergruppenbach and Pleidelsheim and has a length of 13 km. The second driving cycle is an urban profile in the city of Heilbronn with a length of 8 km. The A81 cycle was driven nine times and the Heilbronn cycle eight times. All 17 drives were recorded at different times on different days.

Table 3. Classification of real speed profiles.

Type	Real speed profile
Profile	
Urban	Heilbronn Heilbronn Altitude
Highway	A81 A81 Altitude
Mixed	–

Moreover, besides the speed and altitude measurement, live traffic flow data was retrieved from the real-time traffic information service [12] during every drive.

In the study, the traffic flow data is used as initial global prediction in the algorithm, while the measured speed profile is used as the actual speed profile for local prediction. In contrast to the study in Sect. 3.1, every initial global profile is paired with only one actual speed profile. Furthermore, Heilbronn Altitude and A81 Altitude have the same speed profile as Heilbronn and A81, but the measured altitude profile was considered in the simulation.

4 Results

This section shows the results from the studies explained in the Chaps. 3.1 and 3.2. At first, the studies are evaluated separately and afterwards the findings are discussed in Chap. 4.3.

In this work, the fuel-saving potential is defined as the relative improvement of fuel consumption with (C_{loc}) and without (C_{glb}) the usage of the sensor system:

$$Fuel - SavingPotential = \frac{C_{glb} - C_{loc}}{C_{glb}}. \quad (4)$$

Additionally, the fuel consumption increase over optimal parameters is defined as the relative deterioration of the fuel consumption with (C_{loc}) and without (C_{glb}) the usage of the sensor system compared to the theoretical optimal result C_{opt} of the ECMS:

$$Fuelconsumptionincreaseoveroptimalparameters = \frac{C_{loc,glb} - C_{opt}}{C_{opt}}. \quad (5)$$

Theoretically, global optimal results are achieved, if the initial global speed profile prediction and the actual speed profile show no deviations.

In the following, the average fuel-saving potential and the fuel consumption increase over optimal parameters with all horizon lengths for all emission test cycles (1015, NEDC, UDDS, HWFET, WLTP) and for all real driving cycles (A81, A81 Altitude, Heilbronn, Heilbronn Altitude) are shown in Tables 4 and 5. Figures 6 and 7 display the fuel-saving potential of the emission test cycles and the real driving cycles for a view horizon of 75 m, 135 m and 200 m as box plots.

4.1 Results for Emission Test Cycles

The 1015 and NEDC show the highest fuel-saving potential in this study with average values of 3.48–10.51%. The results of the UDDS show almost the same potential as the NEDC. However, the results of the WLTP and thc HWFET show almost no fuel-saving potential with average values of 0.21–0.73%. For 1015, NEDC and UDDS the fuel-saving potential decrease with longer view and planning horizons. This trend is not seen in the results of HWFET and WLTP. However, for all cycles the absolute fuel consumption decreases with longer view and planning horizons.

4.2 Results for Real Speed Profiles

The results for the real speed profiles confirm the initial findings of emission test cycle results. The two city driving cycles with urban speed profile Heilbronn and Heilbronn Altitude show average fuel-saving potentials between 5.61 and 12.6%. A81 and A81 Altitude show considerably lower average fuel-saving potential results between 1.18 and 2.40%. Heilbronn shows decreasing fuel-saving potential for longer view and planning horizons, whereas Heilbronn Altitude has a higher potential, but no significant result differences regarding horizon length. As in 4.1, absolute fuel consumption decreases with longer view and planning horizons for all driving cycles. Furthermore, considering measured altitude profiles the fuel consumption increase over optimal parameters is higher compared to the speed profiles A81 and Heilbronn.

Table 4 Results of the study on emission test cycles.

Cycle	Potential in %	Horizons in m			Fuel consumption increase over optimal parameters in %	
		View	Driving	Planning	with sensor system	without sensor system
1015	**10.51** ± 1.85	75	50	75	**10.90** ± 1.97	**23.96** ± 2.29
1015	**9.60** ± 1.97	135	50	135	**6.61** ± 1.75	**17.98** ± 2.67
1015	**7.92** ± 1.02	200	50	200	**4.53** ± 1.48	**13.52** ± 1.23
NEDC	**5.64** ± 1.93	75	50	75	**16.54** ± 1.56	**23.55** ± 2.98
NEDC	**5.47** ± 1.34	135	50	135	**11.79** ± 1.45	**18.29** ± 2.62
NEDC	**4.63** ± 1.53	200	50	200	**9.14** ± 1.62	**14.46** ± 2.34
UDDS	**5.08** ± 1.35	75	50	75	**15.24** ± 1.61	**21.42** ± 2.06
UDDS	**4.39** ± 1.44	135	50	135	**8.23** ± 1.15	**13.21** ± 1.52
UDDS	**3.48** ± 1.01	200	50	200	**5.38** ± 0.79	**9.19** ± 1.32
HWFET	**0.21** ± 0.50	75	50	75	**6.85** ± 1.31	**6.64** ± 1.50
HWFET	**0.17** ± 0.56	135	50	135	**5.45** ± 1.16	**5.63** ± 1.48
HWFET	**0.68** ± 0.47	200	50	200	**4.25** ± 1.01	**4.97** ± 1.19
WLTP class 3	**0.73** ± 0.71	75	50	75	**11.64** ± 0.92	**12.46** ± 0.88
WLTP class 3	**0.50** ± 0.66	135	50	135	**8.24** ± 0.93	**8.78** ± 0.70
WLTP class 3	**0.51** ± 0.62	200	50	200	**6.30** ± 0.84	**6.85** ± 0.74

Table 5 Results of the study on real speed profiles.

Cycle	Fuel-Saving Potential in %	Horizons in m			Fuel consumption increase over optimal parameters in %	
		View	Driving	Planning	with sensor system	without sensor system
A81	**2.37** ± 0.48	75	50	75	**1.15** ± 0.81	**3.62** ± 1.02
A81	**2.31** ± 0.32	135	50	135	**1.04** ± 0.75	**3.43** ± 0.86
A81	**2.40** ± 0.48	200	50	200	**0.95** ± 0.72	**3.44** ± 0.95
A81 Altitude	**1.18** ± 1.13	75	50	75	**3.05** ± 0.81	**4.29** ± 1.14
A81 Altitude	**1.70** ± 0.70	135	50	135	**2.65** ± 0.59	**4.42** ± 0.51
A81 Altitude	**1.87** ± 0.78	200	50	200	**2.24** ± 0.57	**4.19** ± 0.63
Heilbronn	**9.11** ± 3.60	75	50	75	**17.63** ± 5.21	**29.41** ± 0.61
Heilbronn	**6.26** ± 1.87	135	50	135	**13.13** ± 3.92	**20.67** ± 1.77
Heilbronn	**5.61** ± 1.62	200	50	200	**10.34** ± 4.47	**16.88** ± 2.73
Heilbronn Altitude	**12.60** ± 3.46	75	50	75	**30.49** ± 12.71	**49.38** ± 14.47
Heilbronn Altitude	**11.99** ± 2.79	135	50	135	**18.90** ± 13.54	**35.41** ± 19.67
Heilbronn Altitude	**12.22** ± 7.83	200	50	200	**13.83** ± 8.12	**30.62** ± 20.90

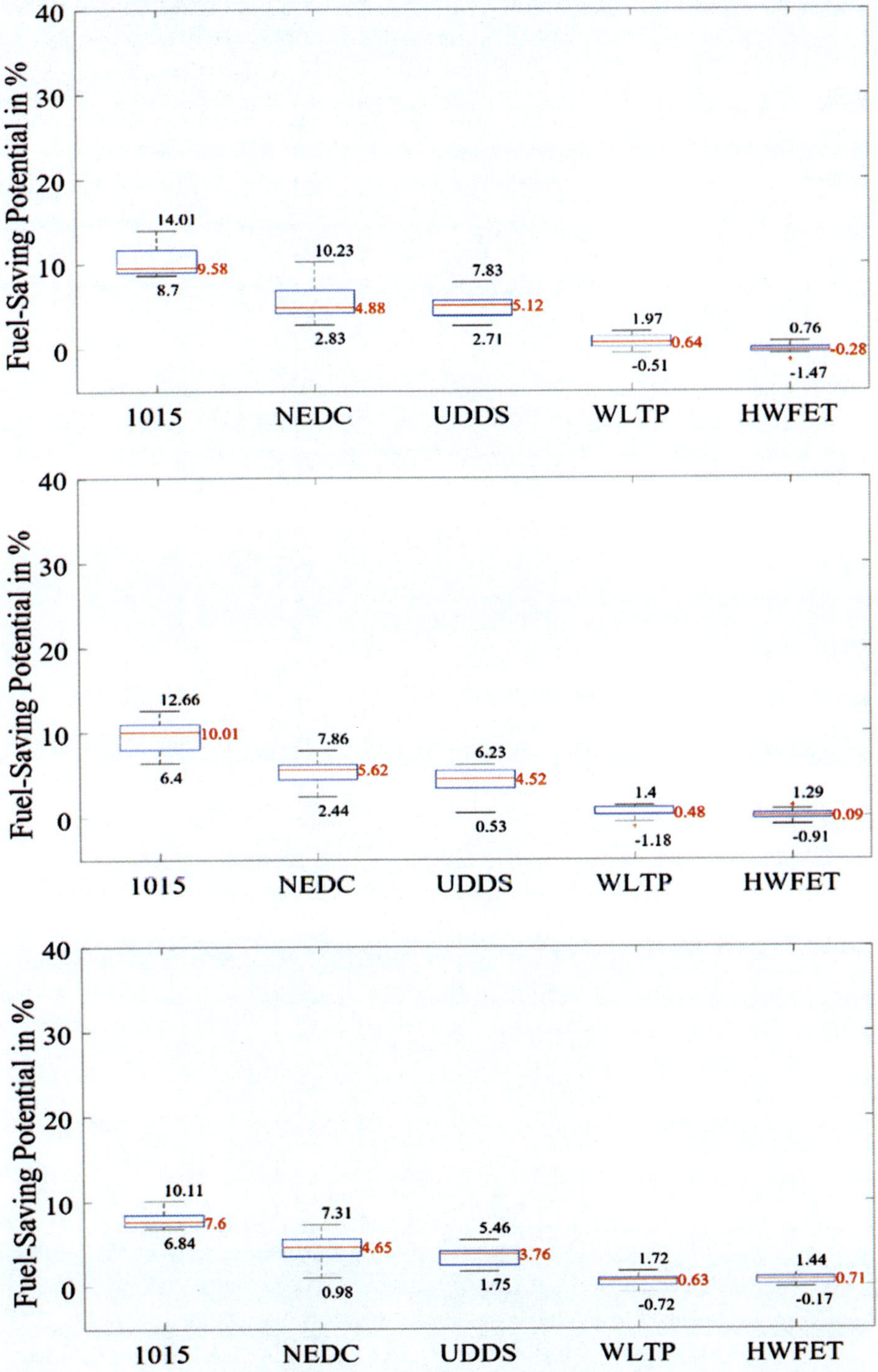

Fig. 6. Fuel-saving potential for emission test cycles 1015, NEDC, UDDS, WLTP and HWFET with 75 m (top), 135 m (mid) and 200 m (bottom) view horizon as box plots.

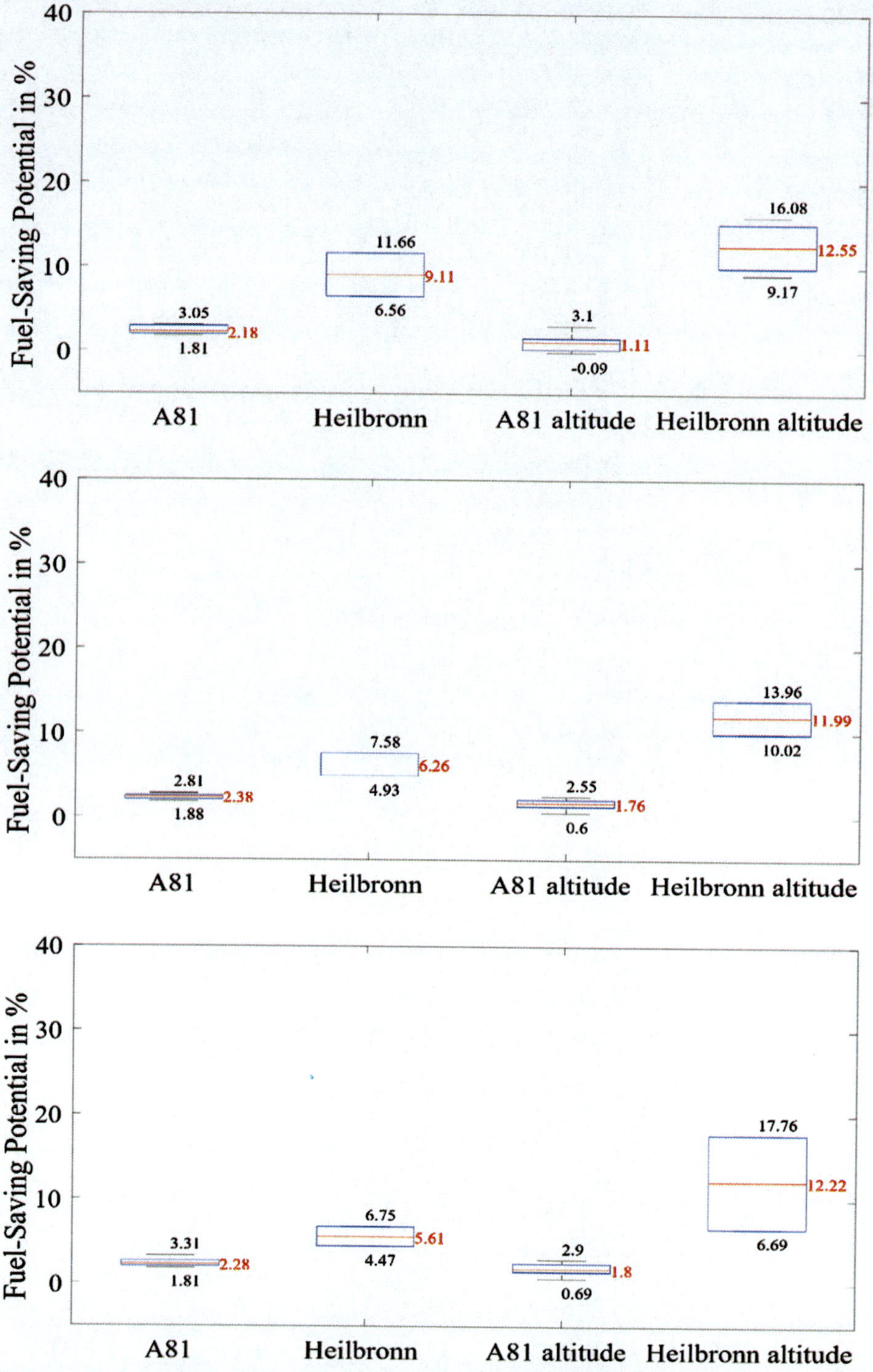

Fig. 7. Fuel-saving potential for real driving cycles A81, Heilbronn, A81 Altitude and Heilbronn Altitude with 75 m (top), 135 m (mid) and 200 m (bottom) view horizon as box plots.

4.3 Findings

Regarding the two studies with application of the developed algorithm on emission test cycles and real driving profiles there are significant findings. Firstly, the Heilbronn Altitude shows the highest fuel-saving potential of both studies with 12.60% at 75 m view and planning horizon and the highest fuel consumption increase over optimal parameters. Nevertheless, the potential values are not significantly higher than that of other cycles. The 1015 and Heilbronn show almost as high fuel-saving potentials as Heilbronn Altitude. Fuel-saving potential on the German highway A81 and A81 Altitude is lower compared to the 1015, NEDC, Heilbronn and Heilbronn Altitude. The WLTP and the HWFET achieve the lowest fuel-saving potentials. Overall, the results show that the fuel-saving potential is higher, if the speed profile has mainly urban characteristics like 1015, NEDC, UDDS, Heilbronn and Heilbronn Altitude. Further analyzation of the results is done by regarding the characteristics of the chosen emission test cycles and real speed profiles.

Mean Profile Speed. Urban speed profiles have significantly lower maximum and average speed than highway or rural profiles. As seen in Fig. 8, the lower the mean profile speed of the cycle, the higher the fuel-saving potential. Considerably high fuel-saving potential over mean profile speed is detected for Heilbronn, 1015 and Heilbronn Altitude, whereas the HWFET, A81 and A81 Altitude show the worst results in this diagram.

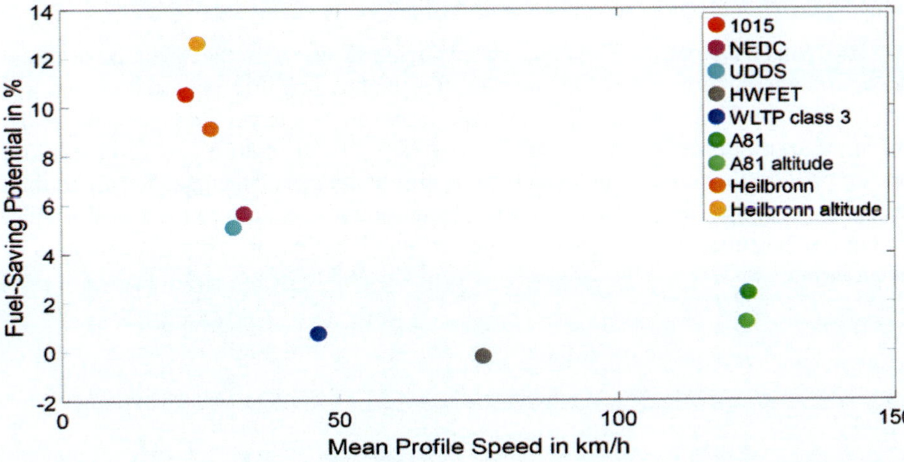

Fig. 8. Illustration of the relationship between fuel-saving potential and mean profile speed.

Number of Standstills. The number of standstills describes the amount of complete stops of the vehicle during the drive. Urban cycles have more stop-and-go traffic than highway profiles. Traffic lights, crossings, interactions with other vehicles and give way signs are only few reasons for the standstills in city traffic. Therefore, there are more acceleration and deceleration phases where surroundings sensor system information is needed the most due to high probability of deviations between initial planned and actual speed profile. Figure 9 shows the relation between the number of stops per kilometer of profile and the fuel-saving potential. Heilbronn, 1015 and Heilbronn Altitude show the best results having the most standstills per kilometer. On the other hand, A81 and A81 Altitude show lower fuel-saving potential having zero standstills per kilometer.

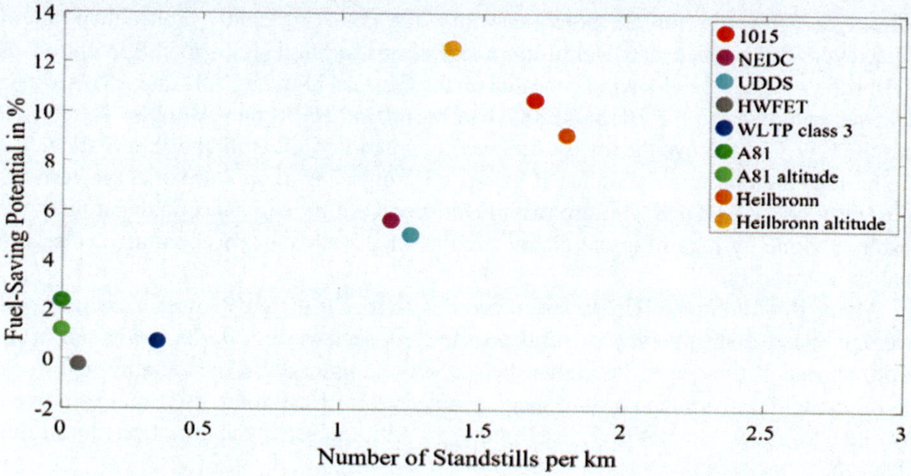

Fig. 9. Illustration of the relationship between fuel-saving potential and number of standstills.

Mean Speed Deviation. The mean speed deviation describes the deviation of initial speed profile prediction and actual speed profile. Traffic flow information does not provide information about standstills. Usually, it only shows a lower average speed at the concerned section because not all cars have to stop. The difference between the initial speed profile prediction and the actual speed profile is therefore bigger for urban than for highway application, where the initial prediction is very close to the actual profile, leaving less potential for fuel-saving using surroundings sensor system information as shown in Fig. 10.

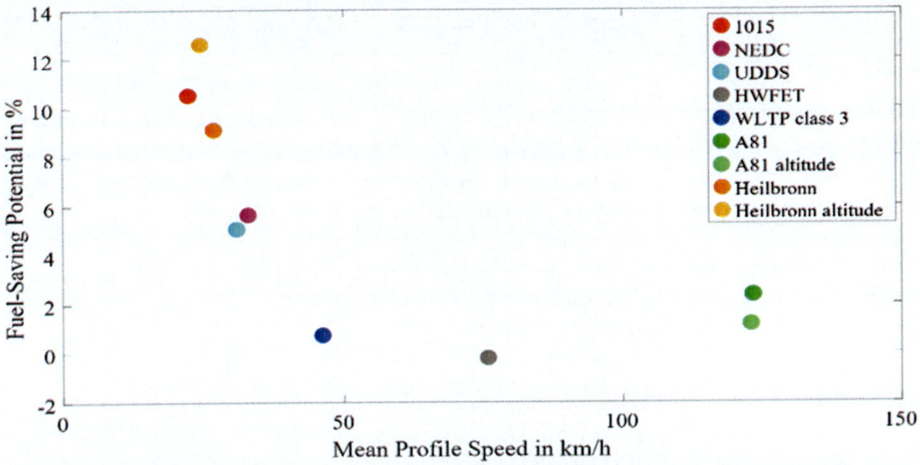

Fig. 10. Illustration of the relationship between fuel-saving potential and mean speed deviations.

5 Conclusion

This paper presents the fuel-saving potential of HEVs using surroundings sensor system information for local online adaptation of the operation strategy. The research is based on a simulation of a P2 HEV extended by a predictive energy management system including the ECMS. The presented algorithm combines an initial global plan, which uses traffic information for offline planning of the drive and a local online adaptation that uses the local speed prediction of the surround sensors to correct SoC-deviations from the initial global plan by adjusting the equivalence factor of the ECMS. The local online adaptation operates a view horizon that describes the distance of the surround sensors' speed prediction, a planning horizon representing the distance in which the equivalence factor is adapted and the driving horizon defined as the distance until the local online adaptation is executed again. In order to determine the fuel-saving potential, the algorithm is applied on emission test cycles and real speed profiles including altitude information in two studies. The results show an average fuel-saving potential of up to 10.51% for emission test cycles and up to 12.60% for real speed profiles. Further evaluation illustrates the relationships between the fuel-saving potential and the mean profile speed, the number of standstills per kilometer and the mean speed deviation of initial global plan and actual speed profile. The findings indicate a higher fuel-saving potential for driving cycles consisting mostly of urban speed profiles, whereas results for highway speed profiles show significantly lower potential for fuel-saving.

Outlook. In future, more mixed speed profiles and urban speed profiles in cities with more traffic volume must be investigated to gain further knowledge about the fuel-saving potential.

The work is based on the assumption, that the surroundings sensor system information is exactly accurate within its view range. However, in reality, sensor data is imperfect and sensor data fusion from different sensors can lead to more deviations. Additionally, sudden traffic situation changes within the driving horizon is not yet considered in the algorithm. Therefore, further investigations need to deal with the inaccuracy of the surroundings sensor system.

The publication [13] uses surroundings sensor system information to develop a car following model, which predicts future speed profiles for multi-lane traffic. The authors aim to further research how the combination of surroundings sensor system information and car following models can be used to increase fuel-saving potential in real driving cycles depending on the individual naturalistic behavior of road users.

Acknowledgement. This work was partially supported by the German Ministry of Education and Research BMBF as part of the project AHeAD (grant 13FH031IX4).

References

1. Kraftfahrt-Bundesamt (2019) Neuzulassungen von Pkw in den Jahren 2009 bis 2018 nach ausgewählten Kraftstoffarten. https://www.kba.de/DE/Statistik/Fahrzeuge/Neuzulassungen/Umwelt/n_umwelt_z.html. Accessed 18 October 2019
2. Paganelli G (2001) General supervisory control policy for the energy optimization of charge-sustaining hybrid electric vehicles. JSAE Rev. https://doi.org/10.1016/S0389-4304(01)00138-2
3. Görke D, Untersuchungen zur kraftstoffoptimalen Betriebsweise von Parallelhybridfahrzeugen und darauf basierende Auslegung regelbasierter Betriebsstrategien. Dissertation. Wissenschaftliche Reihe Fahrzeugtechnik Universität Stuttgart, Wiesbaden
4. Serrao L, Onori S, Rizzoni G (2009) ECMS as a realization of Pontryagin's minimum principle for HEV control. In: 2009 American control conference, St. Louis, MO, USA, 10.06.2009–12.06.2009. IEEE, pp 3964–3969. https://doi.org/10.1109/ACC.2009.5160628
5. Winkler M, Geulen S, Josevski M, Tegethoff M, Abel D, Vöcking B (2015) Online parameter tuning methods for adaptive ECMS control strategies in hybrid electric vehicles. F2014-TMH-032. In: FISITA world automotive congress 2014, Maastricht, The Netherlands, 2–6 June. Curran, Red Hook, NY
6. Musardo C, Rizzoni G, Guezennec Y, Staccia B (2005) A-ECMS. An adaptive algorithm for hybrid electric vehicle energy management. Eur J Control. https://doi.org/10.3166/ejc.11.509-524
7. Onori S, Serrao L, Rizzoni G (2010) Adaptive equivalent consumption minimization strategy for hybrid electric vehicles. In: Proceedings of the ASME dynamic systems and control conference 2010. Presented at …, September 12–15, 2010, Cambridge, Massachusetts, USA, pp 499–505. ASME, New York, NY. https://doi.org/10.1115/DSCC2010-4211
8. Bäker B (2008) Strategies for the control of hybrid propulsion systems. In: Stan C, Cipolla G (eds) Alternative propulsion systems for automobiles. Fachbuch / Haus der Technik, Bd. 87, pp 3–14. Expert-Verl., Renningen
9. ETH Zürich: Downloads – Institute for Dynamic Systems and Control. https://idsc.ethz.ch/research-guzzella-onder/downloads.html. Accessed 18 Oct 2019
10. Sundstrom O, Guzzella L (2009) A generic dynamic programming Matlab function. In: 2009 IEEE control applications (CCA) & intelligent control (ISIC), St. Petersburg, Russia, 8–10 July 2009; [18th IEEE International Conference on Control Applications (CCA) and 24th IEEE International Symposium on Intelligent Control (ISIC); part of the 3rd IEEE Multi-Conference on Systems and Control (MSC 2009)]. 2009 IEEE International Conference on Control Applications (CCA), St. Petersburg, Russia, 8–10 July 2009, pp 1625–1630. IEEE, Piscataway, NJ. https://doi.org/10.1109/CCA.2009.5281131

11. Ambühl D (2009) Energy management strategies for hybrid electric vehicles. A dissertation submitted to ETH Zürich for the degree of Doctor of Science. Dissertation, ETH Zürich (2009)
12. TomTom: Flow segment data traffic API documentation, traffic flow traffic API TomTom developer portal. https://developer.tomtom.com/traffic-api/traffic-api-documentation-traffic-flow/flow-segment-data. Accessed 18 Oct 2019
13. Buyer J, Waldenmayer D, Sußmann N, Zöllner R, Zöllner JM (2019) Interaction-aware approach for online parameter estimation of a multi-lane intelligent driver model. In: IEEE intelligent transportation systems society (ed) ITSC 2019, The 22nd IEEE international conference on intelligent transportation systems, Auckland, NZ, October 27–30, 2019

Quality Assurance of Composite Materials for Powertrain Applications

Andrew Willett(✉)

Toyota Motor Europe, 33 Hoge Wei, Zaventem 1930, Belgium
andrew.willett@toyota-europe.com

Abstract. Unidirectional carbon fibre reinforced plastic (CFRP) composite materials can be used in gas storage systems, lightweight housings for powertrain components (e.g. battery casing, transmission casing), and also for light-weighting of the vehicle body to reduce load on the powertrain. In each case, we would like to optimize costs and minimize the amount of composite required. Due to the complex interaction of many factors, it is often difficult to interpret the performance of composite prototype parts, and implement countermeasures regarding the materials, material synthesis route, and part production methods. One challenge we face is a lack of detailed real information on the incoming material feedstock. In particular, to obtain fibre content and fibre distribution it is necessary to cut the material and inspect it under a microscope. A single material spool may contain hundreds of meters of material, and therefore the extracted and inspected material may not be representative of the entire length of material. In this paper, we present a novel non-destructive and non-contact method for measuring the fibre content and fibre distribution of unidirectional carbon fibre composite materials based on terahertz time domain spectroscopy.

Keywords: Composites · Non-destructive testing · Fibre distribution

1 Introduction

1.1 Materials of Interest

Fibre reinforced composite materials are receiving increased attention for their potential to contribute to vehicle and powertrain weight reduction, and ultimately extending the driving range of electrified vehicles. Thermoplastic composite materials such as woven organosheets and unidirectional tapes are of interest to avoid cure cycles and management of reactive materials within the supply chain. They also enable overmolding processes and welding processes with comparative ease compared to thermoset based materials.

Unidirectional tape materials (Fig. 1) are of particular interest as they enable the orientation of the fibres with respect to expected load paths and maximize opportunity for topology optimization. The width of such tapes can be in the range 5–100 mm, but more commonly 12–50 mm, and thickness in the range of 0.05–0.30 mm. The materials are typically produced continuously in lengths of 250–1000 m. In many cases, the width of the tapes is achieved by slitting them from a substantially wider sheet.

Euroforum Deutschland GmbH (Ed.): *CTI SYMPOSIUM 2019*, pp. 562–574, 2021.
https://doi.org/10.1007/978-3-662-61515-7_49

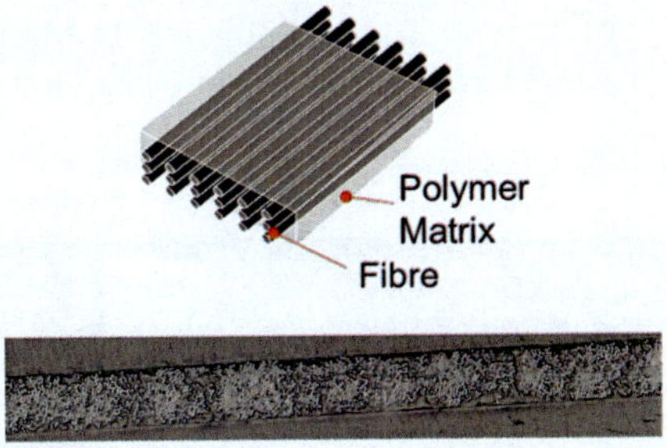

Fig. 1. Tape materials and actual cross-sectional image

1.2 Material Processing

There are various routes for processing these materials into parts. The tape materials normally have a thin resin surface layer. Almost all process routes applied to the tapes to convert them into parts utilize heating, pressure and subsequent cooling in order to bond this thin resin surface layer to other similar tapes or substrates containing a compatible resin system.

Figure 2 shows the laser assisted tape placement process [1] in which a laser is used to generate heat at the so-called nip-point between an incoming material and already placed material in order to weld them together under the action of a pressure roller.

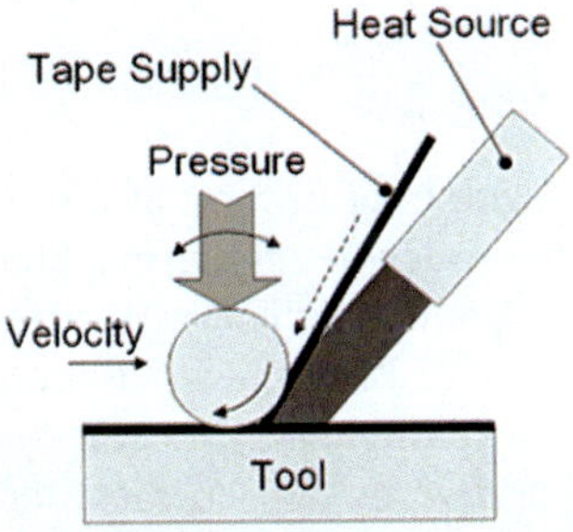

Fig. 2. Laser Assisted Tape Placement

Toyota Motor Europe has been publicly engaged in the collaborative development of new machines for laser assisted tape placement. Within European Framework 7 project STELLAR (grant agreement number: 609067) the Fraunhofer Institute of Production Technology developed the ability to cut the material on the fly, allowing for the lamination of near zero waste composite blanks [2–4].

In a further pre-competitive consortium project led by Aachen Centre for Integrative Lightweight Production (AZL GmbH) and supported by private funding 12 international companies, this technology has been optimized and parallelized to give a new prototype ultrafast tape placement system (Fig. 3) that is capable of depositing and consolidating

500 kg of tape material per hour into net-shape blanks. In March 2019, this collaboration won JEC World Composites global innovation award in the field of industry and machinery [5, 6].

Fig. 3. AZL ultrafast tape placement machine

1.3 Motivation for Quality Assurance

Considering the prospective growth in utilization of thermoplastic composite tape that may arise from the development of such disruptive machine concepts, there is increased interest to be able to assure the quality of the tape feedstock from the perspectives of process development, product optimization and serial production process stability.

Toyota Motor Europe therefore participated in a collaborative project with two major material suppliers to map the critical quality characteristics and create a database of measuring devices capable of measuring those characteristics. Despite this intensive exercise, we feel there is still a need for methods capable of measuring the constituent contents of carbon fibre tapes; in particular the amount of fibre and its distribution within the tape.

Methods for determining the constituent contents are given in ISO 14127: 2008 (Carbon Fibre Reinforced Composites—Determination of the Resin, Fibre, and Void Contents), ASTM D3171-15 (Standard Test Methods for Constituent Content of Composite Materials), and ASTM D2734-16 (Standard Test Methods for Void Content of Reinforced Plastics).

The methods proposed in these standards are destructive in nature, in that a small sample is extracted from the material or part, and the resin is removed by combustion, ignition or chemical digestion. Even within these standards substantial limitations for application exist depending on the exact resin system, fibre system, fillers and binders present in the composite. For example some resin systems cannot be removed or cannot be removed without generation by-products that may in turn deposit on the fibres and corrupt the measurement.

In such cases, the methods for determining the constituent content are generally limited to destructive inspection of cut and polished cross-section of the component. This gives a local value for the constituent content, which differs from others parts of the material that are finally used to make a component.

1.4 Objective of This Paper

It is the objective of this paper to share our first success with terahertz (THz) time domain spectroscopy [7] for mapping the constituent contents of carbon fibre tape materials both non-destructively and without the need for any liquid couplings.

The majority of literature in the field of THz measurement of composites has concentrated on defects in bulk composite objects [8–12] as opposed to the original composite feedstock. In the case of carbon fibre composite bulk objects it is well known that the penetration of THz waves is highly restricted for multi-directional fibre topologies [8–12]. It has therefore been generally challenging to use THz for carbon fibre composites compared to glass fibre composites. In contrast, the application of THz measurement in this paper is successful for carbon fibre composite due to the low thickness and single fibre orientation of the feedstock.

2 Description of Experiments

2.1 Material Selection

Within this work, three commercial tape materials with carbon fibre were selected from an actual tape processing job shop. The materials are representative of actual materials supplied to various industries that are already using or developing tape based parts. In particular the PEEK/CF material is from a supplier that is well known for its high quality impregnation process. Therefore, at the outset of this work we had expected that it is very high quality material.

- PA6/CF—exemplary of automotive industry
- HDPE/CF—exemplary of oil and gas industry
- PEEK/CF—exemplary of aerospace industry

2.2 THz Time Domain Spectroscopy

The measurements presented in this paper were made with the kind collaboration of Protemics GmbH (Aachen, Germany). Preliminary tests were made to check the THz transmission. These preliminary tests showed that it is necessary to polarize the THz radiation perpendicular to the main fibre axis in the tape in order to achieve any appreciable transmission that can be later analyzed. All subsequent tests were therefore made in the configuration shown in Fig. 4. This is a critical point for the success of this measurement.

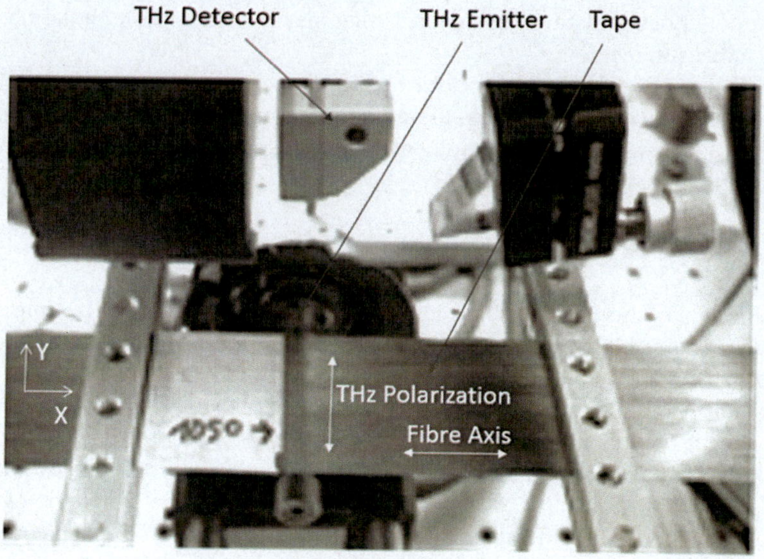

Fig. 4. THz test setup

2.3 Microscopy

After making the THz measurements the tapes have been cross-sectioned (at the same region at which THz measurement has been made), embedded, polished and inspected by microscopy. Using automated image analysis (Fig. 5), the fibres have been isolated and counted in discrete strips of 15um, 30 um, and 50um width. The THz data has then been aligned with the image analysis data as far as possible. Relationships between the transmitted THz signal and the amount of fibre have been investigated.

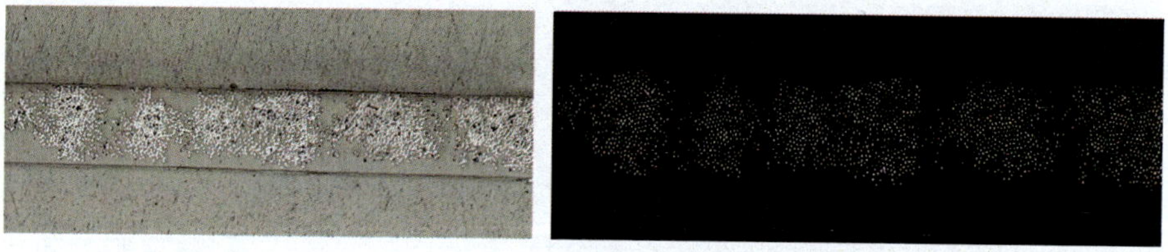

Fig. 5. Automated image analysis (left: original, right: after processing)

3 Results and Discussion

3.1 Typical Areal Scan Results

Figure 6 shows a series of three typical areal scans of the same tape material at the same location for different time delay stage positions. While it is clear there is some discernable fibre topology aligned with the fibre direction (x-axis direction), the actual features are inconsistent between the three images shown. In order to better examine signal variation over different time delay stage positions it is more convenient to use line scans (parallel to y-axis) and time domain scans. It is also substantially quicker to make a line scan rather than a full areal scan.

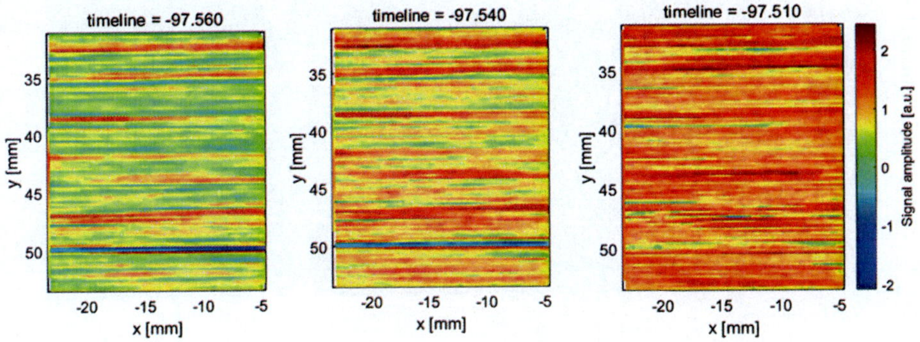

Fig. 6. Typical areal scan of tape material

The areal scans in Fig. 6 serve to show that the measurement is not varying substantially in the x-direction and therefore an arbitrary line scan measurement parallel to the y-axis within this region can be considered to be similar to any other line measured parallel to the y-axis within this region.

3.2 Line Scan and Time Domain Analysis in PA6/CF Tape

Figure 7 shows an exemplary complete line scan for the PA6/CF tape material. From this line scan we have extracted a number of time domain scans at positions labelled A, B, and C. These are shown individually in Fig. 8. The time delay stage position (on the x-axis) refers to the physical position of the time delay stage in the THz measurement system that is used to sample the THz pulse at different time positions.

We have examined various characteristics of these time domain graphs and compared them with the actual fibre count as extracted from microscopy images. Through this activity we could formulate a relationship between the fibre count and the time delay stage position at the maximum peak of the THz pulse. Other factors such as simple maximum amplitude have a comparatively poor correlation with the fibre count.

Figure 9 shows the results of the microscopy image analysis in terms of the number of fibres counted at each position across the tape. In this image "position in Y-axis" refers to the measurement direction and axes shown in the measurement setup Fig. 4. The dark line indicates the actual fibre amount at each position as obtained from microscopy image processing. For the positions A, B, and C, the white dots represent the fibre amount corresponding to the relevant time domain analysis curve in Fig. 8.

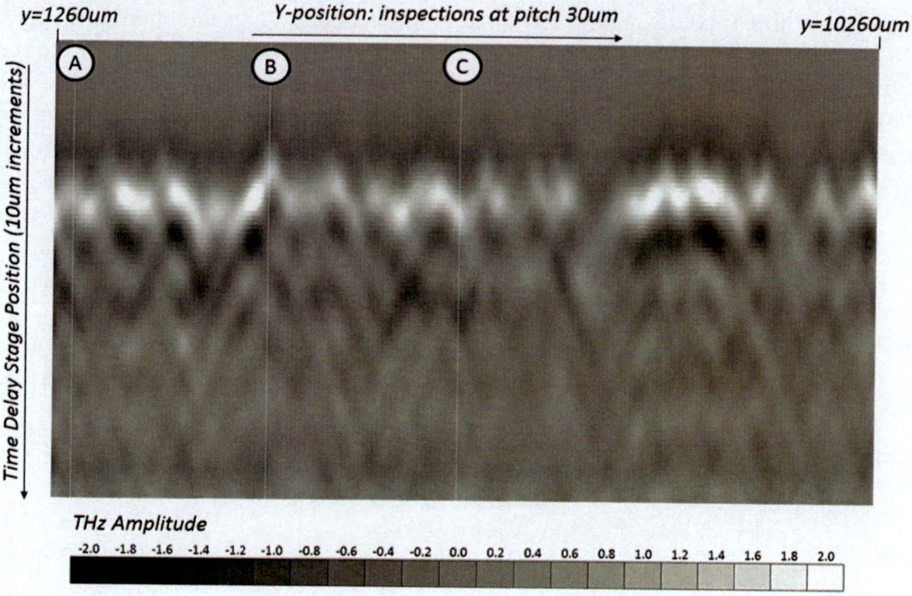

Fig. 7. THz line scan inspection of PA6/CF

3.3 Time Domain Analysis in HDPE/CF and PEEK/CF Tape

Analogous measurements were made for HDPE/CF (as shown in Figs. 10 and 11), and PEEK/CF (as shown in Figs. 12 and 13). Notably, the HDPE/CF has much more homogeneous fibre distribution that the PA6/CF tape. This is also well represented in the low variation of the time delay stage position for the maximum peak of the THz pulse at each of the three analyzed positions A, B, and C.

Despite this small variation, it is still possible to discern a local minimum, maximum and intermediary fibre count values that correspond appropriately to A, B, and C. The HDPE/CF inspection also provides evidence that the technique continues to function for low amounts of carbon fibre. Similarly, the PEEK/CF inspection provides evidence that the technique continues to function for higher amounts of carbon fibre.

3.4 Combination of Results

Figure 14 shows a summary of the results for three positions in each of the materials in terms of the fibre amount and the THz time delay stage position at maximum peak of the THz pulse. We have in fact implemented this for hundreds of measurements, but show here for simplicity a reduced number of measurements.

Within the individual tapes and across the different tape types a consistent relationship can be identified that appears to imply that the carbon fibre amount is the dominant influencing factor compared to the resin system.

The relationship between time delay stage position and number of fibres would seem to saturate as the amount of carbon fibre increases. It may therefore become challenging

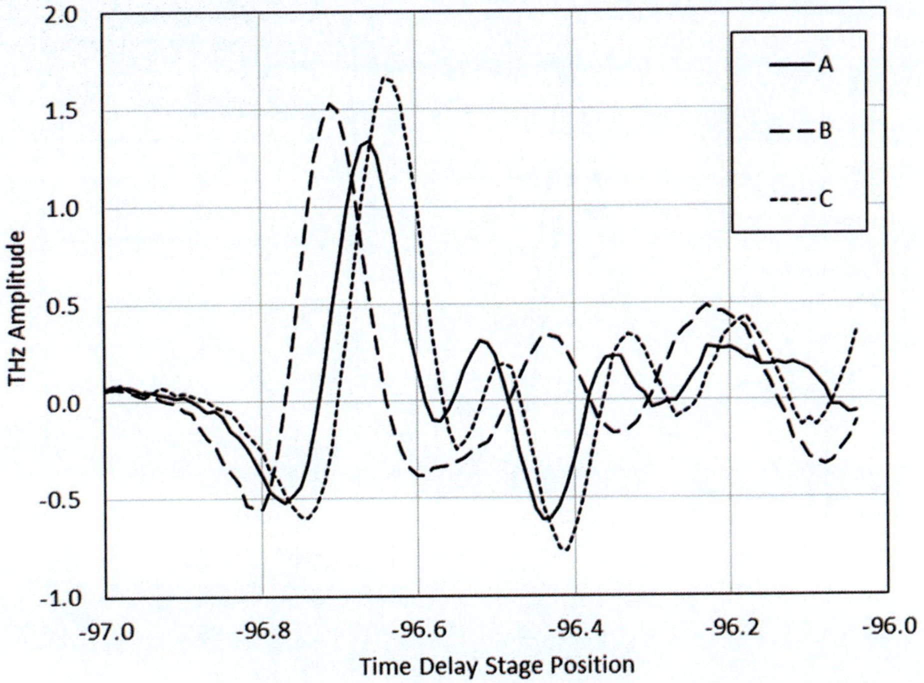

Fig. 8. THz waveform at three selected locations in PA6/CF

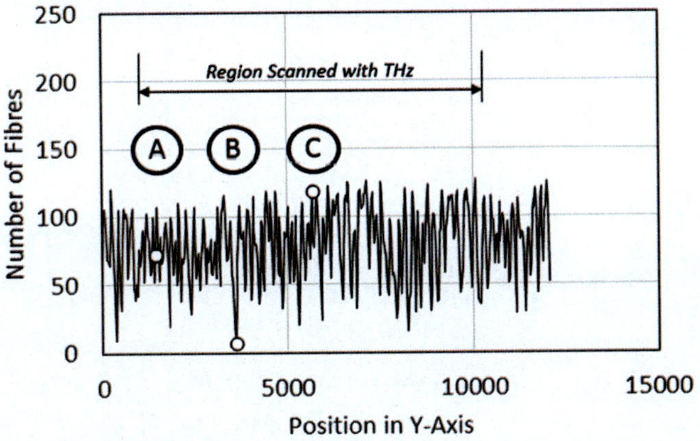

Fig. 9. Actual amount of fibre in 30um windows in PA6/CF (dark line)

to examine tapes with higher amount of fibre than those inspected in this paper without adjusting the inspection system.

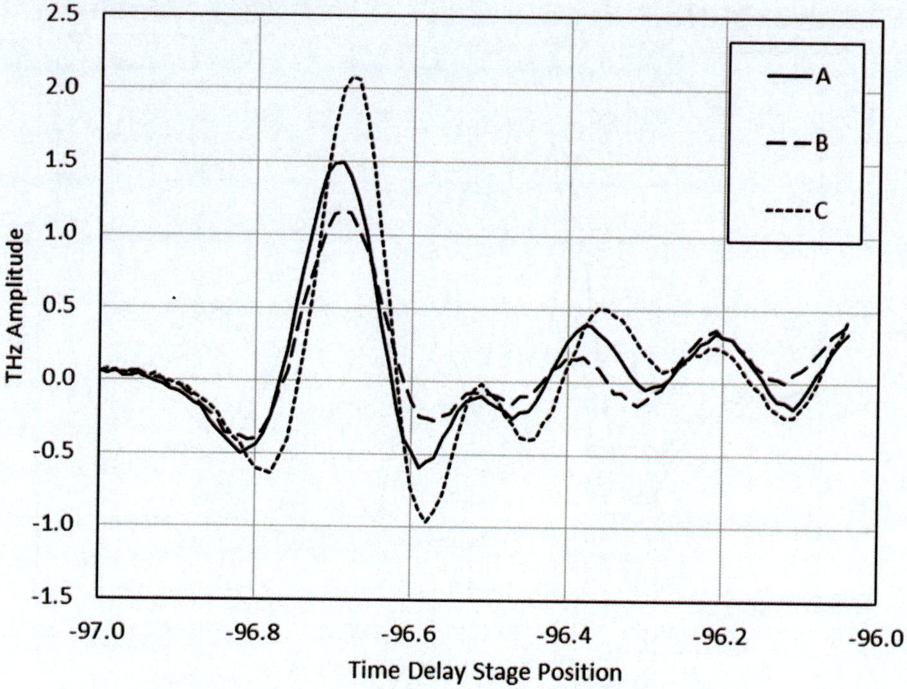

Fig. 10. THz waveform at three selected locations in HDPE/CF

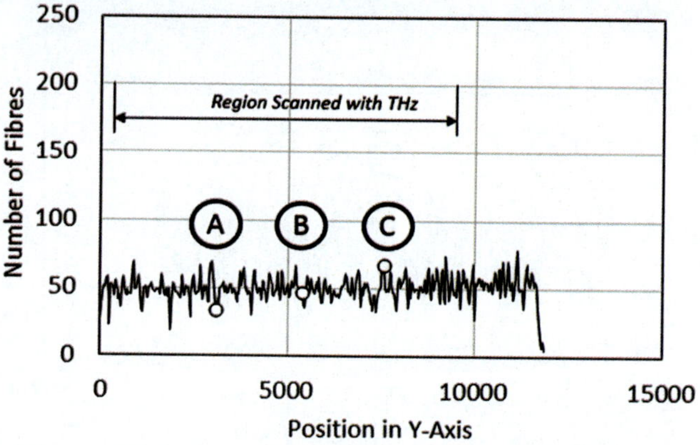

Fig. 11. Actual amount of fibre in 30um windows in HDPE/CF (dark line)

3.5 Implementation Example

Similar relationships to those in Fig. 14 were built up for slice sizes of 15 um, 30 um and 50 um slice sizes. The relationship generated for 50 um slices has then been applied to both PA6/CF tape and PEEK/CF tape to give predicted values for the fibre content. The

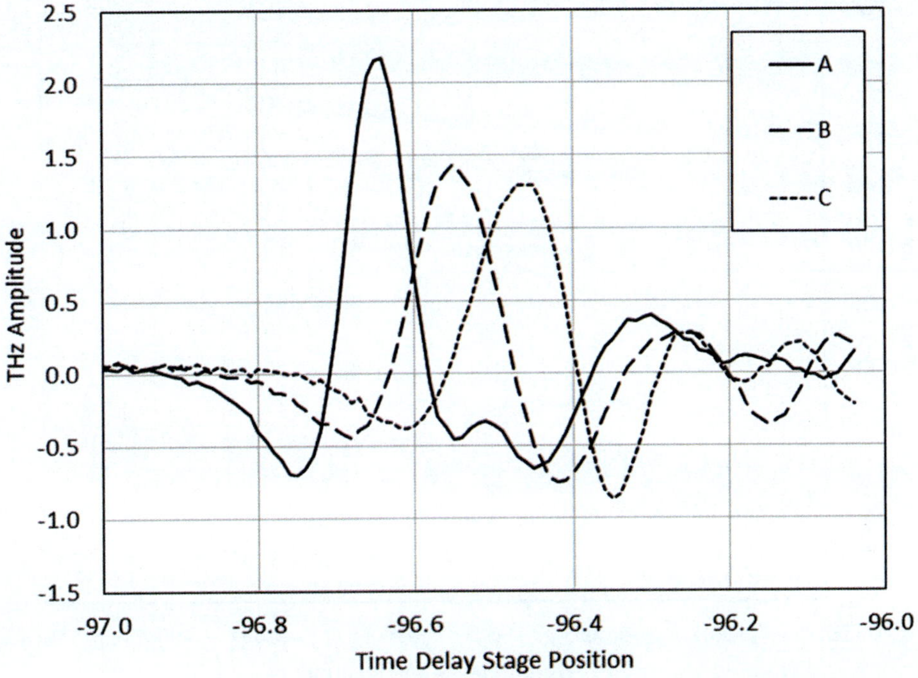

Fig. 12. THz waveform at three selected locations in PEEK/CF

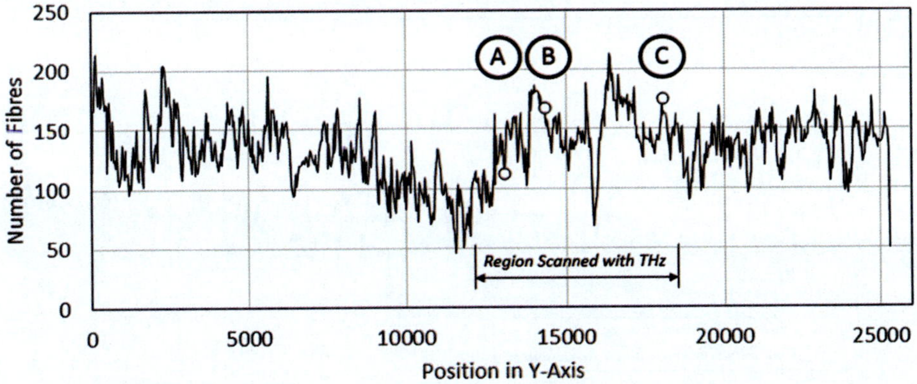

Fig. 13. Actual amount of fibre in 30um windows in PEEK/CF (dark line)

results shown in Figs. 15 and 16 respectively suggest that the technique can be used to non-destructively obtain the approximate carbon fibre content.

There are some instances in Fig. 15 where the predicted value overshoots the actual value or a sudden change is not captured by the THz measurement. We believe this is due to a combination of factors that may include:

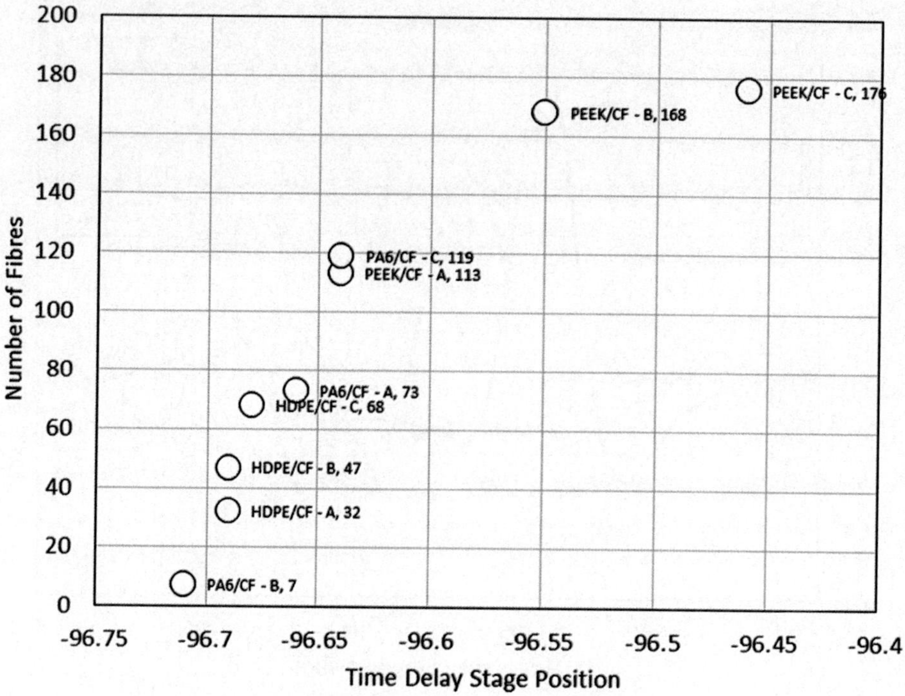

Fig. 14. Exemplary Summary of Relationship

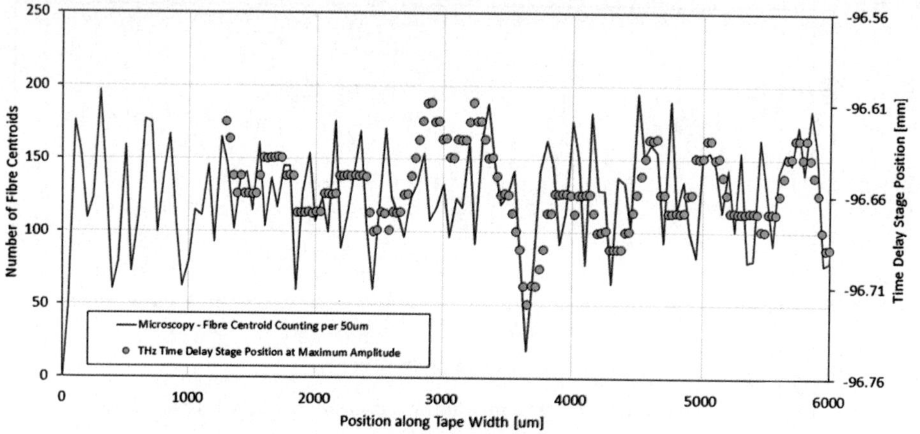

Fig. 15. Example of Application to PA6/CF Tape

- data alignment issues where the probe is not at exactly the same position as that used in the image analysis.
- slight curvature of the microscopy data due to the flexible nature of the material
- the image analysis itself is not capturing all the fibres.

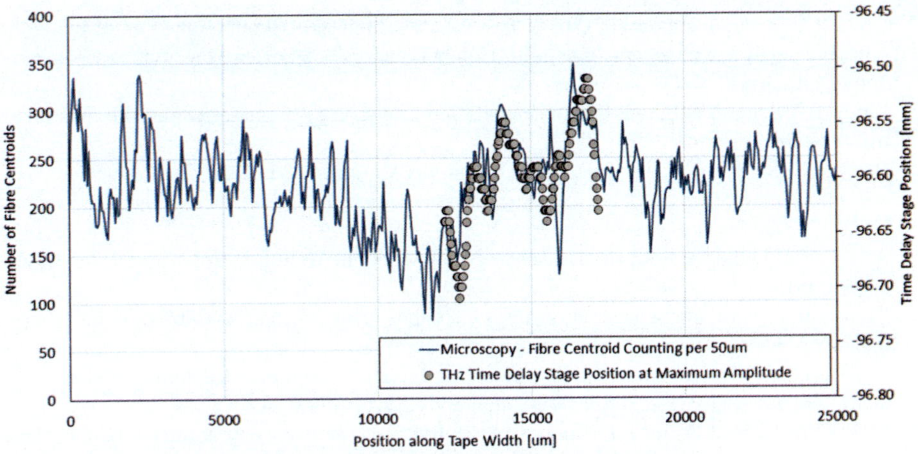

Fig. 16. Example of application to PEEK/CF tape

- the probe may be measuring on the edge of facets such that part of the beam is substantially influenced while another part of the beam is not.

Figure 16 suggests that this technique is more reliable for measurements where the fibre amount is not rapidly changing along the tape width. In both Figs. 15 and 16 only a subset of the originally measured data is shown due to challenges faced with data alignment across the entire dataset.

4 Conclusion

This work has proven the fundamental capability of THz time domain spectroscopy for the material screening of uni-directional carbon fibre composite in the form of thermoplastic tapes. We believe it can be equally applied to unidirectional carbon fibre composite thermoset towpreg. The technique can give both the amount of fibre and also the distribution of fibre. The systematics and method required for this has been published in detail as a patent WO2019/025011A1 [7]. The method shown can equally be applied to glass fibre and natural fibres. The evolution of THz inspection systems is accelerating, and lower cost and higher speed systems are now under investigation for this application.

Further work is necessary to develop robust test protocols and calibration methods. If appropriate calibration methods and hierarchy are developed then it is also possible that THz measurement may be used to support the interpretation of data from other sensor types that may be lower cost, lower resolution, simpler construction but with faster measurement speeds suitable for 100% in-process measurement.

References

1. Brecher C, Kermer-Meyer A, Werner D (2012) Customized solutions for laser-assisted tape placement. JEC Mag 76:70–73

2. Brecher C, Peters T, Emonts M (2015) Cost-effective high speed production of multi-material components by selective tape laying. In: 10th international conference on composite science and technology
3. Lindner F, Schmitt R (2015) Goal-oriented life cycle investigations for composite manufacturing chains. Procedia CIRP 29:438–443
4. Brecher C, Schmitt R, Lindner F, Peters T, Emonts M, Grose BM (2016) Increasing cost and eco efficiency for selective tape placement and forming by adaptive process design. Procedia CIRP 57:769–774
5. AZL GmbH homepage. https://azl-aachen-gmbh.de/jec-innovation-award-winner/. Accessed 31 Oct 2019
6. Composites world homepage. https://www.compositesworld.com/news/azl-production-system-wins-jec-innovation-award. Accessed 31 Oct 2019
7. Willett A, Non-destructive method for detecting defects in unidirectional composite intermediate. International Patent Publication Number WO2019/025011A1
8. Park J, Im K, Hsu D, Jung J, Yang I (2012) Terahertz spectroscopy approach of the fibre orientation influence on CFRP composite solid laminates. J Mech Sci Technol 26(7):2051–2054
9. Ospald F, Zouathi W, Beigang R, Matheis C, Jonuscheit J (2014) Aeronautics composite material inspection with a terahertz time-domain spectroscopy system. Opt Eng SPIE 53(3):1–14
10. Belitskaya A., Baryshev A., Khudchenko A (2016) Space terahertz instrumentation for integrity inspection of non-conducting composites. 19th world conference on non-destructive testing
11. Dong J, Locquet A, Declercq N, Citrin D (2016) Polarization-resolved terahertz imaging of intra- and inter-laminar damages in hybrid fibre-reinforced composite laminate subject to low-velocity impact. Compos B Eng 92:167–174
12. Dong J, Pomarede P, Chehami L, Locquet A, Meraghni F, Declercq N, Citrin D (2018) Visulaziation of subsurface damage in woven carbon fiber-reinforced composites using polarization-sensitive terahertz imaging. NDT&E Int 99:72–79

The Rolling Bearing in the Electrified Power Train—Requirements and Solutions

Thomas M. Wolf(✉)

SKF GmbH, Schweinfurt, Germany
thomas.wolf@skf.com

Abstract. Climate change and global regulations for CO2 emission reduction are nowadays the drivers for the megatrend to E-mobility. Initiated with the hybridization of the power train the industry arrived today at the battery driven electric vehicle. Here one of the most important sub-assemblies is the E-axle, which consists of an electric machine and a reduction gearbox. Market requirements like rising power density, energy efficiency, maintenance free operation and cost efficiency require adequate technical solutions on component level. Especially the rolling bearing is subjected to high demanding operating conditions. High dynamics, poor lubrication, elevated temperature levels, risk of current leakage and others more need to be considered and call for a customized bearing design. For bearing dimensioning and development work on component level an SKF proprietary simulation software toolkit is available. By that it is possible to model the entire E-axle, to analyze the bearing reactions and to optimize the bearing components in terms of service life and efficiency. The simulation of cage deformation at high rotor speed is crucial to develop a high-speed cage design. The potential risk of electrical continuity which damages balls and raceways and reduce grease life, is another issue which asks for a technical solution. With the focus on bearing technology an E-axle and its specific operating conditions will be described, simulation tools and calculation results presented and the current SKF bearing portfolio for E-mobility applications introduced.

Keywords: E-Drive · Bearing technology · Application engineering

Nowadays we are obviously living in the middle of the most significant metamorphosis of the automotive industry—the megatrend to E-mobility! This transformation will result in a lot of changes both for the OEMs and their suppliers. The question will be how the market players will adapt themselves and how they will participate to this novel megatrend. How do the specific technical challenges look like? SKF as bearing company is directly involved in this change process and so we asked ourselves: What are the specific challenges on rolling bearings and how do the solutions look like to be prepared for this megatrend.

Power train (r)evolution

Perhaps it is helpful to start with a view at the roots—the traditional power train. A comparison with the electrified power train displays easily some significant differences. The most evident one is a dramatically reduced number of components. From technical

Euroforum Deutschland GmbH (Ed.): *CTI SYMPOSIUM 2019*, pp. 575–583, 2021.
https://doi.org/10.1007/978-3-662-61515-7_50

point of view the characteristics of power supply in terms of torque and speed generated by an internal combustion engine (ICE) and an E-machine are significantly different. While the maximum crankshaft speed of a conventional ICE will usually not exceed 6500 rpms the rotational speed of an E-machine runs up to 20000 rpms today, with a certain trend to even higher numbers. Moreover the E-machine delivers full torque already at "zero" speed which results in a high degree of vehicle acceleration. This genuine speed and torque behaviour will have consequences on the design of an E-axle and ultimately on the specification of the rolling bearings, too.

Electrification over the last decades

Looking into the past it is amazing to recognize that the electrification of the (car) power train is not new at all. Instead it started with the invention of the starter and the introduction of the alternator many decades ago. And even the idea of a battery driven car is not new, e.g. consider the Lohner Porsche developped in 1899. In the recent past the various levels of hybridization—from microhybrid to plug-in and full hybrid—act as a bridge between the traditional power train and the battery driven electro car (BEV). And as a matter of fact—without the BEVs the ambitious CO_2-emission targets of 95 g CO_2 emission per kilometer by 2021 will not be achievable.

Functional principle and bearing positions

Together with the power electronics and the battery pack the E-axle—consisting of the E-machine and the corresponding reduction gearbox—completes the architecture of a battery driven car. Because of package constraints the E-machine should be as compact as possible and so—for a high output—the E-machine runs at high speed. On the differential shaft however the speed equals wheel speed and so it needs to be low and the torque to be high. This transformation of torque and speed between e-machine and the differential is the task of the reduction gearbox. As a mechanical device both the e-machine and the gearbox features a number of shafts and gear wheels which need to be supported by rolling bearings. Bearing types and their dimensions depend on the e-axle design, the operating conditions and the required service life. In other words: Each bearing position inside the e-axle is subjected to individual loading conditions and needs to get customized accordingly. The rotor of the E-machine is usually supported by two ball bearings which are subjected to high acceleration, high speed, elevated temperature but rather low load. The bearing arrangement is axially preloaded by a disc spring which centers the rotor being beneficial for the NVH behaviour. For the definition of the right initial radial clearance the temperature gradient between rotor and housing is relevant. A temperature difference of up to 50 K is not unusual and so the selected radial clearance class is often according to C4 or even C5 in order to avoid a radial preload at operation.

The likelihood of current leakage is another important aspect. Depending on the characteristic of the power electronics and the design of the e-axle there is a certain risk of capacitive current which damages the ring raceway, balls and the lubricant. In case of grease as lubricant (instead of oil lubrication) the grease life which depends mainly on speed and temperature, needs to be respected as limiting factor of the bearing service life. The usage of high performance greases on polyurea base is often the consequence.

The gearbox input shaft is directly connected to the rotor shaft. Carrying a gear wheel with a helix angle an alternating axial gear mesh load acts during driving and recuperation which is supported by a located bearing. Usually again a ball bearing will be employed. This bearing gets the same dynamics as the rotor bearings but is heavier loaded by the gear mesh forces. Consequently its dynamic capacity rating has to be as high as possible.

For the bearing arrangement on the intermediate shaft two options are possible. Either an opposed bearing arrangement in face-to-face or a located/non-located arrangement consisting of two deep groove ball bearings or one ball—and one cylindrical roller bearing. For energy efficiency—and assembly reasons a located/non-located bearing concept is preferable.

The differential shaft is usually supported by two tapered roller bearings in face-to-face arrangement or—again for the purpose of energy efficiency–two angular contact ball bearings could replace the tapered roller bearings. A located/non-located arrangement is thinkable, too. Basically—due to an friction optimized bearing design—power loss reduction for all bearing positions is the target.

Market requirements on car level

Today the service life of a car is expected to be 300 000 km respectively 8000 h which means an average speed of nearly 40 km/h. The mileage people expect to reach with a BEV is founded on their experience with the ICE, preferably more than 500 km. The basic requirement of "energy efficiency" asks for a light weight design which means growing power density. Moreover NVH requirements for the entire power train are getting more relevant. Last but not least there is certainly a global competition which requires a certain focus on cost at the same time.

Against the background of the given requirements for a BEV it is obvious that they are also valid for the subassemblies and modules belonging to the BEV. Finally it means that those requirements—given by the customer's book of requirement—influence even the design of the rolling bearings being an invisible but one of the most important mechanical component of the electrified power train.

Consequences on the bearing design

To survive in the E-axle the bearing service life has to be beyond the 300 000 km requirement. That means that both the classical subsurface fatigue and the surface fatigue process initiated by poor lubrication and contaminated lubricant needs to be considered and avoided by a suitable and clean lubricant and a robust internal bearing geometry. On the other side a "safe" but over dimensioned bearing will hardly meet the energy—and cost efficiency requirement. The focus on friction and power loss is conflictive to robustness and enlarged service life. This conflict needs to be resolved. As described previously the rotor speed is as high as 20 000 rpm nowadays and might become even higher in the future. This results in a speed factor "n*dm" beyond one million (bearing speed [rpm] * bearing pitch diameter [mm]).

High centrifugal forces acting inside the bearing request a reinforced cage design. Temperature peaks of maximal 170 °C need to be respected when the selection of material and heat treatment takes place. The likelihood of electrical continuity at operation requires a sort of insulated working contacts to prevent a gradual destruction of the raceway topography by thousands of little sparks per second. The growing usage of low viscosity lubricants—or maybe even water based fluids in the future—require an enhanced wear resistance in the working contact in order to prevent the process of wear and surface distress ending up in a premature bearing failure.

... and so the relevant component features are

The following bearing components are involved:

- inner ring
- outer ring
- cage
- ball- or roller set
- seals
- grease

Operating conditions like high rotational acceleration, speed and temperature concern all bearing components. In terms of high dynamics a low frictional torque, low moment of inertia and low density are key factors. Besides the cage needs a high structural strength to withstand the high centrifugal loads acting on the cage prongs. The seal lip/inner ring contact features a low contact force in order to be suitable for high circumferential speed. Balls made of silicon nitride do have—compared to steel balls—a significantly reduced density and a higher Young's modulus which results in lower centrifugal loads, less friction and better wear resistance at elevated operating temperature. And finally a long lasting high operating temperature requires a special heat treatment process in order to avoid retained austenite transformation in the matrix which will end up in an unacceptable dimensional growth of the rings.

Energy efficient rolling bearings are preferable for the employment in e-axles because they assist a higher mechanical efficiency rate. By using friction optimized bearings the power loss can be reduced and consequently the cruising range of a BEV can be enlarged. On bearing level it means that both the load and speed dependent friction torque must be reduced. A bigger raceway radius—a wider osculation—will reduce the size of the contact ellipse and consequently reduce the (sliding) friction. But also other contacts like the ball-cage pocket contact and the seal lip-ring contact are sources of friction and can be optimized. Last but not least the consistency and quantity of the grease inside the bearing influence the speed dependent churning losses and is part of the optimization process, too.

Detrimental high frequency voltage at operation acting between rotor and stator will charge the rotor and cause current leakage through the bearing to the housing. To protect raceways and grease from damage the ideal solution is the utilization of a hybrid ball bearing having balls made of ceramics. This bearing offers a perfect insulation but is comparably expensive. Therefore alternatives like conductive seals and/or ionic grease as well as alternative insulation methods are under development.

The need for down-sizing is equally important. The challenge is to install a maximum of dynamic capacity inside a limited package in order to meet the life requirement anyway. Increasing the capacity and keeping the boundary dimensions requires an optimization process of the internal bearing geometry and/or the usage of special steel. Alternative bearing steel and/or additional heat treatment like carbonitriding will support the trend for down sizing.

The lubrication conditions at operation are relevant for the bearing service life, too. However with the trend to low-viscosity oil—and maybe even water based fluids in the future—the tribology of the working contact is downgraded. In other words—because of a lower lubricant film thickness between rolling elements and raceways the risk of mixed lubrication or even boundary lubrication is rising. Mild wear and surface fatigue as a final consequence can be avoided by the deployment of high quality steel, additional heat treatment and balls of silicon nitride and glassfiber reinforced polymer cage material.

Last but not least the wide field of acoustics is—in context with e-mobility—equally relevant. The fact that the rolling bearing needs to run in a noise sensitive environment needs to be respected in a NVH sensitive specification of the bearing components as well. In particular bearing's running accuracy class, a "quiet" grease type and a favourable cage material are elements which support the requirement of good NVH behaviour at operation.

Simulation tools for application engineering and product development

For the analysis of bearing response and the dimensioning of a set of bearings for a certain e-axle application SKF proprietary software is available. As input the following (customer) data are considered:

- e-drive layout (shaft geometry, axial positions of gears and bearings, etc.)
- gear data (pitch dia., module, helix angle, etc.)
- load cycle (torque, speed, duration)
- lubricant viscosity, temperature and contamination

After modelling the e-axle in the SKF software "SimPro Expert" and translating the customer input the following bearing reactions can be discussed in detail:

- loaded zone (how many balls or rollers are loaded?)
- elastic deformation in the contact, size of pressure ellipse
- Hertzian pressure profiles (e.g. on the most loaded rolling element)
- friction torque and power loss for each bearing and load case
- bearing life calculation according to SKF stressed volume based life model

Considering and assessing all these results simultaneously enables an application oriented and tailor-made bearing design which meets the market requirements in the best way.

Besides another SKF proprietary software called "BEAST" supports mainly product development studies. This software can be regarded as virtual test rig by simulation and optimization of design parameters on bearing components like cages and seals.

Modelling of the reduction gearbox

The reduction gearbox (Fig. 1) consists of three shafts—input, intermediate and differential shaft. The shafts are connected via two pairs of gear wheels to realize a total ratio of about i = 10. As mentioned previously the purpose of the reduction gearbox is to transform the high speed & low torque on the e-rotor and gearbox input shaft to low speed & high torque at the differential/output shaft. Having helical gears there is also an axial load in the gear mesh which needs support by a bearing. Only on the intermediate shaft with two simultaneous gear meshes (driven & driving) a certain compensation of both axial forces—acting in opposite direction—takes place. This fact will reduce the bearing load and is beneficial for the bearing service life. It is noticeable that for a required service life of 300 000 km the corresponding number of revolutions for each shaft is different. In fact the ball bearings on the input shaft experience 1.4 billion revolutions while the bearings on the differential shaft experience "only" 140 million revs.

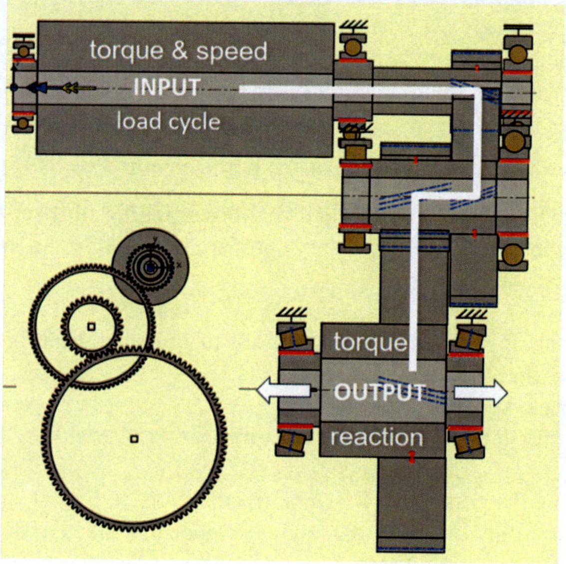

Fig. 1. Modelling of an E-axle in SKF SimPro Expert software

Results like loaded zone, Hertzian pressure and friction respectively power loss

Both the loaded zone and the Hertzian pressure levels (Fig. 2) are significant for the assessment of the bearing performance at operation. Both parameters are decisive for the power loss and the service life of a rolling bearing. In other words—the contribution to the mechanical efficiency rate of the e-axle and bearing's reliability can be estimated by those results. Moreover the knowledge of this bearing response is an excellent prerequisite for the optimization of the internal bearing geometry in order to balance the conflicting requirements of robustness and energy efficiency.

Friction torque and power loss at 30 kW motor power

At operation the bearings are subjected to forces, speed and temperature. Dependent on the lubricant characteristics these external stress factors generate a load—and speed dependent frictional torque. Multiplied by speed the corresponding power loss can be

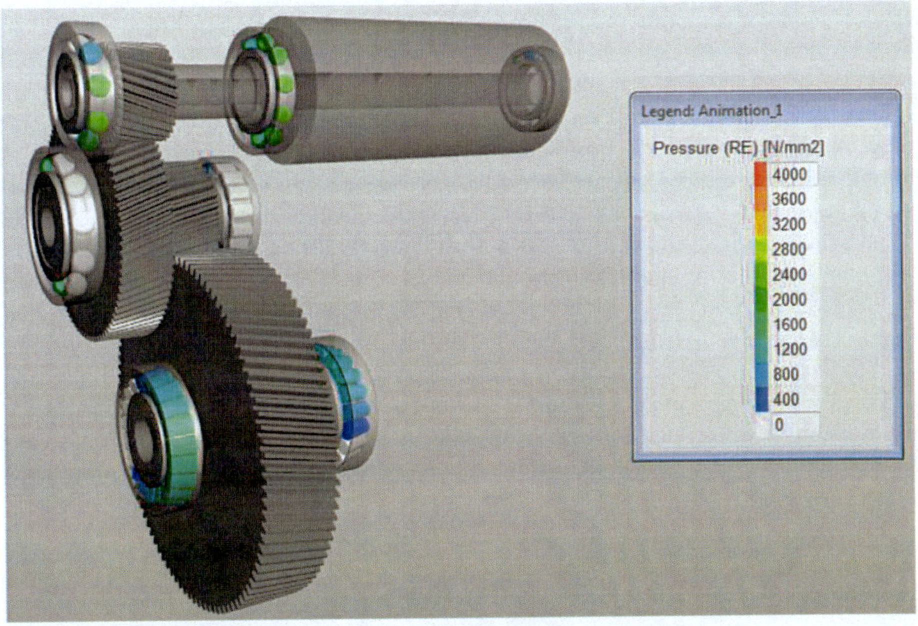

Fig. 2. Bearing reactions displayed by SKF SimPro Expert software

calculated. This power loss multiplied by duration (for instance in hours) indicates the loss of energy consumed by a bearing. It is evident that especially for an E-axle—for the sake of a long mileage—the power loss shall be as low as possible.

By means of the SKF software the friction torque of any bearing for any load case can be calculated.

Example: For an engine power of 30 kW (27Nm at 11000 rpm) on the input shaft the corresponding power losses are

- 240 W on the E-rotor and gearbox input shaft (three ball bearings)
- 51 W on the intermediate shaft (one ball and one roller bearing)
- 24 W on the differential shaft (two tapered roller bearings)

so in total 315 W, which means approximately 1% of the engine power.

As expected—due to the high rotational speed—the biggest power loss acts on the input shaft. On the "slowly" rotating differential the power loss is negligibly low, a further reduction of friction torque by a more sensitive internal geometry (with less robustness) of the differential bearings is actually not reasonable.

SKF software "BEAST" for the development of a ball bearing high speed cage

Ball bearings on the rotor shaft are often subjected to high rotational acceleration and eventually high rotational speed, too. An important criterion is the previously mentioned "speed factor". Speed factors beyond 800 000 request a reinforced cage design, a topic which becomes even more important for a speed factor of up to 1.6 million expected in the near future.

As a consequence of the rotational speed a centrifugal force acts on the rotating components—inner ring, ball set and cage. Usually the cage speed is half of the shaft speed. Apart from the need for an adequate heat dissipation a sufficiently high structural strength of the one-piece polymer cage has to be assured. In fact—due to the Conrad-assembly process—the cage pockets of a deep groove ball bearing are open on one side to make the snap-in of the balls possible. Unfortunately the cage prongs will elastically deform with increasing speed, in particular on their "open" side. This behaviour is called "umbrella effect" and needs to be limited (Fig. 3). If the structural strength is too low there is a risk of touching the outer ring shoulder and—as a worst case scenario—the cage will be ejected from the ball set ending up with a catastrophic bearing failure. By means of BEAST (**bea**ring simulation **t**ool) the reaction on speed of any cage design and—material can be simulated. So the cage design can be optimized in terms of robustness and cost efficiency on a virtual test rig before any investment in tooling for prototypes is necessary.

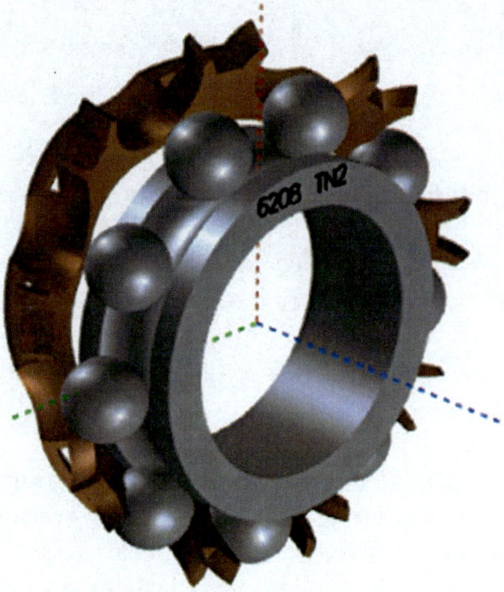

Fig. 3. Cage ejection at high speed (deformation visually increased by factor "4"

SKF Bearing portfolio for the electrified power train

In order to meet all the discussed application requirements acting on the individual bearing positions inside the E-machine and the reduction gearbox the bearing type, the bearing concept and the internal geometry needs to be carefully specified. Nowadays the following bearing types are considered:

- deep groove ball bearings
- cylindrical roller bearings

- angular contact ball bearings
- tapered roller bearings.

Apart from the boundary dimensions which shall be as small as possible the internal geometry is even more relevant. Therefore the geometry inside each bearing is adapted to the predominant operating conditions. For example high speed asks for a friction optimized design and current leakage calls for a hybrid bearing with ceramic balls. High power density on the other side demands a high dynamic capacity ball bearing featuring an increased ball diameter and pitch diameter.

The cylindrical roller bearing is the ideal solution for a non-located bearing position—unifying high compactness, high radial carrying capacity and low power loss.

And finally—for opposed bearing arrangements—usually the tapered roller bearing is the preferred solution. Again the internal geometry can be modified for the sake of power loss reduction and robustness at the same time. And in case of moderate loading conditions the tapered roller bearing can possibly be substituted by a driveline angular contact ball bearing. At the end the SKF bearing portfolio represents a reliable and cost efficient"toolkit" for the electrified power train.

Author Index

Euroforum Deutschland GmbH (Ed.): *CTI SYMPOSIUM 2019*, pp. 585–587, 2021.
https://doi.org/10.1007/978-3-662-61515-7